# Student Solutions Manual

## to accompany

# Chemical Principles

### Sixth Edition

Steven S. Zumdahl

**Thomas J. Hummel**
**Steven S. Zumdahl**

*University of Illinois at Urbana-Champaign*

**BROOKS/COLE**
CENGAGE Learning

Australia • Brazil • Japan • Korea • Mexico • Singapore • Spain • United Kingdom • United States

Vice President and Publisher: Charles Hartford

Senior Marketing Manager: Nicole Moore

Senior Development Editor: Rebecca Berardy Schwartz

Supplements Editor: Kathryn White

Editorial Associate: Henry Cheek

Ancillary Coordinator: Sean McGann

New Title Project Manager: Susan Peltier

Marketing Associate: Kris Bishop

Cover Credit: © 1995-2007 by Michael W. Davidson and the
Florida State University

ISBN-13: 978-0-618-95336-3
ISBN-10: 0-618-95336-1

**Brooks/Cole**
10 Davis Drive
Belmont, CA 94002-3098
USA

Cengage Learning is a leading provider of customized learning
solutions with office locations around the globe, including
Singapore, the United Kingdom, Australia, Mexico, Brazil, and
Japan. Locate your local office at: **www.cengage.com/global**.

Cengage Learning products are represented in Canada by
Nelson Education, Ltd.

To learn more about Brooks/Cole, visit
**www.cengage.com/brookscole**.

Purchase any of our products at your local college store or at
our preferred online store **www.ichapters.com**.

# TABLE OF CONTENTS

iii

# HOW TO USE THIS GUIDE

Solutions to all of the odd-numbered end of chapter exercises are in this manual. This "Solutions Guide" can be very valuable if you use it properly. The way <u>NOT</u> to use it is to look at an exercise in the book and then immediately check the solution, often saying to yourself, "That's easy, I can do it." Developing problem solving skills takes practice. Don't look up a solution to a problem until you have tried to work it on your own. If you are completely stuck, see if you can find a similar problem in the Sample Exercises in the chapter. Only look up the solution as a last resort. If you do this for a problem, look for a similar problem in the end of chapter exercises and try working it. The more problems you do, the easier chemistry becomes. It is also in your self interest to try to work as many problems as possible. Most exams that you will take in chemistry will involve a lot of problem solving. If you have worked several problems similar to the ones on an exam, you will do much better than if the exam is the first time you try to solve a particular type of problem. No matter how much you read and study the text, or how well you think you understand the material, you don't really understand it until you have taken the information in the text and applied the principles to problem solving. You will make mistakes, but the good students learn from their mistakes.

In this manual we have worked problems as in the textbook. We have shown intermediate answers to the correct number of significant figures and used the rounded answer in later calculations. Thus, some of your answers may differ slightly from ours. When we have not followed this convention, we have usually noted this in the solution. The most common exception is when working with the natural logarithm (ln) function, where we usually carried extra significant figures in order to reduce round-off error. In addition, we tried to use constants and conversion factors reported to at least one more significant figure as compared to numbers given in the problem. For some problems, this required the use of more precise atomic masses for H, C, N, and O as given in Chapter 3. This practice of carrying one extra significant figure in constants helps minimize round-off error.

We are grateful to Claire Szoke for her outstanding effort in preparing the manuscript of this manual. We also thank Estelle Lebeau for her careful accuracy review and Jim Madru for his thorough copyediting of the Solutions Manual. We also are grateful to Don DeCoste for his assistance in creating solutions to some of the problems in this Solutions Guide.

TJH
SSZ

# CHAPTER 2

# ATOMS, MOLECULES, AND IONS

## Development of the Atomic Theory

19.     From Avogadro's hypothesis (law), volume ratios are equal to molecule ratios at constant temperature and pressure. Therefore, we can write a balanced equation using the volume data, $Cl_2 + 3 F_2 \rightarrow 2 X$. Two molecules of X contain 6 atoms of F and two atoms of Cl. The formula of X is $ClF_3$ for a balanced reaction.

21.     Hydrazine: $1.44 \times 10^{-1}$ g H/g N;   Ammonia: $2.16 \times 10^{-1}$ g H/g N;   Hydrogen azide: $2.40 \times 10^{-2}$ g H/g N;     Let's try all of the ratios:

$$\frac{0.144}{0.0240} = 6.00; \quad \frac{0.216}{0.0240} = 9.00; \quad \frac{0.216}{0.144} = 1.50 = \frac{3}{2}$$

All the masses of hydrogen in these three compounds can be expressed as simple whole-number ratios. The g H/g N in hydrazine, ammonia, and hydrogen azide are in the ratios 6 : 9 : 1.

23.     To get the atomic mass of H to be 1.00, we divide the mass that reacts with 1.00 g of oxygen by 0.126, that is, 0.126/0.126 = 1.00. To get Na, Mg, and O on the same scale, we do the same division.

$$Na: \frac{2.875}{0.126} = 22.8; \quad Mg: \frac{1.500}{0.126} = 11.9; \quad O: \frac{1.00}{0.126} = 7.94$$

|               | H      | O      | Na    | Mg    |
|---------------|--------|--------|-------|-------|
| Relative value | 1.00   | 7.94   | 22.8  | 11.9  |
| Accepted value | 1.0079 | 15.999 | 22.99 | 24.31 |

The atomic masses of O and Mg are incorrect. The atomic masses of H and Na are close. Something must be wrong about the assumed formulas of the compounds. It turns out that the correct formulas are $H_2O$, $Na_2O$, and $MgO$. The smaller discrepancies result from the error in the assumed atomic mass of H.

1

## The Nature of the Atom

25.        Density of hydrogen nucleus (contains one proton only):

$$V_{nucleus} = \frac{4}{3}\pi r^3 = \frac{4}{3}(3.14)(5 \times 10^{-14}\ cm)^3 = 5 \times 10^{-40}\ cm^3$$

$$d = density = \frac{1.67 \times 10^{-24}\ g}{5 \times 10^{-40}\ cm^3} = 3 \times 10^{15}\ g/cm^3$$

Density of H atom (contains one proton and one electron):

$$V_{atom} = \frac{4}{3}(3.14)(1 \times 10^{-8}\ cm)^3 = 4 \times 10^{-24}\ cm^3$$

$$d = \frac{1.67 \times 10^{-24}\ g + 9 \times 10^{-28}\ g}{4 \times 10^{-24}\ cm^3} = 0.4\ g/cm^3$$

27.        First, divide all charges by the smallest quantity, $6.40 \times 10^{-13}$.

$$\frac{2.56 \times 10^{-12}}{6.40 \times 10^{-13}} = 4.00; \qquad \frac{7.68}{0.640} = 12.00; \qquad \frac{3.84}{0.640} = 6.00$$

Because all charges are whole-number multiples of $6.40 \times 10^{-13}$ zirkombs, the charge on one electron could be $6.40 \times 10^{-13}$ zirkombs. However, $6.40 \times 10^{-13}$ zirkombs could be the charge of two electrons (or three electrons, etc.). All one can conclude is that the charge of an electron is $6.40 \times 10^{-13}$ zirkombs or an integer fraction of $6.40 \times 10^{-13}$.

29.        If the plum pudding model were correct (a diffuse positive charge with electrons scattered throughout), then $\alpha$ particles should have traveled through the thin foil with very minor deflections in their path. This was not the case because a few of the $\alpha$ particles were deflected at very large angles. Rutherford reasoned that the large deflections of these $\alpha$ particles could be caused only by a center of concentrated positive charge that contains most of the atom's mass (the nuclear model of the atom).

## Elements and the Periodic Table

31.        The atomic number of an element is equal to the number of protons in the nucleus of an atom of that element. The mass number is the sum of the number of protons plus neutrons in the nucleus. The atomic mass is the actual mass of a particular isotope (including electrons). As we will see in Chapter 3, the average mass of an atom is taken from a measurement made on a large number of atoms. The average atomic mass value is listed in the periodic table.

33.        a.    The noble gases are He, Ne, Ar, Kr, Xe, and Rn (helium, neon, argon, krypton, xenon, and radon). Radon has only radioactive isotopes. In the periodic table, the whole number enclosed in parentheses is the mass number of the longest-lived isotope of the element.

b.   promethium (Pm) and technetium (Tc)

35.    a.   Five:  F, Cl, Br, I, and At            b.   Six:  Li, Na, K, Rb, Cs, and Fr

                                                        (H is not considered an alkali metal.)

       c.   14:  Ce, Pr, Nd, Pm, Sm, Eu, Gd, Tb, Dy, Ho, Er, Tm, Yb, and Lu

       d.   40:  All elements in the block defined by Sc, Zn, Uub, and Ac are transition metals.

37.    a.   $^{24}_{12}$Mg: 12 protons, 12 neutrons, 12 electrons

       b.   $^{24}_{12}$Mg$^{2+}$: 12 p, 12 n, 10 e            c.   $^{59}_{27}$Co$^{2+}$: 27 p, 32 n, 25 e

       d.   $^{59}_{27}$Co$^{3+}$: 27 p, 32 n, 24 e            e.   $^{59}_{27}$Co: 27 p, 32 n, 27 e

       f.   $^{79}_{34}$Se: 34 p, 45 n, 34 e            g.   $^{79}_{34}$Se$^{2-}$: 34 p, 45 n, 36 e

       h.   $^{63}_{28}$Ni: 28 p, 35 n, 28 e            i.   $^{59}_{28}$Ni$^{2+}$: 28 p, 31 n, 26 e

39.    Atomic number = 63 (Eu);  net charge = +63 − 60 = 3+;  mass number = 63 + 88 = 151;
       symbol: $^{151}_{63}$Eu$^{3+}$

       Atomic number = 50 (Sn);  mass number = 50 + 68 = 118;  net charge = +50 − 48 = 2+;
       symbol: $^{118}_{50}$Sn$^{2+}$.

41.    In ionic compounds, metals lose electrons to form cations, and nonmetals gain electrons to
       form anions. Group 1A, 2A and 3A metals form stable 1+, 2+ and 3+ charged cations,
       respectively. Group 5A, 6A, and 7A nonmetals form 3−, 2− and 1− charged anions,
       respectively.

       a.   Lose 2 e$^-$ to form Ra$^{2+}$.       b.   Lose 3 e$^-$ to form In$^{3+}$.       c.   Gain 3 e$^-$ to form P$^{3-}$.

       d.   Gain 2 e$^-$ to form Te$^{2-}$.       e.   Gain 1 e$^-$ to form Br$^-$.       f.   Lose 1 e$^-$ to form Rb$^+$.

## Nomenclature

43.    AlCl$_3$, aluminum chloride; CrCl$_3$, chromium(III) chloride; ICl$_3$, iodine trichloride; AlCl$_3$ and
       CrCl$_3$ are ionic compounds following the rules for naming ionic compounds. The major
       difference is that CrCl$_3$ contains a transition metal (Cr) that generally exhibits two or more
       stable charges when in ionic compounds. We need to indicate which charged ion we have in
       the compound. This is generally true whenever the metal in the ionic compound is a transition
       metal. ICl$_3$ is made from only nonmetals and is a covalent compound. Predicting formulas for
       covalent compounds is extremely difficult. Because of this, we need to indicate the number of
       each nonmetal in the binary covalent compound. The exception is when there is only one of
       the first species present in the formula; when this is the case, mono- is not used (it is
       assumed).

45.   a.  sulfur difluoride              b.  dinitrogen tetroxide

      c.  iodine trichloride            d.  tetraphosphorus hexoxide

47.   a.  copper(I) iodide          b.  copper(II) iodide          c.  cobalt(II) iodide

      d.  sodium carbonate        e.  sodium hydrogen carbonate or sodium bicarbonate

      f.  tetrasulfur tetranitride      g.  sulfur hexafluoride          h.  sodium hypochlorite

      i.  barium chromate          j.  ammonium nitrate

49.   a.  $SO_2$          b.  $SO_3$          c.  $Na_2SO_3$          d.  $KHSO_3$

      e.  $Li_3N$          f.  $Cr_2(CO_3)_3$          g.  $Cr(C_2H_3O_2)_2$          h.  $SnF_4$

      i.  $NH_4HSO_4$:  composed of $NH_4^+$ and $HSO_4^-$ ions

      j.  $(NH_4)_2HPO_4$      k.  $KClO_4$          l.  $NaH$

      m.  $HBrO$          n.  $HBr$

51.   a.  $Pb(C_2H_3O_2)_2$;  lead(II) acetate          b.  $CuSO_4$;  copper(II) sulfate

      c.  $CaO$;  calcium oxide                    d.  $MgSO_4$;  magnesium sulfate

      e.  $Mg(OH)_2$;  magnesium hydroxide        f.  $CaSO_4$;  calcium sulfate

      g.  $N_2O$;  dinitrogen monoxide or nitrous oxide (common name)

53.   a.  nitric acid, $HNO_3$          b.  perchloric acid, $HClO_4$          c.  acetic acid, $HC_2H_3O_2$

      d.  sulfuric acid, $H_2SO_4$      e.  phosphoric acid, $H_3PO_4$

## Additional Exercises

55.   A compound will always have a constant composition by mass. From the initial data given,
      the mass ratio of H : S : O in sulfuric acid ($H_2SO_4$) is:

$$\frac{2.02}{2.02} : \frac{32.07}{2.02} : \frac{64.00}{2.02} = 1 : 15.9 : 31.7$$

      If we have 7.27 g H, then we will have $7.27 \times 15.9 = 116$ g S and $7.27 \times 31.7 = 230.$ g O in
      the second sample of $H_2SO_4$.

57.   From the $Na_2X$ formula, X has a 2– charge. Because 36 electrons are present, X has 34
      protons, $79 - 34 = 45$ neutrons, and is selenium.

      a.  True. Nonmetals bond together using covalent bonds and are called covalent compounds.

      b.  False. The isotope has 34 protons.

      c.  False. The isotope has 45 neutrons.

      d.  False. The identity is selenium, Se.

59.    From the $XBr_2$ formula, the charge on element X is 2+.  Therefore, the element has 88 protons, which identifies it as radium, Ra.  230 − 88 = 142 neutrons.

61.    In the case of sulfur, $SO_4^{2-}$ is sulfate, and $SO_3^{2-}$ is sulfite.  By analogy:

   $SeO_4^{2-}$: selenate;   $SeO_3^{2-}$: selenite;   $TeO_4^{2-}$: tellurate;   $TeO_3^{2-}$: tellurite

63.    If the formula is InO, then one atomic mass of In would combine with one atomic mass of O, or:

$$\frac{A}{16.00} = \frac{4.784 \text{ g In}}{1.000 \text{ g O}}, \quad A = \text{atomic mass of In} = 76.54$$

   If the formula is $In_2O_3$, then two times the atomic mass of In will combine with three times the atomic mass of O, or:

$$\frac{2A}{(3)16.00} = \frac{4.784 \text{ g In}}{1.000 \text{ g O}}, \quad A = \text{atomic mass of In} = 114.8$$

   The latter number is the atomic mass of In used in the modern periodic table.

65.    The cation has 51 protons and 48 electrons. The number of protons corresponds to the atomic number. Thus this is element 51, antimony. There are 3 fewer electrons than protons. Therefore, the charge on the cation is 3+.  The anion has one-third the number of protons of the cation which corresponds to 17 protons; this is element 17, chlorine. The number of electrons in this anion of chlorine is 17 + 1 = 18 electrons. The anion must have a charge of 1−.

   The formula of the compound formed between $Sb^{3+}$ and $Cl^-$ is $SbCl_3$. The name of the compound is antimony(III) chloride. The Roman numeral is used to indicate the charge of Sb because the predicted charge is not obvious from the periodic table.

67.    Because this is a relatively small number of neutrons, the number of protons will be very close to the number of neutrons present. The heavier elements have significantly more neutrons than protons in their nuclei.  Because this element forms anions, it is a nonmetal and will be a halogen because halogens form stable 1− charged ions in ionic compounds. From the halogens listed, chlorine, with an average atomic mass of 35.45, fits the data. The two isotopes are $^{35}Cl$ and $^{37}Cl$, and the number of electrons in the 1− ion is 18. Note that because the atomic mass of chlorine listed in the periodic table is closer to 35 than 37, we can assume that $^{35}Cl$ is the more abundant isotope. This is discussed in Chapter 3.

## Challenge Problems

69.    a.    Both compounds have $C_2H_6O$ as the formula.  Because they have the same formula, their mass percent composition will be identical.  However, these are different compounds with different properties because the atoms are bonded together differently.  These compounds are called isomers of each other.

       b.    When wood burns, most of the solid material in wood is converted to gases, which escape.  The gases produced are most likely $CO_2$ and $H_2O$.

c.   The atom is not an indivisible particle but is instead composed of other smaller particles, for example, electrons, neutrons, and protons.

d.   The two hydride samples contain different isotopes of either hydrogen and/or lithium. Although the compounds are composed of different isotopes, their properties are similar because different isotopes of the same element have similar properties (except, of course, their mass).

71.     Compound I: $\dfrac{14.0 \text{ g R}}{3.00 \text{ g Q}} = \dfrac{4.67 \text{ g R}}{1.00 \text{ g Q}}$;   Compound II: $\dfrac{7.00 \text{ g R}}{4.50 \text{ g Q}} = \dfrac{1.56 \text{ g R}}{1.00 \text{ g Q}}$

The ratio of the masses of R that combines with 1.00 g Q is $\dfrac{4.67}{1.56} = 2.99 \approx 3$.

As expected from the law of multiple proportions, this ratio is a small whole number.

Because compound I contains three times the mass of R per gram of Q as compared with compound II (RQ), the formula of compound I should be $R_3Q$.

73.     Avogadro proposed that equal volumes of gases (at constant temperature and pressure) contain equal numbers of molecules. In terms of balanced equations, Avogadro's hypothesis (law) implies that volume ratios will be identical to molecule ratios. Assuming one molecule of octane reacting, then 1 molecule of $C_xH_y$ produces 8 molecules of $CO_2$ and 9 molecules of $H_2O$.   $C_xH_y + n\, O_2 \rightarrow 8\, CO_2 + 9\, H_2O$. Because all the carbon in octane ends up as carbon in $CO_2$, octane must contain 8 atoms of C. Similarly, all hydrogen in octane ends up as hydrogen in $H_2O$, so one molecule of octane must contain $9 \times 2 = 18$ atoms of H. Octane formula = $C_8H_{18}$ and the ratio of C:H = 8:18 or 4:9.

# CHAPTER 3

# STOICHIOMETRY

**Atomic Masses and the Mass Spectrometer**

23.    A = 0.0800(45.95269) + 0.0730(46.951764) + 0.7380(47.947947) + 0.0550(48.947841)  +

0.0540(49.944792) = 47.88 amu

This is element Ti (titanium).

25.    186.207 = 0.6260(186.956) + 0.3740(A),  186.207 - 117.0 = 0.3740(A)

$$A = \frac{69.2}{0.3740} = 185 \text{ amu } (A = 184.95 \text{ amu without rounding to proper significant figures})$$

27.    There are three peaks in the mass spectrum, each 2 mass units apart.  This is consistent with two isotopes, differing in mass by two mass units.  The peak at 157.84 corresponds to a $Br_2$ molecule composed of two atoms of the lighter isotope.  This isotope has mass equal to 157.84/2, or 78.92.  This corresponds to $^{79}Br$.  The second isotope is $^{81}Br$ with mass equal to 161.84/2 = 80.92.  The peaks in the mass spectrum correspond to $^{79}Br_2$, $^{79}Br^{81}Br$, and $^{81}Br_2$ in order of increasing mass.  The intensities of the highest and lowest masses tell us the two isotopes are present at about equal abundance.  The actual abundance is 50.68% $^{79}Br$ and 49.32% $^{81}Br$.

29.    GaAs can be either $^{69}GaAs$ or $^{71}GaAs$.  The mass spectrum for GaAs will have two peaks at 144 (= 69 + 75) and 146 (= 71 + 75) with intensities in the ratio of 60 : 40 or 3 : 2.

144        146

Ga$_2$As$_2$ can be $^{69}Ga_2As_2$, $^{69}Ga^{71}GaAs_2$, or $^{71}Ga_2As_2$.  The mass spectrum will have three peaks at 288, 290, and 292 with intensities in the ratio of 36 : 48 : 16 or 9 : 12 : 4.  We get this ratio from the following probability table:

|                    | $^{69}$Ga (0.60) | $^{71}$Ga (0.40) |
|--------------------|------------------|------------------|
| $^{69}$Ga (0.60)   | 0.36             | 0.24             |
| $^{71}$Ga (0.40)   | 0.24             | 0.16             |

## Moles and Molar Masses

31.   a.   $20.0 \text{ mg C}_8\text{H}_{10}\text{N}_4\text{O}_2 \times \dfrac{1 \text{ g}}{1000 \text{ mg}} \times \dfrac{1 \text{ mol}}{194.20 \text{ g}} = 1.03 \times 10^{-4} \text{ mol C}_8\text{H}_{10}\text{N}_4\text{O}_2$

   b.   $2.72 \times 10^{21} \text{ molecules C}_2\text{H}_5\text{OH} \times \dfrac{1 \text{ mol}}{6.022 \times 10^{23} \text{ molecules}} = 4.52 \times 10^{-3} \text{ mol C}_2\text{H}_5\text{OH}$

   c.   $1.50 \text{ g CO}_2 \times \dfrac{1 \text{ mol}}{44.01 \text{ g}} = 3.41 \times 10^{-2} \text{ mol CO}_2$

33.   $4.0 \text{ g H}_2 \times \dfrac{1 \text{ mol H}_2}{2.016 \text{ g H}_2} \times \dfrac{2 \text{ mol H}}{1 \text{ mol H}_2} \times \dfrac{6.022 \times 10^{23} \text{ atoms H}}{1 \text{ mol H}} = 2.4 \times 10^{24} \text{ atoms}$

   $4.0 \text{ g He} \times \dfrac{1 \text{ mol He}}{4.003 \text{ g He}} \times \dfrac{6.022 \times 10^{23} \text{ atoms He}}{1 \text{ mol He}} = 6.0 \times 10^{23} \text{ atoms}$

   $1.0 \text{ mol F}_2 \times \dfrac{2 \text{ mol F}}{1 \text{ mol F}_2} \times \dfrac{6.022 \times 10^{23} \text{ atoms F}}{1 \text{ mol F}} = 1.2 \times 10^{24} \text{ atoms}$

   $44.0 \text{ g CO}_2 \times \dfrac{1 \text{ mol CO}_2}{44.01 \text{ g CO}_2} \times \dfrac{3 \text{ mol atoms}(1 \text{ C} + 2 \text{ O})}{1 \text{ mol CO}_2} \times \dfrac{6.022 \times 10^{23} \text{ atoms}}{1 \text{ mol atoms}}$

   $$= 1.81 \times 10^{24} \text{ atoms}$$

   $146 \text{ g SF}_6 \times \dfrac{1 \text{ mol SF}_6}{146.07 \text{ g SF}_6} \times \dfrac{7 \text{ mol atoms}(1 \text{ S} + 6 \text{ F})}{1 \text{ mol SF}_6} \times \dfrac{6.022 \times 10^{23} \text{ atoms}}{1 \text{ mol atoms}} = 4.21 \times 10^{24} \text{ atoms}$

   The order is:  $4.0 \text{ g He} < 1.0 \text{ mol F}_2 < 44.0 \text{ g CO}_2 < 4.0 \text{ g H}_2 < 146 \text{ g SF}_6$

35.  a.  $2(12.01) + 3(1.008) + 3(35.45) + 2(16.00) = 165.39$ g/mol

b.  $500.0 \text{ g} \times \dfrac{1 \text{ mol}}{165.39 \text{ g}} = 3.023 \text{ mol } C_2H_3Cl_3O_2$

c.  $2.0 \times 10^{-2} \text{ mol} \times \dfrac{165.39 \text{ g}}{\text{mol}} = 3.3 \text{ g } C_2H_3Cl_3O_2$

d.  $5.0 \text{ g } C_2H_3Cl_3O_2 \times \dfrac{1 \text{ mol}}{165.39 \text{ g}} \times \dfrac{6.02 \times 10^{23} \text{ molecules}}{\text{mol}} \times \dfrac{3 \text{ atoms Cl}}{\text{molecule}}$

$= 5.5 \times 10^{22}$ atoms of chlorine

e.  $1.0 \text{ g Cl} \times \dfrac{1 \text{ mol Cl}}{35.45 \text{ g}} \times \dfrac{1 \text{ mol } C_2H_3Cl_3O_2}{3 \text{ mol Cl}} \times \dfrac{165.39 \text{ g } C_2H_3Cl_3O_2}{\text{mol } C_2H_3Cl_3O_2} = 1.6 \text{ g chloral hydrate}$

f.  $500 \text{ molecules} \times \dfrac{1 \text{ mol}}{6.022 \times 10^{23} \text{ molecules}} \times \dfrac{165.39 \text{ g}}{\text{mol}} = 1.373 \times 10^{-19} \text{ g}$

## Percent Composition

37.  Molar mass $= 20(12.01) + 29(1.008) + 19.00 + 3(16.00) = 336.43$ g/mol

Mass % C $= \dfrac{20(12.01) \text{ g C}}{336.43 \text{ g compound}} \times 100 = 71.40\%$ C

Mass % H $= \dfrac{29(1.008) \text{ g H}}{336.43 \text{ g compound}} \times 100 = 8.689\%$ H

Mass % F $= \dfrac{19.00 \text{ g F}}{336.43 \text{ g compound}} \times 100 = 5.648\%$ F

Mass % O $= 100.00 - (71.40 + 8.689 + 5.648) = 14.26\%$ O or:

Mass % O $= \dfrac{3(16.00) \text{ g O}}{336.43 \text{ g compound}} \times 100 = 14.27\%$ O

39.  For each compound, determine the mass of 1 mole of compound, and then calculate the mass percentage of N in 1 mole of that compound.

a.  NO: % N $= \dfrac{14.007 \text{ g N}}{30.006 \text{ g NO}} \times 100 = 46.681\%$ N

b.  $NO_2$: % N = $\dfrac{14.007 \text{ g N}}{46.005 \text{ g } NO_2} \times 100 = 30.447\%$ N

c.  $N_2O_4$: % N = $\dfrac{28.014 \text{ g N}}{92.010 \text{ g } N_2O_4} \times 100 = 30.447\%$ N

d.  $N_2O$: % N = $\dfrac{28.014 \text{ g N}}{44.013 \text{ g } N_2O} \times 100 = 63.649\%$ N

41.  There are 0.390 g Cu for every 100.000 g of fungal laccase.  Let's assume 100.000 g fungal laccase.

Mol fungal laccase = 0.390 g Cu $\times \dfrac{1 \text{ mol Cu}}{63.55 \text{ g Cu}} \times \dfrac{1 \text{ mol fungal laccase}}{4 \text{ mol Cu}} = 1.53 \times 10^{-3}$ mol

$\dfrac{x \text{ g fungal laccase}}{1 \text{ mol fungal laccase}} = \dfrac{100.000 \text{ g}}{1.53 \times 10^{-3} \text{ mol}}$, $x$ = molar mass = $6.54 \times 10^4$ g/mol

## Empirical and Molecular Formulas

43.  a.  Molar mass of $CH_2O$ = 1 mol C$\left(\dfrac{12.011 \text{ g}}{\text{mol C}}\right)$ + 2 mol H$\left(\dfrac{1.0079 \text{ g}}{\text{mol H}}\right)$

$$+ 1 \text{ mol O}\left(\dfrac{15.999 \text{ g}}{\text{mol O}}\right) = 30.026 \text{ g/mol}$$

% C = $\dfrac{12.011 \text{ g C}}{30.026 \text{ g } CH_2O} \times 100 = 40.002\%$ C;  % H = $\dfrac{2.0158 \text{ g H}}{30.026 \text{ g } CH_2O} \times 100 = 6.7135\%$ H

% O = $\dfrac{15.999 \text{ g O}}{30.026 \text{ g } CH_2O} \times 100 = 53.284\%$ O  or % O = 100.000 − (40.002 + 6.7135)

$$= 53.285\%$$

b.  Molar mass of $C_6H_{12}O_6$ = 6(12.011) + 12(1.0079) + 6(15.999) = 180.155 g/mol

% C = $\dfrac{72.066 \text{ g C}}{180.155 \text{ g } C_6H_{12}O_6} \times 100 = 40.002\%$;   % H = $\dfrac{12(1.0079) \text{ g}}{180.155 \text{ g}} \times 100 = 6.7136\%$

% O = 100.00 − (40.002 + 6.7136) = 53.284%

c.  Molar Mass of $HC_2H_3O_2$ = 2(12.011) + 4(1.0079) + 2(15.999) = 60.052 g/mol

$$\% \text{ C} = \frac{24.022 \text{ g}}{60.052 \text{ g}} \times 100 = 40.002\%; \quad \% \text{ H} = \frac{4.0316 \text{ g}}{60.052 \text{ g}} \times 100 = 6.7135\%$$

$$\% \text{ O} = 100.000 - (40.002 + 6.7135) = 53.285\%$$

All three compounds have the same empirical formula, $CH_2O$, and different molecular formulas. The composition of all three in mass percent is also the same (within rounding differences). Therefore, elemental analysis will give us only the empirical formula.

45.    Compound I:  mass O = 0.6498 g $Hg_xO_y$ − 0.6018 g Hg = 0.0480 g O

$$0.6018 \text{ g Hg} \times \frac{1 \text{ mol Hg}}{200.6 \text{ g Hg}} = 3.000 \times 10^{-3} \text{ mol Hg}$$

$$0.0480 \text{ g O} \times \frac{1 \text{ mol O}}{16.00 \text{ g O}} = 3.00 \times 10^{-3} \text{ mol O}$$

The mole ratio between Hg and O is 1 : 1, so the empirical formula of compound I is HgO.

Compound II:  mass Hg = 0.4172 g $Hg_xO_y$ − 0.016 g O = 0.401 g Hg

$$0.401 \text{ g Hg} \times \frac{1 \text{ mol Hg}}{200.6 \text{ g Hg}} = 2.00 \times 10^{-3} \text{ mol Hg}; \quad 0.016 \text{ g O} \times \frac{1 \text{ mol O}}{16.00 \text{ g O}} = 1.0 \times 10^{-3} \text{ mol O}$$

The mole ratio between Hg and O is 2 : 1, so the empirical formula is $Hg_2O$.

47.    First, we will determine composition in mass percent. We assume that all the carbon in the 0.213 g $CO_2$ came from the 0.157 g of the compound and that all the hydrogen in the 0.0310 g $H_2O$ came from the 0.157 g of the compound.

$$0.213 \text{ g CO}_2 \times \frac{12.01 \text{ g C}}{44.01 \text{ g CO}_2} = 0.0581 \text{ g C}; \quad \% \text{ C} = \frac{0.0581 \text{ g C}}{0.157 \text{ g compound}} \times 100 = 37.0\% \text{ C}$$

$$0.0310 \text{ g H}_2O \times \frac{2.016 \text{ g H}}{18.02 \text{ g H}_2O} = 3.47 \times 10^{-3} \text{ g H}; \quad \% \text{ H} = \frac{3.47 \times 10^{-3} \text{ g}}{0.157 \text{ g}} \times 100 = 2.21\% \text{ H}$$

We get the mass percent of N from the second experiment:

$$0.0230 \text{ g NH}_3 \times \frac{14.01 \text{ g N}}{17.03 \text{ g NH}_3} = 1.89 \times 10^{-2} \text{ g N}$$

$$\% \text{ N} = \frac{1.89 \times 10^{-2} \text{ g}}{0.103 \text{ g}} \times 100 = 18.3\% \text{ N}$$

The mass percent of oxygen is obtained by difference:

$$\% \ O = 100.00 - (37.0 + 2.21 + 18.3) = 42.5\% \ O$$

So, out of 100.00 g of compound, there are:

$$37.0 \ \text{g C} \times \frac{1 \ \text{mol C}}{12.01 \ \text{g C}} = 3.08 \ \text{mol C}; \quad 2.21 \ \text{g H} \times \frac{1 \ \text{mol H}}{1.008 \ \text{g H}} = 2.19 \ \text{mol H}$$

$$18.3 \ \text{g N} \times \frac{1 \ \text{mol N}}{14.01 \ \text{g N}} = 1.31 \ \text{mol N}; \quad 42.5 \ \text{g O} \times \frac{1 \ \text{mol O}}{16.00 \ \text{g O}} = 2.66 \ \text{mol O}$$

Lastly, and often the hardest part, we need to find simple whole number ratios. Divide all mole values by the smallest number:

$$\frac{3.08}{1.31} = 2.35; \quad \frac{2.19}{1.31} = 1.67; \quad \frac{1.31}{1.31} = 1.00; \quad \frac{2.66}{1.31} = 2.03$$

Multiplying all these ratios by 3 gives an empirical formula of $C_7H_5N_3O_6$.

49.    Assuming 100.00 g of compound (mass hydrogen = 100.00 g − 49.31 g C − 43.79 g O
$$= 6.90 \ \text{g H}):$$

$$49.31 \ \text{g C} \times \frac{1 \ \text{mol C}}{12.011 \ \text{g C}} = 4.105 \ \text{mol C}; \quad 6.90 \ \text{g H} \times \frac{1 \ \text{mol H}}{1.008 \ \text{g H}} = 6.85 \ \text{mol H}$$

$$43.79 \ \text{g O} \times \frac{1 \ \text{mol O}}{15.999 \ \text{g O}} = 2.737 \ \text{mol O}$$

Dividing all mole values by 2.737 gives:

$$\frac{4.105}{2.737} = 1.500; \quad \frac{6.85}{2.737} = 2.50; \quad \frac{2.737}{2.737} = 1.000$$

Because a whole number ratio is required, the empirical formula is $C_3H_5O_2$.

Empirical formula mass $\approx 3(12.0) + 5(1.0) + 2(16.0) = 73.0 \ \text{g/mol}$

$$\frac{\text{Molar mass}}{\text{Empirical formula mass}} = \frac{146.1}{73.0} = 2.00; \quad \text{molecular formula} = (C_3H_5O_2)_2 = C_6H_{10}O_4$$

51.   First, we will determine composition by mass percent:

$$16.01 \text{ mg CO}_2 \times \frac{1 \text{ g}}{1000 \text{ mg}} \times \frac{12.011 \text{ g C}}{44.009 \text{ g CO}_2} \times \frac{1000 \text{ mg}}{\text{g}} = 4.369 \text{ mg C}$$

$$\% \text{ C} = \frac{4.369 \text{ mg C}}{10.68 \text{ mg compound}} \times 100 = 40.91\% \text{ C}$$

$$4.37 \text{ mg H}_2\text{O} \times \frac{1 \text{ g}}{1000 \text{ mg}} \times \frac{2.016 \text{ g H}}{18.02 \text{ g H}_2\text{O}} \times \frac{1000 \text{ mg}}{\text{g}} = 0.489 \text{ mg H}$$

$$\% \text{ H} = \frac{0.489 \text{ mg}}{10.68 \text{ mg}} \times 100 = 4.58\% \text{ H}; \quad \% \text{ O} = 100.00 - (40.91 + 4.58) = 54.51\% \text{ O}$$

So, in 100.00 g of the compound, we have:

$$40.91 \text{ g C} \times \frac{1 \text{ mol C}}{12.011 \text{ g C}} = 3.406 \text{ mol C}; \quad 4.58 \text{ g H} \times \frac{1 \text{ mol H}}{1.008 \text{ g H}} = 4.54 \text{ mol H}$$

$$54.51 \text{ g O} \times \frac{1 \text{ mol O}}{15.999 \text{ g O}} = 3.407 \text{ mol O}$$

Dividing by the smallest number: $\dfrac{4.54}{3.406} = 1.33 \approx \dfrac{4}{3}$; the empirical formula is $C_3H_4O_3$.

The empirical formula mass of $C_3H_4O_3$ is $\approx 3(12) + 4(1) + 3(16) = 88$ g.

Because $\dfrac{176.1}{88} = 2.0$, the molecular formula is $C_6H_8O_6$.

## Balancing Chemical Equations

53.   When balancing reactions, start with elements that appear in only one of the reactants and one of the products, and then go on to balance the remaining elements.

a.   $C_6H_{12}O_6(s) + O_2(g) \rightarrow CO_2(g) + H_2O(g)$

Balance C atoms:   $C_6H_{12}O_6 + O_2 \rightarrow 6\,CO_2 + H_2O$

Balance H atoms:   $C_6H_{12}O_6 + O_2 \rightarrow 6\,CO_2 + 6\,H_2O$

Lastly, balance O atoms:   $C_6H_{12}O_6(s) + 6\,O_2(g) \rightarrow 6\,CO_2(g) + 6\,H_2O(g)$

b.   $Fe_2S_3(s) + HCl(g) \rightarrow FeCl_3(s) + H_2S(g)$

Balance Fe atoms:  $Fe_2S_3 + HCl \rightarrow 2\ FeCl_3 + H_2S$

Balance S atoms:  $Fe_2S_3 + HCl \rightarrow 2\ FeCl_3 + 3\ H_2S$

There are 6 H and 6 Cl on right, so balance with 6 HCl on left:

   $Fe_2S_3(s) + 6\ HCl(g) \rightarrow 2\ FeCl_3(s) + 3\ H_2S(g)$

c.   $CS_2(l) + NH_3(g) \rightarrow H_2S(g) + NH_4SCN(s)$

C and S are balanced;  balance N:

   $CS_2 + 2\ NH_3 \rightarrow H_2S + NH_4SCN$

H is also balanced.  $CS_2(l) + 2\ NH_3(g) \rightarrow H_2S(g) + NH_4SCN(s)$.

55.   a.   $16\ Cr(s) + 3\ S_8(s) \rightarrow 8\ Cr_2S_3(s)$

b.   $2\ NaHCO_3(s) \rightarrow Na_2CO_3(s) + CO_2(g) + H_2O(g)$

c.   $2\ KClO_3(s) \rightarrow 2\ KCl(s) + 3\ O_2(g)$

d.   $2\ Eu(s) + 6\ HF(g) \rightarrow 2\ EuF_3(s) + 3\ H_2(g)$

e.   $2\ C_6H_6(l) + 15\ O_2(g) \rightarrow 12\ CO_2(g) + 6\ H_2O(g)$

## Reaction Stoichiometry

57.   $1.000\ kg\ Al \times \dfrac{1000\ g\ Al}{kg\ Al} \times \dfrac{1\ mol\ Al}{26.98\ g\ Al} \times \dfrac{3\ mol\ NH_4ClO_4}{3\ mol\ Al} \times \dfrac{117.49\ g\ NH_4ClO_4}{mol\ NH_4ClO_4}$

$$= 4355\ g\ NH_4ClO_4$$

59.   $Fe_2O_3(s) + 2\ Al(s) \rightarrow 2\ Fe(l) + Al_2O_3(s)$

$15.0\ g\ Fe \times \dfrac{1\ mol\ Fe}{55.85\ g\ Fe} = 0.269\ mol\ Fe;\ \ 0.269\ mol\ Fe \times \dfrac{2\ mol\ Al}{2\ mol\ Fe} \times \dfrac{26.98\ g\ Al}{mol\ Al} = 7.26\ g\ Al$

$0.269\ mol\ Fe \times \dfrac{1\ mol\ Fe_2O_3}{2\ mol\ Fe} \times \dfrac{159.70\ g\ Fe_2O_3}{mol\ Fe_2O_3} = 21.5\ g\ Fe_2O_3$

$0.269\ mol\ Fe \times \dfrac{1\ mol\ Al_2O_3}{2\ mol\ Fe} \times \dfrac{101.96\ g\ Al_2O_3}{mol\ Al_2O_3} = 13.7\ g\ Al_2O_3$

61.      $2 LiOH(s) + CO_2(g) \rightarrow Li_2CO_3(aq) + H_2O(l)$

The total volume of air exhaled each minute for the 7 astronauts is $7 \times 20. = 140$ L/min.

$$25,000 \text{ g LiOH} \times \frac{1 \text{ mol LiOH}}{23.95 \text{ g LiOH}} \times \frac{1 \text{ mol CO}_2}{2 \text{ mol LiOH}} \times \frac{44.01 \text{ g CO}_2}{\text{mol CO}_2} \times \frac{100 \text{ g air}}{4.0 \text{ g CO}_2} \times$$

$$\frac{1 \text{ mL air}}{0.0010 \text{ g air}} \times \frac{1 \text{ L}}{1000 \text{ mL}} \times \frac{1 \text{ min}}{140 \text{ L air}} \times \frac{1 \text{ h}}{60 \text{ min}} = 68 \text{ h} = 2.8 \text{ days}$$

63.      $$1.0 \times 10^3 \text{ g phosphorite} \times \frac{75 \text{ g Ca}_3(\text{PO}_4)_2}{100 \text{ g phosphorite}} \times \frac{1 \text{ mol Ca}_3(\text{PO}_4)_2}{310.18 \text{ g Ca}_3(\text{PO}_4)_2} \times$$

$$\frac{1 \text{ mol P}_4}{2 \text{ mol Ca}_3(\text{PO}_4)_2} \times \frac{123.88 \text{ g P}_4}{\text{mol P}_4} = 150 \text{ g P}_4$$

## Limiting Reactants and Percent Yield

65.      The product formed in the reaction is $NO_2$; the other species present in the product represent-tation is excess $O_2$. Therefore, NO is the limiting reactant. In the pictures, 6 NO molecules react with 3 $O_2$ molecules to form 6 $NO_2$ molecules.

$$6 \text{ NO(g)} + 3 \text{ O}_2(g) \rightarrow 6 \text{ NO}_2(g)$$

For smallest whole numbers, the balanced reaction is:

$$2 \text{ NO(g)} + \text{O}_2(g) \rightarrow 2 \text{ NO}_2(g)$$

67.      $$1.50 \text{ g BaO}_2 \times \frac{1 \text{ mol BaO}_2}{169.3 \text{ g BaO}_2} = 8.86 \times 10^{-3} \text{ mol BaO}_2$$

$$25.0 \text{ mL} \times \frac{0.0272 \text{ g HCl}}{\text{mL}} \times \frac{1 \text{ mol HCl}}{36.46 \text{ g HCl}} = 1.87 \times 10^{-2} \text{ mol HCl}$$

The required mole ratio from the balanced reaction is 2 mol HCl to 1 mol $BaO_2$. The actual mole ratio is:

$$\frac{1.87 \times 10^{-2} \text{ mol HCl}}{8.86 \times 10^{-3} \text{ mol BaO}_2} = 2.11$$

Because the actual mole ratio is larger than the required mole ratio, the denominator ($BaO_2$) is the limiting reagent.

$$8.86 \times 10^{-3} \text{ mol BaO}_2 \times \frac{1 \text{ mol H}_2\text{O}_2}{\text{mol BaO}_2} \times \frac{34.02 \text{ g H}_2\text{O}_2}{\text{mol H}_2\text{O}_2} = 0.301 \text{ g H}_2\text{O}_2$$

The amount of HCl reacted is:

$$8.86 \times 10^{-3} \text{ mol BaO}_2 \times \frac{2 \text{ mol HCl}}{\text{mol BaO}_2} = 1.77 \times 10^{-2} \text{ mol HCl}$$

Excess mol HCl $= 1.87 \times 10^{-2}$ mol $- 1.77 \times 10^{-2}$ mol $= 1.0 \times 10^{-3}$ mol HCl

Mass of excess HCl $= 1.0 \times 10^{-3}$ mol HCl $\times \dfrac{36.46 \text{ g HCl}}{\text{mol HCl}} = 3.6 \times 10^{-2}$ g HCl

69.    $2.50$ metric tons $Cu_3FeS_3 \times \dfrac{1000 \text{ kg}}{\text{metric ton}} \times \dfrac{1000 \text{ g}}{\text{kg}} \times \dfrac{1 \text{ mol Cu}_3\text{FeS}_3}{342.71 \text{ g}} \times \dfrac{3 \text{ mol Cu}}{1 \text{ mol Cu}_3\text{FeS}_3} \times$

$$\dfrac{63.55 \text{ g}}{\text{mol Cu}} = 1.39 \times 10^6 \text{ g Cu (theoretical)}$$

$1.39 \times 10^6$ g Cu (theoretical) $\times \dfrac{86.3 \text{ g Cu (actual)}}{100. \text{ g Cu (theoretical)}} = 1.20 \times 10^6$ g Cu $= 1.20 \times 10^3$ kg Cu

$$= 1.20 \text{ metric tons Cu (actual)}$$

71.    An alternative method to solve limiting-reagent problems is to assume that each reactant is limiting and then calculate how much product could be produced from each reactant.  The reactant that produces the smallest amount of product will run out first and is the limiting reagent.

$$5.00 \times 10^6 \text{ g NH}_3 \times \frac{1 \text{ mol NH}_3}{17.03 \text{ g NH}_3} \times \frac{2 \text{ mol HCN}}{2 \text{ mol NH}_3} = 2.94 \times 10^5 \text{ mol HCN}$$

$$5.00 \times 10^6 \text{ g O}_2 \times \frac{1 \text{ mol O}_2}{32.00 \text{ g O}_2} \times \frac{2 \text{ mol HCN}}{3 \text{ mol O}_2} = 1.04 \times 10^5 \text{ mol HCN}$$

$$5.00 \times 10^6 \text{ g CH}_4 \times \frac{1 \text{ mol CH}_4}{16.04 \text{ g CH}_4} \times \frac{2 \text{ mol HCN}}{2 \text{ mol CH}_4} = 3.12 \times 10^5 \text{ mol HCN}$$

$O_2$ is limiting because it produces the smallest amount of HCN.  Although more product could be produced from $NH_3$ and $CH_4$, only enough $O_2$ is present to produce $1.04 \times 10^5$ mol HCN.  The mass of HCN that can be produced is:

$$1.04 \times 10^5 \text{ mol HCN} \times \frac{27.03 \text{ g HCN}}{\text{mol HCN}} = 2.81 \times 10^6 \text{ g HCN}$$

$$5.00 \times 10^6 \text{ g O}_2 \times \frac{1 \text{ mol O}_2}{32.00 \text{ g O}_2} \times \frac{6 \text{ mol H}_2\text{O}}{3 \text{ mol O}_2} \times \frac{18.02 \text{ g H}_2\text{O}}{1 \text{ mol H}_2\text{O}} = 5.63 \times 10^6 \text{ g H}_2\text{O}$$

73.    $P_4(s) + 6 F_2(g) \rightarrow 4 PF_3(g)$;  the theoretical yield of $PF_3$ is:

$$120. \text{ g } PF_3 \text{ (actual)} \times \frac{100.0 \text{ g } PF_3 \text{ (theoretical)}}{78.1 \text{ g } PF_3 \text{ (actual)}} = 154 \text{ g } PF_3 \text{ (theoretical)}$$

$$154 \text{ g } PF_3 \times \frac{1 \text{ mol } PF_3}{87.97 \text{ g } PF_3} \times \frac{6 \text{ mol } F_2}{4 \text{ mol } PF_3} \times \frac{38.00 \text{ g } F_2}{\text{mol } F_2} = 99.8 \text{ g } F_2$$

99.8 g $F_2$ are needed to actually produce 120. g of $PF_3$ if the percent yield is 78.1%.

## Additional Exercises

75.    $17.3 \text{ g H} \times \dfrac{1 \text{ mol H}}{1.008 \text{ g H}} = 17.2 \text{ mol H}; \quad 82.7 \text{ g C} \times \dfrac{1 \text{ mol C}}{12.01 \text{ g C}} = 6.89 \text{ mol C}$

$\dfrac{17.2}{6.89} = 2.50$;  the empirical formula is $C_2H_5$.

The empirical formula mass is ~29 g, so two times the empirical formula would put the compound in the correct range of the molar mass.  Molecular formula = $(C_2H_5)_2 = C_4H_{10}$

$$2.59 \times 10^{23} \text{ atoms H} \times \frac{1 \text{ molecule } C_4H_{10}}{10 \text{ atoms H}} \times \frac{1 \text{ mol } C_4H_{10}}{6.022 \times 10^{23} \text{ molecules}}$$

$$= 4.30 \times 10^{-2} \text{ mol } C_4H_{10}$$

$$4.30 \times 10^{-2} \text{ mol } C_4H_{10} \times \frac{58.12 \text{ g}}{\text{mol } C_4H_{10}} = 2.50 \text{ g } C_4H_{10}$$

77.    $\text{Molar mass } X_2 = \dfrac{0.105 \text{ g}}{8.92 \times 10^{20} \text{ molecules} \times \dfrac{1 \text{ mol}}{6.022 \times 10^{23} \text{ molecules}}} = 70.9 \text{ g/mol}$

The mass of X = 1/2(70.9 g/mol) = 35.5 g/mol.  This is the element chlorine.

Assuming 100.00 g of $MX_3$ compound:

$$54.47 \text{ g Cl} \times \frac{1 \text{ mol}}{35.45 \text{ g}} = 1.537 \text{ mol Cl}$$

$$1.537 \text{ mol Cl} \times \frac{1 \text{ mol M}}{3 \text{ mol Cl}} = 0.5123 \text{ mol M}$$

$$\text{Molar mass of M} = \frac{45.53 \text{ g M}}{0.5123 \text{ mol M}} = 88.87 \text{ g/mol M}$$

M is the element yttrium (Y), and the name of $YCl_3$ is yttrium(III) chloride.

The balanced equation is $2\ Y + 3\ Cl_2 \rightarrow 2\ YCl_3$.

Assuming $Cl_2$ is limiting:

$$1.00\ g\ Cl_2 \times \frac{1\ mol\ Cl_2}{70.90\ g\ Cl_2} \times \frac{2\ mol\ YCl_3}{3\ mol\ Cl_2} \times \frac{195.26\ g\ YCl_3}{1\ mol\ YCl_3} = 1.84\ g\ YCl_3$$

Assuming Y is limiting:

$$1.00\ g\ Y \times \frac{1\ mol\ Y}{88.91\ g\ Y} \times \frac{2\ mol\ YCl_3}{2\ mol\ Y} \times \frac{195.26\ g\ YCl_3}{1\ mol\ YCl_3} = 2.20\ g\ YCl_3$$

Because $Cl_2$, when it all reacts, produces the smaller amount of product, $Cl_2$ is the limiting reagent, and the theoretical yield is $1.84\ g\ YCl_3$.

79.   Mass of $H_2O = 0.755\ g\ CuSO_4{\bullet}xH_2O - 0.483\ g\ CuSO_4 = 0.272\ g\ H_2O$

$$0.483\ g\ CuSO_4 \times \frac{1\ mol\ CuSO_4}{159.62\ g\ CuSO_4} = 0.00303\ mol\ CuSO_4$$

$$0.272\ g\ H_2O \times \frac{1\ mol\ H_2O}{18.02\ g\ H_2O} = 0.0151\ mol\ H_2O$$

$$\frac{0.0151\ mol\ H_2O}{0.00303\ mol\ CuSO_4} = \frac{4.98\ mol\ H_2O}{1\ mol\ CuSO_4} ;\ \text{compound formula} = CuSO_4{\bullet}5H_2O,\ x = 5$$

81.   Consider the case of aluminum plus oxygen.  Aluminum forms $Al^{3+}$ ions;  oxygen forms $O^{2-}$ anions.  The simplest compound of the two elements is $Al_2O_3$.  Similarly, we would expect the formula of a Group 6A element with Al to be $Al_2X_3$.  Assuming this, out of 100.00 g of compound, there are 18.56 g Al and 81.44 g of the unknown element, X.  Let's use this information to determine the molar mass of X, which will allow us to identify X from the periodic table.

$$18.56\ g\ Al \times \frac{1\ mol\ Al}{26.98\ g\ Al} \times \frac{3\ mol\ X}{2\ mol\ Al} = 1.032\ mol\ X$$

81.44 g of X contains 1.032 mol of X.   Molar mass of $X = \dfrac{81.44\ g\ X}{1.032\ mol\ X} = 78.91\ g/mol$

From the periodic table, the unknown element is selenium, and the formula is $Al_2Se_3$.

83.        $1.20\ g\ CO_2 \times \dfrac{1\ mol\ CO_2}{44.01\ g} \times \dfrac{1\ mol\ C}{mol\ CO_2} \times \dfrac{1\ mol\ C_{24}H_{30}N_3O}{24\ mol\ C} \times \dfrac{376.51\ g}{mol\ C_{24}H_{30}N_3O}$

$$= 0.428\ g\ C_{24}H_{30}N_3O$$

$$\frac{0.428\ g\ C_{24}H_{30}N_3O}{1.00\ g\ sample} \times 100 = 42.8\%\ C_{24}H_{30}N_3O\ (LSD)$$

85.     $453 \text{ g Fe} \times \dfrac{1 \text{ mol Fe}}{55.85 \text{ g Fe}} \times \dfrac{1 \text{ mol Fe}_2\text{O}_3}{2 \text{ mol Fe}} \times \dfrac{159.70 \text{ g Fe}_2\text{O}_3}{\text{mol Fe}_2\text{O}_3} = 648 \text{ g Fe}_2\text{O}_3$

Mass % $\text{Fe}_2\text{O}_3 = \dfrac{648 \text{ g Fe}_2\text{O}_3}{752 \text{ g ore}} \times 100 = 86.2\%$

87.     $\dfrac{^{85}\text{Rb atoms}}{^{87}\text{Rb atoms}} = 2.591;$   If we had exactly 100 atoms, $x$ = number of $^{85}$Rb atoms and $100 - x$ = number of $^{87}$Rb atoms.

$\dfrac{x}{100 - x} = 2.591, \;\; x = 259.1 - (2.591)x, \;\; x = \dfrac{259.1}{3.591} = 72.15; \;\; 72.15\% \; ^{85}\text{Rb}$

$0.7215(84.9117) + 0.2785(A) = 85.4678, \;\; A = \dfrac{85.4678 - 61.26}{0.2785} = 86.92 \text{ amu}$

89.     The volume of a gas is proportional to the number of molecules of gas.  Thus the formulas are:

I:  $NH_3$        II:  $N_2H_4$        III:  $HN_3$

The mass ratios are:

I:  $\dfrac{4.634 \text{ g N}}{\text{g H}}$    II:  $\dfrac{6.949 \text{ g N}}{\text{g H}}$    III:  $\dfrac{41.7 \text{ g N}}{\text{g H}}$

If we set the atomic mass of H equal to 1.008, then the atomic mass for nitrogen is:

I:  14.01        II:  14.01        III.  14.0

For example, for compound I:  $\dfrac{A}{3(1.008)} = \dfrac{4.634}{1}, \;\; A = 14.01$

91.     $1.375 \text{ g AgI} \times \dfrac{1 \text{ mol AgI}}{234.8 \text{ g AgI}} = 5.856 \times 10^{-3} \text{ mol AgI} = 5.856 \times 10^{-3} \text{ mol I}$

$1.375 \text{ g AgI} \times \dfrac{126.9 \text{ g I}}{234.8 \text{ g AgI}} = 0.7431 \text{ g I}; \;\; XI_2 \text{ contains } 0.7431 \text{ g I and } 0.257 \text{ g X.}$

$5.856 \times 10^{-3} \text{ mol I} \times \dfrac{1 \text{ mol X}}{2 \text{ mol I}} = 2.928 \times 10^{-3} \text{ mol X}$

Molar mass $= \dfrac{0.257 \text{ g X}}{2.928 \times 10^{-3} \text{ mol X}} = \dfrac{87.8 \text{ g}}{\text{mol}}; \;\;$ atomic mass $= 87.8 \text{ amu}$  (X is Sr.)

93.    Assuming 1 mole of vitamin A (286.4 g vitamin A):

$$\text{mol C} = 286.4 \text{ g vitamin A} \times \frac{0.8396 \text{ g C}}{\text{g vitamin A}} \times \frac{1 \text{ mol C}}{12.011 \text{ g C}} = 20.00 \text{ mol C}$$

$$\text{mol H} = 286.4 \text{ g vitamin A} \times \frac{0.1056 \text{ g H}}{\text{g vitamin A}} \times \frac{1 \text{ mol H}}{1.0079 \text{ g H}} = 30.01 \text{ mol H}$$

Because 1 mole of vitamin A contains 20 mol C and 30 mol H, the molecular formula of vitamin A is $C_{20}H_{30}E$.  To determine E, lets calculate the molar mass of E:

$$286.4 \text{ g} = 20(12.01) + 30(1.008) + \text{molar mass E},  \text{molar mass E} = 16.0 \text{ g/mol}$$

From the periodic table, E = oxygen, and the molecular formula of vitamin A is $C_{20}H_{30}O$.

## Challenge Problems

95.    When the discharge voltage is low, the ions present are in the form of molecules.  When the discharge voltage is increased, the bonds in the molecules are broken, and the ions present are in the form of individual atoms.  Therefore, the high discharge data indicate that the ions $^{16}O^+$, $^{18}O^+$, and $^{40}Ar^+$ are present.  The only combination of these individual ions that can explain the mass data at low discharge is $^{16}O^{16}O^+$ (mass = 32), $^{16}O^{18}O^+$ (mass = 34), and $^{40}Ar^+$ (mass = 40).  Therefore, the gas mixture contains $^{16}O^{16}O$, $^{16}O^{18}O$, and $^{40}Ar$.  To determine the percent composition of each isotope, we use the relative intensity data from the high discharge data to determine the percentage that each isotope contributes to the total relative intensity.  For $^{40}Ar$:

$$\frac{1.0000}{0.7500 + 0.0015 + 1.0000} \times 100 = \frac{1.0000}{1.7515} \times 100 = 57.094\% \ ^{40}Ar$$

For $^{16}O$: $\dfrac{0.7500}{1.7515} \times 100 = 42.82\% \ ^{16}O$;  for $^{18}O$: $\dfrac{0.0015}{1.7515} \times 100 = 8.6 \times 10^{-2}\% \ ^{18}O$

*Note*: $^{18}F$ instead of $^{18}O$ could also explain the data.  However, OF(g) is not a stable compound.  This is why $^{18}O$ is the best choice because $O_2(g)$ does form.

97.    10.00 g $XCl_2$ + excess $Cl_2 \rightarrow$ 12.55 g $XCl_4$; 2.55 g Cl reacted with $XCl_2$ to form $XCl_4$. $XCl_4$ contains 2.55 g Cl and 10.00 g $XCl_2$.  From mole ratios, 10.00 g $XCl_2$ must also contain 2.55 g Cl; mass X in $XCl_2$ = 10.00 – 2.55 = 7.45 g X.

$$2.55 \text{ g Cl} \times \frac{1 \text{ mol Cl}}{35.45 \text{ g Cl}} \times \frac{1 \text{ mol } XCl_2}{2 \text{ mol Cl}} \times \frac{1 \text{ mol X}}{\text{mol } XCl_2} = 3.60 \times 10^{-2} \text{ mol X}$$

So, $3.60 \times 10^{-2}$ mol X  has a mass equal to 7.45 g X.  The molar mass of X is:

$$\frac{7.45 \text{ g X}}{3.60 \times 10^{-2} \text{ mol X}} = 207 \text{ g/mol X}; \quad \text{atomic mass} = 207 \text{ amu, so X is Pb.}$$

99.     For a gas, density and molar mass are directly proportional to each other.

$$\text{Molar mass XH}_n = 2.393(32.00) = \frac{76.58 \text{ g}}{\text{mol}}; \quad 0.803 \text{ g H}_2\text{O} \times \frac{2 \text{ mol H}}{18.02 \text{ g H}_2\text{O}}$$
$$= 8.91 \times 10^{-2} \text{ mol H}$$

$$\frac{8.91 \times 10^{-2} \text{ mol H}}{2.23 \times 10^{-2} \text{ mol XH}_n} = \frac{4 \text{ mol H}}{\text{mol XH}_n}$$

Molar mass X = $76.58 - 4(1.008 \text{ g}) = 72.55$ g/mol;   the element is Ge.

101.    $4.000 \text{ g M}_2\text{S}_3 \rightarrow 3.723 \text{ g MO}_2$

There must be twice as many moles of $MO_2$ as moles of $M_2S_3$ in order to balance M in the reaction.  Setting up an equation for $2(\text{mol M}_2\text{S}_3) = \text{mol MO}_2$ where A = molar mass M:

$$2\left(\frac{4.000 \text{ g}}{2\text{A} + 3(32.07)}\right) = \frac{3.723 \text{ g}}{\text{A} + 2(16.00)}, \quad \frac{8.000}{2\text{A} + 96.21} = \frac{3.723}{\text{A} + 32.00}$$

$(8.000)\text{A} + 256.0 = (7.446)\text{A} + 358.2, \quad (0.554)\text{A} = 102.2, \quad \text{A} = 184 \text{ g/mol; atomic mass}$
$$= 184 \text{ amu}$$

103.    The balanced equations are:

$$\text{C(s)} + 1/2 \text{ O}_2\text{(g)} \rightarrow \text{CO(g) and C(s)} + \text{O}_2\text{(g)} \rightarrow \text{CO}_2\text{(g)}$$

If we have 100.0 mol of products, then we have 72.0 mol $CO_2$, 16.0 mol CO, and 12.0 mol $O_2$.  The initial moles of C equals 72.0 (from $CO_2$) + 16.00 (from CO) = 88.0 mol C and the initial moles of $O_2$ equals 72.0 (from $CO_2$) + 16.0/2 (from CO) + 12.0 (unreacted $O_2$) = 92.0 mol $O_2$.  The initial reaction mixture contained:

$$\frac{92.0 \text{ mol O}_2}{88.0 \text{ mol C}} = 1.05 \text{ mol O}_2/\text{mol C}$$

105.    $LaH_{2.90}$ is the formula.  If only $La^{3+}$ is present, $LaH_3$ would be the formula.  If only $La^{2+}$ is present, $LaH_2$ would be the formula.  Let $x = \text{mol La}^{2+}$ and $y = \text{mol La}^{3+}$:

$$(\text{La}^{2+})_x(\text{La}^{3+})_y\text{H}_{(2x + 3y)} \text{ where } x + y = 1.00 \text{ and } 2x + 3y = 2.90$$

Solving by simultaneous equations:

$$2x + 3y = 2.90$$
$$\underline{-2x - 2y = -2.00}$$
$$y = 0.90 \text{ and } x = 0.10$$

$LaH_{2.90}$ contains $\dfrac{1}{10}$ $La^{2+}$, or 10.% $La^{2+}$, and $\dfrac{9}{10}$ $La^{3+}$, or 90.% $La^{3+}$.

107 .    Let $x$ = mass KCl and $y$ = mass $KNO_3$.  Assuming 100.0 g of mixture, $x + y = 100.0$ g.

Molar mass KCl = 74.55 g/mol;  molar mass $KNO_3$ = 101.11 g/mol

Mol KCl = $\dfrac{x}{74.55}$ ;  mol $KNO_3$ = $\dfrac{y}{101.11}$

Knowing that the mixture is 43.2% K, then in the 100.0 g mixture:

$$39.10\left( \dfrac{x}{74.55} + \dfrac{y}{101.11} \right) = 43.2$$

We have two equations and two unknowns:

$$(0.5245)x + (0.3867)y = 43.2$$
$$x + y = 100.0$$

Solving, $x$ = 32.9 g KCl;  $\dfrac{32.9 \text{ g}}{100.0 \text{ g}} \times 100 = 32.9\%$ KCl

109.    The balanced equations are:

$$4\ NH_3(g) + 5\ O_2(g) \rightarrow 4\ NO(g) + 6\ H_2O(g) \text{ and } 4\ NH_3(g) + 7\ O_2(g) \rightarrow 4\ NO_2(g)$$
$$+ 6\ H_2O(g)$$

Let $4x$ = number of moles of NO formed, and let $4y$ = number of moles of $NO_2$ formed.
Then:

$$4x\ NH_3 + 5x\ O_2 \rightarrow 4x\ NO + 6x\ H_2O \text{ and } 4y\ NH_3 + 7y\ O_2 \rightarrow 4y\ NO_2 + 6y\ H_2O$$

All the $NH_3$ reacted, so $4x + 4y = 2.00$.  $10.00 - 6.75 = 3.25$ mol $O_2$ reacted, so $5x + 7y$
$$= 3.25.$$

Solving by the method of simultaneous equations:

$$20x + 28y = 13.0$$
$$\underline{-20x - 20y = -10.0}$$
$$8y = 3.0, \quad y = 0.38; \quad 4x + 4 \times 0.38 = 2.00, \quad x = 0.12$$

Mol NO = $4x = 4 \times 0.12 = 0.48$ mol NO formed

# CHAPTER 4

# TYPES OF CHEMICAL REACTIONS AND SOLUTION STOICHIOMETRY

## Aqueous Solutions: Strong and Weak Electrolytes

11.   a.   Polarity is a term applied to covalent compounds. Polar covalent compounds have an unequal sharing of electrons in bonds that results in unequal charge distribution in the overall molecule. Polar molecules have a partial negative end and a partial positive end. These are not full charges as in ionic compounds but are charges much smaller in magnitude. Water is a polar molecule and dissolves other polar solutes readily. The oxygen end of water (the partial negative end of the polar water molecule) aligns with the partial positive end of the polar solute, whereas the hydrogens of water (the partial positive end of the polar water molecule) align with the partial negative end of the solute. These opposite charge attractions stabilize polar solutes in water. This process is called hydration. Nonpolar solutes do not have permanent partial negative and partial positive ends; nonpolar solutes are not stabilized in water and do not dissolve.

      b.   KF is a soluble ionic compound, so it is a strong electrolyte. KF(aq) actually exists as separate hydrated $K^+$ ions and hydrated $F^-$ ions in solution. $C_6H_{12}O_6$ is a polar covalent molecule that is a nonelectrolyte. $C_6H_{12}O_6$ is hydrated as described in part a.

      c.   RbCl is a soluble ionic compound, so it exists as separate hydrated $Rb^+$ ions and hydrated $Cl^-$ ions in solution. AgCl is an insoluble ionic compound so the ions stay together in solution and fall to the bottom of the container as a precipitate.

      d.   $HNO_3$ is a strong acid and exists as separate hydrated $H^+$ ions and hydrated $NO_3^-$ ions in solution. CO is a polar covalent molecule and is hydrated as explained in part a.

13.   a.   $Ba(NO_3)_2(aq) \rightarrow Ba^{2+}(aq) + 2\ NO_3^-(aq)$;  picture iv represents the $Ba^{2+}$ and $NO_3^-$ ions present in $Ba(NO_3)_2(aq)$.

      b.   $NaCl(aq) \rightarrow Na^+(aq) + Cl^-(aq)$;  picture ii represents $NaCl(aq)$.

      c.   $K_2CO_3(aq) \rightarrow 2\ K^+(aq) + CO_3^{2-}(aq)$;  picture iii represents $K_2CO_3(aq)$.

      d.   $MgSO_4(aq) \rightarrow Mg^{2+}(aq) + SO_4^{2-}(aq)$;  picture i represents $MgSO_4(aq)$.

23

## Solution Concentration:  Molarity

15.   a.   $2.00 \text{ L} \times \dfrac{0.250 \text{ mol NaOH}}{\text{L}} \times \dfrac{40.00 \text{ g NaOH}}{\text{mol}} = 20.0 \text{ g NaOH}$

Place 20.0 g NaOH in a 2-L volumetric flask; add water to dissolve the NaOH, and fill to the mark with water, mixing several times along the way.

   b.   $2.00 \text{ L} \times \dfrac{0.250 \text{ mol NaOH}}{\text{L}} \times \dfrac{1 \text{ L stock}}{1.00 \text{ mol NaOH}} = 0.500 \text{ L}$

Add 500. mL of 1.00 $M$ NaOH stock solution to a 2-L volumetric flask; fill to the mark with water, mixing several times along the way.

   c.   $2.00 \text{ L} \times \dfrac{0.100 \text{ mol K}_2\text{CrO}_4}{\text{L}} \times \dfrac{194.20 \text{ g K}_2\text{CrO}_4}{\text{mol K}_2\text{CrO}_4} = 38.8 \text{ g K}_2\text{CrO}_4$

Similar to the solution made in part a, instead using 38.8 g $K_2CrO_4$.

   d.   $2.00 \text{ L} \times \dfrac{0.100 \text{ mol K}_2\text{CrO}_4}{\text{L}} \times \dfrac{1 \text{ L stock}}{1.75 \text{ mol K}_2\text{CrO}_4} = 0.114 \text{ L}$

Similar to the solution made in part  b, instead using 114 mL of the 1.75 $M\,K_2CrO_4$ stock solution.

17.   $\text{Mol Na}_2\text{CO}_3 = 0.0700 \text{ L} \times \dfrac{3.0 \text{ mol Na}_2\text{CO}_3}{\text{L}} = 0.21 \text{ mol Na}_2\text{CO}_3$

$Na_2CO_3(s) \rightarrow 2 \text{ Na}^+(aq) + CO_3^{2-}(aq); \text{ mol Na}^+ = 2(0.21) = 0.42 \text{ mol}$

$\text{Mol NaHCO}_3 = 0.0300 \text{ L} \times \dfrac{1.0 \text{ mol NaHCO}_3}{\text{L}} = 0.030 \text{ mol NaHCO}_3$

$NaHCO_3(s) \rightarrow Na^+(aq) + HCO_3^-(aq); \text{ mol Na}^+ = 0.030 \text{ mol}$

$M_{\text{Na}^+} = \dfrac{\text{total mol Na}^+}{\text{total volume}} = \dfrac{0.42 \text{ mol} + 0.030 \text{ mol}}{0.0700 \text{ L} + 0.0300 \text{ L}} = \dfrac{0.45 \text{ mol}}{0.1000 \text{ L}} = 4.5 \ M\,\text{Na}^+$

19.   Stock solution $= \dfrac{10.0 \text{ mg}}{500.0 \text{ mL}} = \dfrac{10.0 \times 10^{-3} \text{ g}}{500.0 \text{ mL}} = \dfrac{2.00 \times 10^{-5} \text{ g steroid}}{\text{mL}}$

$100.0 \times 10^{-6} \text{ L stock} \times \dfrac{1000 \text{ mL}}{\text{L}} \times \dfrac{2.00 \times 10^{-5} \text{ g steroid}}{\text{mL}} = 2.00 \times 10^{-6} \text{ g steroid}$

This is diluted to a final volume of 100.0 mL.

$\dfrac{2.00 \times 10^{-6} \text{ g steroid}}{100.0 \text{ mL}} \times \dfrac{1000 \text{ mL}}{\text{L}} \times \dfrac{1 \text{ mol steroid}}{336.4 \text{ g steroid}} = 5.95 \times 10^{-8} \ M \text{ steroid}$

21.   a.   5.0 ppb Hg in water = $\dfrac{5.0 \text{ ng Hg}}{\text{mL H}_2\text{O}} = \dfrac{5.0 \times 10^{-9} \text{ g Hg}}{\text{mL H}_2\text{O}}$

$\dfrac{5.0 \times 10^{-9} \text{ g Hg}}{\text{mL}} \times \dfrac{1 \text{ mol Hg}}{200.6 \text{ g Hg}} \times \dfrac{1000 \text{ mL}}{\text{L}} = 2.5 \times 10^{-8} \, M \text{ Hg}$

b.   $\dfrac{1.0 \times 10^{-9} \text{ g CHCl}_3}{\text{mL}} \times \dfrac{1 \text{ mol CHCl}_3}{119.4 \text{ g CHCl}_3} \times \dfrac{1000 \text{ mL}}{\text{L}} = 8.4 \times 10^{-9} \, M \text{ CHCl}_3$

c.   10.0 ppm As = $\dfrac{10.0 \text{ μg As}}{\text{mL}} = \dfrac{10.0 \times 10^{-6} \text{ g As}}{\text{mL}}$

$\dfrac{10.0 \times 10^{-6} \text{ g As}}{\text{mL}} \times \dfrac{1 \text{ mol As}}{74.92 \text{ g As}} \times \dfrac{1000 \text{ mL}}{\text{L}} = 1.33 \times 10^{-4} \, M \text{ As}$

d.   $\dfrac{0.10 \times 10^{-6} \text{ g DDT}}{\text{mL}} \times \dfrac{1 \text{ mol DDT}}{354.5 \text{ g DDT}} \times \dfrac{1000 \text{ mL}}{\text{L}} = 2.8 \times 10^{-7} \, M \text{ DDT}$

## Precipitation Reactions

23.   Use the solubility rules in Table 4.1. Some soluble bromides by Rule 2 would be NaBr, KBr, and $NH_4Br$ (there are others). The insoluble bromides by Rule 3 would be AgBr, $PbBr_2$, and $Hg_2Br_2$. Similar reasoning is used for the other parts to this problem.

Sulfates: $Na_2SO_4$, $K_2SO_4$, and $(NH_4)_2SO_4$ (and others) would be soluble, and $BaSO_4$, $CaSO_4$, and $PbSO_4$ (or $Hg_2SO_4$) would be insoluble.

Hydroxides: NaOH, KOH, $Ca(OH)_2$ (and others) would be soluble, and $Al(OH)_3$, $Fe(OH)_3$, and $Cu(OH)_2$ (and others) would be insoluble.

Phosphates: $Na_3PO_4$, $K_3PO_4$, $(NH_4)_3PO_4$ (and others) would be soluble, and $Ag_3PO_4$, $Ca_3(PO_4)_2$, and $FePO_4$ (and others) would be insoluble.

Lead: $PbCl_2$, $PbBr_2$, $PbI_2$, $Pb(OH)_2$, $PbSO_4$, and PbS (and others) would be insoluble. $Pb(NO_3)_2$ would be a soluble $Pb^{2+}$ salt.

25.   For the following answers, the balanced molecular equation is first, followed by the complete ionic equation, and then the net ionic equation.

a.   $(NH_4)_2SO_4(aq) + Ba(NO_3)_2(aq) \rightarrow 2 \, NH_4NO_3(aq) + BaSO_4(s)$

$2 \, NH_4^{+}(aq) + SO_4^{2-}(aq) + Ba^{2+}(aq) + 2 \, NO_3^{-}(aq) \rightarrow 2 \, NH_4^{+}(aq) + 2 \, NO_3^{-}(aq)$
$+ BaSO_4(s)$

$Ba^{2+}(aq) + SO_4^{2-}(aq) \rightarrow BaSO_4(s)$ is the net ionic equation (spectator ions omitted).

b.  $Pb(NO_3)_2(aq) + 2\ NaCl(aq) \rightarrow PbCl_2(s) + 2\ NaNO_3(aq)$

$Pb^{2+}(aq) + 2\ NO_3^-(aq) + 2\ Na^+(aq) + 2\ Cl^-(aq) \rightarrow PbCl_2(s) + 2\ Na^+(aq) + 2\ NO_3^-(aq)$

$Pb^{2+}(aq) + 2\ Cl^-(aq) \rightarrow PbCl_2(s)$

c.  The possible products, potassium phosphate and sodium nitrate, are both soluble in water. Therefore, no reaction occurs.

d.  No reaction occurs because all possible products are soluble.

e.  $CuCl_2(aq) + 2\ NaOH(aq) \rightarrow Cu(OH)_2(s) + 2\ NaCl(aq)$

$Cu^{2+}(aq) + 2\ Cl^-(aq) + 2\ Na^+(aq) + 2\ OH^-(aq) \rightarrow Cu(OH)_2(s) + 2\ Na^+(aq) + 2\ Cl^-(aq)$

$Cu^{2+}(aq) + 2\ OH^-(aq) \rightarrow Cu(OH)_2(s)$

27.  a.  When $CuSO_4(aq)$ is added to $Na_2S(aq)$, the precipitate that forms is $CuS(s)$. Therefore, $Na^+$ (the gray spheres) and $SO_4^{2-}$ (the bluish green spheres) are the spectator ions.

$CuSO_4(aq) + Na_2S(aq) \rightarrow CuS(s) + Na_2SO_4(aq);\ Cu^{2+}(aq) + S^{2-}(aq) \rightarrow CuS(s)$

b.  When $CoCl_2(aq)$ is added to $NaOH(aq)$, the precipitate that forms is $Co(OH)_2(s)$. Therefore, $Na^+$ (the gray spheres) and $Cl^-$ (the green spheres) are the spectator ions.

$CoCl_2(aq) + 2\ NaOH(aq) \rightarrow Co(OH)_2(s) + 2\ NaCl(aq)$

$Co^{2+}(aq) + 2\ OH^-(aq) \rightarrow Co(OH)_2(s)$

c.  When $AgNO_3(aq)$ is added to $KI(aq)$, the precipitate that forms is $AgI(s)$. Therefore, $K^+$ (the red spheres) and $NO_3^-$ (the blue spheres) are the spectator ions.

$AgNO_3(aq) + KI(aq) \rightarrow AgI(s) + KNO_3(aq);\ Ag^+(aq) + I^-(aq) \rightarrow AgI(s)$

29.  Because a precipitate formed with $Na_2SO_4$, the possible cations are $Ba^{2+}$, $Pb^{2+}$, $Hg_2^{2+}$, and $Ca^{2+}$ (from the solubility rules). Because no precipitate formed with KCl, $Pb^{2+}$ and $Hg_2^{2+}$ cannot be present. Because both $Ba^{2+}$ and $Ca^{2+}$ form soluble chlorides and soluble hydroxides, both these cations could be present. Therefore, the cations could be $Ba^{2+}$ and $Ca^{2+}$ (by the solubility rules in Table 4.1). For students who do a more rigorous study of solubility, $Sr^{2+}$ could also be a possible cation (it forms an insoluble sulfate salt, whereas the chloride and hydroxide salts of strontium are soluble).

31.  $2\ AgNO_3(aq) + CaCl_2(aq) \rightarrow 2\ AgCl(s) + Ca(NO_3)_2(aq)$

$$\text{Mol } AgNO_3 = 0.1000\ L \times \frac{0.20\ \text{mol } AgNO_3}{L} = 0.020\ \text{mol } AgNO_3$$

$$\text{Mol CaCl}_2 = 0.1000 \text{ L} \times \frac{0.15 \text{ mol CaCl}_2}{\text{L}} = 0.015 \text{ mol CaCl}_2$$

The required mol $AgNO_3$ to mol $CaCl_2$ ratio is 2 : 1 (from the balanced equation). The actual mole ratio present is $0.020/0.015 = 1.3$ (1.3 : 1). Therefore, $AgNO_3$ is the limiting reagent.

$$\text{Mass AgCl} = 0.020 \text{ mol AgNO}_3 \times \frac{1 \text{ mol AgCl}}{1 \text{ mol AgNO}_3} \times \frac{143.4 \text{ g AgCl}}{\text{mol AgCl}} = 2.9 \text{ g AgCl}$$

The net ionic equation is $Ag^+(aq) + Cl^-(aq) \rightarrow AgCl(s)$. The ions remaining in solution are the unreacted $Cl^-$ ions and the spectator ions, $NO_3^-$ and $Ca^{2+}$ (all $Ag^+$ is used up in forming AgCl). The moles of each ion present initially (before reaction) can be easily determined from the moles of each reactant. 0.020 mol $AgNO_3$ dissolves to form 0.020 mol $Ag^+$ and 0.020 mol $NO_3^-$. 0.015 mol $CaCl_2$ dissolves to form 0.015 mol $Ca^{2+}$ and $2(0.015) = 0.030$ mol $Cl^-$.

Mol unreacted $Cl^-$ = 0.030 mol $Cl^-$ initially – 0.020 mol $Cl^-$ reacted

Mol unreacted $Cl^-$ = 0.010 mol $Cl^-$ unreacted

$$M_{Cl^-} = \frac{0.010 \text{ mol Cl}^-}{\text{total volume}} = \frac{0.010 \text{ mol Cl}^-}{0.1000 \text{ L} + 0.1000 \text{ L}} = 0.050 \ M \ Cl^-$$

The molarity of the spectator ions are:

$$M_{NO_3^-} = \frac{0.020 \text{ mol NO}_3^-}{0.2000 \text{ L}} = 0.10 \ M \ NO_3^-; \quad M_{Ca^{2+}} = \frac{0.015 \text{ mol Ca}^{2+}}{0.2000 \text{ L}} = 0.075 \ M \ Ca^{2+}$$

33.    $2 \text{ AgNO}_3(aq) + \text{Na}_2\text{CrO}_4(aq) \rightarrow \text{Ag}_2\text{CrO}_4(s) + 2 \text{ NaNO}_3(aq)$

$$0.0750 \text{ L} \times \frac{0.100 \text{ mol AgNO}_3}{\text{L}} \times \frac{1 \text{ mol Na}_2\text{CrO}_4}{2 \text{ mol AgNO}_3} \times \frac{161.98 \text{ g Na}_2\text{CrO}_4}{\text{mol Na}_2\text{CrO}_4} = 0.607 \text{ g Na}_2\text{CrO}_4$$

35.    Use aluminum in the formulas to convert from mass of $Al(OH)_3$ to mass of $Al_2(SO_4)_3$ in the mixture.

$$0.107 \text{ g Al(OH)}_3 \times \frac{1 \text{ mol Al(OH)}_3}{78.00 \text{ g}} \times \frac{1 \text{ mol Al}^{3+}}{\text{mol Al(OH)}_3} \times \frac{1 \text{ mol Al}_2(\text{SO}_4)_3}{2 \text{ mol Al}^{3+}} \times$$

$$\frac{342.17 \text{ g Al}_2(\text{SO}_4)_3}{\text{mol Al}_2(\text{SO}_4)_3} = 0.235 \text{ g Al}_2(\text{SO}_4)_3$$

$$\text{Mass \% Al}_2(\text{SO}_4)_3 = \frac{0.235 \text{ g}}{1.45 \text{ g}} \times 100 = 16.2\%$$

37.    All the sulfur in $BaSO_4$ came from the saccharin. The conversion from $BaSO_4$ to saccharin uses the molar masses and formulas of each compound.

$$0.5032 \text{ g BaSO}_4 \times \frac{32.07 \text{ g S}}{233.4 \text{ g BaSO}_4} \times \frac{183.9 \text{ g saccharin}}{32.07 \text{ g S}} = 0.3949 \text{ g saccharin}$$

$$\frac{\text{Average mass}}{\text{Tablet}} = \frac{0.3949 \text{ g}}{10 \text{ tablets}} = \frac{3.949 \times 10^{-2} \text{ g}}{\text{tablet}} = \frac{39.49 \text{ mg}}{\text{tablet}}$$

$$\text{Average mass \%} = \frac{0.3949 \text{ g saccharin}}{0.5894 \text{ g}} \times 100 = 67.00\% \text{ saccharin by mass}$$

39.     $M_2SO_4(aq) + CaCl_2(aq) \rightarrow CaSO_4(s) + 2 \text{ MCl}(aq)$

$$1.36 \text{ g CaSO}_4 \times \frac{1 \text{ mol CaSO}_4}{136.15 \text{ g CaSO}_4} \times \frac{1 \text{ mol M}_2SO_4}{\text{mol CaSO}_4} = 9.99 \times 10^{-3} \text{ mol M}_2SO_4$$

From the problem, 1.42 g $M_2SO_4$ was reacted, so:

$$\text{molar mass} = \frac{1.42 \text{ g M}_2SO_4}{9.99 \times 10^{-3} \text{ mol M}_2SO_4} = 142 \text{ g/mol}$$

$142 \text{ amu} = 2(\text{atomic mass M}) + 32.07 + 4(16.00), \ \text{ atomic mass M} = 23 \text{ amu}$

From periodic table, M is Na(sodium).

## Acid-Base Reactions

41.     a.   Perchloric acid reacted with potassium hydroxide is a possibility.

$HClO_4(aq) + KOH(aq) \rightarrow H_2O(l) + KClO_4(aq)$

b.   Nitric acid reacted with cesium hydroxide is a possibility.

$HNO_3(aq) + CsOH(aq) \rightarrow H_2O(l) + CsNO_3(aq)$

c.   Hydroiodic acid reacted with calcium hydroxide is a possibility.

$2 \text{ HI}(aq) + Ca(OH)_2(aq) \rightarrow 2 \text{ H}_2O(l) + CaI_2(aq)$

43.     If we begin with 50.00 mL of 0.100 $M$ NaOH, then:

$$50.00 \times 10^{-3} \text{ L} \times \frac{0.100 \text{ mol}}{\text{L}} = 5.00 \times 10^{-3} \text{ mol NaOH to be neutralized.}$$

a.   $NaOH(aq) + HCl(aq) \rightarrow NaCl(aq) + H_2O(l)$

$$5.00 \times 10^{-3} \text{ mol NaOH} \times \frac{1 \text{ mol HCl}}{\text{mol NaOH}} \times \frac{1 \text{ L soln}}{0.100 \text{ mol}} = 5.00 \times 10^{-2} \text{ L or 50.0 mL}$$

b.   $2 NaOH(aq) + H_2SO_3(aq) \rightarrow 2 H_2O(l) + Na_2SO_3(aq)$

$$5.00 \times 10^{-3} \text{ mol NaOH} \times \frac{1 \text{ mol H}_2\text{SO}_3}{2 \text{ mol NaOH}} \times \frac{1 \text{ L soln}}{0.100 \text{ mol H}_2\text{SO}_3} = 2.50 \times 10^{-2} \text{ L or } 25.0 \text{ mL}$$

c.   $3 NaOH(aq) + H_3PO_4(aq) \rightarrow Na_3PO_4(aq) + 3 H_2O(l)$

$$5.00 \times 10^{-3} \text{ mol NaOH} \times \frac{1 \text{ mol H}_3\text{PO}_4}{3 \text{ mol NaOH}} \times \frac{1 \text{ L soln}}{0.200 \text{ mol H}_3\text{PO}_4} = 8.33 \times 10^{-3} \text{ L or } 8.33 \text{ mL}$$

d.   $HNO_3(aq) + NaOH(aq) \rightarrow H_2O(l) + NaNO_3(aq)$

$$5.00 \times 10^{-3} \text{ mol NaOH} \times \frac{1 \text{ mol HNO}_3}{\text{mol NaOH}} \times \frac{1 \text{ L soln}}{0.150 \text{ mol HNO}_3} = 3.33 \times 10^{-2} \text{ L or } 33.3 \text{ mL}$$

e.   $HC_2H_3O_2(aq) + NaOH(aq) \rightarrow H_2O(l) + NaC_2H_3O_2(aq)$

$$5.00 \times 10^{-3} \text{ mol NaOH} \times \frac{1 \text{ mol HC}_2\text{H}_3\text{O}_2}{\text{mol NaOH}} \times \frac{1 \text{ L soln}}{0.200 \text{ mol HC}_2\text{H}_3\text{O}_2} = 2.50 \times 10^{-2} \text{ L}$$

$$\text{or } 25.0 \text{ mL}$$

f.   $H_2SO_4(aq) + 2 NaOH(aq) \rightarrow 2 H_2O(l) + Na_2SO_4(aq)$

$$5.00 \times 10^{-3} \text{ mol NaOH} \times \frac{1 \text{ mol H}_2\text{SO}_4}{2 \text{ mol NaOH}} \times \frac{1 \text{ L soln}}{0.300 \text{ mol H}_2\text{SO}_4} = 8.33 \times 10^{-3} \text{ L or } 8.33 \text{ mL}$$

45.    The acid is a diprotic acid ($H_2A$) meaning that it has two $H^+$ ions in the formula to donate to a base. The reaction is $H_2A(aq) + 2 NaOH(aq) \rightarrow 2 H_2O(l) + Na_2A(aq)$, where $A^{2-}$ is what is left over from the acid formula when the two protons ($H^+$ ions) are reacted.

For the HCl reaction, the base has the ability to accept two protons. The most common examples are $Ca(OH)_2$, $Sr(OH)_2$, and $Ba(OH)_2$. A possible reaction would be $2 HCl(aq) + Ca(OH)_2(aq) \rightarrow 2 H_2O(l) + CaCl_2(aq)$.

47.    The pertinent reactions are:

$$2 NaOH(aq) + H_2SO_4(aq) \rightarrow Na_2SO_4(aq) + 2 H_2O(l)$$

$$HCl(aq) + NaOH(aq) \rightarrow NaCl(aq) + H_2O(l)$$

Amount of NaOH added $= 0.0500 \text{ L} \times \dfrac{0.213 \text{ mol}}{\text{L}} = 1.07 \times 10^{-2} \text{ mol NaOH}$

Amount of NaOH neutralized by HCl:

$$0.01321 \text{ L HCl} \times \frac{0.103 \text{ mol HCl}}{\text{L HCl}} \times \frac{1 \text{ mol NaOH}}{\text{mol HCl}} = 1.36 \times 10^{-3} \text{ mol NaOH}$$

The difference, $9.3 \times 10^{-3}$ mol, is the amount of NaOH neutralized by the sulfuric acid.

$$9.3 \times 10^{-3} \text{ mol NaOH} \times \frac{1 \text{ mol H}_2\text{SO}_4}{2 \text{ mol NaOH}} = 4.7 \times 10^{-3} \text{ mol H}_2\text{SO}_4$$

$$\text{Concentration of H}_2\text{SO}_4 = \frac{4.7 \times 10^{-3} \text{ mol}}{0.1000 \text{ L}} = 4.7 \times 10^{-2} \text{ } M \text{ H}_2\text{SO}_4$$

49.    $\text{HC}_2\text{H}_3\text{O}_2(aq) + \text{NaOH}(aq) \rightarrow \text{H}_2\text{O}(l) + \text{NaC}_2\text{H}_3\text{O}_2(aq)$

a.    $16.58 \times 10^{-3}$ L soln $\times \dfrac{0.5062 \text{ mol NaOH}}{\text{L soln}} \times \dfrac{1 \text{ mol acetic acid}}{\text{mol NaOH}}$

$$= 8.393 \times 10^{-3} \text{ mol acetic acid}$$

$$\text{Concentration of acetic acid} = \frac{8.393 \times 10^{-3} \text{ mol}}{0.01000 \text{ L}} = 0.8393 \text{ } M \text{ HC}_2\text{H}_3\text{O}_2$$

b.    If we have 1.000 L of solution:  total mass = 1000. mL $\times \dfrac{1.006 \text{ g}}{\text{mL}} = 1006$ g solution

$$\text{Mass of HC}_2\text{H}_3\text{O}_2 = 0.8393 \text{ mol} \times \frac{60.052 \text{ g}}{\text{mol}} = 50.40 \text{ g HC}_2\text{H}_3\text{O}_2$$

$$\text{Mass \% acetic acid} = \frac{50.40 \text{ g}}{1006 \text{ g}} \times 100 = 5.010\%$$

51.    HCl and $\text{HNO}_3$ are strong acids;  $\text{Ca(OH)}_2$ and RbOH are strong bases. The net ionic equation that occurs is $\text{H}^+(aq) + \text{OH}^-(aq) \rightarrow \text{H}_2\text{O}(l)$.

$$\text{Mol H}^+ = 0.0500 \text{ L} \times \frac{0.100 \text{ mol HCl}}{\text{L}} \times \frac{1 \text{ mol H}^+}{\text{mol HCl}} +$$

$$0.1000 \text{ L} \times \frac{0.200 \text{ mol HNO}_3}{\text{L}} \times \frac{1 \text{ mol H}^+}{\text{mol HNO}_3} = 0.00500 + 0.0200 = 0.0250 \text{ mol H}^+$$

$$\text{Mol OH}^- = 0.5000 \text{ L} \times \frac{0.0100 \text{ mol Ca(OH)}_2}{\text{L}} \times \frac{2 \text{ mol OH}^-}{\text{mol Ca(OH)}_2} +$$

$$0.2000 \text{ L} \times \frac{0.100 \text{ mol RbOH}}{\text{L}} \times \frac{1 \text{ mol OH}^-}{\text{mol RbOH}} = 0.0100 + 0.0200 = 0.0300 \text{ mol OH}^-$$

We have an excess of $\text{OH}^-$, so the solution is basic (not neutral). The moles of excess $\text{OH}^- = 0.0300$ mol $\text{OH}^-$ initially $- 0.0250$ mol $\text{OH}^-$ reacted (with $\text{H}^+$) $= 0.0050$ mol $\text{OH}^-$ excess.

$$M_{\text{OH}^-} = \frac{0.0050 \text{ mol OH}^-}{(0.05000 + 0.1000 + 0.5000 + 0.2000) \text{ L}} = \frac{0.0050 \text{ mol}}{0.8500 \text{ L}} = 5.9 \times 10^{-3} \text{ } M$$

53.    $Ba(OH)_2(aq) + 2 HCl(aq) \rightarrow BaCl_2(aq) + 2 H_2O(l); \quad H^+(aq) + OH^-(aq) \rightarrow H_2O(l)$

$75.0 \times 10^{-3} L \times \dfrac{0.250 \text{ mol HCl}}{L} = 1.88 \times 10^{-2} \text{ mol HCl} = 1.88 \times 10^{-2} \text{ mol H}^+ +$
$1.88 \times 10^{-2} \text{ mol Cl}^-$

$225.0 \times 10^{-3} L \times \dfrac{0.0550 \text{ mol Ba(OH)}_2}{L} = 1.24 \times 10^{-2} \text{ mol Ba(OH)}_2 = 1.24 \times 10^{-2} \text{ mol Ba}^{2+} +$
$2.48 \times 10^{-2} \text{ mol OH}^-$

The net ionic equation requires a 1 : 1 mol ratio between $OH^-$ and $H^+$. The actual mol $OH^-$ to mol $H^+$ ratio is greater than 1 : 1, so $OH^-$ is in excess. Because $1.88 \times 10^{-2}$ mol $OH^-$ will be neutralized by the $H^+$, we have $(2.48 - 1.88) \times 10^{-2} = 0.60 \times 10^{-2}$ mol $OH^-$ in excess.

$$M_{OH^-} = \frac{\text{mol OH}^- \text{ excess}}{\text{total volume}} = \frac{6.0 \times 10^{-3} \text{ mol OH}^-}{0.0750 \text{ L} + 0.2250 \text{ L}} = 2.0 \times 10^{-2} \, M \, OH^-$$

## Oxidation-Reduction Reactions

55.    Apply rules in Table 4.3.

a.    $KMnO_4$ is composed of $K^+$ and $MnO_4^-$ ions. Assign oxygen an oxidation state value of $-2$, which gives manganese a $+7$ oxidation state because the sum of oxidation states for all atoms in $MnO_4^-$ must equal the $1-$ charge on $MnO_4^-$. K, $+1$; O, $-2$; Mn, $+7$.

b.    Assign O a $-2$ oxidation state, which gives nickel a $+4$ oxidation state. Ni, $+4$; O, $-2$.

c.    $K_4Fe(CN)_6$ is composed of $K^+$ cations and $Fe(CN)_6^{4-}$ anions. $Fe(CN)_6^{4-}$ is composed of iron and $CN^-$ anions. For an overall anion charge of $4-$, iron must have a $+2$ oxidation state.

d.    $(NH_4)_2HPO_4$ is made of $NH_4^+$ cations and $HPO_4^{2-}$ anions. Assign $+1$ as oxidation state of H and $-2$ as the oxidation state of O. For N in $NH_4^+$: $x + 4(+1) = +1$, $x = -3 = $ oxidation state of N. For P in $HPO_4^{2-}$: $+1 + y + 4(-2) = -2$, $y = +5 = $ oxidation state of P.

e.    O, $-2$; P, $+3$                       f.    O, $-2$; Fe, $+ 8/3$

g.    O, $-2$; F, $-1$; Xe, $+6$            h.    F, $-1$; S, $+4$

i.    O, $-2$; C, $+2$                        j.    H, $+1$; O, $-2$; C, $0$

57.    a.    $SrCr_2O_7$: Composed of $Sr^{2+}$ and $Cr_2O_7^{2-}$ ions. Sr, $+2$; O, $-2$; Cr, $2x + 7(-2) = -2$, $x = +6$

b.    Cu, $+2$; Cl, $-1$        c.    O, $0$                d.    H, $+1$; O, $-1$

e.    $Mg^{2+}$ and $CO_3^{2-}$ ions present. Mg, $+2$; O, $-2$; C, $+4$;       f.    Ag, $0$

g.    $Pb^{2+}$ and $SO_3^{2-}$ ions present. Pb, $+2$; O, $-2$; S, $+4$;       h.    O, $-2$; Pb, $+4$

i.    $Na^+$ and $C_2O_4^{2-}$ ions present. Na, $+1$; O, $-2$; C, $2x + 4(-2) = -2$, $x = +3$

j.    O, $-2$; C, $+4$

k.   Ammonium ion has a 1+ charge ($NH_4^+$), and sulfate ion has a 2– charge ($SO_4^{2-}$). Therefore, the oxidation state of cerium must be +4 ($Ce^{4+}$).  H, +1;  N, –3;  O, –2;  S, +6

l.   O, –2;  Cr, +3

59.   a.   $Al(s) + 3\ HCl(aq) \rightarrow AlCl_3(aq) + 3/2\ H_2(g)$ or $2\ Al(s) + 6\ HCl(aq) \rightarrow 2\ AlCl_3(aq) + 3\ H_2(g)$

Hydrogen is reduced (goes from the +1 oxidation state to the 0 oxidation state), and aluminum Al is oxidized ($0 \rightarrow +3$).

b.   Balancing S is most complicated because sulfur is in both products. Balance C and H first;  then worry about S.

$CH_4(g) + 4\ S(s) \rightarrow CS_2(l) + 2\ H_2S(g)$

Sulfur is reduced ($0 \rightarrow -2$), and carbon is oxidized ($-4 \rightarrow +4$).

c.   Balance C and H first;  then balance O.

$C_3H_8(g) + 5\ O_2(g) \rightarrow 3\ CO_2(g) + 4\ H_2O(l)$

Oxygen is reduced ($0 \rightarrow -2$), and carbon is oxidized ($-8/3 \rightarrow +4$).

d.   Although this reaction is mass balanced, it is not charge balanced. We need 2 mol of silver on each side to balance the charge.

$Cu(s) + 2\ Ag^+(aq) \rightarrow 2\ Ag(s) + Cu^{2+}(aq)$

Silver is reduced ($+1 \rightarrow 0$), and copper is oxidized ($0 \rightarrow +2$).

61.   a.   Review Section 4.11 of the text for rules on balancing by the half-reaction method. The first step is to separate the reaction into two half-reactions, and then balance each half-reaction separately.

$(Cu \rightarrow Cu^{2+} + 2\ e^-) \times 3$ $\qquad\qquad\qquad\qquad$ $NO_3^- \rightarrow NO + 2\ H_2O$
$\qquad\qquad\qquad\qquad\qquad\qquad\qquad (3\ e^- + 4\ H^+ + NO_3^- \rightarrow NO + 2\ H_2O) \times 2$

Adding the two balanced half-reactions so electrons cancel:

$$3\ Cu \rightarrow 3\ Cu^{2+} + 6\ e^-$$
$$6\ e^- + 8\ H^+ + 2\ NO_3^- \rightarrow 2\ NO + 4\ H_2O$$

$$3\ Cu(s) + 8\ H^+(aq) + 2\ NO_3^-(aq) \rightarrow 3\ Cu^{2+}(aq) + 2\ NO(g) + 4\ H_2O(l)$$

b.   $(2\ Cl^- \rightarrow Cl_2 + 2\ e^-) \times 3$ $\qquad\qquad\qquad\qquad$ $Cr_2O_7^{2-} \rightarrow 2\ Cr^{3+} + 7\ H_2O$
$\qquad\qquad\qquad\qquad\qquad\qquad\qquad 6\ e^- + 14\ H^+ + Cr_2O_7^{2-} \rightarrow 2\ Cr^{3+} + 7\ H_2O$

Add the two balanced half-reactions with six electrons transferred:

$$6\ Cl^- \rightarrow 3\ Cl_2 + 6\ e^-$$
$$6\ e^- + 14\ H^+ + Cr_2O_7^{2-} \rightarrow 2\ Cr^{3+} + 7\ H_2O$$

$$14\ H^+(aq) + Cr_2O_7^{2-}(aq) + 6\ Cl^-(aq) \rightarrow 3\ Cl_2(g) + 2\ Cr^{3+}(aq) + 7\ H_2O(l)$$

c.              $Pb \rightarrow PbSO_4$                                    $PbO_2 \rightarrow PbSO_4$

$Pb + H_2SO_4 \rightarrow PbSO_4 + 2\ H^+$                    $PbO_2 + H_2SO_4 \rightarrow PbSO_4 + 2\ H_2O$

$Pb + H_2SO_4 \rightarrow PbSO_4 + 2\ H^+ + 2\ e^-$      $2\ e^- + 2\ H^+ + PbO_2 + H_2SO_4 \rightarrow PbSO_4 + 2\ H_2O$

Add the two half-reactions with two electrons transferred:

$$2\ e^- + 2\ H^+ + PbO_2 + H_2SO_4 \rightarrow PbSO_4 + 2\ H_2O$$

$$Pb + H_2SO_4 \rightarrow PbSO_4 + 2\ H^+ + 2\ e^-$$

$$\overline{Pb(s) + 2\ H_2SO_4(aq) + PbO_2(s) \rightarrow 2\ PbSO_4(s) + 2\ H_2O(l)}$$

This is the reaction that occurs in an automobile lead storage battery.

d.              $Mn^{2+} \rightarrow MnO_4^-$

$(4\ H_2O + Mn^{2+} \rightarrow MnO_4^- + 8\ H^+ + 5\ e^-) \times 2$

$$NaBiO_3 \rightarrow Bi^{3+} + Na^+$$

$$6\ H^+ + NaBiO_3 \rightarrow Bi^{3+} + Na^+ + 3\ H_2O$$

$$(2\ e^- + 6\ H^+ + NaBiO_3 \rightarrow Bi^{3+} + Na^+ + 3\ H_2O) \times 5$$

$$8\ H_2O + 2\ Mn^{2+} \rightarrow 2\ MnO_4^- + 16\ H^+ + 10\ e^-$$

$$10\ e^- + 30\ H^+ + 5\ NaBiO_3 \rightarrow 5\ Bi^{3+} + 5\ Na^+ + 15\ H_2O$$

$$\overline{8\ H_2O + 30\ H^+ + 2\ Mn^{2+} + 5\ NaBiO_3 \rightarrow 2\ MnO_4^- + 5\ Bi^{3+} + 5\ Na^+ + 15\ H_2O + 16\ H^+}$$

Simplifying:

$$14\ H^+(aq) + 2\ Mn^{2+}(aq) + 5\ NaBiO_3(s) \rightarrow 2\ MnO_4^-(aq) + 5\ Bi^{3+}(aq) + 5\ Na^+(aq) + 7\ H_2O(l)$$

e.              $H_3AsO_4 \rightarrow AsH_3$                             $(Zn \rightarrow Zn^{2+} + 2\ e^-) \times 4$

$H_3AsO_4 \rightarrow AsH_3 + 4\ H_2O$

$8\ e^- + 8\ H^+ + H_3AsO_4 \rightarrow AsH_3 + 4\ H_2O$

$$8\ e^- + 8\ H^+ + H_3AsO_4 \rightarrow AsH_3 + 4\ H_2O$$

$$4\ Zn \rightarrow 4\ Zn^{2+} + 8\ e^-$$

$$\overline{8\ H^+(aq) + H_3AsO_4(aq) + 4\ Zn(s) \rightarrow 4\ Zn^{2+}(aq) + AsH_3(g) + 4\ H_2O(l)}$$

f.              $As_2O_3 \rightarrow H_3AsO_4$

$As_2O_3 \rightarrow 2\ H_3AsO_4$

$(5\ H_2O + As_2O_3 \rightarrow 2\ H_3AsO_4 + 4\ H^+ + 4\ e^-) \times 3$

$$NO_3^- \rightarrow NO + 2\ H_2O$$

$$4\ H^+ + NO_3^- \rightarrow NO + 2\ H_2O$$

$$(3\ e^- + 4\ H^+ + NO_3^- \rightarrow NO + 2\ H_2O) \times 4$$

$$12\ e^- + 16\ H^+ + 4\ NO_3^- \rightarrow 4\ NO + 8\ H_2O$$

$$15\ H_2O + 3\ As_2O_3 \rightarrow 6\ H_3AsO_4 + 12\ H^+ + 12\ e^-$$

$$\overline{7\ H_2O(l) + 4\ H^+(aq) + 3\ As_2O_3(s) + 4\ NO_3^-(aq) \rightarrow 4\ NO(g) + 6\ H_3AsO_4(aq)}$$

g.   $(2 Br^- \rightarrow Br_2 + 2 e^-) \times 5$                              $MnO_4^- \rightarrow Mn^{2+} + 4 H_2O$

$(5 e^- + 8 H^+ + MnO_4^- \rightarrow Mn^{2+} + 4 H_2O) \times 2$

$10 Br^- \rightarrow 5 Br_2 + 10 e^-$

$10 e^- + 16 H^+ + 2 MnO_4^- \rightarrow 2 Mn^{2+} + 8 H_2O$

---

$16 H^+(aq) + 2 MnO_4^-(aq) + 10 Br^-(aq) \rightarrow 5 Br_2(l) + 2 Mn^{2+}(aq) + 8 H_2O(l)$

h.   $CH_3OH \rightarrow CH_2O$                                    $Cr_2O_7^{2-} \rightarrow Cr^{3+}$

$(CH_3OH \rightarrow CH_2O + 2 H^+ + 2 e^-) \times 3$              $14 H^+ + Cr_2O_7^{2-} \rightarrow 2 Cr^{3+} + 7 H_2O$

$6 e^- + 14 H^+ + Cr_2O_7^{2-} \rightarrow 2 Cr^{3+} + 7 H_2O$

$3 CH_3OH \rightarrow 3 CH_2O + 6 H^+ + 6 e^-$

$6 e^- + 14 H^+ + Cr_2O_7^{2-} \rightarrow 2 Cr^{3+} + 7 H_2O$

---

$8 H^+(aq) + 3 CH_3OH(aq) + Cr_2O_7^{2-}(aq) \rightarrow 2 Cr^{3+}(aq) + 3 CH_2O(aq) + 7 H_2O(l)$

63.   a.   $HCl(aq)$ dissociates to $H^+(aq) + Cl^-(aq)$.  For simplicity, let's use $H^+$ and $Cl^-$ separately.

$H^+ \rightarrow H_2$                                       $Fe \rightarrow HFeCl_4$

$(2 H^+ + 2 e^- \rightarrow H_2) \times 3$                  $(H^+ + 4 Cl^- + Fe \rightarrow HFeCl_4 + 3 e^-) \times 2$

$6 H^+ + 6 e^- \rightarrow 3 H_2$

$2 H^+ + 8 Cl^- + 2 Fe \rightarrow 2 HFeCl_4 + 6 e^-$

---

$8 H^+ + 8 Cl^- + 2 Fe \rightarrow 2 HFeCl_4 + 3 H_2$

or       $8 HCl(aq) + 2 Fe(s) \rightarrow 2 HFeCl_4(aq) + 3 H_2(g)$

b.                       $IO_3^- \rightarrow I_3^-$                              $I^- \rightarrow I_3^-$

$3 IO_3^- \rightarrow I_3^-$                              $(3 I^- \rightarrow I_3^- + 2 e^-) \times 8$

$3 IO_3^- \rightarrow I_3^- + 9 H_2O$

$16 e^- + 18 H^+ + 3 IO_3^- \rightarrow I_3^- + 9 H_2O$

$16 e^- + 18 H^+ + 3 IO_3^- \rightarrow I_3^- + 9 H_2O$

$24 I^- \rightarrow 8 I_3^- + 16 e^-$

---

$18 H^+ + 24 I^- + 3 IO_3^- \rightarrow 9 I_3^- + 9 H_2O$

Reducing:  $6 H^+(aq) + 8 I^-(aq) + IO_3^-(aq) \rightarrow 3 I_3^-(aq) + 3 H_2O(l)$

c.   $(Ce^{4+} + e^- \rightarrow Ce^{3+}) \times 97$

$Cr(NCS)_6^{4-} \rightarrow Cr^{3+} + NO_3^- + CO_2 + SO_4^{2-}$

$54 H_2O + Cr(NCS)_6^{4-} \rightarrow Cr^{3+} + 6 NO_3^- + 6 CO_2 + 6 SO_4^{2-} + 108 H^+$

Charge on left $= -4$.  Charge on right $= +3 + 6(-1) + 6(-2) + 108(+1) = +93$.  Add 97 $e^-$ to the product side, and then add the two balanced half-reactions with a common factor of 97 $e^-$ transferred.

$$54\ H_2O + Cr(NCS)_6{}^{4-} \rightarrow Cr^{3+} + 6\ NO_3{}^- + 6\ CO_2 + 6\ SO_4{}^{2-} + 108\ H^+ + 97\ e^-$$
$$97\ e^- + 97\ Ce^{4+} \rightarrow 97\ Ce^{3+}$$

$$97\ Ce^{4+}(aq) + 54\ H_2O(l) + Cr(NCS)_6{}^{4-}(aq) \rightarrow 97\ Ce^{3+}(aq) + Cr^{3+}(aq) + 6\ NO_3{}^-(aq)$$
$$+ 6\ CO_2(g) + 6\ SO_4{}^{2-}(aq) + 108\ H^+(aq)$$

This is very complicated.  A check of the net charge is a good check to see if the equation is balanced.  Left:   charge $= 97(+4) -4 = +384$.  Right:   charge $= 97(+3) + 3 + 6(-1) + 6(-2) + 108(+1) = +384$.

d.              $CrI_3 \rightarrow CrO_4{}^{2-} + IO_4{}^-$                               $Cl_2 \rightarrow Cl^-$
$(16\ H_2O + CrI_3 \rightarrow CrO_4{}^{2-} + 3\ IO_4{}^- + 32\ H^+ + 27\ e^-) \times 2$        $(2\ e^- + Cl_2 \rightarrow 2\ Cl^-) \times 27$

Common factor is a transfer of 54 $e^-$.

$$54\ e^- + 27\ Cl_2 \rightarrow 54\ Cl^-$$
$$32\ H_2O + 2\ CrI_3 \rightarrow 2\ CrO_4{}^{2-} + 6\ IO_4{}^- + 64\ H^+ + 54\ e^-$$

$$32\ H_2O + 2\ CrI_3 + 27\ Cl_2 \rightarrow 54\ Cl^- + 2\ CrO_4{}^{2-} + 6\ IO_4{}^- + 64\ H^+$$

Add 64 $OH^-$ to both sides and convert 64 $H^+$ into 64 $H_2O$.

$$64\ OH^- + 32\ H_2O + 2\ CrI_3 + 27\ Cl_2 \rightarrow 54\ Cl^- + 2\ CrO_4{}^{2-} + 6\ IO_4{}^- + 64\ H_2O$$

Reducing gives:

$$64\ OH^-(aq) + 2\ CrI_3(s) + 27\ Cl_2(g) \rightarrow 54\ Cl^-(aq) + 2\ CrO_4{}^{2-}\ (aq) + 6\ IO_4{}^-(aq)$$
$$+ 32\ H_2O(l)$$

e.                    $Ce^{4+} \rightarrow Ce(OH)_3$
$(e^- + 3\ H_2O + Ce^{4+} \rightarrow Ce(OH)_3 + 3\ H^+) \times 61$

$$Fe(CN)_6{}^{4-} \rightarrow Fe(OH)_3 + CO_3{}^{2-} + NO_3{}^-$$
$$Fe(CN)_6{}^{4-} \rightarrow Fe(OH)_3 + 6\ CO_3{}^{2-} + 6\ NO_3{}^-$$

There are 39 extra O atoms on right.  Add 39 $H_2O$ to left, then add 75 $H^+$ to right to balance $H^+$.

$$39\ H_2O + Fe(CN)_6{}^{4-} \rightarrow Fe(OH)_3 + 6\ CO_3{}^{2-} + 6\ NO_3{}^- + 75\ H^+$$
net charge $= 4-$           net charge $= 57+$

Add 61 $e^-$ to the product side, and then add the two balanced half-reactions with a common factor of 61 $e^-$ transferred.

$$39\ H_2O + Fe(CN)_6{}^{4-} \rightarrow Fe(OH)_3 + 6\ CO_3{}^- + 6\ NO_3{}^- + 75\ H^+ + 61\ e^-$$
$$61\ e^- + 183\ H_2O + 61\ Ce^{4+} \rightarrow 61\ Ce(OH)_3 + 183\ H^+$$

$$222\ H_2O + Fe(CN)_6{}^{4-} + 61\ Ce^{4+} \rightarrow 61\ Ce(OH)_3 + Fe(OH)_3 + 6\ CO_3{}^{2-} + 6\ NO_3{}^- + 258\ H^+$$

Adding 258 OH⁻ to each side, and then reducing gives:

$$258 \, OH^-(aq) + Fe(CN)_6^{4-}(aq) + 61 \, Ce^{4+}(aq) \rightarrow 61 \, Ce(OH)_3(s) + Fe(OH)_3(s)$$
$$+ 6 \, CO_3^{2-}(aq) + 6 \, NO_3^-(aq) + 36 \, H_2O(l)$$

65.     $(H_2C_2O_4 \rightarrow 2 \, CO_2 + 2 \, H^+ + 2 \, e^-) \times 5$          $(5 \, e^- + 8 \, H^+ + MnO_4^- \rightarrow Mn^{2+} + 4 \, H_2O) \times 2$

$$5 \, H_2C_2O_4 \rightarrow 10 \, CO_2 + 10 \, H^+ + 10 \, e^-$$
$$10 \, e^- + 16 \, H^+ + 2 \, MnO_4^- \rightarrow 2 \, Mn^{2+} + 8 \, H_2O$$

$$\overline{6 \, H^+(aq) + 5 \, H_2C_2O_4(aq) + 2 \, MnO_4^-(aq) \rightarrow 10 \, CO_2(g) + 2 \, Mn^{2+}(aq) + 8 \, H_2O(l)}$$

$$0.1058 \text{ g } H_2C_2O_4 \times \frac{1 \text{ mol } H_2C_2O_4}{90.034 \text{ g}} \times \frac{2 \text{ mol } MnO_4^-}{5 \text{ mol } H_2C_2O_4} = 4.700 \times 10^{-4} \text{ mol } MnO_4^-$$

$$\text{Molarity} = \frac{4.700 \times 10^{-4} \text{ mol } MnO_4^-}{28.97 \text{ mL}} \times \frac{1000 \text{ mL}}{L} = 1.622 \times 10^{-2} \, M \, MnO_4^-$$

67.                                              $(Fe^{2+} \rightarrow Fe^{3+} + e^-) \times 5$
$$5 \, e^- + 8 \, H^+ + MnO_4^- \rightarrow Mn^{2+} + 4 \, H_2O$$

$$\overline{8 \, H^+(aq) + MnO_4^-(aq) + 5 \, Fe^{2+}(aq) \rightarrow 5 \, Fe^{3+}(aq) + Mn^{2+}(aq) + 4 \, H_2O(l)}$$

From the titration data we can get the number of moles of $Fe^{2+}$. We then convert this to a mass of iron and calculate the mass percent of iron in the sample.

$$38.37 \times 10^{-3} \text{ L } MnO_4^- \times \frac{0.0198 \text{ mol } MnO_4^-}{L} \times \frac{5 \text{ mol } Fe^{2+}}{\text{mol } MnO_4^-} = 3.80 \times 10^{-3} \text{ mol } Fe^{2+}$$

$$= 3.80 \times 10^{-3} \text{ mol Fe present}$$

$$3.80 \times 10^{-3} \text{ mol Fe} \times \frac{55.85 \text{ g Fe}}{\text{mol Fe}} = 0.212 \text{ g Fe}$$

$$\text{Mass \% Fe} = \frac{0.212 \text{ g}}{0.6128 \text{ g}} \times 100 = 34.6\% \text{ Fe}$$

69.     $Mg(s) + 2 \, HCl(aq) \rightarrow MgCl_2(aq) + H_2(g)$

$$3.00 \text{ g Mg} \times \frac{1 \text{ mol Mg}}{24.31 \text{ g Mg}} \times \frac{2 \text{ mol HCl}}{\text{mol Mg}} \times \frac{1 \text{ L HCl}}{5.0 \text{ mol HCl}} = 0.0494 \text{ L} = 49.4 \text{ mL HCl}$$

## Additional Exercises

71.     Mol $CaCl_2$ present $= 0.230$ L $CaCl_2 \times \dfrac{0.275 \text{ mol } CaCl_2}{\text{L } CaCl_2} = 6.33 \times 10^{-2}$ mol $CaCl_2$

The volume of $CaCl_2$ solution after evaporation is:

$$6.33 \times 10^{-2} \text{ mol } CaCl_2 \times \frac{1 \text{ L } CaCl_2}{1.10 \text{ mol } CaCl_2} = 5.75 \times 10^{-2} \text{ L} = 57.5 \text{ mL } CaCl_2$$

Volume $H_2O$ evaporated $= 230.$ mL $- 57.5$ mL $= 173$ mL $H_2O$ evaporated

73.     a.   $MgCl_2(aq) + 2\ AgNO_3(aq) \rightarrow 2\ AgCl(s) + Mg(NO_3)_2(aq)$

$$0.641 \text{ g } AgCl \times \frac{1 \text{ mol } AgCl}{143.4 \text{ g } AgCl} \times \frac{1 \text{ mol } MgCl_2}{2 \text{ mol } AgCl} \times \frac{95.21 \text{ g}}{\text{mol } MgCl_2} = 0.213 \text{ g } MgCl_2$$

$$\frac{0.213 \text{ g } MgCl_2}{1.50 \text{ g mixture}} \times 100 = 14.2\% \ MgCl_2$$

       b.   $0.213 \text{ g } MgCl_2 \times \dfrac{1 \text{ mol } MgCl_2}{95.21 \text{ g}} \times \dfrac{2 \text{ mol } AgNO_3}{\text{mol } MgCl_2} \times \dfrac{1 \text{ L}}{0.500 \text{ mol } AgNO_3} \times \dfrac{1000 \text{ mL}}{1 \text{ L}}$

$$= 8.95 \text{ mL } AgNO_3$$

75.     a.   $0.308 \text{ g } AgCl \times \dfrac{35.45 \text{ g } Cl}{143.4 \text{ g } AgCl} = 0.0761 \text{ g } Cl;\ \%\ Cl = \dfrac{0.0761 \text{ g}}{0.256 \text{ g}} \times 100 = 29.7\% \ Cl$

Cobalt(III) oxide, $Co_2O_3$:  $2(58.93) + 3(16.00) = 165.86$ g/mol

$$0.145 \text{ g } Co_2O_3 \times \frac{117.86 \text{ g } Co}{165.86 \text{ g } Co_2O_3} = 0.103 \text{ g } Co;\ \%\ Co = \frac{0.103 \text{ g}}{0.416 \text{ g}} \times 100 = 24.8\% \ Co$$

The remainder, $100.0 - (29.7 + 24.8) = 45.5\%$, is water.

Assuming 100.0 g of compound:

$$45.5 \text{ g } H_2O \times \frac{2.016 \text{ g } H}{18.02 \text{ g } H_2O} = 5.09 \text{ g } H;\ \%\ H = \frac{5.09 \text{ g } H}{100.0 \text{ g compound}} \times 100 = 5.09\% \ H$$

$$45.5 \text{ g } H_2O \times \frac{16.00 \text{ g } O}{18.02 \text{ g } H_2O} = 40.4 \text{ g } O;\ \%\ O = \frac{40.4 \text{ g } O}{100.0 \text{ g compound}} \times 100 = 40.4\% \ O$$

The mass percent composition is 24.8% Co, 29.7% Cl, 5.09% H, and 40.4% O.

b.  Out of 100.0 g of compound, there are:

$$24.8 \text{ g Co} \times \frac{1 \text{ mol}}{58.93 \text{ g Co}} = 0.421 \text{ mol Co};\ \ 29.7 \text{ g Cl} \times \frac{1 \text{ mol}}{35.45 \text{ g Cl}} = 0.838 \text{ mol Cl}$$

$$5.09 \text{ g H} \times \frac{1 \text{ mol}}{1.008 \text{ g H}} = 5.05 \text{ mol H};\ \ 40.4 \text{ g O} \times \frac{1 \text{ mol}}{16.00 \text{ g O}} = 2.53 \text{ mol O}$$

Dividing all results by 0.421, we get $CoCl_2 \bullet 6H_2O$ for the empirical formula, which is also the molecular formula.

c.  $CoCl_2 \bullet 6H_2O(aq) + 2 \text{ AgNO}_3(aq) \rightarrow 2 \text{ AgCl}(s) + Co(NO_3)_2(aq) + 6 H_2O(l)$

$CoCl_2 \bullet 6H_2O(aq) + 2 \text{ NaOH}(aq) \rightarrow Co(OH)_2(s) + 2 \text{ NaCl}(aq) + 6 H_2O(l)$

$Co(OH)_2 \rightarrow Co_2O_3$    This is an oxidation-reduction reaction.  Thus we also need to include an oxidizing agent.  The obvious choice is $O_2$.

$4 \text{ Co(OH)}_2(s) + O_2(g) \rightarrow 2 \text{ Co}_2O_3(s) + 4 H_2O(l)$

77.    $Ag^+(aq) + Cl^-(aq) \rightarrow AgCl(s)$;  let $x$ = mol NaCl and $y$ = mol KCl.

$(22.90 \times 10^{-3} \text{ L}) \times 0.1000 \text{ mol/L} = 2.290 \times 10^{-3} \text{ mol Ag}^+ = 2.290 \times 10^{-3} \text{ mol Cl}^-$ total

$x + y = 2.290 \times 10^{-3} \text{ mol Cl}^-$,  $x = 2.290 \times 10^{-3} - y$

Because the molar mass of NaCl is 58.44 g/mol and the molar mass of KCl is 74.55 g/mol:

$(58.44)x + (74.55)y = 0.1586 \text{ g}$

$58.44(2.290 \times 10^{-3} - y) + (74.55)y = 0.1586,\ (16.11)y = 0.0248,\ y = 1.54 \times 10^{-3} \text{ mol KCl}$

$$\text{Mass \% KCl} = \frac{1.54 \times 10^{-3} \text{ mol} \times 74.55 \text{ g/mol}}{0.1586 \text{ g}} \times 100 = 72.4\% \text{ KCl}$$

% NaCl = 100.0 − 72.4 = 27.6% NaCl

79.    $Cr(NO_3)_3(aq) + 3 \text{ NaOH}(aq) \rightarrow Cr(OH)_3(s) + 3 \text{ NaNO}_3(aq)$

$$\text{Mol NaOH used} = 2.06 \text{ g Cr(OH)}_3 \times \frac{1 \text{ mol Cr(OH)}_3}{103.02 \text{ g}} \times \frac{3 \text{ mol NaOH}}{\text{mol Cr(OH)}_3} = 6.00 \times 10^{-2} \text{ mol}$$
to form precipitate

$\text{NaOH}(aq) + \text{HCl}(aq) \rightarrow \text{NaCl}(aq) + H_2O(l)$

Mol NaOH used = $0.1000 \text{ L} \times \dfrac{0.400 \text{ mol HCl}}{\text{L}} \times \dfrac{1 \text{ mol NaOH}}{\text{mol HCl}} = 4.00 \times 10^{-2}$ mol
to react with HCl

$M_{NaOH} = \dfrac{\text{total mol NaOH}}{\text{volume}} = \dfrac{6.00 \times 10^{-2} \text{ mol} + 4.00 \times 10^{-2} \text{ mol}}{0.0500 \text{ L}} = 2.00 \ M \text{ NaOH}$

81.    The amount of KHP used = $0.4016 \text{ g} \times \dfrac{1 \text{ mol}}{204.22 \text{ g}} = 1.967 \times 10^{-3}$ mol KHP.

Because 1 mole of NaOH reacts completely with 1 mole of KHP, the NaOH solution contains $1.967 \times 10^{-3}$ mol NaOH.

Molarity of NaOH = $\dfrac{1.967 \times 10^{-3} \text{ mol}}{25.06 \times 10^{-3} \text{ L}} = \dfrac{7.849 \times 10^{-2} \text{ mol NaOH}}{\text{L}}$

Maximum molarity = $\dfrac{1.967 \times 10^{-3} \text{ mol}}{25.01 \times 10^{-3} \text{ L}} = \dfrac{7.865 \times 10^{-2} \text{ mol NaOH}}{\text{L}}$

Minimum molarity = $\dfrac{1.967 \times 10^{-3} \text{ mol}}{25.11 \times 10^{-3} \text{ L}} = \dfrac{7.834 \times 10^{-2} \text{ mol NaOH}}{\text{L}}$

We can express this as $0.07849 \pm 0.00016 \ M$. An alternate way is to express the molarity as $0.0785 \pm 0.0002 \ M$. This second way shows the actual number of significant figures in the molarity. The advantage of the first method is that it shows that we made all our individual measurements to four significant figures.

83.    Mol $C_6H_8O_7$ = $0.250 \text{ g } C_6H_8O_7 \times \dfrac{1 \text{ mol } C_6H_8O_7}{192.1 \text{ g } C_6H_8O_7} = 1.30 \times 10^{-3}$ mol $C_6H_8O_7$

Let $H_xA$ represent citric acid, where $x$ is the number of acidic hydrogens. The balanced neutralization reaction is:

$H_xA(aq) + x \ OH^-(aq) \rightarrow x \ H_2O(l) + A^{x-}(aq)$

Mol $OH^-$ reacted = $0.0372 \text{ L} \times \dfrac{0.105 \text{ mol } OH^-}{\text{L}} = 3.91 \times 10^{-3}$ mol $OH^-$

$x = \dfrac{\text{mol } OH^-}{\text{mol citric acid}} = \dfrac{3.91 \times 10^{-3} \text{ mol}}{1.30 \times 10^{-3} \text{ mol}} = 3.01$

Therefore, the general acid formula for citric acid is $H_3A$, meaning that citric acid has three acidic hydrogens per citric acid molecule (citric acid is a triprotic acid).

## Challenge Problems

85.     a.    Let $x$ = mass of Mg, so $10.00 - x$ = mass of Zn.  $Ag^+(aq) + Cl^-(aq) \rightarrow AgCl(s)$.

From the given balanced equations, there is a 2 : 1 mole ratio between mol Mg and mol $Cl^-$. The same is true for Zn.  Because mol $Ag^+$ = mol $Cl^-$ present, one can setup an equation relating mol $Cl^-$ present to mol $Ag^+$ added.

$$x \text{ g Mg} \times \frac{1 \text{ mol Mg}}{24.31 \text{ g Mg}} \times \frac{2 \text{ mol Cl}^-}{\text{mol Mg}} + (10.00 - x) \text{ g Zn} \times \frac{1 \text{ mol Zn}}{65.38 \text{ g Zn}} \times \frac{2 \text{ mol Cl}^-}{\text{mol Zn}}$$

$$= 0.156 \text{ L} \times \frac{3.00 \text{ mol Ag}^+}{\text{L}} \times \frac{1 \text{ mol Cl}^-}{\text{mol Ag}^+} = 0.468 \text{ mol Cl}^-$$

$$\frac{2x}{24.31} + \frac{2(10.00 - x)}{65.38} = 0.468, \quad 24.31 \times 65.38 \left( \frac{2x}{24.31} + \frac{20.00 - 2x}{65.38} = 0.468 \right)$$

$(130.8)x + 486.2 - (48.62)x = 743.8$  (carrying 1 extra sig. fig.)

$(82.2)x = 257.6, \quad x = 3.13 \text{ g Mg};$     % Mg $= \dfrac{3.13 \text{ g Mg}}{10.00 \text{ g mixture}} \times 100 = 31.3\% \text{ Mg}$

b.    $0.156 \text{ L} \times \dfrac{3.00 \text{ mol Ag}^+}{\text{L}} \times \dfrac{1 \text{ mol Cl}^-}{\text{mol Ag}^+} = 0.468 \text{ mol Cl}^- = 0.468 \text{ mol HCl added}$

$$M_{HCl} = \frac{0.468 \text{ mol}}{0.0780 \text{ L}} = 6.00 \ M \text{ HCl}$$

87.     $Zn(s) + 2 AgNO_2(aq) \rightarrow 2 Ag(s) + Zn(NO_2)_2(aq)$

Let $x$ = mass of Ag and $y$ = mass of Zn after the reaction has stopped.  Then $x + y = 29.0$ g. Because the moles of Ag produced will equal two times the moles of Zn reacted:

$$(19.0 - y) \text{ g Zn} \times \frac{1 \text{ mol Zn}}{65.38 \text{ g Zn}} \times \frac{2 \text{ mol Ag}}{1 \text{ mol Zn}} = x \text{ g Ag} \times \frac{1 \text{ mol Ag}}{107.9 \text{ g Ag}}$$

Simplifying:

$$3.059 \times 10^{-2}(19.0 - y) = (9.268 \times 10^{-3})x$$

Substituting $x = 29.0 - y$ into the equation gives:

$$3.059 \times 10^{-2}(19.0 - y) = 9.268 \times 10^{-3}(29.0 - y)$$

Solving:

$0.581 - (3.059 \times 10^{-2})y = 0.269 - (9.268 \times 10^{-3})y, \ (2.132 \times 10^{-2})y = 0.312, \ y = 14.6 \text{ g Zn}$

14.6 g Zn are present, and $29.0 - 14.6 = 14.4$ g Ag are also present after the reaction is stopped.

89.      Molar masses:  KCl, $39.10 + 35.45 = 74.55$ g/mol;  KBr, $39.10 + 79.90 = 119.00$ g/mol

AgCl, $107.9 + 35.45 = 143.4$ g/mol;  AgBr, $107.9 + 79.90 = 187.8$ g/mol

Let $x$ = number of moles of KCl in mixture and $y$ = number of moles of KBr in mixture. $Ag^+ + Cl^- \rightarrow AgCl$ and $Ag^+ + Br^- \rightarrow AgBr$;  so, $x$ = moles AgCl and $y$ = moles AgBr.

Setting up two equations from the given information:

$0.1024 \text{ g} = (74.55)x + (119.0)y$ and $0.1889 \text{ g} = (143.4)x + (187.8)y$

Multiply the first equation by $\dfrac{187.8}{119.0}$, and then subtract from the second.

$$
\begin{array}{ll}
0.1889 = & (143.4)x + (187.8)y \\
-0.1616 = & -(117.7)x - (187.8)y \\
\hline
0.0273 = & (25.7)x, \qquad\qquad x = 1.06 \times 10^{-3} \text{ mol KCl}
\end{array}
$$

$1.06 \times 10^{-3} \text{ mol KCl} \times \dfrac{74.55 \text{ KCl}}{\text{mol KCl}} = 0.0790 \text{ g KCl}$

Mass % KCl $= \dfrac{0.0790 \text{ g}}{0.1024 \text{ g}} \times 100 = 77.1\%$,  % KBr $= 100.0 - 77.1 = 22.9\%$

91.      a.  $C_{12}H_{10-n}Cl_n + n\,Ag^+ \rightarrow n\,AgCl$;  molar mass of AgCl $= 143.4$ g/mol

Molar mass of PCB $= 12(12.01) + (10 - n)(1.008) + n(35.45) = 154.20 + (34.44)n$

Because $n$ mol AgCl are produced for every 1 mol PCB reacted, $n(143.4)$ g of AgCl will be produced for every $[154.20 + (34.44)n]$ g of PCB reacted.

$\dfrac{\text{Mass of AgCl}}{\text{Mass of PCB}} = \dfrac{(143.4)n}{154.20 + (34.44)n}$ or $\text{mass}_{AgCl}[154.20 + (34.44)n] = \text{mass}_{PCB}(143.4)n$

b.  $0.4971[154.20 + (34.44)n] = 0.1947(143.4)n, \ 76.65 + (17.12)n = (27.92)n$

$76.65 = (10.80)n, \ n = 7.097$

93.      a.  Flow rate $= 5.00 \times 10^4 \text{ L/s} + 3.50 \times 10^3 \text{ L/s} = 5.35 \times 10^4 \text{ L/s}$

b.  $C_{HCl} = \dfrac{3.50 \times 10^3 (65.0)}{5.35 \times 10^4} = 4.25 \text{ ppm HCl}$

c.  1 ppm = 1 mg/kg $H_2O$ = 1 mg/L   (assuming density = 1.00 g/mL)

$$8.00 \text{ h} \times \frac{60 \text{ min}}{\text{h}} \times \frac{60 \text{ s}}{\text{min}} \times \frac{1.80 \times 10^4 \text{ L}}{\text{s}} \times \frac{4.25 \text{ mg HCl}}{\text{L}} \times \frac{1 \text{ g}}{1000 \text{ mg}} = 2.20 \times 10^6 \text{ g HCl}$$

$$2.20 \times 10^6 \text{ g HCl} \times \frac{1 \text{ mol HCl}}{36.46 \text{ g HCl}} \times \frac{1 \text{ mol CaO}}{2 \text{ mol HCl}} \times \frac{56.08 \text{ g Ca}}{\text{mol CaO}} = 1.69 \times 10^6 \text{ g CaO}$$

d.  The concentration of $Ca^{2+}$ going into the second plant was:

$$\frac{5.00 \times 10^4 (10.2)}{5.35 \times 10^4} = 9.53 \text{ ppm}$$

The second plant used: $1.80 \times 10^4$ L/s × (8.00 × 60 × 60) s = $5.18 \times 10^8$ L of water.

$$1.69 \times 10^6 \text{ g CaO} \times \frac{40.08 \text{ g Ca}^{2+}}{56.08 \text{ g CaO}} = 1.21 \times 10^6 \text{ g Ca}^{2+} \text{ was added to this water.}$$

$$C_{Ca^{2+}} \text{ (plant water)} = 9.53 + \frac{1.21 \times 10^9 \text{ mg}}{5.18 \times 10^8 \text{ L}} = 9.53 + 2.34 = 11.87 \text{ ppm}$$

Because 90.0% of this water is returned, $(1.80 \times 10^4) \times 0.900 = 1.62 \times 10^4$ L/s of water with 11.87 ppm $Ca^{2+}$ is mixed with $(5.35 - 1.80) \times 10^4 = 3.55 \times 10^4$ L/s of water containing 9.53 ppm $Ca^{2+}$.

$$C_{Ca^{2+}} \text{ (final)} = \frac{(1.62 \times 10^4 \text{ L/s})(11.87 \text{ ppm}) + (3.55 \times 10^4 \text{ L/s})(9.53 \text{ ppm})}{1.62 \times 10^4 \text{ L/s} + 3.55 \times 10^4 \text{ L/s}} = 10.3 \text{ ppm}$$

95.   a.  $YBa_2Cu_3O_{6.5}$:

+3 + 2(+2) + 3x + 6.5(−2) = 0

7 + 3x − 13 = 0,  3x = 6,  x = +2     Only $Cu^{2+}$ present.

$YBa_2Cu_3O_7$:

+3 + 2(+2) + 3x + 7(−2) = 0,  x = +2 1/3 or 2.33

This corresponds to two $Cu^{2+}$ and one $Cu^{3+}$ present.

$YBa_2Cu_3O_8$:

+3 + 2(+2) + 3x + 8(−2) = 0,  x = +3;   Only $Cu^{3+}$ present.

b.          $(e^- + Cu^{2+} + I^- \rightarrow CuI) \times 2$                    $2\,e^- + Cu^{3+} + I^- \rightarrow CuI$
                    $3I^- \rightarrow I_3^- + 2\,e^-$                              $3I^- \rightarrow I_3^- + 2\,e^-$

$$2 \, Cu^{2+}(aq) + 5 \, I^-(aq) \rightarrow 2 \, CuI(s) + I_3^-(aq) \qquad Cu^{3+}(aq) + 4 \, I^-(aq) \rightarrow CuI(s) + I_3^-(aq)$$
$$2 \, S_2O_3^{2-} \rightarrow S_4O_6^{2-} + 2 \, e^-$$
$$2 \, e^- + I_3^- \rightarrow 3 \, I^-$$

$$2 \, S_2O_3^{2-}(aq) + I_3^-(aq) \rightarrow 3 \, I^-(aq) + S_4O_6^{2-}(aq)$$

c.  Step II data:  All Cu is converted to $Cu^{2+}$.  *Note*:  Superconductor abbreviated as "123."

$$22.57 \times 10^{-3} \, L \times \frac{0.1000 \, mol \, S_2O_3^{2-}}{L} \times \frac{1 \, mol \, I_3^-}{2 \, mol \, S_2O_3^{2-}} \times \frac{2 \, mol \, Cu^{2+}}{mol \, I_3^-}$$
$$= 2.257 \times 10^{-3} \, mol \, Cu^{2+}$$

$$2.257 \times 10^{-3} \, mol \, Cu \times \frac{1 \, mol \, "123"}{3 \, mol \, Cu} = 7.523 \times 10^{-4} \, mol \, "123"$$

$$\text{Molar mass of } YBa_2Cu_3O_x = \frac{0.5402 \, g}{7.523 \times 10^{-4} \, mol} = 670.2 \, g/mol$$

$$670.2 = 88.91 + 2(137.3) + 3(63.55) + x(16.00), \quad 670.2 = 554.2 + x(16.00)$$

$x = 7.250$;  formula is $YBa_2Cu_3O_{7.25}$.

Check with Step I data:  Both $Cu^{2+}$ and $Cu^{3+}$ present.

$$37.77 \times 10^{-3} \, L \times \frac{0.1000 \, mol \, S_2O_3^{2-}}{L} \times \frac{1 \, mol \, I_3^-}{2 \, mol \, S_2O_3^{2-}} = 1.889 \times 10^{-3} \, mol \, I_3^-$$

We get 1 mol $I_3^-$ per mol $Cu^{3+}$ and 1 mol $I_3^-$ per 2 mol $Cu^{2+}$.  Let $n_{Cu^{3+}}$ = mol $Cu^{3+}$ and $n_{Cu^{2+}}$ = mol $Cu^{2+}$, then:

$$n_{Cu^{3+}} + \frac{n_{Cu^{2+}}}{2} = 1.889 \times 10^{-3} \, mol$$

In addition:  $\dfrac{0.5625 \, g}{670.2 \, g/mol} = 8.393 \times 10^{-4} \, mol \, "123"$

This amount of "123" contains:

$$3(8.393 \times 10^{-4}) = 2.518 \times 10^{-3} \, mol \, Cu \, total = n_{Cu^{3+}} + n_{Cu^{2+}}$$

Solving by simultaneous equations:

$$n_{Cu^{3+}} + n_{Cu^{2+}} = 2.518 \times 10^{-3}$$
$$-n_{Cu^{3+}} - \frac{n_{Cu^{2+}}}{2} = -1.889 \times 10^{-3}$$

$$\frac{n_{Cu^{2+}}}{2} = 6.29 \times 10^{-4}$$

$n_{Cu^{2+}} = 1.26 \times 10^{-3}$ mol $Cu^{2+}$;  $n_{Cu^{3+}} = 2.518 \times 10^{-3} - 1.26 \times 10^{-3} = 1.26 \times 10^{-3}$ mol $Cu^{3+}$

This sample of superconductor contains equal moles of $Cu^{2+}$ and $Cu^{3+}$. Therefore, 1 mole of $YBa_2Cu_3O_x$ contains 1.50 mol $Cu^{2+}$ and 1.50 mol $Cu^{3+}$. Solving for $x$ using oxidation states:

$$+3 + 2(+2) + 1.50(+2) + 1.50(+3) + x(-2) = 0, \quad 14.50 = 2x, \quad x = 7.25$$

The two experiments give the same result, $x = 7.25$ with formula $YBa_2Cu_3O_{7.25}$.

Average oxidation state of Cu:

$$+3 + 2(+2) + 3(x) + 7.25(-2) = 0, \quad 3x = 7.50, \quad x = +2.50$$

As determined from Step I data, this superconductor sample contains equal moles of $Cu^{2+}$ and $Cu^{3+}$, giving an average oxidation state of +2.50.

97.    There are three unknowns so we need three equations to solve for the unknowns. Let $x =$ mass $AgNO_3$, $y =$ mass $CuCl_2$, and $z =$ mass $FeCl_3$. Then $x + y + z = 1.0000$ g. The $Cl^-$ in $CuCl_2$ and $FeCl_3$ will react with the excess $AgNO_3$ to form the precipitate $AgCl(s)$. Assuming silver has an atomic mass of 107.90:

$$\text{Mass of Cl in mixture} = 1.7809 \text{ g AgCl} \times \frac{35.45 \text{ g Cl}}{143.35 \text{ g AgCl}} = 0.4404 \text{ g Cl}$$

$$\text{Mass of Cl from } CuCl_2 = y \text{ g } CuCl_2 \times \frac{2(35.45) \text{ g Cl}}{134.45 \text{ g } CuCl_2} = (0.5273)y$$

$$\text{Mass of Cl from } FeCl_3 = z \text{ g } FeCl_3 \times \frac{3(35.45) \text{ g Cl}}{162.20 \text{ g } FeCl_3} = (0.6557)z$$

The second equation is:  0.4404 g Cl = (0.5273)y + (0.6557)z

Similarly, let's calculate the mass of metals in each salt.

$$\text{Mass of Ag in } AgNO_3 = x \text{ g } AgNO_3 \times \frac{107.9 \text{ g Ag}}{169.91 \text{ g } AgNO_3} = (0.6350)x$$

For $CuCl_2$ and $FeCl_3$, we already calculated the amount of Cl in each initial amount of salt; the remainder must be the mass of metal in each salt.

Mass of Cu in $CuCl_2 = y - (0.5273)y = (0.4727)y$

Mass of Fe in $FeCl_3 = z - (0.6557)z = (0.3443)z$

The third equation is:  $0.4684 \text{ g metals} = (0.6350)x + (0.4727)y + (0.3443)z$

We now have three equations with three unknowns.  Solving:

$$
\begin{array}{rl}
-0.6350 \,(1.0000 = & x \quad + \quad y \quad + \quad z) \\
0.4684 = & (0.6350)x + (0.4727)y + (0.3443)z \\
\hline
-0.1666 = & -(0.1623)y - (0.2907)z
\end{array}
$$

$$
\frac{0.5273}{0.1623} \,[-0.1666 = -(0.1623)y - (0.2907)z]
$$

$$
\begin{array}{rl}
0.4404 = & (0.5273)y + (0.6557)z \\
\hline
-0.1009 = & -(0.2888)z, \quad z = \dfrac{0.1009}{0.2888} = 0.3494 \text{ g } FeCl_3
\end{array}
$$

$0.4404 = (0.5273)y + 0.6557(0.3494), \; y = 0.4007 \text{ g } CuCl_2$

$x = 1.0000 - y - z = 1.0000 - 0.4007 - 0.3494 = 0.2499 \text{ g } AgNO_3$

Mass % $AgNO_3 = \dfrac{0.2499 \text{ g}}{1.0000 \text{ g}} \times 100 = 24.99\% \; AgNO_3$

Mass % $CuCl_2 = \dfrac{0.4007 \text{ g}}{1.0000 \text{ g}} \times 100 = 40.07\% \; CuCl_2; \; \text{mass } \% \; FeCl_3 = 34.94\%$

# CHAPTER 5

# GASES

## Pressure

21.   $4.75 \text{ cm} \times \dfrac{10 \text{ mm}}{\text{cm}} = 47.5 \text{ mm Hg or } 47.5 \text{ torr}; \quad 47.5 \text{ torr} \times \dfrac{1 \text{ atm}}{760 \text{ torr}} = 6.25 \times 10^{-2} \text{ atm}$

$6.25 \times 10^{-2} \text{ atm} \times \dfrac{1.013 \times 10^5 \text{ Pa}}{\text{atm}} = 6.33 \times 10^3 \text{ Pa}$

23.   Suppose we have a column of mercury $1.00 \text{ cm} \times 1.00 \text{ cm} \times 76.0 \text{ cm} = V = 76.0 \text{ cm}^3$:

$\text{mass} = 76.0 \text{ cm}^3 \times 13.59 \text{ g/cm}^3 = 1.03 \times 10^3 \text{ g} \times \dfrac{1 \text{ kg}}{1000 \text{ g}} = 1.03 \text{ kg}$

$F = mg = 1.03 \text{ kg} \times 9.81 \text{ m/s}^2 = 10.1 \text{ kg m/s}^2 = 10.1 \text{ N}$

$\dfrac{\text{Force}}{\text{Area}} = \dfrac{10.1 \text{ N}}{\text{cm}^2} \times \left(\dfrac{100 \text{ cm}}{\text{m}}\right)^2 = 1.01 \times 10^5 \ \dfrac{\text{N}}{\text{m}^2} \text{ or } 1.01 \times 10^5 \text{ Pa}$

(*Note*: $76.0 \text{ cm Hg} = 1 \text{ atm} = 1.01 \times 10^5 \text{ Pa}$.)

To exert the same pressure, a column of water will have to contain the same mass as the 76.0-cm column of mercury.   Thus the column of water will have to be 13.59 times taller or $76.0 \text{ cm} \times 13.59 = 1.03 \times 10^3 \text{ cm} = 10.3 \text{ m}$.

25.   a.   $4.8 \text{ atm} \times \dfrac{760 \text{ mm Hg}}{\text{atm}} = 3.6 \times 10^3 \text{ mm Hg}; \quad$ b.   $3.6 \times 10^3 \text{ mm Hg} \times \dfrac{1 \text{ torr}}{\text{mm Hg}}$

$= 3.6 \times 10^3 \text{ torr}$

c.   $4.8 \text{ atm} \times \dfrac{1.013 \times 10^5 \text{ Pa}}{\text{atm}} = 4.9 \times 10^5 \text{ Pa}; \quad$ d.   $4.8 \text{ atm} \times \dfrac{14.7 \text{ psi}}{\text{atm}} = 71 \text{ psi}$

## Gas Laws

27.    Treat each gas separately, and use the relationship $P_1V_1 = P_2V_2$  (n and T are constant).

For $H_2$:  $P_2 = \dfrac{P_1V_1}{V_2} = 475 \text{ torr} \times \dfrac{2.00 \text{ L}}{3.00 \text{ L}} = 317 \text{ torr}$

For $N_2$:  $P_2 = 0.200 \text{ atm} \times \dfrac{1.00 \text{ L}}{3.00 \text{ L}} = 0.0667 \text{ atm};\ \ 0.0667 \text{ atm} \times \dfrac{760 \text{ torr}}{\text{atm}} = 50.7 \text{ torr}$

$P_{total} = P_{H_2} + P_{N_2} = 317 + 50.7 = 368 \text{ torr}$

29.    $PV = nRT,\ \ \dfrac{nT}{P} = \dfrac{V}{R} = \text{constant},\ \ \dfrac{n_1 T_1}{P_1} = \dfrac{n_2 T_2}{P_2};$  moles × molar mass = mass

$\dfrac{n_1 (\text{molar mass}) T_1}{P_1} = \dfrac{n_2 (\text{molar mass}) T_2}{P_2},\ \ \dfrac{\text{mass}_1 \times T_1}{P_1} = \dfrac{\text{mass}_2 \times T_2}{P_2}$

$\text{mass}_2 = \dfrac{\text{mass}_1 \times T_1 P_2}{T_2 P_1} = \dfrac{1.00 \times 10^3 \text{ g} \times 291 \text{ K} \times 650.\, \text{psi}}{299 \text{ K} \times 2050.\, \text{psi}} = 309 \text{ g}$

31.    $P = P_{CO_2} = \dfrac{n_{CO_2} RT}{V} = \dfrac{\left( 22.0 \text{ g} \times \dfrac{1 \text{ mol}}{44.01 \text{ g}} \right) \times \dfrac{0.08206 \text{ L atm}}{\text{K mol}} \times 300.\, \text{K}}{4.00 \text{ L}} = 3.08 \text{ atm}$

With air present, the partial pressure of $CO_2$ will still be 3.08 atm.  The total pressure will be the sum of the partial pressures.

$P_{total} = P_{CO_2} + P_{air} = 3.08 \text{ atm} + \left( 740.\, \text{torr} \times \dfrac{1 \text{ atm}}{760 \text{ torr}} \right) = 3.08 + 0.974 = 4.05 \text{ atm}$

33.    $n = \dfrac{PV}{RT} = \dfrac{135 \text{ atm} \times 200.0 \text{ L}}{\dfrac{0.08206 \text{ L atm}}{\text{K mol}} \times (273 + 24) \text{ K}} = 1.11 \times 10^3 \text{ mol}$

For He:  $1.11 \times 10^3 \text{ mol} \times \dfrac{4.003 \text{ g He}}{\text{mol}} = 4.44 \times 10^3 \text{ g He}$

For $H_2$:  $1.11 \times 10^3 \text{ mol} \times \dfrac{2.016 \text{ g H}_2}{\text{mol}} = 2.24 \times 10^3 \text{ g H}_2$

35.     $\dfrac{PV}{nT}$ = R;  for a gas at two conditions:

$\dfrac{P_1 V_1}{n_1 T_1} = \dfrac{P_2 V_2}{n_2 T_2}$ ;  because n and V are constant:  $\dfrac{P_1}{T_1} = \dfrac{P_2}{T_2}$

$T_2 = \dfrac{P_2 T_1}{P_1} = \dfrac{2500 \text{ torr} \times 294.2 \text{ K}}{758 \text{ torr}} = 970 \text{ K} = 7.0 \times 10^2 \,^\circ\text{C}$

37.     As $NO_2$ is converted completely into $N_2O_4$, the moles of gas present will decrease by a factor of one-half (from the 2 : 1 mol ratio in the balanced equation).  Using Avogadro's law:

$\dfrac{V_1}{n_1} = \dfrac{V_2}{n_2}$,    $V_2 = V_1 \times \dfrac{n_2}{n_1} = 25.0 \text{ mL} \times \dfrac{1}{2} = 12.5 \text{ mL}$

$N_2O_4(g)$ will occupy one-half the original volume of $NO_2(g)$.

39.     PV = nRT, P is constant.    $\dfrac{nT}{V} = \dfrac{P}{R}$ = constant,  $\dfrac{n_1 T_1}{V_1} = \dfrac{n_2 T_2}{V_2}$

$\dfrac{n_2}{n_1} = \dfrac{T_1 V_2}{T_2 V_1} = \dfrac{294 \text{ K}}{335 \text{ K}} \times \dfrac{4.20 \times 10^3 \text{ m}^3}{4.00 \times 10^3 \text{ m}^3} = 0.921$

41.     At constant T and P, Avogadro's law applies; that is, equal volumes contain equal moles of molecules.  In terms of balanced equations, we can say that mole ratios and volume ratios between the various reactants and products will be equal to each other.  $Br_2 + 3\,F_2 \rightarrow 2\,X$; two moles of X must contain 2 moles of Br and 6 moles of F;  X must have the formula $BrF_3$.

43.     $P_{He} + P_{H_2O} = 1.00 \text{ atm} = 760. \text{ torr} = P_{He} + 23.8 \text{ torr}, \ P_{He} = 736 \text{ torr}$

$n_{He} = 0.586 \text{ g} \times \dfrac{1 \text{ mol}}{4.003 \text{ g}} = 0.146 \text{ mol He}$

$V = \dfrac{n_{He}RT}{P_{He}} = \dfrac{0.146 \text{ mol} \times \dfrac{0.08206 \text{ L atm}}{\text{K mol}} \times 298 \text{ K}}{736 \text{ torr} \times \dfrac{1 \text{ atm}}{760 \text{ torr}}} = 3.69 \text{ L}$

45.     a.   Mole fraction $CH_4 = \chi_{CH_4} = \dfrac{P_{CH_4}}{P_{total}} = \dfrac{0.175 \text{ atm}}{0.175 \text{ atm} + 0.250 \text{ atm}} = 0.412$

$\chi_{O_2} = 1.000 - 0.412 = 0.588$

b.   PV = nRT,  $n_{total} = \dfrac{P_{total} \times V}{RT} = \dfrac{0.425 \text{ atm} \times 10.5 \text{ L}}{\dfrac{0.08206 \text{ L atm}}{\text{K mol}} \times 338 \text{ K}} = 0.161 \text{ mol}$

c.  $\chi_{CH_4} = \dfrac{n_{CH_4}}{n_{total}}$, $n_{CH_4} = \chi_{CH_4} \times n_{total} = 0.412 \times 0.161$ mol $= 6.63 \times 10^{-2}$ mol $CH_4$

$$6.63 \times 10^{-2} \text{ mol } CH_4 \times \frac{16.04 \text{ g } CH_4}{\text{mol } CH_4} = 1.06 \text{ g } CH_4$$

$$n_{O_2} = 0.588 \times 0.161 \text{ mol} = 9.47 \times 10^{-2} \text{ mol } O_2; \ 9.47 \times 10^{-2} \text{ mol } O_2 \times \frac{32.00 \text{ g } O_2}{\text{mol } O_2}$$

$$= 3.03 \text{ g } O_2$$

47.    The partial pressures can be determined by using the mole fractions.

$P_{methane} = P_{total} \times \chi_{methane} = 1.44$ atm $\times 0.915 = 1.32$ atm;  $P_{ethane} = 1.44 - 1.32 = 0.12$ atm

Determining the number of moles of natural gas combusted:

$$n_{natural \ gas} = \frac{PV}{RT} = \frac{1.44 \text{ atm} \times 15.00 \text{ L}}{\dfrac{0.08206 \text{ L atm}}{\text{K mol}} \times 293 \text{ K}} = 0.898 \text{ mol natural gas}$$

$n_{methane} = n_{natural \ gas} \times \chi_{methane} = 0.898$ mol $\times 0.915 = 0.822$ mol methane

$n_{ethane} = 0.898 - 0.822 = 0.076$ mol ethane

$CH_4(g) + 2 \ O_2(g) \rightarrow CO_2(g) + 2 \ H_2O(l)$;   $2 \ C_2H_6 + 7 \ O_2(g) \rightarrow 4 \ CO_2(g) + 6 \ H_2O(l)$

$$0.822 \text{ mol } CH_4 \times \frac{2 \text{ mol } H_2O}{1 \text{ mol } CH_4} \times \frac{18.02 \text{ g } H_2O}{\text{mol } H_2O} = 29.6 \text{ g } H_2O$$

$$0.076 \text{ mol } C_2H_6 \times \frac{6 \text{ mol } H_2O}{2 \text{ mol } C_2H_6} \times \frac{18.02 \text{ g } H_2O}{\text{mol } H_2O} = 4.1 \text{ g } H_2O$$

The total mass of $H_2O$ produced $= 29.6$ g $+ 4.1$ g $= 33.7$ g $H_2O$.

## Gas Density, Molar Mass, and Reaction Stoichiometry

49.    Out of 100.0 g of compound, there are:

$$87.4 \text{ g N} \times \frac{1 \text{ mol N}}{14.01 \text{ g N}} = 6.24 \text{ mol N}; \ \frac{6.24}{6.24} = 1.00$$

$$12.6 \text{ g H} \times \frac{1 \text{ mol H}}{1.008 \text{ g H}} = 12.5 \text{ mol H}; \ \frac{12.5}{6.24} = 2.00$$

Empirical formula is $NH_2$.  $P \times$ (molar mass) $= dRT$, where $d =$ density.

$$\text{Molar mass} = \frac{dRT}{P} = \frac{\dfrac{0.977 \text{ g}}{L} \times \dfrac{0.08206 \text{ L atm}}{\text{K mol}} \times 373 \text{ K}}{710. \text{ torr} \times \dfrac{1 \text{ atm}}{760 \text{ torr}}} = 32.0 \text{ g/mol}$$

Empirical formula mass of $NH_2 = 16.0$ g.  Therefore, the molecular formula is $N_2H_4$.

51.     If $Be^{3+}$, the formula is $Be(C_5H_7O_2)_3$ and molar mass $\approx 13.5 + 15(12) + 21(1) + 6(16)$
        $= 311$ g/mol.  If $Be^{2+}$, the formula is $Be(C_5H_7O_2)_2$ and molar mass $\approx 9.0 + 10(12) + 14(1) +$
        $4(16) = 207$ g/mol.

        Data set I  (molar mass = dRT/P and d = mass/V):

$$\text{molar mass} = \frac{\text{mass} \times RT}{PV} = \frac{0.2022 \text{ g} \times \dfrac{0.08206 \text{ L atm}}{\text{K mol}} \times 286 \text{ K}}{(765.2 \text{ torr} \times \dfrac{1 \text{ atm}}{760 \text{ torr}}) \times (22.6 \times 10^{-3} \text{ L})} = 209 \text{ g/mol}$$

        Data set II:

$$\text{molar mass} = \frac{\text{mass} \times RT}{PV} = \frac{0.2224 \text{ g} \times \dfrac{0.08206 \text{ L atm}}{\text{K mol}} \times 290. \text{ K}}{(764.6 \text{ torr} \times \dfrac{1 \text{ atm}}{760 \text{ torr}}) \times (26.0 \times 10^{-3} \text{ L})} = 202 \text{ g/mol}$$

        These results are close to the expected value of 207 g/mol for $Be(C_5H_7O_2)_2$.  Thus we
        conclude from these data that beryllium is a divalent element with an atomic weight (mass) of
        9.0 g/mol.

53.     $$\text{Molar mass} = \frac{dRT}{P} = \frac{\dfrac{0.70902 \text{ g}}{L} \times \dfrac{0.08206 \text{ L atm}}{\text{K mol}} \times 273.2 \text{ K}}{1.000 \text{ atm}} = 15.90 \text{ g/mol}$$

        15.90 g/mol is the average molar mass of the mixture of methane and helium.  Assume
        100.00 mol of total gas present, and let $x$ = mol of $CH_4$ in the 100.00 mol mixture.  This value
        of $x$ is also equal to the volume percentage of $CH_4$ in 100.00 L of mixture because T and P
        are constant.

        $15.90 = \dfrac{x(16.04) + (100.00 - x)(4.003)}{100.00}$,   $1590. = (16.04)x + 400.3 - (4.003)x$

        $1190. = (12.04)x$,   $x = 98.84\%$ $CH_4$ by volume;   % He $= 100.00 - x = 1.16\%$ He by volume

55.  $$n_{H_2} = \frac{PV}{RT} = \frac{1.0 \text{ atm} \times \left[ 4800 \text{ m}^3 \times \left( \frac{100 \text{ cm}}{\text{m}} \right)^3 \times \frac{1 \text{ L}}{1000 \text{ cm}^3} \right]}{\frac{0.08206 \text{ L atm}}{\text{K mol}} \times 273 \text{ K}} = 2.1 \times 10^5 \text{ mol}$$

$2.1 \times 10^5$ mol $H_2$ are in the balloon. This is 80.% of the total amount of $H_2$ that had to be generated:

$0.80(\text{total mol } H_2) = 2.1 \times 10^5$,  total mol $H_2 = 2.6 \times 10^5$ mol $H_2$

$2.6 \times 10^5 \text{ mol } H_2 \times \frac{1 \text{ mol Fe}}{\text{mol } H_2} \times \frac{55.85 \text{ g Fe}}{\text{mol Fe}} = 1.5 \times 10^7 \text{ g Fe}$

$2.6 \times 10^5 \text{ mol } H_2 \times \frac{1 \text{ mol } H_2SO_4}{\text{mol } H_2} \times \frac{98.09 \text{ g } H_2SO_4}{\text{mol } H_2SO_4} \times \frac{100 \text{ g reagent}}{98 \text{ g } H_2SO_4} = 2.6 \times 10^7$ g of 98% sulfuric acid

57.  Because P and T are constant, V and n are directly proportional. The balanced equation requires 2 L of $H_2$ to react with 1 L of CO (2 : 1 volume ratio due to 2 : 1 mole ratio in the balanced equation). The actual volume ratio present in 1 minute is 16.0 L/25.0 L = 0.640 (0.640 : 1). Because the actual volume ratio present is smaller than the required volume ratio, $H_2$ is the limiting reactant. The volume of $CH_3OH$ produced at STP will be one-half the volume of $H_2$ reacted due to the 1 : 2 mole ratio in the balanced equation. In 1 minute, 16.0 L/2 = 8.00 L $CH_3OH$ are produced (theoretical yield).

$$n_{CH_3OH} = \frac{PV}{RT} = \frac{1.00 \text{ atm} \times 8.00 \text{ L}}{\frac{0.08206 \text{ L atm}}{\text{K mol}} \times 273 \text{ K}} = 0.357 \text{ mol } CH_3OH \text{ in 1 minute}$$

$0.357 \text{ mol } CH_3OH \times \frac{32.04 \text{ g } CH_3OH}{\text{mol } CH_3OH} = 11.4 \text{ g } CH_3OH$ (theoretical yield per minute)

Percent yield $= \frac{\text{actual yield}}{\text{theoretical yield}} \times 100 = \frac{5.30 \text{ g}}{11.4 \text{ g}} \times 100 = 46.5\%$ yield

59.  $150 \text{ g } (CH_3)_2N_2H_2 \times \frac{1 \text{ mol } (CH_3)_2N_2H_2}{60.10 \text{ g}} \times \frac{3 \text{ mol } N_2}{\text{mol } (CH_3)_2N_2H_2} = 7.5 \text{ mol } N_2$ produced

$$P_{N_2} = \frac{nRT}{V} = \frac{7.5 \text{ mol} \times \frac{0.08206 \text{ L atm}}{\text{K mol}} \times 300. \text{ K}}{250 \text{ L}} = 0.74 \text{ atm}$$

We could do a similar calculation for $P_{H_2O}$ and $P_{CO_2}$ and then calculate $P_{total}$ ($= P_{N_2} + P_{H_2O} + P_{CO_2}$). Or we can recognize that 9 total moles of gaseous products form for every mole of $(CH_3)_2N_2H_2$ reacted. This is three times the moles of $N_2$ produced. Therefore, $P_{total}$ will be three times larger than $P_{N_2}$. $P_{total} = 3 \times P_{N_2} = 3 \times 0.74 \text{ atm} = 2.2 \text{ atm}$.

61.      $2 NaClO_3(s) \rightarrow 2 NaCl(s) + 3 O_2(g)$

$P_{total} = P_{O_2} + P_{H_2O}$ , $P_{O_2} = P_{total} - P_{H_2O} = 734$ torr $- 19.8$ torr $= 714$ torr

$$n_{O_2} = \frac{P_{O_2} \times V}{RT} = \frac{\left( 714 \text{ torr} \times \dfrac{1 \text{ atm}}{760 \text{ torr}} \right) \times 0.0572 \text{ L}}{\dfrac{0.08206 \text{ L atm}}{\text{K mol}} \times (273 + 22) \text{ K}} = 2.22 \times 10^{-3} \text{ mol } O_2$$

Mass $NaClO_3$ decomposed $= 2.22 \times 10^{-3}$ mol $O_2 \times \dfrac{2 \text{ mol } NaClO_3}{3 \text{ mol } O_2} \times \dfrac{106.44 \text{ g } NaClO_3}{\text{mol } NaClO_3}$

$$= 0.158 \text{ g } NaClO_3$$

Mass % $NaClO_3 = \dfrac{0.158 \text{ g}}{0.8765 \text{ g}} \times 100 = 18.0\%$

63.      $P_{total} = P_{N_2} + P_{H_2O}$ , $P_{N_2} = 726$ torr $- 23.8$ torr $= 702$ torr $\times \dfrac{1 \text{ atm}}{760 \text{ torr}} = 0.924$ atm

$$n_{N_2} = \frac{P_{N_2} \times V}{RT} = \frac{0.924 \text{ atm} \times 31.8 \times 10^{-3} \text{ L}}{\dfrac{0.08206 \text{ L atm}}{\text{K mol}} \times 298 \text{ K}} = 1.20 \times 10^{-3} \text{ mol } N_2$$

Mass of N in compound $= 1.20 \times 10^{-3}$ mol $N_2 \times \dfrac{28.02 \text{ g } N_2}{\text{mol}} = 3.36 \times 10^{-2}$ g nitrogen

Mass % N $= \dfrac{3.36 \times 10^{-2} \text{ g}}{0.253 \text{ g}} \times 100 = 13.3\%$ N

65.      For $NH_3$:  $P_2 = \dfrac{P_1 V_1}{V_2} = 0.500$ atm $\times \dfrac{2.00 \text{ L}}{3.00 \text{ L}} = 0.333$ atm

For $O_2$:  $P_2 = \dfrac{P_1 V_1}{V_2} = 1.50$ atm $\times \dfrac{1.00 \text{ L}}{3.00 \text{ L}} = 0.500$ atm

After the stopcock is opened, V and T will be constant, so $P \propto n$.  The balanced equation requires:

$$\frac{n_{O_2}}{n_{NH_3}} = \frac{P_{O_2}}{P_{NH_3}} = \frac{5}{4} = 1.25$$

The actual ratio present is:  $\dfrac{P_{O_2}}{P_{NH_3}} = \dfrac{0.500 \text{ atm}}{0.333 \text{ atm}} = 1.50$

The actual ratio is larger than the required ratio, so $NH_3$ in the denominator is limiting. Because equal moles of NO will be produced as $NH_3$ reacted, the partial pressure of NO produced is 0.333 atm (the same as $P_{NH_3}$ reacted).

67.     Rigid container (constant volume): As reactants are converted to products, the moles of gas particles present decrease by one-half. As n decreases, the pressure will decrease (by one-half). Density is the mass per unit volume. Mass is conserved in a chemical reaction, so the density of the gas will not change because mass and volume do not change.

        Flexible container (constant pressure): Pressure is constant because the container changes volume in order to keep a constant pressure. As the moles of gas particles decrease by a factor of 2, the volume of the container will decrease (by one-half). We have the same mass of gas in a smaller volume, so the gas density will increase (is doubled).

69.     $2 NH_3(g) \rightarrow N_2(g) + 3 H_2(g)$; as reactants are converted into products, we go from 2 moles of gaseous reactants to 4 moles of gaseous products (1 mol $N_2$ + 3 mol $H_2$). Because the moles of gas doubles as reactants are converted into products, the volume of the gases will double (at constant P and T).

        $$PV = nRT, P = \left( \frac{RT}{V} \right) n = (constant)n; \text{ pressure is directly related to n at constant T and V.}$$

        As the reaction occurs, the moles of gas will double, so the pressure will double. Because 1 mol of $N_2$ is produced for every 2 mol of $NH_3$ reacted, $P_{N_2} = 1/2 \, P_{NH_3}^o$. Owing to the 3 to 2 mole ratio in the balanced equation, $P_{H_2} = 3/2 \, P_{NH_3}^o$.

        *Note*: $P_{total} = P_{H_2} + P_{N_2} = 3/2 \, P_{NH_3}^o + 1/2 \, P_{NH_3}^o = 2 \, P_{NH_3}^o$. As we said earlier, the total pressure will double from the initial pressure of $NH_3$ as reactants are completely converted into products.

## Kinetic Molecular Theory and Real Gases

71.     V, T, and P are all constant, so n must be constant. Because we have equal moles of gas in each container, gas B molecules must be heavier than gas A molecules.

        a.   Both gas samples have the same number of molecules present (n is constant).

        b.   Because T is constant, $KE_{ave}$ must be the same for both gases ($KE_{ave} = 3/2 \, RT$).

        c.   The lighter gas A molecules will have the faster average velocity.

        d.   The heavier gas B molecules do collide more forcefully, but gas A molecules, with the faster average velocity, collide more frequently. The end result is that P is constant between the two containers.

73.    $(KE)_{avg} = 3/2\ RT$;   KE depends only on temperature.  At each temperature $CH_4$ and $N_2$ will have the same average KE.  For energy units of joules (J), use $R = 8.3145\ J\ K^{-1}\ mol^{-1}$.  To determine average KE per molecule, divide by Avogadro's number, $6.022 \times 10^{23}$ molecules/mol.

At 273 K:  $(KE)_{avg} = \dfrac{3}{2} \times \dfrac{8.3145\ J}{K\ mol} \times 273\ K = 3.40 \times 10^3\ J/mol = 5.65 \times 10^{-21}\ J/molecule$

At 546 K:  $(KE)_{avg} = \dfrac{3}{2} \times \dfrac{8.3145\ J}{K\ mol} \times 546\ K = 6.81 \times 10^3\ J/mol = 1.13 \times 10^{-20}\ J/molecule$

75.    No;  the numbers calculated in Exercise 73 are the average kinetic energies at the various temperatures.  At each temperature, there is a distribution of energies.  Similarly, the numbers calculated in Exercise 74 are a special kind of average velocity.  There is a distribution of velocities as shown in Figs. 5.15 to 5.17 of the text.  Note that the major reason there is a distribution of kinetic energies is because there is a distribution of velocities for any gas sample at some temperature.

77.    a.  They will all have the same average kinetic energy because they are all at the same temperature.  Average kinetic energy depends only on temperature.

       b.  Flask C;  at constant T, $u_{rms} \propto (1/M)^{1/2}$.  In general, the lighter the gas molecules, the greater is the root mean square velocity (at constant T).

       c.  Flask A:  collision frequency is proportional to average velocity $\times$ n/V (as the average velocity doubles, the number of collisions will double, and as the number of molecules in the container doubles, the number of collisions again doubles).  At constant T and V, n is proportional to P, and average velocity is proportional to $(1/M)^{1/2}$.  We use these relationships and the data in the exercise to determine the following relative values.

|   | n (relative) | $u_{avg}$ (relative) | Coll. Freq. (relative) = n $\times$ $u_{avg}$ |
|---|---|---|---|
| A | 1.0 | 1.0 | 1.0 |
| B | 0.33 | 1.0 | 0.33 |
| C | 0.13 | 3.7 | 0.48 |

79.    Graham's law of effusion:   $\dfrac{Rate_1}{Rate_2} = \left(\dfrac{M_2}{M_1}\right)^{1/2}$

Let Freon-12 = gas 1 and Freon-11 = gas 2:

$\dfrac{1.07}{1.00} = \left(\dfrac{137.4}{M_1}\right)^{1/2}$ ,   $1.14 = \dfrac{137.4}{M_1}$,   $M_1 = 121\ g/mol$

The molar mass of $CF_2Cl_2$ is equal to 121 g/mol, so Freon-12 is $CF_2Cl_2$.

81. $\dfrac{Rate_1}{Rate_2} = \left(\dfrac{M_2}{M_1}\right)^{1/2}$;   $rate_1 = \dfrac{24.0 \text{ mL}}{\text{min}}$;   $rate_2 = \dfrac{47.8 \text{ mL}}{\text{min}}$;   $M_2 = \dfrac{16.04 \text{ g}}{\text{mol}}$;   $M_1 = ?$

$\dfrac{24.0}{47.8} = \left(\dfrac{16.04}{M_1}\right)^{1/2} = 0.502$,   $16.04 = (0.502)^2 \times M_1$,   $M_1 = \dfrac{16.04}{0.252} = \dfrac{63.7 \text{ g}}{\text{mol}}$

83.  a.   $PV = nRT$

$P = \dfrac{nRT}{V} = \dfrac{0.5000 \text{ mol} \times \dfrac{0.08206 \text{ L atm}}{\text{K mol}} \times (25.0 + 273.2) \text{ K}}{1.0000 \text{ L}} = 12.24 \text{ atm}$

b.   $\left[P + a\left(\dfrac{n}{V}\right)^2\right](V - nb) = nRT$;   for $N_2$: $a = 1.39$ atm $L^2/mol^2$ and $b = 0.0391$ L/mol

$\left[P + 1.39\left(\dfrac{0.5000}{1.0000}\right)^2 \text{ atm}\right](1.0000 \text{ L} - 0.5000 \times 0.0391 \text{ L}) = 12.24 \text{ L atm}$

$(P + 0.348 \text{ atm})(0.9805 \text{ L}) = 12.24 \text{ L atm}$

$P = \dfrac{12.24 \text{ L atm}}{0.9805 \text{ L}} - 0.348 \text{ atm} = 12.48 - 0.348 = 12.13 \text{ atm}$

c.   The ideal gas law is high by 0.11 atm,  or  $\dfrac{0.11}{12.13} \times 100 = 0.91\%$.

85.   The kinetic molecular theory assumes that gas particles do not exert forces on each other and that gas particles are volumeless.  Real gas particles do exert attractive forces for each other, and real gas particles do have volumes.  A gas behaves most ideally at low pressures and high temperatures.  The effect of attractive forces is minimized at high temperatures because the gas particles are moving very rapidly.  At low pressure, the container volume is relatively large (P and V are inversely related), so the volume of the container taken up by the gas particles is negligible.

87.   The pressure measured for real gases is too low compared to ideal gases. This is due to the attractions gas particles do have for each other; these attractions "hold" them back from hitting the container walls as forcefully. To make up for this slight decrease in pressure for real gases, a factor is added to the measured pressure.  The measured volume is too large. A fraction of the space of the container volume is taken up by the volume of the molecules themselves. Therefore, the actual volume available to real gas molecules is slightly less than the container volume. A term is subtracted from the container volume to correct for the volume taken up by real gas molecules.

89.    The values of $a$ are:  $H_2$, $\dfrac{0.244 \text{ atm L}^2}{\text{mol}^2}$ ;  $CO_2$, 3.59;  $N_2$, 1.39;  $CH_4$, 2.25

Because $a$ is a measure of intermolecular attractions, the attractions are greatest for $CO_2$.

91.    $u_{rms} = \left(\dfrac{3RT}{M}\right)^{1/2} = \left[\dfrac{3\left(\dfrac{8.3145 \text{ kg m}^2}{\text{s}^2 \text{ K mol}}\right)(227 + 273) \text{ K}}{28.02 \times 10^{-3} \text{ kg/mol}}\right]^{1/2} = 667 \text{ m/s}$

$u_{mp} = \left(\dfrac{2RT}{M}\right)^{1/2} = \left[\dfrac{2\left(\dfrac{8.3145 \text{ kg m}^2}{\text{s}^2 \text{ K mol}}\right)(500. \text{ K})}{28.02 \times 10^{-3} \text{ kg/mol}}\right]^{1/2} = 545 \text{ m/s}$

$u_{avg} = \left(\dfrac{8RT}{\pi M}\right)^{1/2} = \left[\dfrac{8\left(\dfrac{8.3145 \text{ kg m}^2}{\text{s}^2 \text{ K mol}}\right)(500. \text{ K})}{\pi(28.02 \times 10^{-3} \text{ kg/mol})}\right]^{1/2} = 615 \text{ m/s}$

93.    The force per impact is proportional to $\Delta(mu) = 2mu$.  Because $m \propto M$, the molar mass, and $u \propto (1/M)^{1/2}$ at constant T,  the force per impact at constant T is proportional to $M \times (1/M)^{1/2} = \sqrt{M}$ .

$\dfrac{\text{Impact force (H}_2)}{\text{Impact force (He)}} = \sqrt{\dfrac{M_{H_2}}{M_{He}}} = \sqrt{\dfrac{2.016}{4.003}} = 0.7097$

95.    $\Delta(mu) = 2mu$ = change in momentum per impact.  Because m is proportional to M, the molar mass, and u is proportional to $(T/M)^{1/2}$:

$\Delta(mu)_{O_2} \propto 2M_{O_2}\left(\dfrac{T}{M_{O_2}}\right)^{1/2}$   and   $\Delta(mu)_{He} \propto 2M_{He}\left(\dfrac{T}{M_{He}}\right)^{1/2}$

$\dfrac{\Delta(mu)_{O_2}}{\Delta(mu)_{He}} = \dfrac{2M_{O_2}\left(\dfrac{T}{M_{O_2}}\right)^{1/2}}{2M_{He}\left(\dfrac{T}{M_{He}}\right)^{1/2}} = \dfrac{M_{O_2}}{M_{He}}\left(\dfrac{M_{He}}{M_{O_2}}\right)^{1/2} = \dfrac{31.998}{4.003}\left(\dfrac{4.003}{31.998}\right)^{1/2} = 2.827$

The change in momentum per impact is 2.827 times larger for $O_2$ molecules than for He atoms.

$$Z_A = A \frac{N}{V}\left(\frac{RT}{2\pi M}\right)^{1/2} = \text{collision rate}$$

$$\frac{Z_{O_2}}{Z_{He}} = \frac{A\left(\dfrac{N}{V}\right)\left(\dfrac{RT}{2\pi M_{O_2}}\right)^{1/2}}{A\left(\dfrac{N}{V}\right)\left(\dfrac{RT}{2\pi M_{He}}\right)^{1/2}} = \frac{\left(\dfrac{1}{M_{O_2}}\right)^{1/2}}{\left(\dfrac{1}{M_{He}}\right)^{1/2}} = 0.3537; \quad \frac{Z_{He}}{Z_{O_2}} = 2.827$$

There are 2.827 times as many impacts per second for He as for $O_2$.

97. Intermolecular collision frequency $= Z = 4\dfrac{N}{V}d^2\left(\dfrac{\pi RT}{M}\right)^{1/2}$, where d = diameter of He atom

$$\frac{n}{V} = \frac{P}{RT} = \frac{3.0 \text{ atm}}{\dfrac{0.08206 \text{ L atm}}{\text{K mol}} \times 300.\text{ K}} = 0.12 \text{ mol/L}$$

$$\frac{N}{V} = \frac{0.12 \text{ mol}}{L} \times \frac{6.022 \times 10^{23} \text{ molecules}}{\text{mol}} \times \frac{1000 \text{ L}}{m^3} = \frac{7.2 \times 10^{25} \text{ molecules}}{m^3}$$

$$Z = 4 \times \frac{7.2 \times 10^{25} \text{ molecules}}{m^3} \times (50. \times 10^{-12} \text{ m})^2 \times \left(\frac{\pi(8.3145)(300.)}{4.00 \times 10^{-3}}\right)^{1/2}$$
$$= 1.0 \times 10^9 \text{ collisions/s}$$

Mean free path $= \lambda = \dfrac{u_{avg}}{Z}$; $u_{avg} = \left(\dfrac{8\,RT}{\pi M}\right)^{1/2} = 1260 \text{ m/s}$; $\lambda = \dfrac{1260 \text{ m/s}}{1.0 \times 10^9 \text{ s}^{-1}} = 1.3 \times 10^{-6} \text{ m}$

## Atmospheric Chemistry

99.  a.  If we have $1.0 \times 10^6$ L of air, then there are $3.0 \times 10^2$ L of CO.

$P_{CO} = \chi_{CO}P_{total}$; $\chi_{CO} = \dfrac{V_{CO}}{V_{total}}$ because $V \propto n$; $P_{CO} = \dfrac{3.0 \times 10^2}{1.0 \times 10^6} \times 628 \text{ torr} = 0.19 \text{ torr}$

b.  $n_{CO} = \dfrac{P_{CO}V}{RT}$;   Assuming 1.0 $m^3$ air, 1 $m^3$ = 1000 L:

$$n_{CO} = \frac{\dfrac{0.19}{760} \text{ atm} \times (1.0 \times 10^3 \text{ L})}{\dfrac{0.08206 \text{ L atm}}{\text{K mol}} \times 273 \text{ K}} = 1.1 \times 10^{-2} \text{ mol CO}$$

$$1.1 \times 10^{-2} \text{ mol} \times \frac{6.02 \times 10^{23} \text{ molecules}}{\text{mol}} = 6.6 \times 10^{21} \text{ CO molecules in 1.0 m}^3 \text{ of air}$$

c.  $\dfrac{6.6 \times 10^{21} \text{ molecules}}{\text{m}^3} \times \left(\dfrac{1 \text{ m}}{100 \text{ cm}}\right)^3 = \dfrac{6.6 \times 10^{15} \text{ molecules CO}}{\text{cm}^3}$

101.    For benzene:

$$89.6 \times 10^{-9} \text{ g} \times \frac{1 \text{ mol}}{78.11 \text{ g}} = 1.15 \times 10^{-9} \text{ mol benzene}$$

$$V_{\text{benzene}} = \frac{n_{\text{benzene}} RT}{P} = \frac{1.15 \times 10^{-9} \text{ mol} \times \dfrac{0.08206 \text{ L atm}}{\text{K mol}} \times 296 \text{ K}}{748 \text{ torr} \times \dfrac{1 \text{ atm}}{760 \text{ torr}}} = 2.84 \times 10^{-8} \text{ L}$$

$$\text{Mixing ratio} = \frac{2.84 \times 10^{-8} \text{ L}}{3.00 \text{ L}} \times 10^6 = 9.47 \times 10^{-3} \text{ ppmv}$$

$$\text{or ppbv} = \frac{\text{vol. of X} \times 10^9}{\text{total vol.}} = \frac{2.84 \times 10^{-8} \text{ L}}{3.00 \text{ L}} \times 10^9 = 9.47 \text{ ppbv}$$

$$\frac{1.15 \times 10^{-9} \text{ mol benzene}}{3.00 \text{ L}} \times \frac{1 \text{ L}}{1000 \text{ cm}^3} \times \frac{6.022 \times 10^{23} \text{ molecules}}{\text{mol}}$$

$$= 2.31 \times 10^{11} \text{ molecules benzene/cm}^3$$

For toluene:

$$153 \times 10^{-9} \text{ g C}_7\text{H}_8 \times \frac{1 \text{ mol}}{92.13 \text{ g}} = 1.66 \times 10^{-9} \text{ mol toluene}$$

$$V_{\text{toluene}} = \frac{n_{\text{toluene}} RT}{P} = \frac{1.66 \times 10^{-9} \text{ mol} \times \dfrac{0.08206 \text{ L atm}}{\text{K mol}} \times 296 \text{ K}}{748 \text{ torr} \times \dfrac{1 \text{ atm}}{760 \text{ torr}}} = 4.10 \times 10^{-8} \text{ L}$$

$$\text{Mixing ratio} = \frac{4.10 \times 10^{-8} \text{ L}}{3.00 \text{ L}} \times 10^6 = 1.37 \times 10^{-2} \text{ ppmv (or 13.7 ppbv)}$$

$$\frac{1.66 \times 10^{-9} \text{ mol toluene}}{3.00 \text{ L}} \times \frac{1 \text{ L}}{1000 \text{ cm}^3} \times \frac{6.022 \times 10^{23} \text{ molecules}}{\text{mol}}$$

$$= 3.33 \times 10^{11} \text{ molecules toluene/cm}^3$$

## Additional Exercises

103.   $0.050 \text{ mL} \times \dfrac{1.149 \text{ g}}{\text{mL}} \times \dfrac{1 \text{ mol O}_2}{32.0 \text{ g}} = 1.8 \times 10^{-3} \text{ mol O}_2$

$$V = \frac{nRT}{P} = \frac{1.8 \times 10^{-3} \text{ mol} \times \dfrac{0.08206 \text{ L atm}}{\text{K mol}} \times 310. \text{ K}}{1.0 \text{ atm}} = 4.6 \times 10^{-2} \text{ L} = 46 \text{ mL}$$

105.   $\text{Mn(s)} + x \text{ HCl(g)} \rightarrow \text{MnCl}_x\text{(s)} + \dfrac{x}{2} \text{ H}_2\text{(g)}$

$$n_{\text{H}_2} = \frac{PV}{RT} = \frac{0.951 \text{ atm} \times 3.22 \text{ L}}{\dfrac{0.08206 \text{ L atm}}{\text{K mol}} \times 373 \text{ K}} = 0.100 \text{ mol H}_2$$

Mol Cl in compound = mol HCl = $0.100 \text{ mol H}_2 \times \dfrac{x \text{ mol Cl}}{\dfrac{x}{2} \text{ mol H}_2} = 0.200 \text{ mol Cl}$

$$\frac{\text{Mol Cl}}{\text{Mol Mn}} = \frac{0.200 \text{ mol Cl}}{2.747 \text{ g Mn} \times \dfrac{1 \text{ mol Mn}}{54.94 \text{ g Mn}}} = \frac{0.200 \text{ mol Cl}}{0.05000 \text{ mol Mn}} = 4.00$$

The formula of compound is $\text{MnCl}_4$.

107.   $PV = nRT$, V and T are constant.   $\dfrac{P_1}{n_1} = \dfrac{P_2}{n_2}$, $\dfrac{P_2}{P_1} = \dfrac{n_2}{n_1}$

We will do this limiting-reagent problem using an alternative method than described in Chapter 3.  Let's calculate the partial pressure of $C_3H_3N$ that can be produced from each of the starting materials assuming each reactant is limiting.  The reactant that produces the smallest amount of product will run out first and is the limiting reagent.

$$P_{C_3H_3N} = 0.500 \text{ MPa} \times \frac{2 \text{ MPa C}_3\text{H}_3\text{N}}{2 \text{ MPa C}_3\text{H}_6} = 0.500 \text{ MPa if C}_3\text{H}_6 \text{ is limiting}$$

$$P_{C_3H_3N} = 0.800 \text{ MPa} \times \frac{2 \text{ MPa C}_3\text{H}_3\text{N}}{2 \text{ MPa NH}_3} = 0.800 \text{ MPa if NH}_3 \text{ is limiting}$$

$$P_{C_3H_3N} = 1.500 \text{ MPa} \times \frac{2 \text{ MPa C}_3\text{H}_3\text{N}}{3 \text{ MPa O}_2} = 1.000 \text{ MPa if O}_2 \text{ is limiting}$$

$C_3H_6$ is limiting.  Although more product could be produced from $NH_3$ and $O_2$, there is only enough $C_3H_6$ to produce 0.500 MPa of $C_3H_3N$.  The partial pressure of $C_3H_3N$ in atmospheres after the reaction is:

$$0.500 \times 10^6 \text{ Pa} \times \frac{1 \text{ atm}}{1.013 \times 10^5 \text{ Pa}} = 4.94 \text{ atm}$$

$$n = \frac{PV}{RT} = \frac{4.94 \text{ atm} \times 150. \text{ L}}{\dfrac{0.08206 \text{ L atm}}{\text{K mol}} \times 298 \text{ K}} = 30.3 \text{ mol } C_3H_3N$$

$$30.3 \text{ mol} \times \frac{53.06 \text{ g}}{\text{mol}} = 1.61 \times 10^3 \text{ g } C_3H_3N \text{ can be produced.}$$

109.    $P_{total} = P_{H_2} + P_{H_2O}$,  $1.032 \text{ atm} = P_{H_2} + 32 \text{ torr} \times \dfrac{1 \text{ atm}}{760 \text{ torr}}$,  $1.032 - 0.042 = 0.990 \text{ atm} = P_{H_2}$

$$n_{H_2} = \frac{P_{H_2}V}{RT} = \frac{0.990 \text{ atm} \times 0.240 \text{ L}}{\dfrac{0.08206 \text{ L atm}}{\text{K mol}} \times 303 \text{ K}} = 9.56 \times 10^{-3} \text{ mol } H_2$$

$$9.56 \times 10^{-3} \text{ mol } H_2 \times \frac{1 \text{ mol Zn}}{\text{mol } H_2} \times \frac{65.38 \text{ g Zn}}{\text{mol Zn}} = 0.625 \text{ g Zn}$$

111.    $P_1V_1 = P_2V_2$;  the total volume is $1.00 \text{ L} + 1.00 \text{ L} + 2.00 \text{ L} = 4.00 \text{ L}$.

For He:  $P_2 = \dfrac{P_1V_1}{V_2} = 200. \text{ torr} \times \dfrac{1.00 \text{ L}}{4.00 \text{ L}} = 50.0 \text{ torr He}$

For Ne:  $P_2 = 0.400 \text{ atm} \times \dfrac{1.00 \text{ L}}{4.00 \text{ L}} = 0.100 \text{ atm}$;  $0.100 \text{ atm} \times \dfrac{760 \text{ torr}}{\text{atm}} = 76.0 \text{ torr Ne}$

For Ar:  $P_2 = 24.0 \text{ kPa} \times \dfrac{2.00 \text{ L}}{4.00 \text{ L}} = 12.0 \text{ kPa}$;  $12.0 \text{ kPa} \times \dfrac{1 \text{ atm}}{101.3 \text{ kPa}} \times \dfrac{760 \text{ torr}}{\text{atm}}$

$$= 90.0 \text{ torr Ar}$$

$P_{total} = 50.0 + 76.0 + 90.0 = 216.0 \text{ torr}$

113.    a.    $156 \text{ mL} \times \dfrac{1.34 \text{ g}}{\text{mL}} = 209 \text{ g HSiCl}_3 = \text{actual yield of HSiCl}_3$

$$n_{HCl} = \frac{PV}{RT} = \frac{10.0 \text{ atm} \times 15.0 \text{ L}}{\dfrac{0.08206 \text{ L atm}}{\text{K mol}} \times 308 \text{ K}} = 5.93 \text{ mol HCl}$$

$$5.93 \text{ mol HCl} \times \frac{1 \text{ mol HSiCl}_3}{3 \text{ mol HCl}} \times \frac{135.45 \text{ g HSiCl}_3}{\text{mol HSiCl}_3} = 268 \text{ g HSiCl}_3$$

$$\text{Percent yield} = \frac{\text{actual yield}}{\text{theoretical yield}} \times 100 = \frac{209 \text{ g}}{268 \text{ g}} \times 100 = 78.0\%$$

b.   $209 \text{ g HiSCl}_3 \times \dfrac{1 \text{ mol HSiCl}_3}{135.45 \text{ g HSiCl}_3} \times \dfrac{1 \text{ mol SiH}_4}{4 \text{ mol HSiCl}_3} = 0.386 \text{ mol SiH}_4$

This is the theoretical yield. If the percent yield is 93.1%, then the actual yield is:

$0.386 \text{ mol SiH}_4 \times 0.931 = 0.359 \text{ mol SiH}_4$

$$V_{\text{SiH}_4} = \frac{nRT}{P} = \frac{0.359 \text{ mol} \times \dfrac{0.08206 \text{ L atm}}{\text{K mol}} \times 308 \text{ K}}{10.0 \text{ atm}} = 0.907 \text{ L} = 907 \text{ mL SiH}_4$$

115.   We will apply Boyle's law to solve.   $PV = nRT = \text{constant}, \ P_1V_1 = P_2V_2$

Let condition (1) correspond to He from the tank that can be used to fill balloons.  We must leave 1.0 atm of He in the tank, so $P_1 = 200. - 1.00 = 199$ atm and $V_1 = 15.0$ L.  Condition (2) will correspond to the filled balloons with $P_2 = 1.00$ atm and $V_2 = N(2.00 \text{ L})$, where N is the number of filled balloons, each at a volume of 2.00 L.

199 atm $\times$ 15.0 L = 1.00 atm $\times$ N(2.00 L),  N = 1492.5;  we can't fill 0.5 of a balloon, so N = 1492 balloons, or to 3 significant figures, 1490 balloons.

117.   $n_{\text{Ar}} = \dfrac{228 \text{ g}}{39.95 \text{ g/mol}} = 5.71 \text{ mol Ar}; \ \chi_{\text{CH}_4} = \dfrac{n_{\text{CH}_4}}{n_{\text{CH}_4} + n_{\text{Ar}}}, \ 0.650 = \dfrac{n_{\text{CH}_4}}{n_{\text{CH}_4} + 5.71}$

$0.650(n_{\text{CH}_4} + 5.71) = n_{\text{CH}_4}, \ 3.71 = (0.350)n_{\text{CH}_4}, \ n_{\text{CH}_4} = 10.6 \text{ mol CH}_4$

$\text{KE}_{\text{avg}} = \dfrac{3}{2}RT \text{ for 1 mol}$

Thus $\text{KE}_{\text{total}} = (10.6 + 5.71 \text{ mol}) \times 3/2 \times 8.3145 \text{ J K}^{-1} \text{ mol}^{-1} \times 298 \text{ K} = 6.06 \times 10^4 \text{ J} = 60.6 \text{ kJ}$

119.   Mol of He removed $= \dfrac{PV}{RT} = \dfrac{1.00 \text{ atm} \times (1.75 \times 10^{-3} \text{ L})}{\dfrac{0.08206 \text{ L atm}}{\text{K mol}} \times 298 \text{ K}} = 7.16 \times 10^{-5} \text{ mol He}$

In the original flask, $7.16 \times 10^{-5}$ mol of He exerted a partial pressure of $1.960 - 1.710 = 0.250$ atm.

$$V = \frac{nRT}{V} = \frac{(7.16 \times 10^{-5} \text{ mol}) \times 0.08206 \text{ L atm K}^{-1} \text{ mol}^{-1} \times 298 \text{ K}}{0.250 \text{ atm}} = 7.00 \times 10^{-3} \text{ L}$$
$$= 7.00 \text{ mL}$$

121.   a.   $2 \text{ CH}_4(g) + 2 \text{ NH}_3(g) + 3 \text{ O}_2(g) \rightarrow 2 \text{ HCN}(g) + 6 \text{ H}_2\text{O}(g)$

b.   Volumes of gases are proportional to moles at constant T and P.  Using the balanced equation, methane and ammonia are in stoichiometric amounts and oxygen is in excess. In 1 second:

$$n_{CH_4} = \frac{PV}{RT} = \frac{1.00 \text{ atm} \times 20.0 \text{ L}}{0.08206 \text{ L atm K}^{-1} \text{ mol}^{-1} \times 423 \text{ K}} = 0.576 \text{ mol CH}_4$$

$$\frac{0.576 \text{ mol CH}_4}{s} \times \frac{2 \text{ mol HCN}}{2 \text{ mol CH}_4} \times \frac{27.03 \text{ g HCN}}{\text{mol HCN}} = 15.6 \text{ g HCN/s}$$

## Challenge Problems

123. Initially we have 1.00 mol $CH_4$ (16.0 g/mol = molar mass) and 2.00 mol $O_2$ (32.0 g/mol = molar mass).

$$CH_4(g) + a\,O_2(g) \rightarrow b\,CO(g) + c\,CO_2(g) + d\,H_2O(g)$$

$b + c = 1.00$ (C balance); $2a = b + 2c + d$ (O balance)

$2d = 4$ (H balance), $d = 2 = 2.00$ mol $H_2O$

$$V_{initial} = \frac{nRT}{P} = \frac{3.00 \text{ mol} \times 0.08206 \text{ L atm K}^{-1} \text{ mol}^{-1} \times 425 \text{ K}}{1.00 \text{ atm}} = 104.6 \text{ L (1 extra sig .fig.)}$$

$$\text{Density}_{initial} = \frac{80.0 \text{ g}}{104.6 \text{ L}} = 0.7648 \text{ g/L (1 extra sig. fig.)}$$

Because mass is constant:

$$\text{mass} = V_{initial} \times d_{initial} = V_{final} \times d_{final}, \; V_{final} = V_{initial} \times \frac{d_{initial}}{d_{final}} = 104.6 \text{ L} \times \frac{0.7648 \text{ g/L}}{0.7282 \text{ g/L}}$$

$V_{final} = 109.9 \text{ L (1 extra sig. fig.)}$

$$n_{final} = \frac{PV}{RT} = \frac{1.00 \text{ atm} \times 109.9 \text{ L}}{\dfrac{0.08206 \text{ L atm}}{\text{K mol}} \times 425 \text{ K}} = 3.15 \text{ total moles of gas}$$

Assuming an excess of $O_2$ is present after reaction, an expression for the total moles of gas present at completion is:

$b + c + 2.00 + (2.00 - a) = 3.15$; *Note*: $d = 2.00$ mol $H_2O$ was determined previously.

Because $b + c = 1.00$, solving gives $a = 1.85$ mol $O_2$ reacted. Indeed, $O_2$ is in excess. From the O balance equation:

$2a = 3.70 = b + 2c + 2.00$, $b + 2c = 1.70$

Because $b + c = 1.00$, solving gives $b = 0.30$ mol CO and $c = 0.70$ mol $CO_2$.

The fraction of methane that reacts to form CO is 0.30 mol CO/1.00 mol $CH_4$ = 0.30 (or 30.% by moles of the reacted methane forms CO).

125.    $Cr(s) + 3 HCl(aq) \rightarrow CrCl_3(aq) + 3/2 H_2(g)$;   $Zn(s) + 2 HCl(aq) \rightarrow ZnCl_2(aq) + H_2(g)$

Mol $H_2$ produced $= n = \dfrac{PV}{RT} = \dfrac{\left(750.\ \text{torr} \times \dfrac{1\ \text{atm}}{760\ \text{torr}}\right) \times 0.225\ \text{L}}{\dfrac{0.08206\ \text{L atm}}{\text{K mol}} \times (273 + 27)\ \text{K}} = 9.02 \times 10^{-3}\ \text{mol}\ H_2$

$9.02 \times 10^{-3}$ mol $H_2$ = mol $H_2$ from Cr reaction  +  mol $H_2$ from Zn reaction

From the balanced equation:  $9.02 \times 10^{-3}$ mol $H_2$ = mol Cr × (3/2)  +  mol Zn × 1

Let $x$ = mass of Cr and $y$ = mass of Zn, then:

$$x + y = 0.362\ \text{g and}\ 9.02 \times 10^{-3} = \frac{(1.5)x}{52.00} + \frac{y}{65.38}$$

We have two equations and two unknowns.  Solving by simultaneous equations:

$$9.02 \times 10^{-3} = (0.02885)x + (0.01530)y$$
$$\underline{-0.01530 \times 0.362 = -(0.01530)x - (0.01530)y}$$
$$3.48 \times 10^{-3} = (0.01355)x, \qquad x = \text{mass of Cr} = \frac{3.48 \times 10^{-3}}{0.01355} = 0.257\ \text{g}$$

$y$ = mass of Zn = 0.362 g − 0.257 g = 0.105 g Zn;  mass % Zn = $\dfrac{0.105\ \text{g}}{0.362\ \text{g}} \times 100$
$= 29.0\%$ Zn

127.    Molar mass $= \dfrac{dRT}{P}$ , P and molar mass are constant; $dT = \dfrac{P \times \text{molar mass}}{R}$ = constant

d = constant(1/T) or $d_1T_1 = d_2T_2$, where T is in kelvin (K).

T = $x$ + °C; 1.2930($x$ + 0.0) = 0.9460($x$ + 100.0)

(1.2930)$x$ = (0.9460)$x$ + 94.60, (0.3470)$x$ = 94.60, $x$ = 272.6

From these data, absolute zero would be −272.6°C.  The actual value is −273.15°C.

129.    $\dfrac{PV}{nRT} = 1 + \beta P$;  $\dfrac{n}{V}$ × molar mass = d

$\dfrac{\text{molar mass}}{RT} \times \dfrac{P}{d} = 1 + \beta P$,  $\dfrac{P}{d} = \dfrac{RT}{\text{molar mass}} + \dfrac{\beta RTP}{\text{molar mass}}$

This is in the equation for a straight line:  $y = b + mx$.  If we plot P/d versus P and extrapolate to P = 0, we get a $y$ intercept = $b$ = 1.398 = RT/molar mass.

At 0.00°C, molar mass = $\dfrac{0.08206 \times 273.15}{1.398} = 16.03$ g/mol.

131.    Figure 5.16 shows the effect of temperature on the Maxwell-Boltzmann distribution of velocities of molecules. Note that as temperature increases, the probability that a gas particle has the most probable velocity decreases. Thus, since the probability of the gas particle with the most probable velocity decreased by one-half, then the temperature must be higher than 300. K.

The equation that determines the probability that a gas molecule has a certain velocity is:

$$f(u) = 4\pi \left( \frac{m}{2\pi k_B T} \right)^{3/2} u^2 e^{-mu^2/2k_B T}$$

Let $T_x$ = the unknown temperature, then:

$$\frac{f(u_{mp,x})}{f(u_{mp,300})} = \frac{1}{2} = \frac{4\pi \left( \dfrac{m}{2\pi k_B T_x} \right)^{3/2} u_{mp,x}^2 \, e^{-mu_{mp,x}^2/2k_B T_x}}{4\pi \left( \dfrac{m}{2\pi k_B T_{300}} \right)^{3/2} u_{mp,300}^2 \, e^{-mu_{mp,300}^2/2k_B T_{300}}}$$

Because $u_{mp} = \sqrt{\dfrac{2k_B T}{m}}$ , the equation reduces to:

$$\frac{1}{2} = \frac{\left( \dfrac{1}{T_x} \right)^{3/2} (T_x)}{\left( \dfrac{1}{T_{300}} \right)^{3/2} (T_{300})} = \left( \frac{T_{300}}{T_x} \right)^{1/2}$$

Note that the overall exponent term cancels from the expression when $2k_B T/m$ is substituted for $u_{mp}^2$ in the exponent term; the temperatures cancel. Solving for $T_x$:

$$\frac{1}{2} = \left( \frac{300.\,K}{T_x} \right), \; T_x = 1.20 \times 10^3 \, K$$

As expected, $T_x$ is higher than 300. K.

133.    From the problem, we want $Z_A/Z = 1.00 \times 10^{18}$ where $Z_A$ is the collision frequency of the gas particles with the walls of the container and Z is the intermolecular collision frequency.

From the text:  $\dfrac{Z_A}{Z} = \dfrac{A \dfrac{N}{V} \sqrt{\dfrac{RT}{2\pi M}}}{4 \dfrac{N}{V} d^2 \sqrt{\dfrac{\pi RT}{M}}} = 1.00 \times 10^{18}, \; 1.00 \times 10^{18} = \dfrac{A}{4 d^2 \pi \sqrt{2}}$

If $\ell$ = length of the cube edge container, then the area A of one cube face is $\ell^2$ and the total area in the cube is $6\ell^2$ (6 faces/cube). He diameter = d = 2(3.2 × 10⁻¹¹ m) = 6.4 × 10⁻¹¹ m.

Solving the expression for A, and then for $\ell$ gives $\ell = 0.11$ m $= 1.1$ dm.

Volume $= \ell^3 = (1.1 \text{ dm})^3 = 1.3 \text{ dm}^3 = 1.3 \text{ L}$

135.   The reactions are:

$$C(s) + 1/2 \, O_2(g) \rightarrow CO(g) \text{ and } C(s) + O_2(g) \rightarrow CO_2(g)$$

$$PV = nRT, \; P = n\left(\frac{RT}{V}\right) = n(\text{constant})$$

Because the pressure has increased by 17.0%, the number of moles of gas has also increased by 17.0%.

$$n_{final} = (1.170)n_{initial} = 1.170(5.00) = 5.85 \text{ mol gas} = n_{O_2} + n_{CO} + n_{CO_2}$$

$$n_{CO} + n_{CO_2} = 5.00 \quad \text{(balancing moles of C).} \quad \text{Solving by simultaneous equations:}$$

$$n_{O_2} + n_{CO} + n_{CO_2} = 5.85$$
$$-(n_{CO} + n_{CO_2} = 5.00)$$
$$\overline{\phantom{n_{O_2} + n_{CO} + n_{CO_2}}}$$
$$n_{O_2} \phantom{+ n_{CO} + n_{CO_2}} = 0.85$$

If all C were converted to $CO_2$, no $O_2$ would be left.  If all C were converted to CO, we would get 5 mol CO and 2.5 mol excess $O_2$ in the reaction mixture.  In the final mixture, moles of CO equals twice the moles of $O_2$ present ($n_{CO} = 2n_{O_2}$).

$$n_{CO} = 2n_{O_2} = 1.70 \text{ mol CO}; \; 1.70 + n_{CO_2} = 5.00, \; n_{CO_2} = 3.30 \text{ mol } CO_2$$

$$\chi_{CO} = \frac{1.70}{5.85} = 0.291; \quad \chi_{CO_2} = \frac{3.30}{5.85} = 0.564; \quad \chi_{O_2} = \frac{0.85}{5.85} = 0.145 \approx 0.15$$

137.   a.   The reaction is:  $CH_4(g) + 2 \, O_2(g) \rightarrow CO_2(g) + 2 \, H_2O(g)$

$$PV = nRT, \; \frac{PV}{n} = RT = \text{constant}, \; \frac{P_{CH_4}V_{CH_4}}{n_{CH_4}} = \frac{P_{air}V_{air}}{n_{air}}$$

The balanced equation requires 2 mol $O_2$ for every mol of $CH_4$ that reacts.  For three times as much oxygen, we would need 6 mol $O_2$ per mol of $CH_4$ reacted ($n_{O_2} = 6n_{CH_4}$). Air is 21% mole percent $O_2$, so $n_{O_2} = (0.21)n_{air}$.  Therefore, the moles of air we would need to delivery the excess $O_2$ are:

$$n_{O_2} = (0.21)n_{air} = 6n_{CH_4}, \; n_{air} = 29n_{CH_4}, \; \frac{n_{air}}{n_{CH_4}} = 29$$

In 1 minute:

$$V_{air} = V_{CH_4} \times \frac{n_{air}}{n_{CH_4}} \times \frac{P_{CH_4}}{P_{air}} = 200. \text{ L} \times 29 \times \frac{1.50 \text{ atm}}{1.00 \text{ atm}} = 8.7 \times 10^3 \text{ L air/min}$$

b.  If $x$ mol of $CH_4$ were reacted, then $6x$ mol $O_2$ were added, producing $(0.950)x$ mol $CO_2$ and $(0.050)x$ mol of $CO$. In addition, $2x$ mol $H_2O$ must be produced to balance the hydrogens.

$$CH_4(g) + 2\ O_2(g) \rightarrow CO_2(g) + 2\ H_2O(g);\ \ CH_4(g) + 3/2\ O_2(g) \rightarrow CO(g) + 2\ H_2O(g)$$

Amount $O_2$ reacted:

$$(0.950)x\ \text{mol}\ CO_2 \times \frac{2\ \text{mol}\ O_2}{\text{mol}\ CO_2} = (1.90)x\ \ \text{mol}\ O_2$$

$$(0.050)x\ \text{mol}\ CO \times \frac{1.5\ \text{mol}\ O_2}{\text{mol}\ CO} = (0.075)x\ \ \text{mol}\ O_2$$

Amount of $O_2$ left in reaction mixture $= (6.00)x\ - (1.90)x - (0.075)x = (4.03)x\ \ \text{mol}\ O_2$

Amount of $N_2 = (6.00)x\ \ \text{mol}\ O_2 \times \frac{79\ \text{mol}\ N_2}{21\ \text{mol}\ O_2} = (22.6)x \approx 23x\ \ \text{mol}\ N_2$

The reaction mixture contains:

$$(0.950)x\ \ \text{mol}\ CO_2 + (0.050)x\ \ \text{mol}\ CO + (4.03)x\ \ \text{mol}\ O_2 + (2.00)x\ \ \text{mol}\ H_2O$$
$$+ 23x\ \ \text{mol}\ N_2 = (30.)x\ \ \text{mol of gas total}$$

$$\chi_{CO} = \frac{(0.050)x}{(30.)x} = 0.0017; \ \ \chi_{CO_2} = \frac{(0.950)x}{(30.)x} = 0.032; \ \ \chi_{O_2} = \frac{(4.03)x}{(30.)x} = 0.13$$

$$\chi_{H_2O} = \frac{(2.00)x}{(30.)x} = 0.067; \ \ \ \ \chi_{N_2} = \frac{23x}{(30.)x} = 0.77$$

c.  The partial pressures are determined by $P = \chi P_{total}$. Because $P_{total} = 1.00$ atm, $P_{CO} = 0.0017$ atm, $P_{CO_2} = 0.032$ atm, $P_{O_2} = 0.13$ atm, $P_{H_2O} = 0.067$ atm, and $P_{N_2} = 0.77$ atm.

139.    Each stage will give an enrichment of:

$$\frac{\text{Diffusion rate}\ ^{12}CO_2}{\text{Diffusion rate}\ ^{13}CO_2} = \left(\frac{M_{^{13}CO_2}}{M_{^{12}CO_2}}\right)^{1/2} = \left(\frac{45.001}{43.998}\right) = 1.0113$$

Because $^{12}CO_2$ moves slightly faster, each successive stage will have less $^{13}CO_2$.

$$\frac{99.90\ ^{12}CO_2}{0.10\ ^{13}CO_2} \times 1.0113^N = \frac{99.990\ ^{12}CO_2}{0.010\ ^{13}CO_2}$$

$$1.0113^N = \frac{9,999.0}{999.00} = 10.009 \ \ \ \ \text{(carrying extra significant figures)}$$

$N \log(1.0113) = \log(10.009)$, $N = \dfrac{1.000391}{4.88 \times 10^{-3}} = 2.05 \times 10^2 \approx 2.1 \times 10^2$ stages are needed.

141.   a.   Average molar mass of air = $0.790 \times 28.02$ g/mol + $0.210 \times 32.00$ g/mol = 28.9 g/mol; molar mass of helium = 4.003 g/mol

A given volume of air at a given set of conditions has a larger density than helium at those conditions.  We need to heat the air to a temperature greater than 25°C in order to lower the air density (by driving air out of the hot air balloon) until the density is the same as that for helium (at 25°C and 1.00 atm).

b.   To provide the same lift as the helium balloon (assume V = 1.00 L), the mass of air in the hot-air balloon (V = 1.00 L) must be the same as that in the helium balloon.  Let MM = molar mass:

$$P \bullet MM = dRT, \quad mass = \frac{MM \bullet PV}{RT}; \quad \text{solving: mass He} = 0.164 \text{ g}$$

$$\text{Mass air} = 0.164 \text{ g} = \frac{28.9 \text{ g/mol} \times 1.00 \text{ atm} \times 1.00 \text{ L}}{\dfrac{0.08206 \text{ L atm}}{\text{K mol}} \times \text{T}}, \quad \text{T} = 2150 \text{ K} \quad \text{(a very high temperature)}$$

143.   d = molar mass(P/RT);  at constant P and T, the density of gas is directly proportional to the molar mass of the gas. Thus the molar mass of the gas has a value which is 1.38 times that of the molar mass of $O_2$.

Molar mass = 1.38(32.00 g/mol) = 44.2 g/mol

Because $H_2O$ is produced when the unknown binary compound is combusted, the unknown must contain hydrogen. Let $A_xH_y$ be the formula for unknown compound.

$$\text{Mol } A_xH_y = 10.0 \text{ g } A_xH_y \times \frac{1 \text{ mol } A_xH_y}{44.2 \text{ g}} = 0.226 \text{ mol } A_xH_y$$

$$\text{Mol H} = 16.3 \text{ g } H_2O \times \frac{1 \text{ mol } H_2O}{18.02 \text{ g}} \times \frac{2 \text{ mol H}}{\text{mol } H_2O} = 1.81 \text{ mol H}$$

$$\frac{1.81 \text{ mol H}}{0.226 \text{ mol } A_xH_y} = 8 \text{ mol H/mol } A_xH_y ; \quad A_xH_y = A_xH_8$$

The mass of the $x$ moles of A in the $A_xH_8$ formula is:

44.2 g – 8(1.008 g) = 36.1 g

From the periodic table and by trial and error, some possibilities for $A_xH_8$ are $ClH_8$, $F_2H_8$, $C_3H_8$, and $Be_4H_8$.  $C_3H_8$ and $Be_4H_8$ fit the data best and because $C_3H_8$ (propane) is a known substance, $C_3H_8$ is the best possible identity from the data in this problem.

# CHAPTER 6

# CHEMICAL EQUILIBRIUM

## Characteristics of Chemical Equilibrium

11.     $2 NOCl(g) \rightleftharpoons 2 NO(g) + Cl_2(g)$   $K = 1.6 \times 10^{-5}$ mol/L

The expression for K is the product concentrations divided by the reactant concentrations. When K has a value much less than one, the product concentrations are relatively small, and the reactant concentrations are relatively large.

$2 NO(g) \rightleftharpoons N_2(g) + O_2(g)$   $K = 1 \times 10^{31}$

When K has a value much greater than one, the product concentrations are relatively large, and the reactant concentrations are relatively small. In both cases, however, the rate of the forward reaction equals the rate of the reverse reaction at equilibrium (this is a definition of equilibrium).

13.     No, it doesn't matter which direction the equilibrium position is reached. Both experiments will give the same equilibrium position because both experiments started with stoichiometric amounts of reactants or products.

15.     When equilibrium is reached, there is no net change in the amount of reactants and products present because the rates of the forward and reverse reactions are equal to each other. The first diagram has 4 $A_2B$ molecules, 2 $A_2$ molecules, and 1 $B_2$ molecule present. The second diagram has 2 $A_2B$ molecules, 4 $A_2$ molecules, and 2 $B_2$ molecules. Therefore, the first diagram cannot represent equilibrium because there was a net change in reactants and products. Is the second diagram the equilibrium mixture? That depends on whether there is a net change between reactants and products when going from the second diagram to the third diagram. The third diagram contains the same number and type of molecules as the second diagram, so the second diagram is the first illustration that represents equilibrium.

The reaction container initially contained only $A_2B$. From the first diagram, 2 $A_2$ molecules and 1 $B_2$ molecule are present (along with 4 $A_2B$ molecules). From the balanced reaction, these 2 $A_2$ molecules and 1 $B_2$ molecule were formed when 2 $A_2B$ molecules decomposed. Therefore, the initial number of $A_2B$ molecules present equals $4 + 2 = 6$ molecules $A_2B$.

## The Equilibrium Constant

17.     K and $K_p$ are equilibrium constants as determined by the law of mass action. For K, concentration units of mol/L are used, and for $K_p$, partial pressures in units of atm are used (generally). Q is called the reaction quotient. Q has the exact same form as K or $K_p$, but instead of equilibrium concentrations, initial concentrations are used to calculate the Q value. We use Q to determine if a reaction is at equilibrium.  When Q = K (or when $Q_p = K_p$), the reaction is at equilibrium. When $Q \neq K$, the reaction is not at equilibrium, and one can deduce the net change that must occur for the system to get to equilibrium.

19.     Solids and liquids do not appear in equilibrium expressions. Only gases and dissolved solutes appear in equilibrium expressions.

a.   $K = \dfrac{[H_2O]}{[NH_3]^2[CO_2]}$;  $K_p = \dfrac{P_{H_2O}}{P_{NH_3}^2 \times P_{CO_2}}$     b.   $K = [N_2][Br_2]^3$;  $K_p = P_{N_2} \times P_{Br_2}^3$

c.   $K = [O_2]^3$;  $K_p = P_{O_2}^3$     d.   $K = \dfrac{[H_2O]}{[H_2]}$;  $K_p = \dfrac{P_{H_2O}}{P_{H_2}}$

21.     $K_p = K(RT)^{\Delta n}$, where $\Delta n$ = sum of gaseous product coefficients – sum of gaseous reactant coefficients.  For this reaction, $\Delta n = 3 - 1 = 2$.

$$K = \frac{[CO][H_2]^2}{[CH_3OH]} = \frac{(0.24)(1.1)^2}{(0.15)} = 1.9 \text{ mol}^2/\text{L}^2$$

$$K_p = K(RT)^2 = 1.9(0.08206 \text{ L atm K}^{-1} \text{ mol}^{-1} \times 600. \text{ K})^2 = 4.6 \times 10^3 \text{ atm}^2$$

23.     $[N_2O] = \dfrac{2.00 \times 10^{-2} \text{ mol}}{2.00 \text{ L}}$;  $[N_2] = \dfrac{2.80 \times 10^{-4} \text{ mol}}{2.00 \text{ L}}$;  $[O_2] = \dfrac{2.50 \times 10^{-5} \text{ mol}}{2.00 \text{ L}}$

$$K = \frac{[N_2O]^2}{[N_2]^2[O_2]} = \frac{\left(\dfrac{2.00 \times 10^{-2}}{2.00}\right)^2}{\left(\dfrac{2.80 \times 10^{-4}}{2.00}\right)^2\left(\dfrac{2.50 \times 10^{-5}}{2.00}\right)} = \frac{(1.00 \times 10^{-2})^2}{(1.40 \times 10^{-4})^2(1.25 \times 10^{-5})}$$

$$= 4.08 \times 10^8 \text{ L/mol}$$

If the given concentrations represent equilibrium concentrations, then they should give a value of K = $4.08 \times 10^8$.

$$\frac{(0.200)^2}{(2.00 \times 10^{-4})^2(0.00245)} = 4.08 \times 10^8$$

Because the given concentrations when plugged into the equilibrium constant expression give a value equal to K ($4.08 \times 10^8$), this set of concentrations is a system at equilibrium.

25.    $K_p = \dfrac{P_{H_2}^4}{P_{H_2O}^4}$;  $P_{total} = P_{H_2O} + P_{H_2}$,  36.3 torr = 15.0 torr + $P_{H_2}$,  $P_{H_2}$ = 21.3 torr

Because 1 atm = 760 torr:  $K_p = \dfrac{\left(21.3\ torr \times \dfrac{1\ atm}{760\ torr}\right)^4}{\left(15.0\ torr \times \dfrac{1\ atm}{760\ torr}\right)^4} = 4.07$

*Note*: Solids and pure liquids are not included in K expressions.

27.    $PCl_5(g) \rightleftharpoons PCl_3(g) + Cl_2(g)$    $K_p = \dfrac{P_{PCl_3} \times P_{Cl_2}}{P_{PCl_5}}$

To determine $K_p$, we must determine the equilibrium partial pressures of each gas. Initially, $P_{PCl_5}$ = 0.50 atm and $P_{PCl_3}$ = $P_{Cl_2}$ = 0 atm. To reach equilibrium, some of the $PCl_5$ reacts to produce some $PCl_3$ and $Cl_2$, all in a 1 : 1 mole ratio. We must determine the change in partial pressures necessary to reach equilibrium. Because moles $\propto$ P at constant V and T, if we let $x$ = atm of $PCl_5$ that reacts to reach equilibrium, this will produce $x$ atm of $PCl_3$ and $x$ atm of $Cl_2$ at equilibrium. The equilibrium partial pressures of each gas will be the initial partial pressure of each gas plus the change necessary to reach equilibrium. The equilibrium partial pressures are:

   $P_{PCl_5}$ = 0.50 atm $- x$,  $P_{PCl_3}$ = $P_{Cl_2}$ = $x$

Now we solve for $x$ using the information in the problem:

   $P_{total} = P_{PCl_5} + P_{PCl_3} + P_{Cl_2}$,  0.84 atm = 0.50 $- x + x + x$,  0.84 atm = 0.50 + $x$,

   $x$ = 0.34 atm

The equilibrium partial pressures are:

   $P_{PCl_5}$ = 0.50 $-$ 0.34 = 0.16 atm,  $P_{PCl_3}$ = $P_{Cl_2}$ = 0.34 atm

   $K_p = \dfrac{P_{PCl_3} \times P_{Cl_2}}{P_{PCl_5}} = \dfrac{(0.34)(0.34)}{(0.16)} = 0.72$ atm

   $K = \dfrac{K_p}{(RT)^{\Delta n}}$, $\Delta n = 2 - 1 = 1$;  $K_p = \dfrac{0.72}{(0.08206)(523)} = 0.017$ mol/L

29.    When solving equilibrium problems, a common method to summarize all the information in the problem is to set up a table. We commonly call this table an ICE table because it summarizes *i*nitial concentrations, *c*hanges that must occur to reach equilibrium, and *e*quilibrium concentrations (the sum of the initial and change columns). For the change column, we will generally use the variable *x*, which will be defined as the amount of reactant (or product) that must react to reach equilibrium. In this problem, the reaction must shift

right to reach equilibrium because there are no products present initially. Therefore, $x$ is defined as the amount of reactant $SO_3$ that reacts to reach equilibrium, and we use the coefficients in the balanced equation to relate the net change in $SO_3$ to the net change in $SO_2$ and $O_2$. The general ICE table for this problem is:

$$2\ SO_3(g) \rightleftharpoons 2\ SO_2(g) + O_2(g) \qquad K = \frac{[SO_2]^2[O_2]}{[SO_3]^2}$$

| | | | |
|---|---|---|---|
| Initial | 12.0 mol/3.0 L | 0 | 0 |
| | Let $x$ mol/L of $SO_3$ react to reach equilibrium. | | |
| Change | $-x$ $\rightarrow$ | $+x$ | $+x/2$ |
| Equil. | $4.0 - x$ | $x$ | $x/2$ |

From the problem, we are told that the equilibrium $SO_2$ concentration is 3.0 mol/3.0 L = 1.0 $M$ ([$SO_2$]$_e$ = 1.0 $M$). From the ICE table setup, [$SO_2$]$_e$ = $x$, so $x$ = 1.0. Solving for the other equilibrium concentrations: [$SO_3$]$_e$ = 4.0 − $x$ = 4.0 − 1.0 = 3.0 $M$; [$O_2$] = $x/2$ = 1.0/2 = 0.50 $M$.

$$K = \frac{[SO_2]^2[O_2]}{[SO_3]^2} = \frac{(1.0\ M)^2\,(0.50\ M)}{(3.0\ M)^2} = 0.056\ \text{mol/L}$$

Alternate method: Fractions in the change column can be avoided (if you want) be defining $x$ differently. If we were to let $2x$ mol/L of $SO_3$ react to reach equilibrium, then the ICE table setup is:

$$2\ SO_3(g) \rightleftharpoons 2\ SO_2(g) + O_2(g) \qquad K = \frac{[SO_2]^2[O_2]}{[SO_3]^2}$$

| | | | |
|---|---|---|---|
| Initial | 4.0 $M$ | 0 | 0 |
| | Let $2x$ mol/L of $SO_3$ react to reach equilibrium. | | |
| Change | $-2x$ $\rightarrow$ | $+2x$ | $+x$ |
| Equil. | $4.0 - 2x$ | $2x$ | $x$ |

Solving: $2x$ = [$SO_2$]$_e$ = 1.0 $M$, $x$ = 0.50 $M$; [$SO_3$]$_e$ = 4.0 − 2(0.50) = 3.0 $M$; [$O_2$]$_e$ = $x$ = 0.50 $M$

These are exactly the same equilibrium concentrations as solved for previously, thus K will be the same (as it must be). The moral of the story is to define $x$ in a manner that is most comfortable for you. Your final answer is independent of how you define $x$ initially.


## Equilibrium Calculations

31.    $CaCO_3(s) \rightleftharpoons CaO(s) + CO_2(g)$    $K_p = P_{CO_2} = 1.04$ atm

We only need to calculate the initial partial pressure of $CO_2$ and compare this value to 1.04 atm. At this temperature, all $CO_2$ will be in the gas phase.

a.   $PV = nRT, \ Q = P_{CO_2} = \dfrac{n_{CO_2}RT}{V} = \dfrac{\dfrac{58.4 \text{ g } CO_2}{44.01 \text{ g/mol}} \times \dfrac{0.08206 \text{ L atm}}{K \text{ mol}} \times 1173 \text{ K}}{50.0 \text{ L}} =$

$$2.55 \text{ atm} > K_p$$

Reaction will shift to the left because $Q > K_p$; the mass of CaO will decrease.

b.   $Q = \ P_{CO_2} = \dfrac{(23.76)(0.08206)(1173)}{(44.01)(50.0)} = 1.04 \text{ atm} = K_p$

At equilibrium because $Q = K_p$; mass of CaO will not change.

c.   Mass of $CO_2$ is the same as in part b.   $P = 1.04 \text{ atm} = K_P$.   At equilibrium; mass of CaO will not change.

d.   $Q = P_{CO_2} = \dfrac{(4.82)(0.08206)(1173)}{(44.01)(50.0)} = 0.211 \text{ atm} < K_p$

Reaction will shift to the right because $Q < K_p$; the mass of CaO will increase.

33.     $H_2O(g) + Cl_2O(g) \rightleftharpoons 2 \ HOCl(g)$        $K = 0.090 = \dfrac{[HOCl]^2}{[H_2O][Cl_2O]}$

a.   The initial concentrations of $H_2O$ and $Cl_2O$ are:

$$\dfrac{1.0 \text{ g } H_2O}{1.0 \text{ L}} \times \dfrac{1 \text{ mol}}{18.0 \text{ g}} = 5.6 \times 10^{-2} \text{ mol/L}; \quad \dfrac{2.0 \text{ g } Cl_2O}{1.0 \text{ L}} \times \dfrac{1 \text{ mol}}{86.9 \text{ g}} = 2.3 \times 10^{-2} \text{ mol/L}$$

Because only reactants are present initially, the reaction must proceed to the right to reach equilibrium.  Summarizing the problem in a table:

|          | $H_2O(g)$ | + | $Cl_2O(g)$ | $\rightleftharpoons$ | $2 \ HOCl(g)$ |
|----------|-----------|---|------------|------|------------|
| Initial  | $5.6 \times 10^{-2} \ M$ | | $2.3 \times 10^{-2} \ M$ | | $0$ |
| | | $x$ mol/L of $H_2O$ reacts to reach equilibrium | | | |
| Change   | $-x$ | | $-x$ | $\rightarrow$ | $+2x$ |
| Equil.   | $5.6 \times 10^{-2} - x$ | | $2.3 \times 10^{-2} - x$ | | $2x$ |

$$K = 0.090 = \dfrac{(2x)^2}{(5.6 \times 10^{-2} - x)(2.3 \times 10^{-2} - x)}$$

$$1.16 \times 10^{-4} - (7.11 \times 10^{-3})x + (0.090)x^2 = 4x^2$$

$$(3.91)x^2 + (7.11 \times 10^{-3})x - 1.16 \times 10^{-4} = 0 \quad \text{(We carried extra significant figures.)}$$

Solving using the quadratic formula (see Appendix 1 of the text):

$$x = \frac{-7.11 \times 10^{-3} \pm (5.06 \times 10^{-5} + 1.81 \times 10^{-3})^{1/2}}{7.82} = 4.6 \times 10^{-3} \; M \text{ or } -6.4 \times 10^{-3} \; M$$

A negative answer makes no physical sense; we can't have less than nothing. Thus $x = 4.6 \times 10^{-3} \; M$.

$[HOCl] = 2x = 9.2 \times 10^{-3} \; M; \quad [Cl_2O] = 2.3 \times 10^{-2} - x = 0.023 - 0.0046 = 1.8 \times 10^{-2} \; M$

$[H_2O] = 5.6 \times 10^{-2} - x = 0.056 - 0.0046 = 5.1 \times 10^{-2} \; M$

b. 

|  | $H_2O(g)$ | $+$ | $Cl_2O(g)$ | $\rightleftharpoons$ | $2\,HOCl(g)$ |
|---|---|---|---|---|---|
| Initial | 0 |  | 0 |  | 1.0 mol/2.0 L $= 0.50\,M$ |

$2x$ mol/L of HOCl reacts to reach equilibrium

| Change | $+x$ |  | $+x$ | $\leftarrow$ | $-2x$ |
| Equil. | $x$ |  | $x$ |  | $0.50 - 2x$ |

$$K = 0.090 = \frac{[HOCl]^2}{[H_2O][Cl_2O]} = \frac{(0.50 - 2x)^2}{x^2}$$

The expression is a perfect square, so we can take the square root of each side:

$$0.30 = \frac{0.50 - 2x}{x}, \quad (0.30)x = 0.50 - 2x, \; (2.30)x = 0.50$$

$x = 0.217 \; M$ (We carried extra significant figures.)

$x = [H_2O] = [Cl_2O] = 0.217 = 0.22 \; M; \quad [HOCl] = 0.50 - 2x = 0.50 - 0.434 = 0.07 \; M$

35. 

|  | $2\,SO_2(g)$ | $+$ | $O_2(g)$ | $\rightleftharpoons$ | $2\,SO_3(g)$ | $K_p = 0.25$ |
|---|---|---|---|---|---|---|
| Initial | 0.50 atm |  | 0.50 atm |  | 0 |  |

$2x$ atm of $SO_2$ reacts to reach equilibrium

| Change | $-2x$ |  | $-x$ | $\rightarrow$ | $+2x$ |
| Equil. | $0.50 - 2x$ |  | $0.50 - x$ |  | $2x$ |

$$K_p = 0.25 = \frac{P_{SO_3}^2}{P_{SO_2}^2 \times P_{O_2}} = \frac{(2x)^2}{(0.50 - 2x)^2 (0.50 - x)}$$

This will give a cubic equation.  Graphing calculators can be used to solve this expression. If you don't have a graphing calculator, an alternative method for solving a cubic equation is to use the method of successive approximations (see Appendix 1 of the text).  The first step is to guess a value for $x$.  Because the value of K is small (K < 1), not much of the forward reaction will occur to reach equilibrium.  This tells us that $x$ is small.  Let's guess that $x =$

0.050 atm.   Now we take this estimated value for $x$ and substitute it into the equation everywhere that $x$ appears except for one.   For equilibrium problems, we will substitute the estimated value for $x$ into the denominator, and then solve for the numerator value of $x$.   We continue this process until the estimated value of $x$ and the calculated value of $x$ converge on the same number.   This is the same answer we would get if we were to solve the cubic equation exactly.   Applying the method of successive approximations and carrying extra significant figures:

$$\frac{4x^2}{[0.50-2(0.050)]^2-[0.50-(0.050)]} = \frac{4x^2}{(0.40)^2(0.45)} = 0.25, \ x = 0.067$$

$$\frac{4x^2}{[0.50-2(0.067)]^2\,[0.50-(0.067)]} = \frac{4x^2}{(0.366)^2(0.433)} = 0.25, \ x = 0.060$$

$$\frac{4x^2}{(0.38)^2(0.44)} = 0.25, \ x = 0.063; \quad \frac{4x^2}{(0.374)^2(0.437)} = 0.25, \quad x = 0.062$$

The next trial gives the same value for $x = 0.062$ atm.   We are done except for determining the equilibrium concentrations.   They are:

$$P_{SO_2} = 0.50 - 2x = 0.50 - 2(0.062) = 0.376 = 0.38 \text{ atm}$$

$$P_{O_2} = 0.50 - x = 0.438 = 0.44 \text{ atm}; \quad P_{SO_3} = 2x = 0.124 = 0.12 \text{ atm}$$

37.   The assumption comes from the value of K being much less than 1. For these reactions, the equilibrium mixture will not have a lot of products present; mostly reactants are present at equilibrium. If we define the change that must occur in terms of $x$ as the amount (molarity or partial pressure) of a reactant that must react to reach equilibrium, then $x$ must be a small number because K is a very small number. We want to know the value of $x$ in order to solve the problem, so we don't assume $x = 0$. Instead, we concentrate on the equilibrium row in the ICE table. Those reactants (or products) that have equilibrium concentrations in the form of $0.10 - x$ or $0.25 + x$ or $3.5 - 3x$, etc., is where an important assumption can be made. The assumption is that because K $\ll$ 1, $x$ will be small ($x \ll 1$), and when we add $x$ or subtract $x$ from some initial concentration, it will make little or no difference. That is, we assume that $0.10 - x \approx 0.10$ or $0.25 + x \approx 0.25$ or $3.5 - 3x \approx 3.5$, etc.; we assume that the initial concentration of a substance is equal to the final concentration. This assumption makes the math much easier and usually gives a value of $x$ that is well within 5% of the true value of $x$ (we get about the same answer with a lot less work).

We check the assumptions for validity using the 5% rule. From doing a lot of these calculations, it is found that when an assumption such as $0.20 - x \approx 0.20$ is made, if $x$ is less than 5% of the number the assumption was made against, then our final answer is within acceptable error limits of the true value of $x$ (as determined when the equation is solved exactly). For our example above ($0.20 - x \approx 0.20$), if $(x/0.20) \times 100 \leq 5\%$, then our assumption is valid by the 5% rule. If the error is greater than 5%, then we must solve the equation exactly or use a math trick called the method of successive approximations. See Appendix 1 for details regarding the method of successive approximations, as well as for a review in solving quadratic equations exactly.

39.                     $2 CO_2(g) \;\rightleftharpoons\; 2 CO(g) \;+\; O_2(g)$   $K = \dfrac{[CO]^2[O_2]}{[CO_2]^2} = 2.0 \times 10^{-6}$

Initial        2.0 mol/5.0 L          0                0
               $2x$ mol/L of $CO_2$ reacts to reach equilibrium
Change      $-2x$          $\rightarrow$      $+2x$             $+x$
Equil.      $0.40 - 2x$                  $2x$               $x$

$K = 2.0 \times 10^{-6} = \dfrac{[CO]^2[O_2]}{[CO_2]^2} = \dfrac{(2x)^2(x)}{(0.40 - 2x)^2}$ ; assuming $2x \ll 0.40$ (K is small, so $x$ is
small.)

$2.0 \times 10^{-6} \approx \dfrac{4x^3}{(0.40)^2}$ , $2.0 \times 10^{-6} = \dfrac{4x^3}{0.16}$ , $x = 4.3 \times 10^{-3}\ M$

Checking assumption: $\dfrac{2(4.3 \times 10^{-3})}{0.40} \times 100 = 2.2\%$; assumption is valid by the 5% rule.

$[CO_2] = 0.40 - 2x = 0.40 - 2(4.3 \times 10^{-3}) = 0.39\ M$

$[CO] = 2x = 2(4.3 \times 10^{-3}) = 8.6 \times 10^{-3}\ M$;   $[O_2] = x = 4.3 \times 10^{-3}\ M$

41.     $CaCO_3(s) \;\rightleftharpoons\; CaO(s) + CO_2(g)$   $K_p = 1.16 = P_{CO_2}$

Some of the 20.0 g of $CaCO_3$ will react to reach equilibrium. The amount that reacts is the
quantity of $CaCO_3$ required to produce a $CO_2$ pressure of 1.16 atm (from the $K_p$ expression).

$n_{CO_2} = \dfrac{P_{CO_2} V}{RT} = \dfrac{1.16\ \text{atm} \times 10.0\ \text{L}}{\dfrac{0.08206\ \text{L atm}}{\text{K mol}} \times 1073\ \text{K}} = 0.132\ \text{mol}\ CO_2$

Mass $CaCO_3$ reacted = 0.132 mol $CO_2 \times \dfrac{1\ \text{mol}\ CaCO_3}{\text{mol}\ CO_2} \times \dfrac{100.09\ \text{g}}{\text{mol}\ CaCO_3} = 13.2\ \text{g}\ CaCO_3$

Mass % $CaCO_3$ reacted = $\dfrac{13.2\ \text{g}}{20.0\ \text{g}} \times 100 = 66.0\%$

43.     a.   $K_p = K(RT)^{\Delta n} = \dfrac{4.5 \times 10^9\ \text{L}}{\text{mol}} \left( \dfrac{0.08206\ \text{L atm}}{\text{K mol}} \times 373\ \text{K} \right)^{-1}$ , where $\Delta n = 1 - 2 = -1$

$K_p = 1.5 \times 10^8\ \text{atm}^{-1}$

b.   $K_p$ is so large that at equilibrium we will have almost all $COCl_2$.  Assume $P_{total} \approx$
$P_{COCl_2} \approx 5.0$ atm.

$$CO(g) + Cl_2(g) \rightleftharpoons COCl_2(g) \qquad K_p = 1.5 \times 10^8$$

| | | | |
|---|---|---|---|
| Initial | 0 | 0 | 5.0 atm |

$x$ atm $COCl_2$ reacts to reach equilibrium

| | | | |
|---|---|---|---|
| Change | $+x$ | $+x$ | $\leftarrow \quad -x$ |
| Equil. | $x$ | $x$ | $5.0 - x$ |

$$K_p = 1.5 \times 10^8 = \frac{5.0 - x}{x^2} \approx \frac{5.0}{x^2} \qquad \text{(Assuming } 5.0 - x \approx 5.0.\text{)}$$

Solving: $x = 1.8 \times 10^{-4}$ atm. Check assumptions: $5.0 - x = 5.0 - 1.8 \times 10^{-4} = 5.0$ atm. Assumptions are good (well within the 5% rule).

$$P_{CO} = P_{Cl_2} = 1.8 \times 10^{-4} \text{ atm and } P_{COCl_2} = 5.0 \text{ atm}$$

## Le Chatelier's Principle

45.     Only statement d is correct. Addition of a catalyst has no effect on the equilibrium position; the reaction just reaches equilibrium more quickly. Statement a is false for reactants that are either solids or liquids (adding more of these has no effect on the equilibrium). Statement b is false always. If temperature remains constant, then the value of K is constant. Statement c is false for exothermic reactions where an increase in temperature decreases the value of K.

47.     a.   No effect; adding more of a pure solid or pure liquid has no effect on the equilibrium position.

    b.   Shifts left; $HF(g)$ will be removed by reaction with the glass. As $HF(g)$ is removed, the reaction will shift left to produce more $HF(g)$.

    c.   Shifts right; as $H_2O(g)$ is removed, the reaction will shift right to produce more $H_2O(g)$.

49.     $H^+ + OH^- \rightarrow H_2O$; sodium hydroxide (NaOH) will react with the $H^+$ on the product side of the reaction. This effectively removes $H^+$ from the equilibrium, which will shift the reaction to the right to produce more $H^+$ and $CrO_4^{2-}$. Because more $CrO_4^{2-}$ is produced, the solution turns yellow.

51.     a.   Right          b.   Right          c.   No effect; $He(g)$ is neither a reactant nor a product.

    d.   Left; because the reaction is exothermic, heat is a product:

    $$CO(g) + H_2O(g) \rightarrow H_2(g) + CO_2(g) + \text{heat}$$

    Increasing T will add heat. The equilibrium shifts to the left to use up the added heat.

    e.   No effect; because the moles of gaseous reactants equals the moles of gaseous products (2 mol versus 2 mol), a change in volume will have no effect on the equilibrium.

53.     a.   Left                    b.   Right                    c.   Left

d.  No effect (reactant and product concentrations are unchanged)

e.  No effect;  because there are equal numbers of product and reactant gas molecules, a change in volume has no effect on the equilibrium position.

f.  Right;  a decrease in temperature will shift the equilibrium to the right because heat is a product in this reaction (as is true in all exothermic reactions).

55.  An endothermic reaction, where heat is a reactant, will shift right to products with an increase in temperature. The amount of $NH_3(g)$ will increase as the reaction shifts right, so the smell of ammonia will increase.


## Additional Exercises

57.  a.   $N_2(g) + O_2(g) \rightleftharpoons 2\ NO(g)$    $K_p = 1 \times 10^{-31} = \dfrac{P_{NO}^2}{P_{N_2} \times P_{O_2}} = \dfrac{P_{NO}^2}{(0.8)(0.2)}$

$P_{NO} = 1 \times 10^{-16}$ atm

In 1.0 cm$^3$ of air:  $n_{NO} = \dfrac{PV}{RT} = \dfrac{(1 \times 10^{-16}\ \text{atm})(1.0 \times 10^{-3}\ \text{L})}{\left(\dfrac{0.08206\ \text{L atm}}{\text{K mol}}\right)(298\ \text{K})} = 4 \times 10^{-21}$ mol NO

$\dfrac{4 \times 10^{-21}\ \text{mol NO}}{\text{cm}^3} \times \dfrac{6.02 \times 10^{23}\ \text{molecules}}{\text{mol NO}} = \dfrac{2 \times 10^3\ \text{molecules NO}}{\text{cm}^3}$

b.  There is more NO in the atmosphere than we would expect from the value of K.  The answer must lie in the rates of the reaction.  At 25°C, the rates of both reactions:

$$N_2 + O_2 \rightarrow 2\ NO \ \text{ and } \ 2\ NO \rightarrow N_2 + O_2$$

are so slow that they are essentially zero.  Very strong bonds must be broken; the activation energy is very high.  Therefore, the reaction essentially doesn't occur at low temperatures.  Nitric oxide, however, can be produced in high-energy or high-temperature environments because the production of NO is endothermic.  In nature, some NO is produced by lightning and the primary manmade source is automobiles.  At these high temperatures, K will increase, and the rates of the reaction will also increase, resulting in a higher production of NO.  Once the NO gets into a more normal temperature environment, it doesn't go back to $N_2$ and $O_2$ because of the slow rate.

59.      $O(g) + NO(g) \rightleftharpoons NO_2(g)$                        $K = 1/6.8 \times 10^{-49} = 1.5 \times 10^{48}$

$NO_2(g) + O_2(g) \rightleftharpoons NO(g) + O_3(g)$           $K = 1/5.8 \times 10^{-34} = 1.7 \times 10^{33}$

---

$O_2(g) + O(g) \rightleftharpoons O_3(g)$                         $K = (1.5 \times 10^{48})(1.7 \times 10^{33}) = 2.6 \times 10^{81}$ L/mol

61.                          $3 H_2(g)$    +    $N_2(g)$    $\rightleftharpoons$    $2 NH_3(g)$

|  | | | |
|---|---|---|---|
| Initial | $[H_2]_0$ | $[N_2]_0$ | 0 |

$x$ mol/L of $N_2$ reacts to reach equilibrium

| Change | $-3x$ | $-x$ | $\rightarrow$ | $+2x$ |
|---|---|---|---|---|
| Equil. | $[H_2]_0 - 3x$ | $[N_2]_0 - x$ | | $2x$ |

From the problem:

$[NH_3]_e = 4.0\ M = 2x,\ x = 2.0\ M;\ [H_2]_e = 5.0\ M = [H_2]_0 - 3x;\ [N_2]_e = 8.0\ M = [N_2]_0 - x$

$5.0\ M = [H_2]_0 - 3(2.0\ M),\ [H_2]_0 = 11.0\ M;\ 8.0\ M = [N_2]_0 - 2.0\ M,\ [N_2]_0 = 10.0\ M$

63.    a.                $PCl_5(g) \rightleftharpoons PCl_3(g) + Cl_2(g)$    $K_p = (P_{PCl_3} \times P_{Cl_2})/P_{PCl_5}$

| | | | | |
|---|---|---|---|---|
| Initial | $P_0$ | 0 | 0 | $P_0$ = initial $PCl_5$ pressure |
| Change | $-x$ | $\rightarrow$ $+x$ | $+x$ | |
| Equil. | $P_0 - x$ | $x$ | $x$ | |

$P_{total} = P_0 - x + x + x = P_0 + x = 358.7$ torr

$$P_0 = \frac{n_{PCl_5}RT}{V} = \frac{\dfrac{2.4156\ g}{208.22\ g/mol} \times \dfrac{0.08206\ L\ atm}{K\ mol} \times 523.2\ K}{2.000\ L} = 0.2490\ atm\ (or\ 189.2\ torr)$$

$x = P_{total} - P_0 = 358.7 - 189.2 = 169.5$ torr

$$P_{PCl_3} = P_{Cl_2} = 169.5\ torr \times \frac{1\ atm}{760\ torr} = 0.2230\ atm$$

$$P_{PCl_5} = 189.2 - 169.5 = 19.7\ torr \times \frac{1\ atm}{760\ torr} = 0.0259\ atm$$

$$K_p = \frac{(0.2230)^2}{0.0259} = 1.92\ atm$$

b.    $P_{Cl_2} = \dfrac{n_{Cl_2}RT}{V} = \dfrac{0.250 \times 0.08206 \times 523.2}{2.000} = 5.37\ atm\ Cl_2\ added$

$$PCl_5(g) \quad \rightleftharpoons \quad PCl_3(g) \quad + \quad Cl_2(g)$$

| | | | | |
|---|---|---|---|---|
| Initial | 0.0259 atm | | 0.2230 atm | 0.2230 atm | (from a) |

Adding 0.250 mol $Cl_2$ increases $P_{Cl_2}$ by 5.37 atm.

| | | | | | |
|---|---|---|---|---|---|
| Initial' | 0.0259 | | 0.2230 | 5.59 | |
| Change | +0.2230 | $\leftarrow$ | −0.2230 | −0.2230 | React completely |
| After | 0.2489 | | 0 | 5.37 | New initial |
| Change | $-x$ | $\rightarrow$ | $+x$ | $+x$ | |
| Equil. | $0.2489 - x$ | | $x$ | $5.37 + x$ | |

$$\frac{(5.37 + x)(x)}{(0.2489 - x)} = 1.92, \quad x^2 + (7.29)x - 0.478 = 0$$

Solving using the quadratic formula: $x = 0.0650$ atm

$$P_{PCl_3} = 0.0650 \text{ atm}; \quad P_{PCl_5} = 0.2489 - 0.0650 = 0.1839 \text{ atm}; \quad P_{Cl_2} = 5.37 + 0.0650$$
$$= 5.44 \text{ atm}$$

65.    $N_2O_4(g) \rightleftharpoons 2 NO_2(g)$ $\qquad K_p = \dfrac{P_{NO_2}^2}{P_{N_2O_4}} = \dfrac{(1.20)^2}{0.34} = 4.2 \text{ atm}$

Doubling the volume decreases each partial pressure by a factor of 2 ($P = nRT/V$).
$P_{NO_2} = 0.600$ atm and $P_{N_2O_4} = 0.17$ atm are the new partial pressures.

$$Q = \frac{(0.600)^2}{0.17} = 2.1, \quad Q < K; \quad \text{equilibrium will shift to the right.}$$

$$N_2O_4(g) \quad \rightleftharpoons \quad 2 NO_2(g)$$

| | | |
|---|---|---|
| Initial | 0.17 atm | 0.600 atm |
| Equil. | $0.17 - x$ | $0.600 + 2x$ |

$$K_p = 4.2 = \frac{(0.600 + 2x)^2}{(0.17 - x)}, \quad 4x^2 + (6.6)x - 0.354 = 0 \quad \text{(carrying extra sig. figs.)}$$

Solving using the quadratic formula: $x = 0.052$ atm

$$P_{NO_2} = 0.600 + 2(0.052) = 0.704 \text{ atm}; \quad P_{N_2O_4} = 0.17 - 0.052 = 0.12 \text{ atm}$$

67.    a.                $2 NaHCO_3(s) \rightleftharpoons Na_2CO_3(s) + CO_2(g) + H_2O(g)$        $K_p = 0.25$

Initial          $-$                  $-$                0          0

NaHCO$_3$(s) decomposes to form $x$ atm each of $CO_2(g)$ and $H_2O(g)$ at equilibrium.

Change        $-$          $\rightarrow$       $-$              $+x$        $+x$

Equil.          $-$                  $-$                $x$          $x$

$$K_p = 0.25 = P_{CO_2} \times P_{H_2O},\ \ 0.25 = x^2,\ x = P_{CO_2} = P_{H_2O} = 0.50 \text{ atm}$$

b.    $n_{CO_2} = \dfrac{P_{CO_2}V}{RT} = \dfrac{(0.50 \text{ atm})(1.00 \text{ L})}{(0.08206 \text{ L atm K}^{-1} \text{ mol}^{-1})(398 \text{ K})} = 1.5 \times 10^{-2} \text{ mol } CO_2$

Mass of Na$_2$CO$_3$ produced:

$$1.5 \times 10^{-2} \text{ mol } CO_2 \times \frac{1 \text{ mol Na}_2CO_3}{\text{mol } CO_2} \times \frac{106.0 \text{ g Na}_2CO_3}{\text{mol Na}_2CO_3} = 1.6 \text{ g Na}_2CO_3$$

Mass of NaHCO$_3$ reacted:

$$1.5 \times 10^{-2} \text{ mol } CO_2 \times \frac{2 \text{ mol NaHCO}_3}{\text{mol } CO_2} \times \frac{84.01 \text{ g NaHCO}_3}{\text{mol}} = 2.5 \text{ g NaHCO}_3$$

Mass of NaHCO$_3$ remaining $= 10.0 - 2.5 = 7.5$ g

c.    $10.0 \text{ g NaHCO}_3 \times \dfrac{1 \text{ mol NaHCO}_3}{84.01 \text{ g NaHCO}_3} \times \dfrac{1 \text{ mol } CO_2}{2 \text{ mol NaHCO}_3} = 5.95 \times 10^{-2} \text{ mol } CO_2$

When all the NaHCO$_3$ has just been consumed, we will have $5.95 \times 10^{-2}$ mol CO$_2$ gas at a pressure of 0.50 atm (from a).

$$V = \frac{nRT}{P} = \frac{(5.95 \times 10^{-2} \text{ mol})(0.08206 \text{ L atm K}^{-1} \text{ mol}^{-1})(398 \text{ K})}{0.50 \text{ atm}} = 3.9 \text{ L}$$

69.            $NH_3(g) + H_2S(g) \rightleftharpoons NH_4HS(s)$    $K = 400. = \dfrac{1}{[NH_3][H_2S]}$

Initial    $\dfrac{2.00 \text{ mol}}{5.00 \text{ L}}$        $\dfrac{2.00 \text{ mol}}{5.00 \text{ L}}$        $-$

$x$ mol/L of NH$_3$ reacts to reach equilibrium

Change    $-x$            $-x$            $-$

Equil.    $0.400 - x$        $0.400 - x$        $-$

$K = 400. = \dfrac{1}{(0.400 - x)(0.400 - x)}$,    $0.400 - x = \left(\dfrac{1}{400.}\right)^{1/2} = 0.0500$,    $x = 0.350\ M$

Mol $NH_4HS(s)$ produced $= 5.00 \text{ L} \times \dfrac{0.350 \text{ mol } NH_3}{\text{L}} \times \dfrac{1 \text{ mol } NH_4HS}{\text{mol } NH_3} = 1.75 \text{ mol}$

Total mol $NH_4HS(s) = 2.00$ mol initially $+ 1.75$ mol produced $= 3.75$ mol total

$3.75 \text{ mol } NH_4HS \times \dfrac{51.12 \text{ g } NH_4HS}{\text{mol } NH_4HS} = 192 \text{ g } NH_4HS$

$[H_2S]_e = 0.400 \ M - x = 0.400 \ M - 0.350 \ M = 0.050 \ M \ H_2S$

$P_{H_2S} = \dfrac{n_{H_2S}RT}{V} = \dfrac{n_{H_2S}}{V} \times RT = \dfrac{0.050 \text{ mol}}{\text{L}} \times \dfrac{0.08206 \text{ L atm}}{\text{K mol}} \times 308.2 \text{ K} = 1.3 \text{ atm}$

71.    $5.63 \text{ g } C_5H_6O_3 \times \dfrac{1 \text{ mol } C_5H_6O_3}{114.10 \text{ g}} = 0.0493 \text{ mol } C_5H_6O_3$ initially

Total moles of gas $= n_{total} = \dfrac{P_{total}V}{RT} = \dfrac{1.63 \text{ atm} \times 2.50 \text{ L}}{\dfrac{0.08206 \text{ L atm}}{\text{K mol}} \times 473 \text{ K}} = 0.105 \text{ mol}$
at equilibrium

$$C_5H_6O_3(g) \ \rightleftharpoons \ C_2H_6(g) \ + \ 3 \ CO(g)$$

Initial         0.0493 mol              0              0
                Let $x$ mol $C_5H_6O_3$ react to reach equilibrium.
Change          $-x$            $\rightarrow$   $+x$          $+3x$
Equil.          $0.0493 - x$            $x$            $3x$

0.105 mol total $= 0.0493 - x + x + 3x = 0.0493 + 3x, \ x = 0.0186$ mol

$K = \dfrac{[C_2H_6][CO]^3}{[C_5H_6O_3]} = \dfrac{\left[\dfrac{0.0186 \text{ mol } C_2H_6}{2.50 \text{ L}}\right]\left[\dfrac{3(0.0186) \text{ mol CO}}{2.50 \text{ L}}\right]^3}{\left[\dfrac{(0.0493 - 0.0186) \text{ mol } C_5H_6O_3}{2.50 \text{ L}}\right]} = 6.74 \times 10^{-6} \text{ mol}^3/\text{L}^3$

## Challenge Problems

73.    a.                    $2 \ NO(g) \ + \ Br_2(g) \ \rightleftharpoons \ 2 \ NOBr(g)$

       Initial        98.4 torr        41.3 torr            0
                      $2x$ torr of NO reacts to reach equilibrium
       Change         $-2x$            $-x$    $\rightarrow$   $+2x$
       Equil.         $98.4 - 2x$      $41.3 - x$           $2x$

$$P_{total} = P_{NO} + P_{Br_2} + P_{NOBr} = (98.4 - 2x) + (41.3 - x) + 2x = 139.7 - x$$

$$P_{total} = 110.5 = 139.7 - x, \ x = 29.2 \text{ torr}; \quad P_{NO} = 98.4 - 2(29.2) = 40.0 \text{ torr} = 0.0526 \text{ atm}$$

$$P_{Br_2} = 41.3 - 29.2 = 12.1 \text{ torr} = 0.0159 \text{ atm}; \quad P_{NOBr} = 2(29.2) = 58.4 \text{ torr} = 0.0768 \text{ atm}$$

$$K_p = \frac{P_{NOBr}^2}{P_{NO}^2 \times P_{Br_2}} = \frac{(0.0768 \text{ atm})^2}{(0.0526 \text{ atm})^2 (0.0159 \text{ atm})} = 134 \text{ atm}^{-1}$$

b.                    $2 \text{ NO(g)} \quad + \quad Br_2(g) \quad \rightleftharpoons \quad 2 \text{ NOBr(g)}$

| | | | |
|---|---|---|---|
| Initial | 0.30 atm | 0.30 atm | 0 |
| | 2x atm of NO reacts to reach equilibrium | | |
| Change | −2x | −x      → | +2x |
| Equil. | 0.30 − 2x | 0.30 − x | 2x |

This would yield a cubic equation, which can be difficult to solve unless you have a graphing calculator. Because $K_p$ is pretty large, let's approach equilibrium in two steps: Assume the reaction goes to completion, and then solve the back equilibrium problem.

$2 \text{ NO} \quad + \quad Br_2 \quad \rightleftharpoons \quad 2 \text{ NOBr}$

| | | | | |
|---|---|---|---|---|
| Before | 0.30 atm | 0.30 atm | 0 | |
| | Let 0.30 atm NO react completely. | | | |
| Change | −0.30 | −0.15   → | +0.30 | React completely |
| After | 0 | 0.15 | 0.30 | New initial |
| | 2y atm of NOBr reacts to reach equilibrium | | | |
| Change | +2y | +y   ← | −2y | |
| Equil. | 2y | 0.15 + y | 0.30 − 2y | |

$$K_p = 134 = \frac{(0.30 - 2y)^2}{(2y)^2 (0.15 + y)}, \ \frac{(0.30 - 2y)^2}{(0.15 + y)} = 134 \times 4y^2 = 536y^2$$

If $y \ll 0.15$: $\dfrac{(0.30)^2}{0.15} \approx 536y^2$ and $y = 0.034$; assumptions are poor ($y$ is 23% of 0.15).

Use 0.034 as an approximation for $y$, and solve by successive approximations (see Appendix 1):

$$\frac{(0.30 - 0.068)^2}{0.15 + 0.034} = 536y^2, \ y = 0.023; \quad \frac{(0.30 - 0.046)^2}{0.15 + 0.023} = 536y^2, \ y = 0.026$$

$$\frac{(0.30 - 0.052)^2}{0.15 + 0.026} = 536y^2, \ y = 0.026 \text{ atm} \quad \text{(We have converged on the correct answer.)}$$

So: $P_{NO} = 2y = 0.052 \text{ atm}; \quad P_{Br_2} = 0.15 + y = 0.18 \text{ atm}; \quad P_{NOBr} = 0.30 - 2y = 0.25 \text{ atm}$

75.   $P_4(g) \rightleftharpoons 2\,P_2(g)$   $K_p = 0.100 = \dfrac{P_{P_2}^2}{P_{P_4}}$;   $P_{P_4} + P_{P_2} = P_{total} = 1.00$ atm,   $P_{P_4} = 1.00 - P_{P_2}$

Let $y = P_{P_2}$ at equilibrium, then $K_p = \dfrac{y^2}{1.00 - y} = 0.100$

Solving: $y = 0.270$ atm $= P_{P_2}$;   $P_{P_4} = 1.00 - 0.270 = 0.73$ atm

To solve for the fraction dissociated, we need the initial pressure of $P_4$.

$$P_4(g) \quad\rightleftharpoons\quad 2\,P_2(g)$$

| | | |
|---|---|---|
| Initial | $P_0$ | 0 |

$P_0$ = initial pressure of $P_4$

$x$ atm of $P_4$ reacts to reach equilibrium

| | | | |
|---|---|---|---|
| Change | $-x$ | $\rightarrow$ | $+2x$ |
| Equil. | $P_0 - x$ | | $2x$ |

$P_{total} = P_0 - x + 2x = 1.00$ atm $= P_0 + x$

Solving: $0.270$ atm $= P_{P_2} = 2x$,  $x = 0.135$ atm;  $P_0 = 1.00 - 0.135 = 0.87$ atm

Fraction dissociated $= \dfrac{x}{P_0} = \dfrac{0.135}{0.87} = 0.16$, or 16% of $P_4$ is dissociated to reach equilibrium.

77.   $d = \text{density} = \dfrac{P \times (\text{molar mass})}{RT} = \dfrac{P_{O_2}(\text{molar mass}_{O_2}) + P_{O_3}(\text{molar mass}_{O_3})}{RT}$

$0.168$ g/L $= \dfrac{P_{O_2}(32.00\text{ g/mol}) + P_{O_3}(48.00\text{ g/mol})}{\dfrac{0.08206\text{ L atm}}{\text{K mol}} \times 448\text{ K}}$,   $(32.00)P_{O_2} + (48.00)P_{O_3} = 6.18$

$P_{total} = P_{O_2} + P_{O_3} = 128\text{ torr} \times \dfrac{1\text{ atm}}{760\text{ torr}} = 0.168$ atm

We have two equations in two unknowns.  Solving using simultaneous equations:

$$(32.00)P_{O_2} + (48.00)P_{O_3} = 6.18$$
$$-(32.00)P_{O_2} - (32.00)P_{O_3} = -5.38$$

$$\overline{\hspace{4cm}(16.00)P_{O_3} = 0.80}$$

$P_{O_3} = \dfrac{0.80}{16.00} = 0.050$ atm and  $P_{O_2} = 0.118$ atm

$K_p = \dfrac{P_{O_3}^2}{P_{O_2}^3} = \dfrac{(0.050)^2}{(0.118)^3} = 1.5$ atm$^{-1}$

79.    $4.72 \text{ g CH}_3\text{OH} \times \dfrac{1 \text{ mol}}{32.04 \text{ g}} = 0.147 \text{ mol CH}_3\text{OH initially}$

Graham's law of effusion:  $\dfrac{\text{Rate}_1}{\text{Rate}_2} = \sqrt{\dfrac{M_2}{M_1}}$

$\dfrac{\text{Rate}_{H_2}}{\text{Rate}_{CH_3OH}} = \sqrt{\dfrac{M_{CH_3OH}}{M_{H_2}}} = \sqrt{\dfrac{32.04}{2.016}} = 3.987$

The effused mixture has 33.0 times as much $H_2$ as $CH_3OH$.  When the effusion rate ratio is multiplied by the equilibrium mole ratio of $H_2$ to $CH_3OH$, the effused mixture will have 33.0 times as much $H_2$ as $CH_3OH$.  Let $n_{H_2}$ and $n_{CH_3OH}$ equal the equilibrium moles of $H_2$ and $CH_3OH$, respectively.

$33.0 = 3.987 \times \dfrac{n_{H_2}}{n_{CH_3OH}}, \quad \dfrac{n_{H_2}}{n_{CH_3OH}} = 8.28$

$$CH_3OH(g) \quad \rightleftharpoons \quad CO(g) \quad + \; 2\,H_2(g)$$

| | | | |
|---|---|---|---|
| Initial | 0.147 mol | 0 | 0 |
| Change | $-x$ | $\rightarrow$ $+x$ | $+2x$ |
| Equil. | $0.147 - x$ | $x$ | $2x$ |

From the ICE table,  $8.28 = \dfrac{n_{H_2}}{n_{CH_3OH}} = \dfrac{2x}{0.147 - x}$

Solving:  $x = 0.118$ mol

$K = \dfrac{[CO][H_2]^2}{[CH_3OH]} = \dfrac{\left(\dfrac{0.118 \text{ mol}}{1.00 \text{ L}}\right)\left(\dfrac{2(0.118 \text{ mol})}{1.00 \text{ L}}\right)^2}{\dfrac{(0.147 - 0.118) \text{ mol}}{1.00 \text{ L}}} = 0.23 \text{ mol}^2/\text{L}^2$

81.    a.  $P_{PCl_5} = \dfrac{n_{PCl_5}RT}{V} = \dfrac{0.100 \text{ mol} \times \dfrac{0.08206 \text{ L atm}}{\text{K mol}} \times 480.\,K}{12.0 \text{ L}} = 0.328 \text{ atm}$

$$PCl_5(g) \quad \rightleftharpoons \quad PCl_3(g) \quad + \quad Cl_2(g) \qquad K_p = 0.267 \text{ atm}$$

| | | | |
|---|---|---|---|
| Initial | 0.328 atm | 0 | 0 |
| Change | $-x$ | $\rightarrow$ $+x$ | $+x$ |
| Equil. | $0.328 - x$ | $x$ | $x$ |

$K_p = \dfrac{x^2}{0.328 - x} = 0.267, \; x^2 + (0.267)x - 0.08758 = 0$  (carrying extra sig. figs.)

Solving using the quadratic formula:  $x = 0.191$ atm

$P_{PCl_3} = P_{Cl_2} = 0.191 \text{ atm}; \; P_{PCl_5} = 0.328 - 0.191 = 0.137 \text{ atm}$

b.                    $PCl_5(g) \rightleftharpoons PCl_3(g) + Cl_2(g)$

| | | | | |
|---|---|---|---|---|
| Initial | $P_0$ | | 0 | 0 | $P_0$ = initial pressure of $PCl_5$ |
| Change | $-x$ | $\rightarrow$ | $+x$ | $+x$ |
| Equil. | $P_0 - x$ | | $x$ | $x$ |

$P_{total} = 2.00$ atm $= (P_0 - x) + x + x = P_0 + x$,  $P_0 = 2.00 - x$

$$K_p = \frac{x^2}{P_0 - x} = 0.267; \quad \frac{x^2}{2.00 - 2x} = 0.267, \quad x^2 = 0.534 - (0.534)x$$

$x^2 + (0.534)x - 0.534 = 0;$      solving using the quadratic formula:

$$x = \frac{-0.534 \pm \sqrt{(0.534)^2 + 4(0.534)}}{2} = 0.511 \text{ atm}$$

$P_0 = 2.00 - x = 2.00 - 0.511 = 1.49$ atm;  the initial pressure of $PCl_5$ was 1.49 atm.

$$n_{PCl_5} = \frac{P_{PCl_5} V}{RT} = \frac{(1.49 \text{ atm})(5.00 \text{ L})}{(0.08206 \text{ L atm K}^{-1} \text{ mol}^{-1})(480. \text{ K})} = 0.189 \text{ mol } PCl_5$$

0.189 mol $PCl_5$ × 208.22 g $PCl_5$/mol = 39.4 g $PCl_5$ was initially introduced.

83.                    $N_2(g) + O_2(g) \rightleftharpoons 2 NO(g)$   Let:

                                                            equilibrium $P_{O_2} = p$

Equil.        $(3.7)p$        $p$                $x$        equilibrium $P_{N_2} = (78/21)P_{O_2} = (3.7)p$

                                                            equilibrium $P_{NO} = x$

                                                            equilibrium $P_{NO_2} = y$

$$K_p = 1.5 \times 10^{-4} = \frac{P_{NO}^2}{P_{O_2} \times P_{N_2}}$$

                    $N_2 + 2 O_2 \rightleftharpoons 2 NO_2$

Equil.        $(3.7)p$        $p$                $y$

$$K_p = 1.0 \times 10^{-5} = \frac{P_{NO_2}^2}{P_{O_2}^2 \times P_{N_2}}$$

We want $P_{NO_2} = P_{NO}$ at equilibrium, so $x = y$.

Taking the ratio of the two $K_p$ expressions:

$$\frac{\dfrac{P_{NO}^2}{P_{O_2} \times P_{N_2}}}{\dfrac{P_{NO_2}^2}{P_{O_2}^2 \times P_{N_2}}} = \frac{1.5 \times 10^{-4}}{1.0 \times 10^{-5}}; \ \text{because} \ P_{NO} = P_{NO_2}: \ P_{O_2} = \frac{1.5 \times 10^{-4}}{1.0 \times 10^{-5}} = 15 \ \text{atm}$$

Air is 21 mol % $O_2$, so:

$$P_{O_2} = (0.21)P_{total}, \quad P_{total} = \frac{15 \ \text{atm}}{0.21} = 71 \ \text{atm}$$

To solve for the equilibrium concentrations of all gases (not required to answer the question), solve one of the $K_p$ expressions where $p = P_{O_2} = 15$ atm.

$$1.5 \times 10^{-4} = \frac{x^2}{15[3.7(15)]}, \ x = P_{NO} = P_{NO_2} = 0.35 \ \text{atm}$$

Equilibrium pressures:

$$P_{O_2} = 15 \ \text{atm}; \quad P_{N_2} = 3.7(15) = 55.5 = 56 \ \text{atm}; \quad P_{NO} = P_{NO_2} = 0.35 \ \text{atm}$$

85.                    $SO_3(g) \ \rightleftharpoons \ SO_2(g) \ + \ 1/2 \ O_2(g)$

| | | | |
|---|---|---|---|
| Initial | $P_0$ | 0 | 0 | $P_0 =$ initial pressure of $SO_3$ |
| Change | $-x$ | $\rightarrow$ $+x$ | $+x/2$ |
| Equil. | $P_0 - x$ | $x$ | $x/2$ |

Average molar mass of the mixture is:

$$\text{average molar mass} = \frac{dRT}{P} = \frac{(1.60 \ \text{g/L})(0.08206 \ \text{L atm K}^{-1} \ \text{mol}^{-1})(873 \ \text{K})}{1.80 \ \text{atm}} = 63.7 \ \text{g/mol}$$

The average molar mass is determined by:

$$\text{average molar mass} = \frac{n_{SO_3}(80.07 \ \text{g/mol}) + n_{SO_2}(64.07 \ \text{g/mol}) + n_{O_2}(32.00 \ \text{g/mol})}{n_{total}}$$

Because $\chi_A =$ mol fraction of component $A = n_A/n_{total} = P_A/P_{total}$:

$$63.7 \ \text{g/mol} = = \frac{P_{SO_3}(80.07) + P_{SO_2}(64.07) + P_{O_2}(32.00)}{P_{total}}$$

$$P_{total} = P_0 - x + x + x/2 = P_0 + x/2 = 1.80 \ \text{atm}, \ P_0 = 1.80 - x/2$$

$$63.7 = \frac{(P_0 - x)(80.07) + x(64.07) + \frac{x}{2}(32.00)}{1.80}$$

$$63.7 = \frac{(1.80 - 3/2\,x)(80.07) + x(64.07) + \frac{x}{2}(32.00)}{1.80}$$

$$115 = 144 - (120.1)x + (64.07)x + (16.00)x, \quad (40.0)x = 29, \quad x = 0.73 \text{ atm}$$

$$P_{SO_3} = P_0 - x = 1.80 - (3/2)x = 0.71 \text{ atm}; \quad P_{SO_2} = 0.73 \text{ atm}; \quad P_{O_2} = x/2 = 0.37 \text{ atm}$$

$$K_p = \frac{P_{SO_2} \times P_{O_2}^{1/2}}{P_{SO_3}} = \frac{(0.73)(0.37)^{1/2}}{(0.71)} = 0.63 \text{ atm}^{1/2}$$

87.    $$N_2(g) + 3\,H_2\,(g) \rightleftharpoons 2\,NH_3(g) \qquad K_p = \frac{P_{NH_3}^2}{P_{N_2} \times P_{H_2}^3} = 6.5 \times 10^{-3}$$

<u>1.0 atm</u>      $N_2(g)$     +     $3\,H_2(g)$     $\rightleftharpoons$     $2\,NH_3(g)$

| | | | |
|---|---|---|---|
| Initial | 0.25 atm | 0.75 atm | 0 |
| Equil. | $0.25 - x$ | $0.75 - 3x$ | $2x$ |

$$\frac{(2x)^2}{(0.75 - 3x)^3(0.25 - x)} = 6.5 \times 10^{-3}; \text{ using successive approximations:}$$

$$x = 1.2 \times 10^{-2} \text{ atm}; \quad P_{NH_3} = 2x = 0.024 \text{ atm}$$

<u>10 atm</u>      $N_2(g)$     +     $3\,H_2(g)$     $\rightleftharpoons$     $2\,NH_3(g)$

| | | | |
|---|---|---|---|
| Initial | 2.5 atm | 7.5 atm | 0 |
| Equil. | $2.5 - x$ | $7.5 - 3x$ | $2x$ |

$$\frac{(2x)^2}{(7.5 - 3x)^3(2.5 - x)} = 6.5 \times 10^{-3}; \text{ using successive approximations:}$$

$$x = 0.69 \text{ atm}; \quad P_{NH_3} = 1.4 \text{ atm}$$

<u>100 atm</u>      Using the same setup as above: $\dfrac{4x^2}{(75 - 3x)^3(25 - x)} = 6.5 \times 10^{-3}$

Solving by successive approximations: $x = 16 \text{ atm}; \quad P_{NH_3} = 32 \text{ atm}$

<u>1000 atm</u>

$$N_2(g) \quad + \quad 3\,H_2(g) \quad \rightleftharpoons \quad 2\,NH_3(g)$$

| | | | |
|---|---|---|---|
| Initial | 250 atm | 750 atm | 0 |

Let 250 atm $N_2$ react completely.

| | | | |
|---|---|---|---|
| New initial | 0 | 0 | $5.0 \times 10^2$ |
| Equil. | $x$ | $3x$ | $5.0 \times 10^2 - 2x$ |

$$\frac{(5.0 \times 10^2 - 2x)^2}{(3x)^3 x} = 6.5 \times 10^{-3}; \text{ Using successive approximations:}$$

$$x = 32 \text{ atm}; \quad P_{NH_3} = 5.0 \times 10^2 - 2x = 440 \text{ atm}$$

The results are plotted as $\log P_{NH_3}$ versus $\log P_{total}$. Notice that as $P_{total}$ increases, a larger fraction of $N_2$ and $H_2$ is converted to $NH_3$, that is, as $P_{total}$ increases (V decreases), the reaction shifts further to the right, as predicted by LeChatelier's principle.

89.     $$\frac{2.00 \text{ g}}{165 \text{ g/mol}} = 0.0121 \text{ mol XY (initially)}$$

$(0.350)(0.0121 \text{ mol}) = 4.24 \times 10^{-3} \text{ mol XY dissociated}$

$$XY \quad \rightarrow \quad X \quad + \quad Y$$

| | | | |
|---|---|---|---|
| Initial | 0.0121 mol | 0 | 0 |
| Change | $-0.00424$ | $\rightarrow +0.00424$ | $+0.00424$ |
| Equil. | 0.0079 mol | 0.00424 mol | 0.00424 mol |

Total moles of gas = $0.0079 + 0.00424 + 0.00424 = 0.0164$ mol

$V \propto n$, so: $\dfrac{V_{final}}{V_{initial}} = \dfrac{n_{final}}{n_{initial}} = \dfrac{0.0164\ mol}{0.0121\ mol} = 1.36$

$V_{initial} = \dfrac{nRT}{P} = \dfrac{(0.0121\ mol)(0.008206\ L\ atm\ K^{-1}\ mol^{-1})(298\ K)}{0.967\ atm} = 0.306\ L$

$V_{final} = 0.306\ L(1.36) = 0.416\ L$;   Because mass is conserved in a chemical reaction:

$$density\ (final) = \frac{mass}{volume} = \frac{2.00\ g}{0.416\ L} = 4.81\ g/L$$

$$K = \frac{[X][Y]}{[XY]} = \frac{\left(\dfrac{0.00424\ mol}{0.416\ L}\right)^2}{\left(\dfrac{0.0079\ mol}{0.416\ L}\right)} = 5.5 \times 10^{-3}\ mol/L$$

# CHAPTER 7

# ACIDS AND BASES

## Nature of Acids and Bases

17. An acid is a proton ($H^+$) donor, and a base is a proton acceptor. A conjugate acid-base pair differs by only a proton ($H^+$) in the formulas.

| | Acid | Base | Conjugate Base of Acid | Conjugate Acid of Base |
|---|---|---|---|---|
| a. | $H_2CO_3$ | $H_2O$ | $HCO_3^-$ | $H_3O^+$ |
| b. | $C_5H_5NH^+$ | $H_2O$ | $C_5H_5N$ | $H_3O^+$ |
| c. | $C_5H_5NH^+$ | $HCO_3^-$ | $C_5H_5N$ | $H_2CO_3$ |

19. The dissociation reaction (the $K_a$ reaction) of an acid in water commonly omits water as a reactant. We will follow this practice. All dissociation reactions produce $H^+$ and the conjugate base of the acid that is dissociated.

a. $HC_2H_3O_2(aq) \rightleftharpoons H^+(aq) + C_2H_3O_2^-(aq)$ $\qquad K_a = \dfrac{[H^+][C_2H_3O_2^-]}{[HC_2H_3O_2]}$

b. $Co(H_2O)_6^{3+}(aq) \rightleftharpoons H^+(aq) + Co(H_2O)_5(OH)^{2+}(aq)$ $\qquad K_a = \dfrac{[H^+][Co(H_2O)_5(OH)^{2+}]}{[Co(H_2O)_6^{3+}]}$

c. $CH_3NH_3^+(aq) \rightleftharpoons H^+(aq) + CH_3NH_2(aq)$ $\qquad K_a = \dfrac{[H^+][CH_3NH_2]}{[CH_3NH_3^+]}$

21. The beaker on the left represents a strong acid in solution; the acid HA is 100% dissociated into the $H^+$ and $A^-$ ions. The beaker on the right represents a weak acid in solution; only a little bit of the acid HB dissociates into ions, so the acid exists mostly as undissociated HB molecules in water.

a. $HNO_2$: weak acid beaker
b. $HNO_3$: strong acid beaker
c. HCl: strong acid beaker
d. HF: weak acid beaker
e. $HC_2H_3O_2$: weak acid beaker

23.    The $K_a$ value is directly related to acid strength. As $K_a$ increases, acid strength increases. For water, use $K_w$ when comparing the acid strength of water to other species. The $K_a$ values are:

HClO$_4$:  strong acid ($K_a \gg 1$);  HClO$_2$:  $K_a = 1.2 \times 10^{-2}$

NH$_4^+$:  $K_a = 5.6 \times 10^{-10}$;  H$_2$O:  $K_a = K_w = 1.0 \times 10^{-14}$

From the $K_a$ values, the ordering is:  HClO$_4$ > HClO$_2$ > NH$_4^+$ > H$_2$O

25.    a.    H$_2$SO$_4$ is a strong acid and water is a very weak acid with $K_a = K_w = 1.0 \times 10^{-14}$. H$_2$SO$_4$ is a much stronger acid than H$_2$O.

b.    H$_2$O, $K_a = K_w = 1.0 \times 10^{-14}$; HOCl, $K_a = 3.5 \times 10^{-8}$; HOCl is a stronger acid than H$_2$O because $K_a$ for HOCl > $K_a$ for H$_2$O.

c.    NH$_4^+$ , $K_a = 5.6 \times 10^{-10}$; HC$_2$H$_2$ClO$_2$, $K_a = 1.35 \times 10^{-3}$; HC$_2$H$_2$ClO$_2$ is a stronger acid than NH$_4^+$ because $K_a$ for HC$_2$H$_2$ClO$_2$ > $K_a$ for NH$_4^+$.

27.    a.    H$_2$O and CH$_3$CO$_2^-$

b.    An acid-base reaction can be thought of as a competition between two opposing bases. Because this equilibrium lies far to the left ($K_a < 1$), CH$_3$CO$_2^-$ is a stronger base than H$_2$O.

c.    The acetate ion is a better base than water and produces basic solutions in water.  When we put acetate ion into solution as the only major basic species, the reaction is:

$$CH_3CO_2^- + H_2O \rightleftharpoons CH_3CO_2H + OH^-$$

Now the competition is between CH$_3$CO$_2^-$ and OH$^-$ for the proton.  Hydroxide ion is the strongest base possible in water.  The preceding equilibrium lies far to the left resulting in a $K_b$ value of less than 1.  Those species we specifically call weak bases ($10^{-14} < K_b < 1$) lie between H$_2$O and OH$^-$ in base strength.  Weak bases are stronger bases than water but are weaker bases than OH$^-$.

29.    In deciding whether a substance is an acid or a base, strong or weak, you should keep in mind a couple ideas:

1.    There are only a few common strong acids and strong bases all of which should be memorized. Common strong acids = HCl, HBr, HI, HNO$_3$, HClO$_4$, and H$_2$SO$_4$.  Common strong bases = LiOH, NaOH, KOH, RbOH, CsOH, Ca(OH)$_2$, Sr(OH)$_2$, and Ba(OH)$_2$.

2.    All other acids and bases are weak and will have $K_a$ and $K_b$ values of less than 1 but greater than $K_w$ ($10^{-14}$).  Reference Table 7.2 for $K_a$ values for some weak acids and Table 7.3 for $K_b$ values for some weak bases.  There are too many weak acids and weak bases to memorize them all.  Therefore, use the tables of $K_a$ and $K_b$ values to help you identify weak acids and weak bases.  Appendix 5 contains more complete tables of $K_a$ and $K_b$ values.

a.   weak acid ($K_a = 4.0 \times 10^{-4}$)          b.   strong acid
c.   weak base ($K_b = 4.38 \times 10^{-4}$)          d.   strong base
e.   weak base ($K_b = 1.8 \times 10^{-5}$)          f.   weak acid ($K_a = 7.2 \times 10^{-4}$)
g.   weak acid ($K_a = 1.8 \times 10^{-4}$)          h.   strong base
i.   strong acid

## Autoionization of Water and pH Scale

31.    a.   $H_2O(l) \rightleftharpoons H^+(aq) + OH^-(aq)$     $K_w = 2.92 \times 10^{-14} = [H^+][OH^-]$

In pure water: $[H^+] = [OH^-]$, $2.92 \times 10^{-14} = [H^+]^2$, $[H^+] = 1.71 \times 10^{-7}\,M = [OH^-]$

b.   $pH = -\log[H^+] = -\log(1.71 \times 10^{-7}) = 6.767$

c.   $[H^+] = K_w/[OH^-] = (2.92 \times 10^{-14})/0.10 = 2.9 \times 10^{-13}\,M$;  $pH = -\log(2.9 \times 10^{-13}) = 12.54$

33.    a.   $[H^+] = 10^{-pH}$, $[H^+] = 10^{-7.40} = 4.0 \times 10^{-8}\,M$

$pOH = 14.00 - pH = 14.00 - 7.40 = 6.60$; $[OH^-] = 10^{-pOH} = 10^{-6.60} = 2.5 \times 10^{-7}\,M$

or $[OH^-] = \dfrac{K_w}{[H^+]} = \dfrac{1.0 \times 10^{-14}}{4.0 \times 10^{-8}} = 2.5 \times 10^{-7}\,M$;  this solution is basic since $pH > 7.00$.

b.   $[H^+] = 10^{-15.3} = 5 \times 10^{-16}\,M$; $pOH = 14.00 - 15.3 = -1.3$; $[OH^-] = 10^{-(-1.3)} = 20\,M$; basic

c.   $[H^+] = 10^{-(-1.0)} = 10\,M$; $pOH = 14.0 - (-1.0) = 15.0$; $[OH^-] = 10^{-15.0} = 1 \times 10^{-15}\,M$; acidic

d.   $[H^+] = 10^{-3.20} = 6.3 \times 10^{-4}\,M$; $pOH = 14.00 - 3.20 = 10.80$; $[OH^-] = 10^{-10.80} = 1.6 \times 10^{-11}\,M$; acidic

e.   $[OH^-] = 10^{-5.0} = 1 \times 10^{-5}\,M$; $pH = 14.0 - pOH = 14.0 - 5.0 = 9.0$; $[H^+] = 10^{-9.0} = 1 \times 10^{-9}\,M$; basic

f.   $[OH^-] = 10^{-9.60} = 2.5 \times 10^{-10}\,M$; $pH = 14.00 - 9.60 = 4.40$; $[H^+] = 10^{-4.40} = 4.0 \times 10^{-5}\,M$; acidic

## Solutions of Acids

35.    Strong acids are assumed to completely dissociate in water, for example, $HCl(aq) + H_2O(l) \rightarrow H_3O^+(aq) + Cl^-(aq)$ or $HCl(aq) \rightarrow H^+(aq) + Cl^-(aq)$.

a.   A 0.10 $M$ HCl solution gives 0.10 $M$ $H^+$ and 0.10 $M$ $Cl^-$ because HCl completely dissociates.  The amount of $H^+$ from $H_2O$ will be insignificant.

$pH = -\log[H^+] = -\log(0.10) = 1.00$

b. 5.0 $M$ $H^+$ is produced when 5.0 $M$ $HClO_4$ completely dissociates. The amount of $H^+$ from $H_2O$ will be insignificant. pH = $-\log(5.0)$ = $-0.70$ (Negative pH values just indicate very concentrated acid solutions.)

c. $1.0 \times 10^{-11}$ $M$ $H^+$ is produced when $1.0 \times 10^{-11}$ $M$ HI completely dissociates. If you take the negative log of $1.0 \times 10^{-11}$, this gives pH = 11.00. This is impossible! We dissolved an acid in water and got a basic pH. What we must consider in this problem is that water by itself donates $1.0 \times 10^{-7}$ $M$ $H^+$. We can normally ignore the small amount of $H^+$ from $H_2O$ except when we have a very dilute solution of an acid (as in the case here). Therefore, the pH is that of neutral water (pH = 7.00) because the amount of HI present is insignificant.

37.   HCl is a strong acid. $[H^+]$ = $10^{-1.50}$ = $3.16 \times 10^{-2}$ $M$ (carrying one extra sig. fig.)

$$M_1V_1 = M_2V_2, \quad V_1 = \frac{M_2V_2}{M_1} = \frac{3.16 \times 10^{-2} \text{ mol/L} \times 1.6 \text{ L}}{12 \text{ mol/L}} = 4.2 \times 10^{-3} \text{ L}$$

4.2 mL of 12 $M$ HCl with enough water added to make 1600 mL of solution will result in a solution having $[H^+]$ = $3.2 \times 10^{-2}$ $M$ and pH = 1.50.

39.   a. Major species: $H^+(aq)$, $Br^-(aq)$, and $H_2O(l)$. (HBr is a strong acid.) $[H^+]$ = 0.250 $M$

pH = $-\log[H^+]$ = $-\log(0.250)$ = 0.602

b. $H^+(aq)$, $ClO_4^-(aq)$, and $H_2O(l)$. ($HClO_4$ is a strong acid.) pH = 0.602

c. $H^+(aq)$, $NO_3^-(aq)$, and $H_2O(l)$. ($HNO_3$ is a strong acid.) pH = 0.602

d. $HNO_2$ ($K_a = 4.0 \times 10^{-4}$) and $H_2O$ ($K_a = K_w = 1.0 \times 10^{-14}$) are the major species. $HNO_2$ is much stronger acid than $H_2O$, so it is the major source of $H^+$. However, $HNO_2$ is a weak acid ($K_a < 1$), so it only partially dissociates in water. We must solve an equilibrium problem to determine $[H^+]$. In this Solutions Guide, we will summarize the *i*nitial, *c*hange and *e*quilibrium concentrations into one table called an ICE table. Solving the weak acid problem:

$$HNO_2(aq) \;\rightleftharpoons\; H^+(aq) \quad + \quad NO_2^-(aq)$$

| | | | |
|---|---|---|---|
| Initial | 0.250 $M$ | ~0 | 0 |
| | $x$ mol/L $HNO_2$ dissociates to reach equilibrium | | |
| Change | $-x$ | $\rightarrow$  $+x$ | $+x$ |
| Equil. | 0.250 $- x$ | $x$ | $x$ |

$$K_a = \frac{[H^+][NO_2^-]}{[HNO_2]}, \quad 4.0 \times 10^{-4} = \frac{x^2}{0.250 - x}; \text{ if we assume } x \ll 0.250, \text{ then:}$$

$$4.0 \times 10^{-4} \approx \frac{x^2}{0.250}, \quad x = \sqrt{4.0 \times 10^{-4}(0.250)} = 0.010 \ M$$

We must check the assumption: $\dfrac{x}{0.250} \times 100 = \dfrac{0.010}{0.250} \times 100 = 4.0\%$

All the assumptions are good. The $H^+$ contribution from water ($10^{-7}$ $M$) is negligible and $x$ is small compared to 0.250 (percent error = 4.0%). If the percent error is less than 5% for an assumption, we will consider it a valid assumption (called the 5% rule). Finishing the problem: $x = 0.010$ $M = [H^+]$; $pH = -\log[H^+] = -\log(0.010) = 2.00$

e. $CH_3CO_2H$ ($K_a = 1.8 \times 10^{-5}$) and $H_2O$ ($K_a = K_w = 1.0 \times 10^{-14}$) are the major species. $CH_3CO_2H$ is the major source of $H^+$. Solving the weak acid problem:

$$CH_3CO_2H \rightleftharpoons H^+ + CH_3CO_2^-$$

Initial      0.250 $M$      ~0      0
$x$ mol/L $CH_3CO_2H$ dissociates to reach equilibrium
Change      $-x$      $\rightarrow$ $+x$      $+x$
Equil.      0.250 $- x$      $x$      $x$

$$K_a = \frac{[H^+][CH_3CO_2^-]}{[CH_3CO_2H]}, \quad 1.8 \times 10^{-5} = \frac{x^2}{0.250 - x} \approx \frac{x^2}{0.250} \quad (\text{assuming } x \ll 0.250)$$

$x = 2.1 \times 10^{-3}$ $M$; checking assumption: $\dfrac{2.1 \times 10^{-3}}{0.250} \times 100 = 0.84\%$. Assumptions good.

$[H^+] = x = 2.1 \times 10^{-3}$ $M$; $pH = -\log(2.1 \times 10^{-3}) = 2.68$

f. HCN ($K_a = 6.2 \times 10^{-10}$) and $H_2O$ are the major species. HCN is the major source of $H^+$.

$$HCN \rightleftharpoons H^+ + CN^-$$

Initial      0.250 $M$      ~0      0
$x$ mol/L HCN dissociates to reach equilibrium
Change      $-x$      $\rightarrow$ $+x$      $+x$
Equil.      0.250 $- x$      $x$      $x$

$$K_a = 6.2 \times 10^{-10} = \frac{[H^+][CN^-]}{[HCN]} = \frac{x^2}{0.250 - x} \approx \frac{x^2}{0.250} \quad (\text{assuming } x \ll 0.250)$$

$x = [H^+] = 1.2 \times 10^{-5}$ $M$; checking assumption: $x$ is 0.0048% of 0.250.

Assumptions good. $pH = -\log(1.2 \times 10^{-5}) = 4.92$

41. a. Major species: HOCl ($K_a = 3.5 \times 10^{-8}$) and water; major source of $H^+$ = HOCl. Because $K_a$ for HOCl is less than 1, HOCl is a weak acid and we must solve an equilibrium problem to determine $[H^+]$. The setup is:

$$HOCl\ (aq) \rightleftharpoons H^+(aq) + OCl^-(aq)$$

| | | | |
|---|---|---|---|
| Initial | 0.20 $M$ | ~0 | 0 |
| | $x$ mol/L HOCl dissociates to reach equilibrium | | |
| Change | $-x$ $\rightarrow$ | $+x$ | $+x$ |
| Equil. | $0.20 - x$ | $x$ | $x$ |

$$K_a = 3.5 \times 10^{-8} = \frac{[H^+][OCl^-]}{[HOCl]} = \frac{x^2}{0.20 - x} \approx \frac{x^2}{0.20} \quad (\text{assuming } x \ll 0.20)$$

$$x = [H^+] = 8.4 \times 10^{-5}\ M$$

We have made two assumptions that we must check.

1. $0.20 - x \approx 0.20$

   $(x/0.20) \times 100 = (8.4 \times 10^{-5}/0.20) \times 100 = 0.042\%$.  Great assumption.  If the percent error in the assumption is less than 5%, then the assumption is valid.

2. HOCl is the major source of $H^+$, that is, we can ignore $1.0 \times 10^{-7}\ M$ $H^+$ already present in neutral $H_2O$.

   $[H^+]$ from HOCl = $8.4 \times 10^{-5} \gg 1.0 \times 10^{-7}$; this assumption is valid.

In future problems we will always begin the problem solving process by making these assumptions, and we will always check them.  However, we may not explicitly state that the assumptions are valid.  We will <u>always</u> state when the assumptions are <u>not</u> valid and we have to use other techniques to solve the problem.  Remember, anytime we make an assumption, we must check its validity before the solution to the problem is complete.  Answering the question:

$$[H^+] = [OCl^-] = 8.4 \times 10^{-5}\ M;\ [OH^-] = K_w/[H^+] = 1.2 \times 10^{-10}\ M$$

$$[HOCl] = 0.20 - x = 0.20\ M;\ pH = -\log[H^+] = -\log(8.4 \times 10^{-5}) = 4.08$$

b.  $HOC_6H_5$ ($K_a = 1.6 \times 10^{-10}$) is the dominant producer of $H^+$.  Solving the weak acid problem:

$$HOC_6H_5 \rightleftharpoons H^+ + OC_6H_5^- \quad K_a = 1.6 \times 10^{-10}$$

| | | | |
|---|---|---|---|
| Initial | 1.5 $M$ | ~0 | 0 |
| | $x$ mol/L $HOC_6H_5$ dissociates to reach equilibrium | | |
| Change | $-x$ $\rightarrow$ | $+x$ | $+x$ |
| Equil. | $1.5 - x$ | $x$ | $x$ |

$$K_a = 1.6 \times 10^{-10} = \frac{[H^+][OC_6H_5^-]}{[HOC_6H_5]} = \frac{x^2}{1.5 - x} \approx \frac{x^2}{1.5} \quad (\text{assuming } x \ll 1.5)$$

$x = [H^+] = 1.5 \times 10^{-5} M$; assumptions good: $1.0 \times 10^{-7} \ll 1.5 \times 10^{-5} \ll 1.5$

$[H^+] = [OC_6H_5^-] = 1.5 \times 10^{-5} M$; $[OH^-] = 6.7 \times 10^{-10} M$

$[HOC_6H_5] = 1.5 - x = 1.5 M$; $pH = -\log(1.5 \times 10^{-5}) = 4.82$

c.    This is a weak acid in water.  Solving the weak acid problem:

$$HF \quad \rightleftharpoons \quad H^+ \quad + \quad F^- \qquad K_a = 7.2 \times 10^{-4}$$

| | | | |
|---|---|---|---|
| Initial | 0.020 $M$ | ~0 | 0 |

$x$ mol/L HF dissociates to reach equilibrium

| | | | |
|---|---|---|---|
| Change | $-x$ | $\rightarrow$   $+x$ | $+x$ |
| Equil. | $0.020 - x$ | $x$ | $x$ |

$$K_a = 7.2 \times 10^{-4} = \frac{[H^+][F^-]}{[HF]} = \frac{x^2}{0.020 - x} \approx \frac{x^2}{0.020} \quad (\text{assuming } x \ll 0.020)$$

$x = [H^+] = 3.8 \times 10^{-3} M$;  Check assumptions:

$$\frac{x}{0.020} \times 100 = \frac{3.8 \times 10^{-3}}{0.020} \times 100 = 19\%$$

The assumption $x \ll 0.020$ is not good ($x$ is more than 5% of 0.020).  We must solve $x^2/(0.020 - x) = 7.2 \times 10^{-4}$ exactly by using either the quadratic formula or the method of successive approximations (see Appendix 1 of the text).  Using successive approximations, we let 0.016 $M$ be a new approximation for [HF].  That is, in the denominator try $x = 0.0038$ (the value of $x$ we calculated making the normal assumption) so that $0.020 - 0.0038 = 0.016$; then solve for a new value of $x$ in the numerator.

$$\frac{x^2}{0.020 - x} \approx \frac{x^2}{0.016} = 7.2 \times 10^{-4}, \quad x = 3.4 \times 10^{-3}$$

We use this new value of $x$ to further refine our estimate of [HF], that is, $0.020 - x = 0.020 - 0.0034 = 0.0166$ (carrying an extra sig. fig.).

$$\frac{x^2}{0.020 - x} \approx \frac{x^2}{0.0166} = 7.2 \times 10^{-4}, \quad x = 3.5 \times 10^{-3}$$

We repeat, until we get a self-consistent answer.  This would be the same answer we would get solving exactly using the quadratic equation.  In this case it is, $x = 3.5 \times 10^{-3}$.  Thus:

$[H^+] = [F^-] = x = 3.5 \times 10^{-3} M$; $[OH^-] = K_w/[H^+] = 2.9 \times 10^{-12} M$

$[HF] = 0.020 - x = 0.020 - 0.0035 = 0.017 M$; $pH = 2.46$

*Note*: When the 5% assumption fails, use whichever method you are most comfortable with to solve exactly.  The method of successive approximations is probably fastest when the percent error is less than ~25% (unless you have a graphing calculator).

43.     Major species: $HIO_3$, $H_2O$;  major source of $H^+$: $HIO_3$ (a weak acid, $K_a = 0.17$)

$$HIO_3 \quad \rightleftharpoons \quad H^+ \quad + \quad IO_3^-$$

Initial      0.010 $M$              ~0                0

$x$ mol/L $HIO_3$ dissociates to reach equilibrium

Change      $-x$          $\rightarrow$      $+x$              $+x$

Equil.      $0.010 - x$              $x$              $x$

$$K_a = 0.17 = \frac{[H^+][IO_3^-]}{[HIO_3]} = \frac{x^2}{0.010-x} \approx \frac{x^2}{0.010}, \quad x = 0.041; \text{ check assumption.}$$

Assumption is horrible ($x$ is more than 400% of 0.010). When the assumption is this poor, it is generally quickest to solve exactly using the quadratic formula (see Appendix 1 in text). Using the quadratic formula and carrying extra significant figures:

$$0.17 = \frac{x^2}{0.010-x}, \quad x^2 = 0.17(0.010 - x), \quad x^2 + (0.17)x - 1.7 \times 10^{-3} = 0$$

$$x = \frac{-0.17 \pm [(0.17)^2 - 4(1)(-1.7 \times 10^{-3})]^{1/2}}{2(1)} = \frac{-0.17 \pm 0.189}{2}, \quad x = 9.5 \times 10^{-3} \, M$$

$$(x \text{ must be positive})$$

$$x = 9.5 \times 10^{-3} \, M = [H^+]; \quad pH = -\log(9.5 \times 10^{-3}) = 2.02$$

45.     This is a weak acid in water. We must solve a weak acid problem. Let $HBz = C_6H_5CO_2H$.

$$0.56 \text{ g HBz} \times \frac{1 \text{ mol HBz}}{122.1 \text{ g}} = 4.6 \times 10^{-3} \text{ mol}; \quad [HBz]_0 = 4.6 \times 10^{-3} \, M$$

$$HBz \quad \rightleftharpoons \quad H^+ \quad + \quad Bz^-$$

Initial      $4.6 \times 10^{-3} \, M$       ~0                0

$x$ mol/L HBz dissociates to reach equilibrium

Change      $-x$          $\rightarrow$      $+x$              $+x$

Equil.      $4.6 \times 10^{-3} - x$      $x$              $x$

$$K_a = 6.4 \times 10^{-5} = \frac{[H^+][Bz^-]}{[HBz]} = \frac{x^2}{(4.6 \times 10^{-3} - x)} \approx \frac{x^2}{4.6 \times 10^{-3}}$$

$$x = [H^+] = 5.4 \times 10^{-4}; \quad \text{check assumptions: } \frac{x}{4.6 \times 10^{-3}} \times 100 = \frac{5.4 \times 10^{-4}}{4.6 \times 10^{-3}} \times 100 = 12\%$$

Assumption is not good ($x$ is 12% of $4.6 \times 10^{-3}$). When assumption(s) fail, we must solve exactly using the quadratic formula or the method of successive approximations (see Appendix 1 of text). Using successive approximations:

$$\frac{x^2}{(4.6 \times 10^{-3}) - (5.4 \times 10^{-4})} = 6.4 \times 10^{-5}, \ x = 5.1 \times 10^{-4}$$

$$\frac{x^2}{(4.6 \times 10^{-3}) - (5.1 \times 10^{-4})} = 6.4 \times 10^{-5}, \ x = 5.1 \times 10^{-4} \ M \ \ \text{(consistent answer)}$$

Thus: $x = [H^+] = [Bz^-] = [C_6H_5CO_2^-] = 5.1 \times 10^{-4} \ M$

$[HBz] = [C_6H_5CO_2H] = 4.6 \times 10^{-3} - x = 4.1 \times 10^{-3} \ M$

$pH = -\log(5.1 \times 10^{-4}) = 3.29; \ pOH = 14.00 - pH = 10.71; \ [OH^-] = 10^{-10.71} = 1.9 \times 10^{-11} \ M$

47. $\ \ [HC_9H_7O_4] = \dfrac{2 \text{ tablets} \times \dfrac{0.325 \text{ g } HC_9H_7O_4}{\text{tablet}} \times \dfrac{1 \text{ mol } HC_9H_7O_4}{180.15 \text{ g}}}{0.237 \text{ L}} = 0.0152 \ M$

$$HC_9H_7O_4 \ \rightleftharpoons \ H^+ \ + \ C_9H_7O_4^-$$

Initial $\quad$ 0.0152 $M$ $\qquad$ ~0 $\qquad$ 0
$\qquad \qquad$ $x$ mol/L $HC_9H_7O_4$ dissociates to reach equilibrium
Change $\quad$ $-x$ $\qquad \rightarrow \ +x \qquad \quad +x$
Equil. $\quad$ 0.0152 $- x$ $\qquad x \qquad \qquad x$

$K_a = 3.3 \times 10^{-4} = \dfrac{[H^+][C_9H_7O_4^-]}{[HC_9H_7O_4]} = \dfrac{x^2}{0.0152 - x} \approx \dfrac{x^2}{0.0152}, \ x = 2.2 \times 10^{-3} \ M$

Assumption that $0.0152 - x \approx 0.0152$ fails the 5% rule: $\dfrac{2.2 \times 10^{-3}}{0.0152} \times 100 = 14\%$

Using successive approximations or the quadratic equation gives an exact answer of
$x = 2.1 \times 10^{-3} \ M$.

$[H^+] = x = 2.1 \times 10^{-3} \ M; \ \ pH = -\log(2.1 \times 10^{-3}) = 2.68$

49. $\quad$ a. $\quad$ HCl is a strong acid. It will produce 0.10 $M$ $H^+$. HOCl is a weak acid. Let's consider
$\qquad \quad$ the equilibrium:

$$HOCl \ \ \rightleftharpoons \ \ H^+ \ + \ OCl^- \qquad K_a = 3.5 \times 10^{-8}$$

Initial $\quad$ 0.10 $M$ $\qquad \quad$ 0.10 $M$ $\qquad$ 0
$\qquad \qquad$ $x$ mol/L HOCl dissociates to reach equilibrium
Change $\quad$ $-x$ $\qquad \rightarrow \quad +x \qquad \quad +x$
Equil. $\quad$ 0.10 $- x$ $\qquad \quad$ 0.10 $+ x \qquad x$

$K_a = 3.5 \times 10^{-8} = \dfrac{[H^+][OCl^-]}{[HOCl]} = \dfrac{(0.10 + x)(x)}{0.10 - x} \approx x, \ x = 3.5 \times 10^{-8} \ M$

Assumptions are great ($x$ is 0.000035% of 0.10). We are really assuming that HCl is the only important source of $H^+$, which it is. The $[H^+]$ contribution from HOCl, $x$, is negligible. Therefore, $[H^+] = 0.10\ M$; pH = 1.00.

b.  $HNO_3$ is a strong acid, giving an initial concentration of $H^+$ equal to 0.050 $M$. Consider the equilibrium:

$$HC_2H_3O_2 \quad \rightleftharpoons \quad H^+ \quad + \quad C_2H_3O_2^- \qquad K_a = 1.8 \times 10^{-5}$$

Initial      0.50 $M$                0.050 $M$              0
$\qquad\qquad$ $x$ mol/L $HC_2H_3O_2$ dissociates to reach equilibrium
Change      $-x$              $\rightarrow$     $+x$           $+x$
Equil.       $0.50 - x$                0.050 + $x$       $x$

$$K_a = 1.8 \times 10^{-5} = \frac{[H^+][C_2H_3O_2^-]}{[HC_2H_3O_2]} = \frac{(0.050 + x)x}{(0.50 - x)} \approx \frac{(0.050)x}{0.50}$$

$x = 1.8 \times 10^{-4}$; assumptions are good (well within the 5% rule).

$[H^+] = 0.050 + x = 0.050\ M$ and pH = 1.30

51.  $\qquad\qquad\qquad$ HX $\quad \rightleftharpoons \quad$ H$^+$ $\quad + \quad$ X$^-$

Initial      I                 ~0              0          where I = $[HX]_0$
$\qquad\qquad$ $x$ mol/L HX dissociates to reach equilibrium
Change      $-x$         $\rightarrow$     $+x$           $+x$
Equil.       I $- x$              $x$              $x$

From the problem, $x = 0.25$(I) and I $- x = 0.30\ M$.

I $- 0.25$(I) $= 0.30\ M$,  I $= 0.40\ M$  and  $x = 0.25(0.40\ M) = 0.10\ M$

$$K_a = \frac{[H^+][X^-]}{[HX]} = \frac{x^2}{I - x} = \frac{(0.10)^2}{0.30} = 0.033$$

53.  pH = 2.77,  $[H^+] = 10^{-2.77} = 1.7 \times 10^{-3}\ M$

$\qquad\qquad$ HOCN $\quad \rightleftharpoons \quad$ H$^+$ $\quad + \quad$ OCN$^-$

Initial  0.0100                ~0              0
Equil.   0.0100 $- x$           $x$              $x$

$x = [H^+] = [OCN^-] = 1.7 \times 10^{-3}\ M$;  $[HOCN] = 0.0100 - x = 0.0100 - 0.0017 = 0.0083\ M$

$$K_a = \frac{[H^+][OCN^-]}{[HOCN]} = \frac{(1.7 \times 10^{-3})^2}{8.3 \times 10^{-3}} = 3.5 \times 10^{-4}$$

55.     Major species: HCOOH and $H_2O$; major source of $H^+$: HCOOH

$$HCOOH \rightleftharpoons H^+ + HCOO^-$$

Initial         C            ~0        0         where C = $[HCOOH]_0$
                $x$ mol/L HCOOH dissociates to reach equilibrium
Change         $-x$         $\rightarrow$ $+x$    $+x$
Equil.         $C - x$                   $x$       $x$

$$K_a = 1.8 \times 10^{-4} = \frac{[H^+][HCOO^-]}{[HCOOH]} = \frac{x^2}{C-x}, \text{ where } x = [H^+]$$

$$1.8 \times 10^{-4} = \frac{[H^+]^2}{C - [H^+]}; \text{ because pH} = 2.70: [H^+] = 10^{-2.70} = 2.0 \times 10^{-3} \, M$$

$$1.8 \times 10^{-4} = \frac{(2.0 \times 10^{-3})^2}{C - (2.0 \times 10^{-3})}, \quad C - (2.0 \times 10^{-3}) = \frac{4.0 \times 10^{-6}}{1.8 \times 10^{-4}}, \quad C = 2.4 \times 10^{-2} \, M$$

A 0.024 $M$ formic acid solution will have pH = 2.70.

## Solutions of Bases

57.     $NO_3^-$: $K_b \ll K_w$ because $HNO_3$ is a strong acid.  All conjugate bases of strong acids have no base strength in water.  $H_2O$: $K_b = K_w = 1.0 \times 10^{-14}$; $NH_3$: $K_b = 1.8 \times 10^{-5}$; $C_5H_5N$: $K_b = 1.7 \times 10^{-9}$

Base strength = $NH_3 > C_5H_5N > H_2O > NO_3^-$  (As $K_b$ increases, base strength increases.)

59.     a.  $C_6H_5NH_2$       b.  $C_6H_5NH_2$       c.  $OH^-$       d.  $CH_3NH_2$

The base with the largest $K_b$ value is the strongest base  ($K_{b, C_6H_5NH_2} = 3.8 \times 10^{-10}$,

$K_{b, CH_3NH_2} = 4.4 \times 10^{-4}$ ).  $OH^-$ is the strongest base possible in water.

61.     $NaOH(aq) \rightarrow Na^+(aq) + OH^-(aq)$; NaOH is a strong base that completely dissociates into $Na^+$ and $OH^-$.  The initial concentration of NaOH will equal the concentration of $OH^-$ donated by NaOH.

a.  $[OH^-] = 0.10 \, M$; pOH = $-\log[OH^-] = -\log(0.10) = 1.00$

pH = 14.00 − pOH = 14.00 − 1.00 = 13.00

Note that $H_2O$ is also present, but the amount of $OH^-$ produced by $H_2O$ will be insignificant compared to the 0.10 $M$ $OH^-$ produced from the NaOH.

b.  The $[OH^-]$ concentration donated by the NaOH is $1.0 \times 10^{-10}$ $M$. Water by itself donates $1.0 \times 10^{-7}$ $M$. In this exercise, water is the major $OH^-$ contributor, and $[OH^-] = 1.0 \times 10^{-7}$ $M$.

$pOH = -\log(1.0 \times 10^{-7}) = 7.00$; $pH = 14.00 - 7.00 = 7.00$

c.  $[OH^-] = 2.0$ $M$; $pOH = -\log(2.0) = -0.30$; $pH = 14.00 - (-0.30) = 14.30$

63.  $pH = 10.50$; $pOH = 14.00 - 10.50 = 3.50$; $[OH^-] = 10^{-3.50} = 3.2 \times 10^{-4}$ $M$

$Ba(OH)_2(aq) \rightarrow Ba^{2+}(aq) + 2\,OH^-(aq)$;  $Ba(OH)_2$ donates 2 mol $OH^-$ per mol $Ba(OH)_2$.

$$[Ba(OH)_2] = 3.2 \times 10^{-4}\ M\ OH^- \times \frac{1\,M\ Ba(OH)_2}{2\,M\ OH^-} = 1.6 \times 10^{-4}\ M\,Ba(OH)_2$$

A $1.6 \times 10^{-4}$ $M\,Ba(OH)_2$ solution will produce a $pH = 10.50$ solution.

65.  Neutrally charged organic compounds containing at least one nitrogen atom generally behave as weak bases. The nitrogen atom has an unshared pair of electrons around it. This lone pair of electrons is used to form a bond to $H^+$.

67.  This is a solution of a weak base in water. We must solve a weak base equilibrium problem.

$$C_2H_5NH_2\ +\ H_2O\ \rightleftharpoons\ C_2H_5NH_3^+\ +\ OH^-\qquad K_b = 5.6 \times 10^{-4}$$

| | | | |
|---|---|---|---|
| Initial | 0.20 $M$ | 0 | ~0 |
| | $x$ mol/L $C_2H_5NH_2$ reacts with $H_2O$ to reach equilibrium | | |
| Change | $-x$ | $\rightarrow$  $+x$ | $+x$ |
| Equil. | $0.20 - x$ | $x$ | $x$ |

$$K_b = \frac{[C_2H_5NH_3^+][OH^-]}{[C_2H_5NH_2]} = \frac{x^2}{0.20 - x} \approx \frac{x^2}{0.20}\quad \text{(assuming } x \ll 0.20\text{)}$$

$x = 1.1 \times 10^{-2}$; checking assumption: $\dfrac{1.1 \times 10^{-2}}{0.20} \times 100 = 5.5\%$

The assumption fails the 5% rule. We must solve exactly using either the quadratic equation or the method of successive approximations (see Appendix 1 of the text). Using successive approximations and carrying extra significant figures:

$$\frac{x^2}{0.20 - 0.011} = \frac{x^2}{0.189} = 5.6 \times 10^{-4}, \quad x = 1.0 \times 10^{-2}\ M \quad \text{(consistent answer)}$$

$$x = [OH^-] = 1.0 \times 10^{-2}\ M; \quad [H^+] = \frac{K_w}{[OH^-]} = \frac{1.0 \times 10^{-14}}{1.0 \times 10^{-2}} = 1.0 \times 10^{-12}\ M; \quad pH = 12.00$$

69.    $\dfrac{5.0 \times 10^{-3} \text{ g}}{0.0100 \text{ L}} \times \dfrac{1 \text{ mol}}{299.4 \text{ g}} = 1.7 \times 10^{-3}$  $M = [\text{codeine}]_0$;  let cod = codeine ($C_{18}H_{21}NO_3$).

Solving the weak base equilibrium problem:

$$\text{cod} + H_2O \rightleftharpoons \text{codH}^+ + OH^- \qquad K_b = 10^{-6.05} = 8.9 \times 10^{-7}$$

Initial       $1.7 \times 10^{-3}$ $M$                       0                            ~0
              $x$ mol/L codeine reacts with $H_2O$ to reach equilibrium
Change        $-x$                       $\rightarrow$     $+x$                       $+x$
Equil.        $1.7 \times 10^{-3} - x$                      $x$                         $x$

$K_b = 8.9 \times 10^{-7} = \dfrac{x^2}{(1.7 \times 10^{-3} - x)} \approx \dfrac{x^2}{1.7 \times 10^{-3}}$ ,  $x = 3.9 \times 10^{-5}$ ;   assumptions good.

$[OH^-] = 3.9 \times 10^{-5}$ $M$;  $[H^+] = K_w/[OH^-] = 2.6 \times 10^{-10}$ $M$;  pH $= -\log[H^+] = 9.59$

71.    To solve for percent ionization, just solve the weak base equilibrium problem.

a.        $NH_3 + H_2O \rightleftharpoons NH_4^+ + OH^- \qquad K_b = 1.8 \times 10^{-5}$

Initial   0.10 $M$                    0              ~0
Equil.    0.10 $- x$                   $x$             $x$

$K_b = 1.8 \times 10^{-5} = \dfrac{x^2}{0.10 - x} \approx \dfrac{x^2}{0.10}$ ,  $x = [OH^-] = 1.3 \times 10^{-3}$ $M$;  assumptions good.

Percent ionization $= \dfrac{[OH^-]}{[NH_3]_0} \times 100 = \dfrac{1.3 \times 10^{-3} \ M}{0.10 \ M} \times 100 = 1.3\%$

b.            $NH_3 + H_2O \rightleftharpoons NH_4^+ + OH^-$

Initial   0.010 $M$                   0              ~0
Equil.    0.010 $- x$                  $x$             $x$

$1.8 \times 10^{-5} = \dfrac{x^2}{0.010 - x} \approx \dfrac{x^2}{0.010}$ ,  $x = [OH^-] = 4.2 \times 10^{-4}$ $M$;  assumptions good.

Percent ionization $= \dfrac{4.2 \times 10^{-4}}{0.010} \times 100 = 4.2\%$

*Note*:  For the same base, the percent ionization increases as the initial concentration of base decreases.

c.        $CH_3NH_2 + H_2O \rightleftharpoons CH_3NH_3^+ + OH^- \qquad K_b = 4.38 \times 10^{-4}$

Initial   0.10 $M$                    0              ~0
Equil.    0.10 $- x$                   $x$             $x$

$$4.38 \times 10^{-4} = \frac{x^2}{0.10-x} \approx \frac{x^2}{0.10} \ , \ x = 6.6 \times 10^{-3}; \quad \text{assumption fails the 5\% rule (} x \text{ is}$$

6.6% of 0.10).  Using successive approximations and carrying extra significant figures:

$$\frac{x^2}{0.10-0.0066} = \frac{x^2}{0.093} = 4.38 \times 10^{-4}, \ x = 6.4 \times 10^{-3} \quad \text{(consistent answer)}$$

$$\text{Percent ionization} = \frac{6.4 \times 10^{-3}}{0.10} \times 100 = 6.4\%$$

73.   Using the $K_b$ reaction to solve where PT = p-toluidine ($CH_3C_6H_4NH_2$):

$$PT \quad + \quad H_2O \quad \rightleftharpoons \quad PTH^+ \quad + \quad OH^-$$

Initial          0.016 $M$                              0          ~0
                 $x$ mol/L of PT reacts with $H_2O$ to reach equilibrium
Change        $-x$                        $\rightarrow$          $+x$          $+x$
Equil.         $0.016 - x$                                      $x$            $x$

$$K_b = \frac{[PTH^+][OH^-]}{[PT]} = \frac{x^2}{0.016-x}$$

Since pH = 8.60:  pOH = 14.00 − 8.60 = 5.40 and $[OH^-] = x = 10^{-5.40} = 4.0 \times 10^{-6} \ M$

$$K_b = \frac{(4.0 \times 10^{-6})^2}{0.016-(4.0 \times 10^{-6})} = 1.0 \times 10^{-9}$$

## Polyprotic Acids

75.   $H_3C_6H_5O_7(aq) \rightleftharpoons H_2C_6H_5O_7^-(aq) + H^+(aq)$          $K_{a_1} = \dfrac{[H_2C_6H_5O_7^-][H^+]}{[H_3C_6H_5O_7]}$

     $H_2C_6H_5O_7^-(aq) \rightleftharpoons HC_6H_5O_7^{2-}(aq) + H^+(aq)$          $K_{a_2} = \dfrac{[HC_6H_5O_7^{2-}][H^+]}{[H_2C_6H_5O_7^-]}$

     $HC_6H_5O_7^{2-}(aq) \rightleftharpoons C_6H_5O_7^{3-}(aq) + H^+(aq)$          $K_{a_3} = \dfrac{[C_6H_5O_7^{3-}][H^+]}{[HC_6H_5O_7^{2-}]}$

77.    The reactions are:

$$H_3AsO_4 \rightleftharpoons H^+ + H_2AsO_4^- \quad K_{a_1} = 5 \times 10^{-3}$$

$$H_2AsO_4^- \rightleftharpoons H^+ + HAsO_4^{2-} \quad K_{a_2} = 8 \times 10^{-8}$$

$$HAsO_4^{2-} \rightleftharpoons H^+ + AsO_4^{3-} \quad K_{a_3} = 6 \times 10^{-10}$$

We will deal with the reactions in order of importance, beginning with the largest $K_a$, $K_{a_1}$.

$$H_3AsO_4 \quad \rightleftharpoons \quad H^+ \quad + \quad H_2AsO_4^- \qquad K_{a_1} = 5 \times 10^{-3} = \frac{[H^+][H_2AsO_4^-]}{[H_3AsO_4]}$$

| | | | |
|---|---|---|---|
| Initial | 0.20 M | ~0 | 0 |
| Equil. | 0.20 - x | x | x |

$$5 \times 10^{-3} = \frac{x^2}{0.20 - x} \approx \frac{x^2}{0.20}, \quad x = 3 \times 10^{-2} \ M; \quad \text{assumption fails the 5\% rule.}$$

Solving by the method of successive approximations:

$$5 \times 10^{-3} = x^2/(0.20 - 0.03), \quad x = 3 \times 10^{-2} \ \text{(consistent answer)}$$

$[H^+] = [H_2AsO_4^-] = 3 \times 10^{-2} \ M; \ [H_3AsO_4] = 0.20 - 0.03 = 0.17 \ M$

Because $K_{a_2} = \dfrac{[H^+][HAsO_4^{2-}]}{[H_2AsO_4^-]} = 8 \times 10^{-8}$ is much smaller than the $K_{a_1}$ value, very little of

$H_2AsO_4^-$ (and $HAsO_4^{2-}$) dissociates compared to $H_3AsO_4$. Therefore, $[H^+]$ and $[H_2AsO_4^-]$ will not change significantly by the $K_{a_2}$ reaction. Using the previously calculated concentrations of $H^+$ and $H_2AsO_4^-$ to calculate the concentration of $HAsO_4^{2-}$:

$$8 \times 10^{-8} = \frac{(3 \times 10^{-2})[HAsO_4^{2-}]}{3 \times 10^{-2}}, \quad [HAsO_4^{2-}] = 8 \times 10^{-8} \ M$$

The assumption that the $K_{a_2}$ reaction does not change $[H^+]$ and $[H_2AsO_4^-]$ is good. We repeat the process using $K_{a_3}$ to get $[AsO_4^{3-}]$.

$$K_{a_3} = 6 \times 10^{-10} = \frac{[H^+][AsO_4^{3-}]}{[HAsO_4^{2-}]} = \frac{(3 \times 10^{-2})[AsO_4^{3-}]}{8 \times 10^{-8}}$$

$[AsO_4^{3-}] = 1.6 \times 10^{-15} \approx 2 \times 10^{-15} \ M; \quad \text{assumption good.}$

So in 0.20 M analytical concentration of $H_3AsO_4$:

$[H_3AsO_4] = 0.17 \ M; \ [H^+] = [H_2AsO_4^-] = 3 \times 10^{-2} \ M;$

$[HAsO_4^{2-}] = 8 \times 10^{-8} \ M; \ [AsO_4^{3-}] = 2 \times 10^{-15} \ M; \ [OH^-] = K_w/[H^+] = 3 \times 10^{-13} \ M$

79.    For $H_2C_6H_6O_6$.  $K_{a_1} = 7.9 \times 10^{-5}$ and $K_{a_2} = 1.6 \times 10^{-12}$.  Because $K_{a_1} \gg K_{a_2}$, the amount of $H^+$ produced by the $K_{a_2}$ reaction will be negligible.

$$[H_2C_6H_6O_6]_0 = \frac{0.500 \text{ g} \times \dfrac{1 \text{ mol } H_2C_6H_6O_6}{176.12 \text{ g}}}{0.2000 \text{ L}} = 0.0142 \ M$$

$$H_2C_6H_6O_6(aq) \ \rightleftharpoons \ HC_6H_6O_6^-(aq) \ + \ H^+(aq) \qquad K_{a_1} = 7.9 \times 10^{-5}$$

| | | | |
|---|---|---|---|
| Initial | 0.0142 $M$ | 0 | ~0 |
| Equil. | 0.0142 – $x$ | $x$ | $x$ |

$$K_{a_1} = 7.9 \times 10^{-5} = \frac{x^2}{0.0142 - x} \approx \frac{x^2}{0.0142}, \ x = 1.1 \times 10^{-3}; \quad \text{assumption fails the 5\% rule.}$$

Solving by the method of successive approximations:

$$7.9 \times 10^{-5} = \frac{x^2}{0.0142 - 1.1 \times 10^{-3}}, \ x = 1.0 \times 10^{-3} \ M \text{ (consistent answer)}$$

Because $H^+$ produced by the $K_{a_2}$ reaction will be negligible, $[H^+] = 1.0 \times 10^{-3}$ and pH = 3.00.

81.    The dominant $H^+$ producer is the strong acid $H_2SO_4$.  A 2.0 $M$ $H_2SO_4$ solution produces 2.0 $M$ $HSO_4^-$ and 2.0 $M$ $H^+$.  However, $HSO_4^-$ is a weak acid that could also add $H^+$ to the solution.

$$HSO_4^- \ \rightleftharpoons \ H^+ \ + \ SO_4^{2-}$$

| | | | |
|---|---|---|---|
| Initial | 2.0 $M$ | 2.0 $M$ | 0 |
| | $x$ mol/L $HSO_4^-$ dissociates to reach equilibrium | | |
| Change | $-x$ | $\rightarrow$ $+x$ | $+x$ |
| Equil. | 2.0 – $x$ | 2.0 + $x$ | $x$ |

$$K_{a_2} = 1.2 \times 10^{-2} = \frac{[H^+][SO_4^{2-}]}{[HSO_4^-]} = \frac{(2.0 + x)x}{2.0 - x} \approx \frac{2.0(x)}{2.0}, \ x = 1.2 \times 10^{-2} \ M$$

Because $x$ is 0.60% of 2.0, the assumption is valid by the 5% rule.  The amount of additional $H^+$ from $HSO_4^-$ is $1.2 \times 10^{-2}$ $M$.  The total amount of $H^+$ present is:

$$[H^+] = 2.0 + (1.2 \times 10^{-2}) = 2.0 \ M; \ \text{pH} = -\log(2.0) = -0.30$$

*Note*: In this problem $H^+$ from $HSO_4^-$ could have been ignored.  However, this is not usually the case in more dilute solutions of $H_2SO_4$.

## Acid-Base Properties of Salts

83.    a.    These are strong acids like HCl, HBr, HI, $HNO_3$, $H_2SO_4$, or $HClO_4$.

     b.    These are salts of the conjugate acids of the bases in Table 7.3. These conjugate acids are all weak acids. $NH_4Cl$, $CH_3NH_3NO_3$, and $C_2H_5NH_3Br$ are three examples. Note that the anions used to form these salts are conjugate bases of strong acids; this is so because they have no acidic or basic properties in water (with the exception of $HSO_4^-$, which has weak acid properties).

     c.    These are strong ases like LiOH, NaOH, KOH, RbOH, CsOH, $Ca(OH)_2$, $Sr(OH)_2$, and $Ba(OH)_2$.

     d.    These are salts of the conjugate bases of the neutrally charged weak acids in Table 7.2. The conjugate bases of weak acids are weak bases themselves. Three examples are $NaClO_2$, $KC_2H_3O_2$, and $CaF_2$. The cations used to form these salts are $Li^+$, $Na^+$, $K^+$, $Rb^+$, $Cs^+$, $Ca^{2+}$, $Sr^{2+}$, or $Ba^{2+}$ because these cations have no acidic or basic properties in water. Notice that these are the cations of the strong bases you should memorize.

     e.    There are two ways to make a neutral salt. The easiest way is to combine a conjugate base of a strong acid (except for $HSO_4^-$) with one of the cations from a strong base. These ions have no acidic/basic properties in water, so salts of these ions are neutral. Three examples are NaCl, $KNO_3$, and $SrI_2$. Another type of strong electrolyte that can produce neutral solutions are salts that contain an ion with weak acid properties combined with an ion of opposite charge having weak base properties. If the $K_a$ for the weak acid ion is equal to the $K_b$ for the weak base ion, then the salt will produce a neutral solution. The most common example of this type of salt is ammonium acetate ($NH_4C_2H_3O_2$). For this salt, $K_a$ for $NH_4^+ = K_b$ for $C_2H_3O_2^- = 5.6 \times 10^{-10}$. This salt at any concentration produces a neutral solution.

85.    One difficult aspect of acid-base chemistry is recognizing what types of species are present in solution, that is, whether a species is a strong acid, strong base, weak acid, weak base, or a neutral species.  Below are some ideas and generalizations to keep in mind that will help in recognizing types of species present.

     a.    Memorize the following strong acids:  HCl, HBr, HI, $HNO_3$, $HClO_4$, and $H_2SO_4$

     b.    Memorize the following strong bases:  LiOH, NaOH, KOH, RbOH, $Ca(OH)_2$, $Sr(OH)_2$, and $Ba(OH)_2$

     c.    Weak acids have a $K_a$ value of less than 1 but greater than $K_w$.  Some weak acids are listed in Table 7.2 of the text.  Weak bases have a $K_b$ value of less than 1 but greater than $K_w$.  Some weak bases are listed in Table 7.3 of the text.

     d.    Conjugate bases of weak acids are weak bases, that is, all have a $K_b$ value of less than 1 but greater than $K_w$.  Some examples of these are the conjugate bases of the weak acids listed in Table 7.2 of the text.

e.    Conjugate acids of weak bases are weak acids, that is, all have a $K_a$ value of less than 1 but greater than $K_w$.  Some examples of these are the conjugate acids of the weak bases listed in Table 7.3 of the text.

f.    Alkali metal ions ($Li^+$, $Na^+$, $K^+$, $Rb^+$, $Cs^+$) and some alkaline earth metal ions ($Ca^{2+}$, $Sr^{2+}$, $Ba^{2+}$) have no acidic or basic properties in water.

g.    Conjugate bases of strong acids ($Cl^-$, $Br^-$, $I^-$, $NO_3^-$, $ClO_4^-$, $HSO_4^-$) have no basic properties in water ($K_b \ll K_w$), and only $HSO_4^-$ has any acidic properties in water.

Let's apply these ideas to this problem to see what types of species are present.

a.    HI:        Strong acid;  HF: weak acid ($K_a = 7.2 \times 10^{-4}$)

    NaF:    $F^-$ is the conjugate base of the weak acid HF, so $F^-$ is a weak base.  The $K_b$ value for $F^- = K_w/K_{a, HF} = 1.4 \times 10^{-11}$.  $Na^+$ has no acidic or basic properties.

    NaI:    Neutral (pH = 7.0);  $Na^+$ and $I^-$ have no acidic/basic properties.

In order of increasing pH, we place the compounds from most acidic (lowest pH) to most basic (highest pH).  Increasing pH:  HI < HF < NaI < NaF.

b.    $NH_4Br$:    $NH_4^+$ is a weak acid ($K_a = 5.6 \times 10^{-10}$), and $Br^-$ is a neutral species.

    HBr:        Strong acid

    KBr:        Neutral;  $K^+$ and $Br^-$ have no acidic/basic properties.

    $NH_3$:        Weak base, $K_b = 1.8 \times 10^{-5}$

Increasing pH:  HBr  <  $NH_4Br$  <  KBr  <  $NH_3$
                     Most                    Most
                     acidic                  basic

c.    $C_6H_5NH_3NO_3$:        $C_6H_5NH_3^+$ is a weak acid ($K_a = K_w/K_{b, C_6H_5NH_2} =$
                              $1.0 \times 10^{-14}/3.8 \times 10^{-10} = 2.6 \times 10^{-5}$), and $NO_3^-$ is a neutral species.

    $NaNO_3$:        Neutral;  $Na^+$ and $NO_3^-$ have no acidic/basic properties.

    NaOH:        Strong base

    $HOC_6H_5$:        Weak acid ($K_a = 1.6 \times 10^{-10}$)

    $KOC_6H_5$:        $OC_6H_5^-$ is a weak base ($K_b = K_w/K_{a, HOC_6H_5} = 6.3 \times 10^{-5}$), and $K^+$ is a neutral species.

    $C_6H_5NH_2$:        Weak base ($K_b = 3.8 \times 10^{-10}$)

    $HNO_3$:        Strong acid

This is a little more difficult than the previous parts of this problem because two weak acids and two weak bases are present.  Between the weak acids, $C_6H_5NH_3^+$ is a stronger weak acid than $HOC_6H_5$ since the $K_a$ value for $C_6H_5NH_3^+$ is larger than the $K_a$ value for $HOC_6H_5$.  Between the two weak bases, because the $K_b$ value for $OC_6H_5^-$ is larger than the $K_b$ value for $C_6H_5NH_2$, $OC_6H_5^-$ is a stronger weak base than $C_6H_5NH_2$.

Increasing pH:  $HNO_3 < C_6H_5NH_3NO_3 < HOC_6H_5 < NaNO_3 < C_6H_5NH_2 < KOC_6H_5 < NaOH$
          Most acidic                                                                      Most basic

87.    Reference Table 7.6 of the text and the solution to Exercise 85 for some generalizations on acid-base properties of salts.

a.    $KCl \rightarrow K^+ + Cl^-$  neutral;  $K^+$ and $Cl^-$ have no effect on pH.

b.    $C_2H_5NH_3CN \rightarrow C_2H_5NH_3^+ + CN^-$  basic;  $C_2H_5NH_3^+$ is a weak acid
      $(K_a = K_w/K_{b,C_2H_5NH_2} = 1.0 \times 10^{-14}/5.6 \times 10^{-4} = 1.8 \times 10^{-11})$, and $CN^-$ is a weak base
      $(K_b = K_w/K_{a,HCN} = 1.0 \times 10^{-14}/6.2 \times 10^{-10} = 1.6 \times 10^{-5})$. Because $K_{b,CN^-} > K_{a,C_2H_5NH_3^+}$,
      the solution of $C_2H_5NH_3CN$ will be basic.

c.    $C_5H_5NHF \rightarrow C_5H_5NH^+ + F^-$  acidic;  $C_5H_5NH^+$ is a weak acid
      $(K_a = K_w/K_{b,C_5H_5N} = 5.9 \times 10^{-6})$, and $F^-$ is a weak base  $(K_b = K_w/K_{a,HF} = 1.4 \times 10^{-11})$.
      Because $K_{a,C_5H_5NH^+} > K_{b,F^-}$, the solution of $C_5H_5NHF$ will be acidic.

d.    $NH_4C_2H_3O_2 \rightarrow NH_4^+ + C_2H_3O_2^-$  neutral;  $NH_4^+$ is a weak acid $(K_a = 5.6 \times 10^{-10})$, and
      $C_2H_3O_2^-$ is a weak base $(K_b = K_w/K_{a,HC_2H_3O_2} = 5.6 \times 10^{-10})$.  Because $K_{a,NH_4^+} =$
      $K_{b,C_2H_3O_2^-}$, the solution of $NH_4C_2H_3O_2$ will have pH = 7.00.

e.    $NaHSO_3 \rightarrow Na^+ + HSO_3^-$  acidic;  $Na^+$ has no acidic/basic properties.  $HSO_3^-$ is a weak
      acid $(K_{a_2} = 1.0 \times 10^{-7})$, and $HSO_3^-$ is also a weak base $(K_b = K_w/K_{a_1,H_2SO_3} = 1.0 \times$
      $10^{-14}/1.5 \times 10^{-2} = 6.7 \times 10^{-13})$.  $HSO_3^-$ is a stronger acid than a base because $K_a > K_b$.
      Therefore, the solution is acidic.

f.    $NaHCO_3 \rightarrow Na^+ + HCO_3^-$  basic; ignore $Na^+$;  $HCO_3^-$ is a weak acid $(K_{a_2} = 4.8 \times 10^{-11})$,
      and $HCO_3^-$ is a weak base $(K_b = K_w/K_{a_1,H_2CO_3} = 2.3 \times 10^{-8})$.  $HCO_3^-$ is a stronger base
      than an acid because $K_b > K_a$.  Therefore, the solution is basic.

89.    a.    $KNO_2 \rightarrow K^+ + NO_2^-$:  $NO_2^-$ is a weak base.  Ignore $K^+$.

$$NO_2^- + H_2O \rightleftharpoons HNO_2 + OH^- \qquad K_b = \frac{K_w}{K_a} = \frac{1.0 \times 10^{-14}}{4.0 \times 10^{-4}} = 2.5 \times 10^{-11}$$

Initial    0.12 $M$                    0          ~0
Equil.    0.12 – $x$                  $x$          $x$

$$K_b = 2.5 \times 10^{-11} = \frac{[OH^-][HNO_2]}{[NO_2^-]} = \frac{x^2}{0.12 - x} \approx \frac{x^2}{0.12}$$

$x = [OH^-] = 1.7 \times 10^{-6} M$;  pOH = 5.77;  pH = 8.23;  assumptions good.

b.   $NaOCl \rightarrow Na^+ + OCl^-$:  $OCl^-$ is a weak base. Ignore $Na^+$.

$$OCl^- + H_2O \rightleftharpoons HOCl + OH^- \qquad K_b = \frac{K_w}{K_a} = \frac{1.0 \times 10^{-14}}{3.5 \times 10^{-8}} = 2.9 \times 10^{-7}$$

| | | | |
|---|---|---|---|
| Initial | 0.45 M | 0 | ~0 |
| Equil. | 0.45 − x | x | x |

$$K_b = 2.9 \times 10^{-7} = \frac{[HOCl][OH^-]}{[OCl^-]} = \frac{x^2}{0.45 - x} \approx \frac{x^2}{0.45}$$

$x = [OH^-] = 3.6 \times 10^{-4}$ $M$;  pOH = 3.44;  pH = 10.56;   assumptions good.

c.   $NH_4ClO_4 \rightarrow NH_4^+ + ClO_4^-$:  $NH_4^+$ is a weak acid.  $ClO_4^-$ is the conjugate base of a strong acid.  $ClO_4^-$ has no basic (or acidic) properties.

$$NH_4^+ \rightleftharpoons NH_3 + H^+ \qquad K_a = \frac{K_w}{K_b} = \frac{1.0 \times 10^{-14}}{1.8 \times 10^{-5}} = 5.6 \times 10^{-10}$$

| | | | |
|---|---|---|---|
| Initial | 0.40 M | 0 | ~0 |
| Equil. | 0.40 − x | x | x |

$$K_a = 5.6 \times 10^{-10} = \frac{[NH_3][H^+]}{[NH_4^+]} = \frac{x^2}{0.40 - x} \approx \frac{x^2}{0.40}$$

$x = [H^+] = 1.5 \times 10^{-5}$ $M$;  pH = 4.82;   assumptions good.

91.   $NaN_3 \rightarrow Na^+ + N_3^-$;  Azide ($N_3^-$) is a weak base since it is the conjugate base of a weak acid.  All conjugate bases of weak acids are weak bases ($K_w < K_b < 1$).  Ignore $Na^+$.

$$N_3^- + H_2O \rightleftharpoons HN_3 + OH^- \qquad K_b = \frac{K_w}{K_a} = \frac{1.0 \times 10^{-14}}{1.9 \times 10^{-5}} = 5.3 \times 10^{-10}$$

| | | | |
|---|---|---|---|
| Initial | 0.010 M | 0 | ~0 |
| | x mol/L of $N_3^-$ reacts with $H_2O$ to reach equilibrium | | |
| Change | −x | → +x | +x |
| Equil. | 0.010 − x | x | x |

$$K_b = \frac{[HN_3][OH^-]}{[N_3^-]}, \quad 5.3 \times 10^{-10} = \frac{x^2}{0.010 - x} \approx \frac{x^2}{0.010} \quad \text{(assuming } x \ll 0.010)$$

$x = [OH^-] = 2.3 \times 10^{-6}$ $M$;  $[H^+] = \dfrac{1.0 \times 10^{-14}}{2.3 \times 10^{-6}} = 4.3 \times 10^{-9}$ $M$;  assumptions good.

$[HN_3] = [OH^-] = 2.3 \times 10^{-6}$ $M$;  $[Na^+] = 0.010$ $M$;  $[N_3^-] = 0.010 - 2.3 \times 10^{-6} = 0.010$ $M$

93.    All these salts contain $Na^+$, which has no acidic/basic properties and a conjugate base of a weak acid (except for NaCl, where $Cl^-$ is a neutral species).  All conjugate bases of weak acids are weak bases since $K_b$ values for these species are between $K_w$ and 1.  To identify the species, we will use the data given to determine the $K_b$ value for the weak conjugate base. From the $K_b$ value and data in Table 7.2 of the text, we can identify the conjugate base present by calculating the $K_a$ value for the weak acid.  We will use $A^-$ as an abbreviation for the weak conjugate base.

$$A^- \;+\; H_2O \;\rightleftharpoons\; HA \;+\; OH^-$$

Initial          0.100 mol/1.00 L                0              ~0
                 $x$ mol/L $A^-$ reacts with $H_2O$ to reach equilibrium
Change         $-x$                    $\rightarrow$        $+x$           $+x$
Equil.         $0.100 - x$                              $x$             $x$

$K_b = \dfrac{[HA][OH^-]}{[A^-]} = \dfrac{x^2}{0.100 - x}$ ;  from the problem, pH = 8.07:

pOH = 14.00 − 8.07 = 5.93;  $[OH^-] = x = 10^{-5.93} = 1.2 \times 10^{-6} \; M$

$K_b = \dfrac{(1.2 \times 10^{-6})^2}{0.100 - (1.2 \times 10^{-6})} = 1.4 \times 10^{-11} = K_b$ value for the conjugate base of a weak acid.

The $K_a$ value for the weak acid equals $K_w/K_b$:  $K_a = \dfrac{1.0 \times 10^{-14}}{1.4 \times 10^{-11}} = 7.1 \times 10^{-4}$

From Table 7.2 of the text, this $K_a$ value is closest to HF.  Therefore, the unknown salt is NaF.

95.    Major species:  $Co(H_2O)_6^{3+}$ ($K_a = 1.0 \times 10^{-5}$), $Cl^-$ (neutral), and $H_2O$ ($K_w = 1.0 \times 10^{-14}$); $Co(H_2O)_6^{3+}$ will determine the pH since it is a stronger acid than water.  Solving the weak acid problem in the usual manner:

$$Co(H_2O)_6^{3+} \;\rightleftharpoons\; Co(H_2O)_5(OH)^{2+} \;+\; H^+ \qquad K_a = 1.0 \times 10^{-5}$$

Initial          0.10 $M$                        0                         ~0
Equil.         0.10 - $x$                        $x$                        $x$

$K_a = 1.0 \times 10^{-5} = \dfrac{x^2}{0.10 - x} \approx \dfrac{x^2}{0.10}$ ,  $x = [H^+] = 1.0 \times 10^{-3} \; M$

pH = −log$(1.0 \times 10^{-3})$ = 3.00;  assumptions good.

97.    Major species:  $NH_4^+$, $OCl^-$, and $H_2O$;  $K_a$ for $NH_4^+ = (1.0 \times 10^{-14})/(1.8 \times 10^{-5}) = 5.6 \times 10^{-10}$ and $K_b$ for $OCl^- = (1.0 \times 10^{-14})/(3.5 \times 10^{-8}) = 2.9 \times 10^{-7}$ .

Because $OCl^-$ is a better base than $NH_4^+$ is an acid, the solution will be basic. The dominant equilibrium is the best acid ($NH_4^+$) reacting with the best base ($OCl^-$) present.

$$NH_4^+ \quad + \quad OCl^- \quad \rightleftharpoons \quad NH_3 \quad + \quad HOCl$$

Initial      0.50 $M$          0.50 $M$              0              0
Change       $-x$              $-x$         $\rightarrow$   $+x$           $+x$
Equil.       $0.50 - x$        $0.50 - x$            $x$            $x$

$$K = K_{a,\,NH_4^+} \times \frac{1}{K_{a,\,HOCl}} = (5.6 \times 10^{-10})/(3.5 \times 10^{-8}) = 0.016$$

$$K = 0.016 = \frac{[NH_3][HOCl]}{[NH_4^+][OCl^-]} = \frac{x(x)}{(0.50 - x)(0.50 - x)}$$

$$\frac{x^2}{(0.50 - x)^2} = 0.016, \quad \frac{x}{0.50 - x} = (0.016)^{1/2} = 0.13, \quad x = 0.058\ M$$

To solve for the $H^+$, use any pertinent $K_a$ or $K_b$ value.  Using $K_a$ for $NH_4^+$:

$$K_{a,\,NH_4^+} = 5.6 \times 10^{-10} = \frac{[NH_3][H^+]}{[NH_4^+]} = \frac{(0.058)[H^+]}{0.50 - 0.058}, \quad [H^+] = 4.3 \times 10^{-9}\ M,\ \ pH = 8.37$$

## Solutions of Dilute Acids and Bases

99.                    $HBrO \quad \rightleftharpoons \quad H^+ \quad + \quad BrO^- \quad K_a = 2 \times 10^{-9}$

Initial      $1.0 \times 10^{-6}\ M$              ~0                 0
             $x$ mol/L HBrO dissociates to reach equilibrium
Change       $-x$             $\rightarrow$        $+x$              $+x$
Equil.       $1.0 \times 10^{-6} - x$             $x$                $x$

$$K_a = 2 \times 10^{-9} = \frac{x^2}{(1.0 \times 10^{-6} - x)} \approx \frac{x^2}{1.0 \times 10^{-6}}, \quad x = [H^+] = 4 \times 10^{-8}\ M;\ \ pH = 7.4$$

Let's check the assumptions.  This answer is impossible!  We can't add a small amount of an acid to a neutral solution and get a basic solution.  The highest pH possible for an acid in water is 7.0.  In the correct solution we would have to take into account the autoionization of water.

101.                  $HCN \quad \rightleftharpoons \quad H^+ \quad + \quad CN^- \quad K_a = 6.2 \times 10^{-10}$

Initial   $5.0 \times 10^{-4}\ M$          ~0             0
Equil.    $5.0 \times 10^{-4} - x$         $x$            $x$

$$K_a = \frac{x^2}{(5.0 \times 10^{-4} - x)} \approx \frac{x^2}{5.0 \times 10^{-4}} = 6.2 \times 10^{-10}, \ x = 5.6 \times 10^{-7}; \ \text{check assumptions.}$$

The assumption that the $H^+$ contribution from water is negligible is poor. Whenever the calculated pH is greater than 6.0 for a weak acid, the water contribution to $[H^+]$ must be considered. From Section 7.9 in text:

$$\text{if} \ \frac{[H^+]^2 - K_w}{[H^+]} \ll [HCN]_0 = 5.0 \times 10^{-4}, \text{ then we can use } [H^+] = (K_a[HCN]_o + K_w)^{1/2}.$$

Using this formula: $[H^+] = [(6.2 \times 10^{-10})(5.0 \times 10^{-4}) + (1.0 \times 10^{-14})]^{1/2}, \ [H^+] = 5.7 \times 10^{-7} \ M$

Checking assumptions: $\dfrac{[H^+]^2 - K_w}{[H^+]} = 5.5 \times 10^{-7} \ll 5.0 \times 10^{-4}$

Assumptions good. $pH = -\log(5.7 \times 10^{-7}) = 6.24$

103.    We can't neglect the $[H^+]$ contribution from $H_2O$ since this is a very dilute solution of the strong acid. Following the strategy developed in Section 7.10 of the text, we first determine the charge balance equation and then manipulate this equation to get into one unknown.

[Positive charge] = [negative charge]

$$[H^+] = [Cl^-] + [OH^-] = 7.0 \times 10^{-7} + \frac{K_w}{[H^+]} \ \text{(because } [Cl^-] = 7.0 \times 10^{-7} \text{ and } [OH^-] = \frac{K_w}{[H^+]})$$

$$\frac{[H^+]^2 - K_w}{[H^+]} = 7.0 \times 10^{-7}, \ [H^+]^2 - (7.0 \times 10^{-7})[H^+] - 1.0 \times 10^{-14} = 0$$

Using the quadratic formula to solve:

$$[H^+] = \frac{-(-7.0 \times 10^{-7}) \pm [(-7.0 \times 10^{-7})^2 - 4(1)(-1.0 \times 10^{-14})]^{1/2}}{2(1)}$$

$$[H^+] = 7.1 \times 10^{-7} \ M; \ pH = -\log(7.1 \times 10^{-7}) = 6.15$$

## Additional Exercises

105.    a.    $NH_3 + H_3O^+ \rightleftharpoons NH_4^+ + H_2O$

$$K_{eq} = \frac{[NH_4^+]}{[NH_3][H^+]} = \frac{1}{K_a \text{ for } NH_4^+} = \frac{K_b}{K_w} = \frac{1.8 \times 10^{-5}}{1.0 \times 10^{-14}} = 1.8 \times 10^9$$

b.    $NO_2^- + H_3O^+ \rightleftharpoons H_2O + HNO_2 \quad K_{eq} = \frac{[HNO_2]}{[NO_2^-][H^+]} = \frac{1}{K_a} = \frac{1}{4.0 \times 10^{-4}} = 2.5 \times 10^3$

c.   $NH_4^+ + CH_3CO_2^- \rightleftharpoons NH_3 + CH_3CO_2H$     $K_{eq} = \dfrac{[NH_3][CH_3CO_2H]}{[NH_4^+][CH_3CO_2^-]} \times \dfrac{[H^+]}{[H^+]}$

$$K_{eq} = \frac{K_a \text{ for } NH_4^+}{K_a \text{ for } CH_3CO_2H} = \frac{K_w}{(K_b \text{ for } NH_3)(K_a \text{ for } CH_3CO_2H)}$$

$$K_{eq} = \frac{1.0 \times 10^{-14}}{(1.8 \times 10^{-5})(1.8 \times 10^{-5})} = 3.1 \times 10^{-5}$$

d.   $H_3O^+ + OH^- \rightleftharpoons 2\,H_2O$    $K_{eq} = \dfrac{1}{K_w} = 1.0 \times 10^{14}$

e.   $NH_4^+ + OH^- \rightleftharpoons NH_3 + H_2O$    $K_{eq} = \dfrac{1}{K_b \text{ for } NH_3} = 5.6 \times 10^4$

f.   $HNO_2 + OH^- \rightleftharpoons H_2O + NO_2^-$

$$K_{eq} = \frac{[NO_2^-]}{[HNO_2][OH^-]} \times \frac{[H^+]}{[H^+]} = \frac{K_a \text{ for } HNO_2}{K_w} = \frac{4.0 \times 10^{-4}}{1.0 \times 10^{-14}} = 4.0 \times 10^{10}$$

107.   The light bulb is bright because a strong electrolyte is present; that is, a solute is present that dissolves to produce a lot of ions in solution. The pH meter value of 4.6 indicates that a weak acid is present. (If a strong acid were present, the pH would be close to zero.) Of the possible substances, only HCl (strong acid), NaOH (strong base), and $NH_4Cl$ are strong electrolytes. Of these three substances, only $NH_4Cl$ contains a weak acid (the HCl solution would have a pH close to zero, and the NaOH solution would have a pH close to 14.0). $NH_4Cl$ dissociates into $NH_4^+$ and $Cl^-$ ions when dissolved in water. $Cl^-$ is the conjugate base of a strong acid, so it has no basic (or acidic properties) in water. $NH_4^+$, however, is the conjugate acid of the weak base $NH_3$, so $NH_4^+$ is a weak acid and would produce a solution with a pH = 4.6 when the concentration is ~1.0 $M$. $NH_4Cl$ is the solute.

109.   $CaO(s) + H_2O(l) \rightarrow Ca(OH)_2(aq);\ \ Ca(OH)_2(aq) \rightarrow Ca^{2+}(aq) + 2\,OH^-(aq)$

$$[OH^-] = \frac{0.25\,g\,CaO \times \dfrac{1\,mol\,CaO}{56.08\,g} \times \dfrac{1\,mol\,Ca(OH)_2}{1\,mol\,CaO} \times \dfrac{2\,mol\,OH^-}{mol\,Ca(OH)_2}}{1.5\,L} = 5.9 \times 10^{-3}\,M$$

$pOH = -\log(5.9 \times 10^{-3}) = 2.23,\ \ pH = 14.00 - 2.23 = 11.77$

111.   $$\frac{30.0 \text{ mg papH}^+\text{Cl}^-}{\text{mL soln}} \times \frac{1000 \text{ mL}}{\text{L}} \times \frac{1 \text{ g}}{1000 \text{ mg}} \times \frac{1 \text{ mol papH}^+\text{Cl}^-}{378.85 \text{ g}} \times \frac{1 \text{ mol papH}^+}{\text{mol papH}^+\text{Cl}^-}$$

$$= 0.0792 \ M$$

$$\text{papH}^+ \ \rightleftharpoons \ \text{pap} + \text{H}^+ \qquad K_a = \frac{K_w}{K_{b, \text{ pap}}} = \frac{2.1 \times 10^{-14}}{8.33 \times 10^{-9}} = 2.5 \times 10^{-6}$$

Initial   0.0792 $M$         0     ~0
Equil.   0.0792 − $x$       $x$     $x$

$$K_a = 2.5 \times 10^{-6} = \frac{x^2}{0.0792 - x} \approx \frac{x^2}{0.0792} \ , \quad x = [\text{H}^+] = 4.4 \times 10^{-4} \ M$$

$$\text{pH} = -\log(4.4 \times 10^{-4}) = 3.36; \quad \text{assumptions good.}$$

113.   a.   $$\text{Fe(H}_2\text{O)}_6^{3+} + \text{H}_2\text{O} \ \rightleftharpoons \ \text{Fe(H}_2\text{O)}_5(\text{OH})^{2+} \ + \ \text{H}_3\text{O}^+$$

Initial     0.10 $M$                 0            ~0
Equil.     0.10 − $x$                $x$             $x$

$$K_a = \frac{[\text{H}_3\text{O}^+][\text{Fe(H}_2\text{O)}_5(\text{OH})^{2+}]}{[\text{Fe(H}_2\text{O)}_6^{3+}]}, \quad 6.0 \times 10^{-3} = \frac{x^2}{0.10 - x} \approx \frac{x^2}{0.10}$$

$x = 2.4 \times 10^{-2} \ M$; assumption is poor (24% error).

Using successive approximations:

$$\frac{x^2}{0.10 - 0.024} = 6.0 \times 10^{-3}, \ x = 0.021$$

$$\frac{x^2}{0.10 - 0.021} = 6.0 \times 10^{-3}, \ x = 0.022; \quad \frac{x^2}{0.10 - 0.022} = 6.0 \times 10^{-3}, \ x = 0.022$$

$x = [\text{H}^+] = 0.022 \ M$; pH = 1.66

b.   $$\frac{[\text{Fe(H}_2\text{O)}_5(\text{OH})^{2+}]}{[\text{Fe(H}_2\text{O)}_6^{3+}]} = \frac{0.0010}{0.9990} \ ; \ K_a = 6.0 \times 10^{-3} = \frac{[\text{H}^+](0.0010)}{0.9990}$$

Solving:  $[\text{H}^+] = 6.0 \ M$; pH = −log(6.0) = −0.78

c.   Because of the lower charge, $\text{Fe}^{2+}$(aq) will not be as strong an acid as $\text{Fe}^{3+}$(aq). A solution of iron(II) nitrate will be less acidic (have a higher pH) than a solution with the same concentration of iron(III) nitrate.

115.    0.50 $M$ HA, $K_a = 1.0 \times 10^{-3}$;  0.20 $M$ HB, $K_a = 1.0 \times 10^{-10}$;  0.10 $M$ HC, $K_a = 1.0 \times 10^{-12}$

Major source of $H^+$ is HA because its $K_a$ value is significantly larger than other $K_a$ values.

$$HA \quad \rightleftharpoons \quad H^+ \quad + \quad A^-$$

Initial      0.50 $M$           ~0           0
Equil.       0.50 − $x$          $x$           $x$

$$K_a = \frac{x^2}{0.50 - x}, \quad 1.0 \times 10^{-3} \approx \frac{x^2}{0.50}, \quad x = 0.022 \, M = [H^+], \quad \frac{0.022}{0.50} \times 100 = 4.4\% \text{ error}$$

Assumption good.  Let's check out the assumption that only HA is an important source of $H^+$.

For HB:  $1.0 \times 10^{-10} = \dfrac{(0.022)[B^-]}{(0.20)}$,  $[B^-] = 9.1 \times 10^{-10} \, M$

At <u>most</u>, HB will produce an additional $9.1 \times 10^{-10} \, M \, H^+$.  Even less will be produced by HC.  Thus our original assumption was good.  $[H^+] = 0.022 \, M$.

117.    Since $NH_3$ is so concentrated, we need to calculate the $OH^-$ contribution from the weak base $NH_3$.

$$NH_3 \quad + \quad H_2O \quad \rightleftharpoons \quad NH_4^+ \quad + \quad OH^- \qquad K_b = 1.8 \times 10^{-5}$$

Initial   15.0 $M$                          0              0.0100 $M$    (Assume no volume change.)
Equil.    15.0 − $x$                        $x$             0.0100 + $x$

$$K_b = 1.8 \times 10^{-5} = \frac{x(0.0100 + x)}{15.0 - x} \approx \frac{x(0.0100)}{15.0}, \quad x = 0.027; \text{ assumption is horrible}$$
$$(x \text{ is } 270\% \text{ of } 0.0100).$$

Using the quadratic formula:

$$(1.8 \times 10^{-5})(15.0 - x) = (0.0100)x + x^2, \quad x^2 + (0.0100)x - 2.7 \times 10^{-4} = 0$$

$$x = 1.2 \times 10^{-2} \, M, \quad [OH^-] = (1.2 \times 10^{-2}) + 0.0100 = 0.022 \, M$$

119.    a.    The initial concentrations are halved since equal volumes of the two solutions are mixed.

$$HC_2H_3O_2 \quad \rightleftharpoons \quad H^+ \quad + \quad C_2H_3O_2^-$$

Initial      0.100 $M$            $5.00 \times 10^{-4} \, M$        0
Equil.       0.100 − $x$          $5.00 \times 10^{-4} + x$        $x$

$$K_a = 1.8 \times 10^{-5} = \frac{x(5.00 \times 10^{-4} + x)}{0.100 - x} \approx \frac{x(5.00 \times 10^{-4})}{0.100}$$

$x = 3.6 \times 10^{-3}$; assumption is horrible.  Using the quadratic formula:

$$x^2 + (5.18 \times 10^{-4})x - 1.8 \times 10^{-6} = 0$$

$$x = 1.1 \times 10^{-3} \, M; \quad [H^+] = 5.00 \times 10^{-4} + x = 1.6 \times 10^{-3} \, M; \quad pH = 2.80$$

b.   $x = [C_2H_3O_2^-] = 1.1 \times 10^{-3} \ M$

121.   Because the values of $K_{a_1}$ and $K_{a_2}$ are fairly close to each other, we should consider the amount of $H^+$ produced by the $K_{a_1}$ and $K_{a_2}$ reactions.

$$H_3C_6H_5O_7 \quad \rightleftharpoons \quad H_2C_6H_5O_7^- \quad + \quad H^+ \quad K_{a_1} = 8.4 \times 10^{-4}$$

Initial     0.15 $M$                                    0                   ~0
Equil.      0.15 - $x$                                  $x$                  $x$

$8.4 \times 10^{-4} = \dfrac{x^2}{0.15 - x} \approx \dfrac{x^2}{0.15}$ ,   $x = 1.1 \times 10^{-2}$;   assumption fails the 5% rule.

Solving more exactly using the method of successive approximations:

$$8.4 \times 10^{-4} = \frac{x^2}{(0.15 - 1.1 \times 10^{-2})} , \quad x = 1.1 \times 10^{-2} \ M \ \text{(consistent answer)}$$

Now let's solve for the $H^+$ contribution from the $K_{a_2}$ reaction.

$$H_2C_6H_5O_7^- \quad \rightleftharpoons \quad HC_6H_5O_7^{2-} \quad + \quad H^+ \qquad K_{a_2} = 1.8 \times 10^{-5}$$

Initial      $1.1 \times 10^{-2} \ M$                          0            $1.1 \times 10^{-2} \ M$
Equil.       $1.1 \times 10^{-2} - x$                         $x$           $1.1 \times 10^{-2} + x$

$1.8 \times 10^{-5} = \dfrac{x(1.1 \times 10^{-2} + x)}{(1.1 \times 10^{-2} - x)} \approx \dfrac{x(1.1 \times 10^{-2})}{1.1 \times 10^{-2}}$ ,   $x = 1.8 \times 10^{-5} \ M$;   assumption good
                                                                                                                                    (0.2% error).

At most, $1.8 \times 10^{-5} \ M \ H^+$ will be added from the $K_{a_2}$ reaction.

$[H^+]_{total} = (1.1 \times 10^{-2}) + (1.8 \times 10^{-5}) = 1.1 \times 10^{-2} \ M$

Note that the $H^+$ contribution from the $K_{a_2}$ reaction was negligible compared to the $H^+$ contribution from the $K_{a_1}$ reaction even though the two $K_a$ values only differed by a factor of 50. Therefore, the $H^+$ contribution from the $K_{a_3}$ reaction will also be negligible since $K_{a_3} < K_{a_2}$.

Solving:  pH $= -\log(1.1 \times 10^{-2}) = 1.96$

## Challenge Problems

123.   a.   $HCO_3^- + HCO_3^- \rightleftharpoons H_2CO_3 + CO_3^{2-}$

$$K_{eq} = \frac{[H_2CO_3][CO_3^{2-}]}{[HCO_3^-][HCO_3^-]} \times \frac{[H^+]}{[H^+]} = \frac{K_{a_2}}{K_{a_1}} = \frac{4.8 \times 10^{-11}}{4.3 \times 10^{-7}} = 1.1 \times 10^{-4}$$

b.  $[H_2CO_3] = [CO_3^{2-}]$ since the reaction in part a is the principal equilibrium reaction.

c.  $H_2CO_3 \rightleftharpoons 2\,H^+ + CO_3^{2-}$    $K_{eq} = \dfrac{[H^+]^2[CO_3^{2-}]}{[H_2CO_3]} = K_{a_1} \times K_{a_2}$

   Because $[H_2CO_3] = [CO_3^{2-}]$ from part b, $[H^+]^2 = K_{a_1} \times K_{a_2}$.

   $[H^+] = (K_{a_1} \times K_{a_2})^{1/2}$, or taking the $-$log of both sides: $pH = \dfrac{pK_{a_1} + pK_{a_2}}{2}$

d.  $[H^+] = [(4.3 \times 10^{-7}) \times (4.8 \times 10^{-11})]^{1/2}$,  $[H^+] = 4.5 \times 10^{-9}\ M$;  $pH = 8.35$

125.           $HC_2H_3O_2 \rightleftharpoons H^+ + C_2H_3O_2^-$        $K_a = 1.8 \times 10^{-5}$

Initial  1.00 $M$        ~0          0
Equil.   $1.00 - x$        $x$          $x$

$1.8 \times 10^{-5} = \dfrac{x^2}{1.00 - x} \approx \dfrac{x^2}{1.00}$,   $x = [H^+] = 4.24 \times 10^{-3}\ M$ (using one extra sig. fig.)

$pH = -\log(4.24 \times 10^{-3}) = 2.37$;   assumptions good.

We want to double the pH to $2(2.37) = 4.74$ by addition of the strong base NaOH. As is true with all strong bases, they are great at accepting protons. In fact, they are so good that we can assume they accept protons 100% of the time. The best acid present will react the strong base. This is $HC_2H_3O_2$. The initial reaction that occurs when the strong base is added is:

   $HC_2H_3O_2 + OH^- \rightarrow C_2H_3O_2^- + H_2O$

Note that this reaction has the net effect of converting $HC_2H_3O_2$ into its conjugate base, $C_2H_3O_2^-$.

For a pH = 4.74, let's calculate the ratio of $[C_2H_3O_2^-]/[HC_2H_3O_2]$ necessary to achieve this pH.

$HC_2H_3O_2 \rightleftharpoons H^+ + C_2H_3O_2^-$   $K_a = \dfrac{[H^+][C_2H_3O_2^-]}{[HC_2H_3O_2]}$

When pH = 4.74,  $[H^+] = 10^{-4.74} = 1.8 \times 10^{-5}$.

$K_a = 1.8 \times 10^{-5} = \dfrac{(1.8 \times 10^{-5})[C_2H_3O_2^-]}{[HC_2H_3O_2]}$,  $\dfrac{[C_2H_3O_2^-]}{[HC_2H_3O_2]} = 1.0$

For a solution having pH = 4.74, we need to have equal concentrations (equal moles) of $C_2H_3O_2^-$ and $HC_2H_3O_2$. Therefore, we need to add an amount of NaOH that will convert one-half of the $HC_2H_3O_2$ into $C_2H_3O_2^-$. This amount is 0.50 $M$ NaOH.

$$HC_2H_3O_2 \; + \; OH^- \; \rightarrow \; C_2H_3O_2^- \; + \; H_2O$$

Before        1.00 $M$        0.50 $M$        0
Change       −0.50         −0.50    →   +0.50
After          0.50 $M$        0            0.50 $M$
completion

From the preceding stoichiometry problem, adding enough NaOH(s) to produce a 0.50 $M$ $OH^-$ solution will convert one-half the $HC_2H_3O_2$ into $C_2H_3O_2^-$;  this results in a solution with pH = 4.74.

$$\text{Mass NaOH} = 1.00\text{ L} \; \times \; \frac{0.50 \text{ mol NaOH}}{\text{L}} \times \frac{40.00 \text{ g NaOH}}{\text{mol}} = 20. \text{ g NaOH}$$

127.    Major species: $BH^+$, $X^-$, and $H_2O$;  because $BH^+$ is the best acid and $X^-$ is the best base in solution, the principal equilibrium is:

$$BH^+ \; + \; X^- \; \rightleftharpoons \; B \; + \; HX$$

Initial   0.100 $M$     0.100 $M$        0           0
Equil.   0.100 − $x$    0.100 − $x$      $x$          $x$

$$K = \frac{K_{a,\,BH^+}}{K_{a,\,HX}} = \frac{[B][HX]}{[BH^+][X^-]} \text{, where } [B] = [HX] \text{ and } [BH^+] = [X^-]$$

To solve for the $K_a$ of HX, let's use the equilibrium expression to derive a general expression that relates pH to the $pK_a$ for $BH^+$ and to the $pK_a$ for HX.

$$\frac{K_{a,\,BH^+}}{K_{a,\,HX}} = \frac{[HX]^2}{[X^-]^2}; \quad K_{a,HX} = \frac{[H^+][X^-]}{[HX]}, \quad \frac{[HX]}{[X^-]} = \frac{[H^+]}{K_{a,\,HX}}$$

$$\frac{K_{a,\,BH^+}}{K_{a,\,HX}} = \frac{[HX]^2}{[X^-]^2} = \left(\frac{[H^+]}{K_{a,\,HX}}\right)^2, \quad [H^+]^2 = K_{a,\,BH^+} \times K_{a,\,HX}$$

Taking the −log of both sides:  $pH = \dfrac{pK_{a,\,BH^+} + pK_{a,\,HX}}{2}$

This is a general equation that applies to all BHX type salts.  Solving the problem:

$$K_b \text{ for B} = 1.0 \times 10^{-3}; \; K_a \text{ for } BH^+ = \frac{K_w}{K_b} = 1.0 \times 10^{-11}$$

$$pH = 8.00 = \frac{11.00 + pK_{a,\,HX}}{2}, \quad pK_{a,\,HX} = 5.00 \text{ and } K_a \text{ for HX} = 10^{-5.00} = 1.0 \times 10^{-5}$$

129.   0.0500 $M$ $HCO_2H$ (HA), $K_a = 1.77 \times 10^{-4}$; 0.150 $M$ $CH_3CH_2CO_2H$ (HB), $K_a = 1.34 \times 10^{-5}$

Because two comparable weak acids are present, each contributes to the total pH.

Charge balance:  $[H^+] = [A^-] + [B^-] + [OH^-] = [A^-] + [B^-] + K_w/[H^+]$

Mass balance for HA and HB:   $0.0500 = [HA] + [A^-]$ and $0.150 = [HB] + [B^-]$

$$\frac{[H^+][A^-]}{[HA]} = 1.77 \times 10^{-4}; \quad \frac{[H^+][B^-]}{[HB]} = 1.34 \times 10^{-5}$$

We have five equations and five unknowns.  Manipulate the equations to solve.

$[H^+] = [A^-] + [B^-] + K_w/[H^+];$  $[H^+]^2 = [H^+][A^-] + [H^+][B^-] + K_w$

$[H^+][A^-] = (1.77 \times 10^{-4})[HA] = (1.77 \times 10^{-4})(0.0500 - [A^-])$

If $[A^-] \ll 0.0500$, then $[H^+][A^-] \approx (1.77 \times 10^{-4})(0.0500) = 8.85 \times 10^{-6}$.

Similarly, assume $[H^+][B^-] \approx (1.34 \times 10^{-5})(0.150) = 2.01 \times 10^{-6}$.

$[H^+]^2 = 8.85 \times 10^{-6} + 2.01 \times 10^{-6} + 1.00 \times 10^{-14}$,  $[H^+] = 3.30 \times 10^{-3}$ mol/L

Check assumptions:  $[H^+][A^-] \approx 8.85 \times 10^{-6}$,  $[A^-] \approx \dfrac{8.85 \times 10^{-6}}{3.30 \times 10^{-3}} \approx 2.68 \times 10^{-3}$

Assumed $0.0500 - [A^-] \approx 0.0500$.  This assumption is borderline ($2.68 \times 10^{-3}$ is 5.4% of 0.0500).   The HB assumption is good (0.4% error).

Using successive approximations to refine the $[H^+][A^-]$ value:

$[H^+] = 3.22 \times 10^{-3}$ $M$,  pH = $-\log(3.22 \times 10^{-3}) = 2.492$

*Note*:  If we treat each acid separately:

$H^+$ from HA = $2.9 \times 10^{-3}$
$H^+$ from HB = $1.4 \times 10^{-3}$

_____

$4.3 \times 10^{-3}$ $M$ = $[H^+]_{total}$

This assumes the acids did not suppress each other's ionization. They do, and we expect the $[H^+]$ to be less than $4.3 \times 10^{-3}$ $M$. We get such an answer.

131.              HA       $\rightleftharpoons$       $H^+$  +  $A^-$    $K_a = 5.00 \times 10^{-10}$

| | | | | | |
|---|---|---|---|---|---|
| Initial | $[HA]_0$ | | $\sim 0$ | 0 | |
| Change | $-x$ | $\rightarrow$ | $+x$ | $+x$ | |
| Equil. | $[HA]_0 - x$ | | $x$ | $x$ | |

From the problem:  pH = 5.650, so $[H^+] = x = 10^{-5.650} = 2.24 \times 10^{-6} M$

$$5.00 \times 10^{-10} = \frac{x^2}{[HA]_0 - x} = \frac{(2.24 \times 10^{-6})^2}{([HA]_0 - 2.24 \times 10^{-6})}, \quad [HA]_0 = 1.00 \times 10^{-2} M$$

After the water is added, the pH of the solution is between 6 and 7, so the water contribution to the $[H^+]$ must be considered.  The general expression for a very dilute weak acid solution is:

$$K_a = \frac{[H^+]^2 - K_w}{[HA]_0 - \dfrac{[H^+]^2 - K_w}{[H^+]}}$$

pH = 6.650;  $[H^+] = 10^{-6.650} = 2.24 \times 10^{-7} M$

Let V = volume of water added:

$$5.00 \times 10^{-10} = \frac{(2.24 \times 10^{-7})^2 - (1.00 \times 10^{-14})}{(1.00 \times 10^{-2})\left(\dfrac{0.0500}{0.0500 + V}\right) - \dfrac{(2.24 \times 10^{-7})^2 - (1.00 \times 10^{-14})}{2.24 \times 10^{-7}}}$$

Solving,  V = 6.16 L of water were added.

133.    $\dfrac{0.135\ \text{mol}\ CO_2}{2.50\ L} = 5.40 \times 10^{-2}\ \text{mol}\ CO_2/L = 5.40 \times 10^{-2}\ M\ H_2CO_3;\ \ 0.105\ M\ CO_3^{2-}$

The best acid ($H_2CO_3$) reacts with the best base present ($CO_3^{2-}$) for the principal equilibrium.

$$H_2CO_3 + CO_3^{2-} \rightarrow 2\ HCO_3^-\quad K = \frac{K_{a_1,\ H_2CO_3}}{K_{a_2,\ H_2CO_3}} = \frac{4.3 \times 10^{-7}}{4.8 \times 10^{-11}} = 9.0 \times 10^3$$

Because K >> 1, assume all $CO_2$ ($H_2CO_3$) is converted into $HCO_3^-$; that is, $5.40 \times 10^{-2}$ mol/L $CO_3^{2-}$ is converted into $HCO_3^-$.

$[HCO_3^-] = 2(5.40 \times 10^{-2}) = 0.108\ M;\ \ [CO_3^{2-}] = 0.105 - 0.0540 = 0.051\ M$

*Note*: If we solve for the $[H_2CO_3]$ using these concentrations, we get $[H_2CO_3] = 2.5 \times 10^{-5}\ M$; our assumption that the reaction goes to completion is good ($2.5 \times 10^{-5}$ is 0.05% of 0.051). Whenever K >> 1, always assume the reaction goes to completion.

To solve for the $[H^+]$ in equilibrium with $HCO_3^-$ and $CO_3^{2-}$, use the $K_a$ expression for $HCO_3^-$.

$HCO_3^- \rightleftharpoons H^+ + CO_3^{2-}\quad K_{a2} = 4.8 \times 10^{-11}$

$$4.8 \times 10^{-11} = \frac{[H^+][CO_3^{2-}]}{[HCO_3^-]} = \frac{[H^+](0.051)}{0.108}$$

$[H^+] = 1.0 \times 10^{-10};\ \ pH = 10.00;\ \ \text{assumptions good.}$

135.    Major species: $H_2O$, $Na^+$, and $NO_2^-$; $NO_2^-$ is a weak base. $NO_2^- + H_2O \rightleftharpoons HNO_2 + OH^-$; because this is a very dilute solution of a weak base, the $OH^-$ contribution from $H_2O$ must be considered. The weak base equations for dilute solutions are analogous to the weak acid equations derived in Section 7.9 of the text.

For $A^-$ type bases ($A^- + H_2O \rightleftharpoons HA + OH^-$), the general equation is:

$$K_b = \frac{[OH^-]^2 - K_w}{[A^-]_0 - \dfrac{[OH^-]^2 - K_w}{[OH^-]}}$$

When $[A^-]_0 \gg \dfrac{[OH^-]^2 - K_w}{[OH^-]}$, then $K_b = \dfrac{[OH^-]^2 - K_w}{[A^-]_0}$ and:

$$[OH^-] = (K_b[A^-]_0 + K_w)^{1/2}$$

Try: $[OH^-] = \left( \dfrac{1.0 \times 10^{-14}}{4.0 \times 10^{-4}} \times (6.0 \times 10^{-4}) + (1.0 \times 10^{-14}) \right)^{1/2} = 1.6 \times 10^{-7} \ M$

Checking assumption: $6.0 \times 10^{-4} \gg \dfrac{(1.6 \times 10^{-7})^2 - (1.0 \times 10^{-14})}{1.6 \times 10^{-7}} = 9.8 \times 10^{-8}$

Assumption good. $[OH^-] = 1.6 \times 10^{-7} \ M$; pOH = 6.80; pH = 7.20

137.    Major species: $H_2O$, $NH_3$, $H^+$, and $Cl^-$; The $H^+$ from the strong acid will react with the best base present ($NH_3$). Because strong acids are great at donating protons, the reaction between $H^+$ and $NH_3$ essentially goes to completion, that is, until one or both of the reactants runs out. The reaction is:

$$NH_3 + H^+ \rightarrow NH_4^+$$

Because equal volumes of $1.0 \times 10^{-4} \ M$ $NH_3$ and $1.0 \times 10^{-4} \ M$ $H^+$ are mixed, both reactants are in stoichiometric amounts, and both reactants will run out at the same time. After reaction, only $NH_4^+$ and $Cl^-$ remain. $Cl^-$ has no basic properties since it is the conjugate base of a strong acid. Therefore, the only species with acid-base properties is $NH_4^+$, a weak acid. The initial concentration of $NH_4^+$ will be exactly one-half of $1.0 \times 10^{-4} \ M$ since equal volumes of $NH_3$ and HCl were mixed. Now we must solve the weak acid problem involving $5.0 \times 10^{-5} \ M$ $NH_4^+$.

$$\text{NH}_4^+ \quad \rightleftharpoons \quad \text{H}^+ \quad + \quad \text{NH}_3 \qquad K_a = \frac{K_w}{K_b} = 5.6 \times 10^{-10}$$

| | | | |
|---|---|---|---|
| Initial | $5.0 \times 10^{-5}\ M$ | ~0 | 0 |
| Equil. | $5.0 \times 10^{-5} - x$ | $x$ | $x$ |

$$K_a = \frac{x^2}{(5.0 \times 10^{-5} - x)} \approx \frac{x^2}{5.0 \times 10^{-5}} = 5.6 \times 10^{-10},\ \ x = 1.7 \times 10^{-7}\ M;\ \ \text{check assumptions.}$$

We cannot neglect $[\text{H}^+]$ that comes from $\text{H}_2\text{O}$.  As discussed in Section 7.9 of the text, assume $5.0 \times 10^{-5} \gg ([\text{H}^+]^2 - K_w)/[\text{H}^+]$.  If this is the case, then:

$$[\text{H}^+] = (K_a[\text{HA}]_0 + K_w)^{1/2} = 1.9 \times 10^{-7}\ M;\ \ \text{checking assumption:}$$

$$\frac{[\text{H}^+]^2 - K_w}{[\text{H}^+]} = 1.4 \times 10^{-7} \ll 5.0 \times 10^{-5}\ \ \text{(assumption good)}$$

So: $[\text{H}^+] = 1.9 \times 10^{-7}\ M;\ \text{pH} = 6.72$

# CHAPTER 8

# APPLICATIONS OF AQUEOUS EQUILIBRIA

## Buffers

15. A buffer solution is one that resists a change in its pH when either hydroxide ions or protons ($H^+$) are added. Any solution that contains a weak acid and its conjugate base or a weak base and its conjugate acid is classified as a buffer. The pH of a buffer depends on the [base]/[acid] ratio. When $H^+$ is added to a buffer, the weak base component of the buffer reacts with the $H^+$ and forms the acid component of the buffer. Even though the concentrations of the acid and base components of the buffer change some, the ratio of [base]/[acid] does not change that much. This translates into a pH that doesn't change much. When $OH^-$ is added to a buffer, the weak acid component is converted into the base component of the buffer. Again, the [base]/[acid] ratio does not change a lot (unless a large quantity of $OH^-$ is added), so the pH does not change much.

$$H^+(aq) + CO_3^{2-}(aq) \rightarrow HCO_3^-(aq); \quad OH^-(aq) + HCO_3^-(aq) \rightarrow H_2O(l) + CO_3^{2-}(aq)$$

17. $pH = pK_a + \log\dfrac{[base]}{[acid]}$; when [acid] > [base], then $\dfrac{[base]}{[acid]} < 1$ and $\log\left(\dfrac{[base]}{[acid]}\right) < 0$.

From the Henderson-Hasselbalch equation, if the log term is negative, then $pH < pK_a$. When one has more acid than base in a buffer, the pH will be on the acidic side of the $pK_a$ value; that is, the pH is at a value lower than the $pK_a$ value. When one has more base than acid in a buffer ([conjugate base] > [weak acid]), then the log term in the Henderson-Hasselbalch equation is positive, resulting in $pH > pK_a$. When one has more base than acid in a buffer, the pH is on the basic side of the $pK_a$ value; that is, the pH is at a value greater than the $pK_a$ value. The other scenario you can run across in a buffer is when [acid] = [base]. Here, the log term is equal to zero, and $pH = pK_a$.

19. a. This is a weak acid problem. Let $HC_3H_5O_2$ = HOPr and $C_3H_5O_2^-$ = OPr$^-$.

|  | HOPr(aq) | $\rightleftharpoons$ | H$^+$(aq) | + | OPr$^-$(aq) | $K_a = 1.3 \times 10^{-5}$ |
|---|---|---|---|---|---|---|
| Initial | 0.100 $M$ | | ~0 | | 0 | |
| | $x$ mol/L HOPr dissociates to reach equilibrium | | | | | |
| Change | $-x$ | $\rightarrow$ | $+x$ | | $+x$ | |
| Equil. | $0.100 - x$ | | $x$ | | $x$ | |

$$K_a = 1.3 \times 10^{-5} = \frac{[H^+][OPr^-]}{[HOPr]} = \frac{x^2}{0.100 - x} \approx \frac{x^2}{0.100}$$

$x = [H^+] = 1.1 \times 10^{-3} \ M$; $pH = 2.96$; assumptions good by the 5% rule.

b.  This is a weak base problem.

$$OPr^-(aq) + H_2O(l) \rightleftharpoons HOPr(aq) + OH^-(aq) \quad K_b = \frac{K_w}{K_a} = 7.7 \times 10^{-10}$$

Initial      0.100 $M$                                    0              ~0
             $x$ mol/L OPr$^-$ reacts with H$_2$O to reach equilibrium
Change       $-x$                              $\rightarrow$   $+x$          $+x$
Equil.       0.100 $- x$                              $x$            $x$

$$K_b = 7.7 \times 10^{-10} = \frac{[HOPr][OH^-]}{[OPr^-]} = \frac{x^2}{0.100 - x} \approx \frac{x^2}{0.100}$$

$x = [OH^-] = 8.8 \times 10^{-6} \, M;$  pOH = 5.06;  pH = 8.94;  assumptions good.

c.  Pure H$_2$O, $[H^+] = [OH^-] = 1.0 \times 10^{-7} \, M;$  pH = 7.00

d.  This solution contains a weak acid and its conjugate base.  This is a buffer solution.  We will solve for the pH through the weak acid equilibrium reaction.

$$HOPr(aq) \rightleftharpoons H^+(aq) + OPr^-(aq) \quad K_a = 1.3 \times 10^{-5}$$

Initial      0.100 $M$              ~0              0.100 $M$
             $x$ mol/L HOPr dissociates to reach equilibrium
Change       $-x$          $\rightarrow$   $+x$            $+x$
Equil.       0.100 $- x$              $x$              0.100 $+ x$

$$1.3 \times 10^{-5} = \frac{(0.100 + x)(x)}{0.100 - x} \approx \frac{(0.100)(x)}{0.100} = x = [H^+]$$

$[H^+] = 1.3 \times 10^{-5} \, M;$  pH = 4.89;  assumptions good.

Alternately, we can use the Henderson-Hasselbalch equation to calculate the pH of buffer solutions.

$$pH = pK_a + \log\frac{[base]}{[acid]} = pK_a + \log\left(\frac{0.100}{0.100}\right) = pK_a = -\log(1.3 \times 10^{-5}) = 4.89$$

The Henderson-Hasselbalch equation will be valid when an assumption of the type 0.1 + $x \approx 0.1$ that we just made in this problem is valid.  From a practical standpoint, this will almost always be true for useful buffer solutions.  If the assumption is not valid, the solution will have such a low buffering capacity it will not be of any use to control the pH. *Note*:  The Henderson-Hasselbalch equation can <u>only</u> be used to solve for the pH of buffer solutions.

21.    a.   OH⁻ will react completely with the best acid present, HOPr.

$$\text{HOPr} \quad + \quad \text{OH}^- \quad \rightarrow \quad \text{OPr}^- \quad + \quad \text{H}_2\text{O}$$

| | HOPr | OH⁻ | | OPr⁻ | H₂O | |
|---|---|---|---|---|---|---|
| Before | 0.100 $M$ | 0.020 $M$ | | 0 | | |
| Change | −0.020 | −0.020 | $\rightarrow$ | +0.020 | | Reacts completely |
| After | 0.080 | 0 | | 0.020 | | |

A buffer solution results after the reaction.  Using the Henderson-Hasselbalch equation:

$$\text{pH} = \text{p}K_a + \log \frac{[\text{base}]}{[\text{acid}]} = 4.89 + \log \frac{(0.020)}{(0.080)} = 4.29; \text{ assumptions good.}$$

b.   We have a weak base and a strong base present at the same time.  The amount of OH⁻
     added by the weak base will be negligible.  To prove it, let's consider the weak base
     equilibrium:

$$\text{OPr}^- \quad + \quad \text{H}_2\text{O} \quad \rightleftharpoons \quad \text{HOPr} \quad + \quad \text{OH}^- \qquad K_b = 7.7 \times 10^{-10}$$

| | OPr⁻ | | HOPr | OH⁻ |
|---|---|---|---|---|
| Initial | 0.100 $M$ | | 0 | 0.020 $M$ |
| | | | | |
| Change | −$x$ | $\rightarrow$ | +$x$ | +$x$ |
| Equil. | 0.100 − $x$ | | $x$ | 0.020 + $x$ |

$x$ mol/L OPr⁻ reacts with H₂O to reach equilibrium

[OH⁻] = 0.020 + $x \approx$ 0.020 $M$;  pOH = 1.70;  pH = 12.30;  assumption good.

*Note*:  The OH⁻ contribution from the weak base OPr⁻ was negligible ($x$ = 3.9 × 10⁻⁹ $M$ as
compared to 0.020 $M$ OH⁻ from the strong base).  The pH can be determined by only
considering the amount of strong base present.

c.   This is a strong base in water.  [OH⁻] = 0.020 $M$;  pOH = 1.70;  pH = 12.30

d.   OH⁻ will react completely with HOPr, the best acid present.

$$\text{HOPr} \quad + \quad \text{OH}^- \quad \rightarrow \quad \text{OPr}^- \quad + \quad \text{H}_2\text{O}$$

| | HOPr | OH⁻ | | OPr⁻ | H₂O | |
|---|---|---|---|---|---|---|
| Before | 0.100 $M$ | 0.020 $M$ | | 0.100 $M$ | | |
| Change | −0.020 | −0.020 | $\rightarrow$ | +0.020 | | Reacts completely |
| After | 0.080 | 0 | | 0.120 | | |

Using the Henderson-Hasselbalch equation to solve for the pH of the resulting buffer
solution:

$$\text{pH} = \text{p}K_a + \log \frac{[\text{base}]}{[\text{acid}]} = 4.89 + \log \frac{(0.120)}{(0.080)} = 5.07; \text{ assumptions good.}$$

23.    Major species: HF, F$^-$, K$^+$, and H$_2$O. K$^+$ has no acidic or basic properties. This is a solution containing a weak acid and its conjugate base. This is a buffer solution. One appropriate equilibrium reaction you can use is the K$_a$ reaction of HF, which contains both HF and F$^-$. However, you could also use the K$_b$ reaction for F$^-$ and come up with the same answer. Alternately, you could use the Henderson-Hasselblach equation to solve for the pH. For this problem, we will use the K$_a$ reaction and set up an ICE table to solve for the pH.

$$\text{HF} \quad \rightleftharpoons \quad \text{F}^- \quad + \quad \text{H}^+$$

| | | | |
|---|---|---|---|
| Initial | 0.60 $M$ | 1.00 $M$ | ~0 |

$x$ mol/L HF dissociates to reach equilibrium

| | | | | |
|---|---|---|---|---|
| Change | $-x$ | $\rightarrow$ | $+x$ | $+x$ |
| Equil. | 0.60 $- x$ | | 1.00 $+ x$ | $x$ |

$$K_a = 7.2 \times 10^{-4} = \frac{[\text{F}^-][\text{H}^+]}{[\text{HF}]} = \frac{(1.00 + x)(x)}{0.60 - x} \approx \frac{(1.00)(x)}{0.60} \quad \text{(assuming } x << 0.60\text{)}$$

$$x = [\text{H}^+] = 0.60 \times (7.2 \times 10^{-4}) = 4.3 \times 10^{-4} \ M; \text{ assumptions good.}$$

$$\text{pH} = -\log(4.3 \times 10^{-4}) = 3.37$$

25.    Major species after NaOH added: HF, F$^-$, K$^+$, Na$^+$, OH$^-$, and H$_2$O. The OH$^-$ from the strong base will react with the best acid present (HF). Any reaction involving a strong base is assumed to go to completion. Because all species present are in the same volume of solution, we can use molarity units to do the stoichiometry part of the problem (instead of moles). The stoichiometry problem is:

$$\text{OH}^- \quad + \quad \text{HF} \quad \rightarrow \quad \text{F}^- \quad + \text{H}_2\text{O}$$

| | | | | |
|---|---|---|---|---|
| Before | 0.10 mol/1.00 L | 0.60 $M$ | 1.00 $M$ | |
| Change | $-0.10\ M$ | $-0.10\ M$ | $\rightarrow$ $+0.10\ M$ | Reacts completely |
| After | 0 | 0.50 | 1.10 | |

After all the OH$^-$ reacts, we are left with a solution containing a weak acid (HF) and its conjugate base (F$^-$). This is what we call a buffer problem. We will solve this buffer problem using the K$_a$ equilibrium reaction. One could also use the K$_b$ equilibrium reaction or use the Henderson-Hasselbalch equation to solve for the pH.

$$\text{HF} \quad \rightleftharpoons \quad \text{F}^- \quad + \quad \text{H}^+$$

| | | | |
|---|---|---|---|
| Initial | 0.50 $M$ | 1.10 $M$ | ~0 |

$x$ mol/L HF dissociates to reach equilibrium

| | | | | |
|---|---|---|---|---|
| Change | $-x$ | $\rightarrow$ | $+x$ | $+x$ |
| Equil. | 0.50 $- x$ | | 1.10 $+ x$ | $x$ |

$$K_a = 7.2 \times 10^{-4} = \frac{(1.10 + x)(x)}{0.50 - x} \approx \frac{(1.10)(x)}{0.50}, \ x = [\text{H}^+] = 3.3 \times 10^{-4} \ M; \ \text{pH} = 3.48;$$

assumptions good.

*Note*: The added NaOH to this buffer solution changes the pH only from 3.37 to 3.48. If the NaOH were added to 1.0 L of pure water, the pH would change from 7.00 to 13.00.

Major species after HCl added:  HF, F⁻, H⁺, K⁺, Cl⁻, and $H_2O$; the added H⁺ from the strong
acid will react completely with the best base present (F⁻).

$$H^+ \quad + \quad F^- \quad \rightarrow \quad HF$$

| | H⁺ | F⁻ | HF | |
|---|---|---|---|---|
| Before | $\dfrac{0.20 \text{ mol}}{1.00 \text{ L}}$ | 1.00 $M$ | 0.60 $M$ | |
| Change | $-0.20\ M$ | $-0.20\ M \rightarrow$ | $+0.20\ M$ | Reacts completely |
| After | 0 | 0.80 | 0.80 | |

After all the H⁺ has reacted, we have a buffer solution (a solution containing a weak acid and
its conjugate base). Solving the buffer problem:

$$HF \quad \rightleftharpoons \quad F^- \quad + \quad H^+$$

| | HF | F⁻ | H⁺ |
|---|---|---|---|
| Initial | 0.80 $M$ | 0.80 $M$ | 0 |
| Equil. | $0.80 - x$ | $0.80 + x$ | $x$ |

$$K_a = 7.2 \times 10^{-4} = \frac{(0.80 + x)(x)}{0.80 - x} \approx \frac{(0.80)(x)}{0.80}, \ x = [H^+] = 7.2 \times 10^{-4}\ M; \ pH = 3.14;$$

assumptions good.

*Note*: The added HCl to this buffer solution changes the pH only from 3.37 to 3.14. If the HCl
were added to 1.0 L of pure water, the pH would change from 7.00 to 0.70.

27.  $[HC_7H_5O_2] = \dfrac{21.5 \text{ g } HC_7H_5O_2 \times \dfrac{1 \text{ mol } HC_7H_5O_2}{122.12 \text{ g}}}{0.2000 \text{ L}} = 0.880\ M$

$[C_7H_5O_2^-] = \dfrac{37.7 \text{ g } NaC_7H_5O_2 \times \dfrac{1 \text{ mol } NaC_7H_5O_2}{144.10 \text{ g}} \times \dfrac{1 \text{ mol } C_7H_5O_2^-}{\text{mol } NaC_7H_5O_2}}{0.2000 \text{ L}} = 1.31\ M$

We have a buffer solution since we have both a weak acid and its conjugate base present at
the same time. One can use the $K_a$ reaction or the $K_b$ reaction to solve. We will use the $K_a$
reaction for the acid component of the buffer.

$$HC_7H_5O_2 \quad \rightleftharpoons \quad H^+ \quad + \quad C_7H_5O_2^-$$

| | | | |
|---|---|---|---|
| Initial | 0.880 $M$ | ~0 | 1.31 $M$ |
| | $x$ mol/L of $HC_7H_5O_2$ dissociates to reach equilibrium | | |
| Change | $-x$ | $\rightarrow +x$ | $+x$ |
| Equil. | $0.880 - x$ | $x$ | $1.31 + x$ |

$$K_a = 6.4 \times 10^{-5} = \frac{x(1.31 + x)}{0.880 - x} \approx \frac{x(1.31)}{0.880}, \ x = [H^+] = 4.3 \times 10^{-5}\ M$$

$$pH = -\log(4.3 \times 10^{-5}) = 4.37; \ \text{assumptions good.}$$

Alternatively, we can use the Henderson-Hasselbalch equation to calculate the pH of buffer solutions.

$$pH = pK_a + \log \frac{[base]}{[acid]} = pK_a + \log \frac{[C_7H_5O_2^-]}{[HC_7H_5O_2]}$$

$$pH = -\log(6.4 \times 10^{-5}) + \log\left(\frac{1.31}{0.880}\right) = 4.19 + 0.173 = 4.36$$

Within round-off error, this is the same answer we calculated solving the equilibrium problem using the $K_a$ reaction.

The Henderson-Hasselbalch equation will be valid when an assumption of the type $1.31 + x \approx 1.31$ that we just made in this problem is valid. From a practical standpoint, this will almost always be true for useful buffer solutions. If the assumption is not valid, the solution will have such a low buffering capacity that it will be of no use to control the pH. *Note*: The Henderson-Hasselbalch equation can <u>only</u> be used to solve for the pH of buffer solutions.

29.    $C_5H_5NH^+ \rightleftharpoons H^+ + C_5H_5N$   $K_a = \dfrac{K_w}{K_b} = \dfrac{1.0 \times 10^{-14}}{1.7 \times 10^{-9}} = 5.9 \times 10^{-6}$

$$pK_a = -\log(5.9 \times 10^{-6}) = 5.23$$

We will use the Henderson-Hasselbalch equation to calculate the concentration ratio necessary for each buffer.

$$pH = pK_a + \log \frac{[base]}{[acid]}, \ \ pH = 5.23 + \log \frac{[C_5H_5N]}{[C_5H_5NH^+]}$$

a.   $4.50 = 5.23 + \log \dfrac{[C_5H_5N]}{[C_5H_5NH^+]}$         b.   $5.00 = 5.23 + \log \dfrac{[C_5H_5N]}{[C_5H_5NH^+]}$

$\log \dfrac{[C_5H_5N]}{[C_5H_5NH^+]} = -0.73$             $\log \dfrac{[C_5H_5N]}{[C_5H_5NH^+]} = -0.23$

$\dfrac{[C_5H_5N]}{[C_5H_5NH^+]} = 10^{-0.73} = 0.19$        $\dfrac{[C_5H_5N]}{[C_5H_5NH^+]} = 10^{-0.23} = 0.59$

c.   $5.23 = 5.23 + \log \dfrac{[C_5H_5N]}{[C_5H_5NH^+]}$         d.   $5.50 = 5.23 + \log \dfrac{[C_5H_5N]}{[C_5H_5NH^+]}$

$\dfrac{[C_5H_5N]}{[C_5H_5NH^+]} = 10^{0.0} = 1.0$         $\dfrac{[C_5H_5N]}{[C_5H_5NH^+]} = 10^{0.27} = 1.9$

31.   When $H^+$ is added, it converts $C_2H_3O_2^-$ into $HC_2H_3O_2$:  $C_2H_3O_2^- + H^+ \rightarrow HC_2H_3O_2$.  From this reaction, the moles of $HC_2H_3O_2$ produced must equal the moles of $H^+$ added and the total concentration of acetate ion + acetic acid must equal 1.0 $M$ (assuming no volume change). Summarizing for each solution:

$$[C_2H_3O_2^-] + [HC_2H_3O_2] = 1.0 \ M \text{ and } [HC_2H_3O_2] = [H^+] \text{ added}$$

a.   $pH = pK_a + \log\dfrac{[C_2H_3O_2^-]}{[HC_2H_3O_2]}$;  for $pH = pK_a$, $[C_2H_3O_2^-] = [HC_2H_3O_2]$.

For this to be true, $[C_2H_3O_2^-] = [HC_2H_3O_2] = 0.50 \ M = [H^+]$ added, which means that 0.50 mol of HCl must be added to 1.0 L of the initial solution to produce a solution with $pH = pK_a$.

b.   $4.20 = 4.74 + \log\dfrac{[C_2H_3O_2^-]}{[HC_2H_3O_2]}$,  $\dfrac{[C_2H_3O_2^-]}{[HC_2H_3O_2]} = 10^{-0.54} = 0.29$

$[C_2H_3O_2^-] = 0.29[HC_2H_3O_2]$;  $0.29[HC_2H_3O_2] + [HC_2H_3O_2] = 1.0 \ M$

$[HC_2H_3O_2] = 0.78 \ M = [H^+]$ added

0.78 mol of HCl must be added to produce a solution with $pH = 4.20$.

c.   $5.00 = 4.74 + \log\dfrac{[C_2H_3O_2^-]}{[HC_2H_3O_2]}$,  $\dfrac{[C_2H_3O_2^-]}{[HC_2H_3O_2]} = 10^{0.26} = 1.8$

$[C_2H_3O_2^-] = 1.8[HC_2H_3O_2]$;  $1.8[HC_2H_3O_2] + [HC_2H_3O_2] = 1.0 \ M$

$[HC_2H_3O_2] = 0.36 \ M = [H^+]$ added

0.36 mol of HCl must be added to produce a solution with $pH = 5.00$.

33.   $pH = pK_a + \log\dfrac{[C_2H_3O_2^-]}{[HC_2H_3O_2]}$;  $pK_a = -\log(1.8 \times 10^{-5}) = 4.74$

Because the buffer components, $C_2H_3O_2^-$ and $HC_2H_3O_2$, are both in the same volume of water, the concentration ratio of $[C_2H_3O_2^-]/[HC_2H_3O_2]$ will equal the mole ratio of mol $C_2H_3O_2^-$/mol $HC_2H_3O_2$.

$5.00 = 4.74 + \log\dfrac{\text{mol } C_2H_3O_2^-}{\text{mol } HC_2H_3O_2}$;  mol $HC_2H_3O_2 = 0.5000 \ L \times \dfrac{0.200 \ \text{mol}}{L} = 0.100$ mol

$0.26 = \log\dfrac{\text{mol } C_2H_3O_2^-}{0.100 \ \text{mol}}$,  $\dfrac{\text{mol } C_2H_3O_2^-}{0.100 \ \text{mol}} = 10^{0.26} = 1.8$,  mol $C_2H_3O_2^- = 0.18$ mol

Mass $NaC_2H_3O_2 = 0.18$ mol $NaC_2H_3O_2 \times \dfrac{82.03 \ g}{\text{mol}} = 15$ g $NaC_2H_3O_2$

35.    a.   $pK_b$ for $C_6H_5NH_2 = -\log(3.8 \times 10^{-10}) = 9.42$; $pK_a$ for $C_6H_5NH_3^+ = 14.00 - 9.42 = 4.58$

$$pH = pK_a + \log\frac{[C_6H_5NH_2]}{[C_6H_5NH_3^+]}, \quad 4.20 = 4.58 + \log\frac{0.50\ M}{[C_6H_5NH_3^+]}$$

$$-0.38 = \log\frac{0.50\ M}{[C_6H_5NH_3^+]}, \quad [C_6H_5NH_3^+] = [C_6H_5NH_3Cl] = 1.2\ M$$

   b.   $4.0\text{ g NaOH} \times \dfrac{1\text{ mol NaOH}}{40.00\text{ g}} \times \dfrac{1\text{ mol OH}^-}{\text{mol NaOH}} = 0.10\text{ mol OH}^-$; $[OH^-] = \dfrac{0.10\text{ mol}}{1.0\text{ L}} = 0.10\ M$

$$C_6H_5NH_3^+ \quad + \quad OH^- \quad \rightarrow \quad C_6H_5NH_2 \quad + \quad H_2O$$

| | | | | |
|---|---|---|---|---|
| Before | 1.2 M | 0.10 M | 0.50 M | |
| Change | −0.10 | −0.10 | → +0.10 | |
| After | 1.1 | 0 | 0.60 | |

A buffer solution exists.  $pH = 4.58 + \log\left(\dfrac{0.60}{1.1}\right) = 4.32$

37.    a.   $pH = pK_a + \log\dfrac{[\text{base}]}{[\text{acid}]}$, $\quad 7.40 = -\log(4.3 \times 10^{-7}) + \log\dfrac{[HCO_3^-]}{[H_2CO_3]} = 6.37 + \log\dfrac{[HCO_3^-]}{[H_2CO_3]}$

$$\frac{[HCO_3^-]}{[H_2CO_3]} = 10^{1.03} = 11; \quad \frac{[H_2CO_3]}{[HCO_3^-]} = \frac{[CO_2]}{[HCO_3^-]} = \frac{1}{11} = 0.091$$

   b.   $7.15 = -\log(6.2 \times 10^{-8}) + \log\dfrac{[HPO_4^{2-}]}{[H_2PO_4^-]}$, $\quad 7.15 = 7.21 + \log\dfrac{[HPO_4^{2-}]}{[H_2PO_4^-]}$

$$\frac{[HPO_4^{2-}]}{[H_2PO_4^-]} = 10^{-0.06} = 0.9, \quad \frac{[H_2PO_4^-]}{[HPO_4^{2-}]} = \frac{1}{0.9} = 1.1 \approx 1$$

   c.   A best buffer has approximately equal concentrations of weak acid and conjugate base, so $pH \approx pK_a$ for a best buffer.   The $pK_a$ value for a $H_3PO_4/H_2PO_4^-$ buffer is $-\log(7.5 \times 10^{-3})$ $= 2.12$.  A pH of 7.15 is too high for a $H_3PO_4/H_2PO_4^-$ buffer to be effective.  At this high of pH, there would be so little $H_3PO_4$ present that we could hardly consider it a buffer; this solution would not be effective in resisting pH changes, especially when strong base is added.

39.    a.   No; a solution of a strong acid ($HNO_3$) and its conjugate base ($NO_3^-$) is not generally considered a buffer solution.

   b.   No; two acids are present ($HNO_3$ and HF), so it is not a buffer solution.

   c.   $H^+$ reacts completely with $F^-$. Since equal volumes are mixed, the initial concentrations in the mixture are $0.10\ M$ $HNO_3$ and $0.20\ M$ NaF.

|          | $H^+$    | $+$ | $F^-$    | $\rightarrow$ | $HF$  |                    |
|----------|----------|-----|----------|---------------|-------|--------------------|
| Before   | $0.10\ M$ |     | $0.20\ M$ |               | $0$   |                    |
| Change   | $-0.10$  |     | $-0.10$  | $\rightarrow$ | $+0.10$ | Reacts completely |
| After    | $0$      |     | $0.10$   |               | $0.10$ |                   |

After $H^+$ reacts completely, a buffer solution results; that is, a weak acid (HF) and its conjugate base ($F^-$) are both present in solution in large quantities.

d.  No; a strong acid ($HNO_3$) and a strong base (NaOH) do not form buffer solutions. They will neutralize each other to form $H_2O$.

41.   A best buffer has large and equal quantities of weak acid and conjugate base.  Because [acid] = [base] for a best buffer,  $pH = pK_a + \log\dfrac{[base]}{[acid]} = pK_a + 0 = pK_a$  ($pH \approx pK_a$ for a best buffer).

The best acid choice for a pH = 7.00 buffer would be the weak acid with a $pK_a$ close to 7.0 or $K_a \approx 1 \times 10^{-7}$.  HOCl is the best choice in Table 7.2 ($K_a = 3.5 \times 10^{-8}$; $pK_a = 7.46$).  To make this buffer, we need to calculate the [base]/[acid] ratio.

$$7.00 = 7.46 + \log\frac{[base]}{[acid]}, \quad \frac{[OCl^-]}{[HOCl]} = 10^{-0.46} = 0.35$$

Any $OCl^-$/HOCl buffer in a concentration ratio of 0.35 : 1 will have a pH = 7.00.  One possibility is [NaOCl] = 0.35 $M$ and [HOCl] = 1.0 $M$.

43.   To solve for [KOCl], we need to use the equation derived in Section 8.3 of the text on the exact treatment of buffered solutions.  The equation is:

$$K_a = \frac{[H^+]\left([A^-]_0 + \dfrac{[H^+]^2 - K_w}{[H^+]}\right)}{[HA]_0 - \dfrac{[H^+]^2 - K_w}{[H^+]}}$$

Because pH = 7.20,  $[H^+] = 10^{-7.20} = 6.3 \times 10^{-8}\ M$.

$$K_a = 3.5 \times 10^{-8} = \frac{6.3 \times 10^{-8}\left([OCl^-] + \dfrac{(6.3 \times 10^{-8})^2 - (1.0 \times 10^{-14})}{6.3 \times 10^{-8}}\right)}{1.0 \times 10^{-6} - \dfrac{(6.3 \times 10^{-8})^2 - (1.0 \times 10^{-14})}{6.3 \times 10^{-8}}}$$

$$3.5 \times 10^{-8} = \frac{6.3 \times 10^{-8}([OCl^-] - 9.57 \times 10^{-8})}{(1.0 \times 10^{-6}) + (9.57 \times 10^{-8})} \quad \text{(Carrying extra sig. figs.)}$$

$$3.83 \times 10^{-14} = 6.3 \times 10^{-8}([OCl^-] - 9.57 \times 10^{-8}), \quad [OCl^-] = [KOCl] = 7.0 \times 10^{-7}\ M$$

45.    Using regular procedures, pH = $pK_a$ = $-\log(1.6 \times 10^{-7})$ = 6.80 since $[A^-]_0$ = $[HA]_0$ in this buffer solution.  However, the pH is very close to that of neutral water, so maybe we need to consider the $H^+$ contribution from water.  Another problem with this answer is that $x$ (= $[H^+]$) is not small as compared with $[HA]_0$ and $[A^-]_0$, which was assumed when solving using the regular procedures.  Because the concentrations of the buffer components are less than $10^{-6}$ M, let us use the expression for the exact treatment of buffers to solve.

$$K_a = 1.6 \times 10^{-7} = \frac{[H^+]\left([A^-]_0 + \dfrac{[H^+]^2 - K_w}{[H^+]}\right)}{[HA]_0 - \dfrac{[H^+]^2 - K_w}{[H^+]}} =$$

$$\frac{[H^+]\left(5.0 \times 10^{-7} + \dfrac{[H^+]^2 - (1.0 \times 10^{-14})}{[H^+]}\right)}{5.0 \times 10^{-7} - \dfrac{[H^+]^2 - (1.0 \times 10^{-14})}{[H^+]}}$$

Solving exactly requires solving a cubic equation.  Instead, we will use the method of successive approximations where our initial guess for $[H^+]$ = $1.6 \times 10^{-7}$ M (the value obtained using the regular procedures).

$$1.6 \times 10^{-7} = \frac{[H^+]\left(5.0 \times 10^{-7} + \dfrac{(1.6 \times 10^{-7})^2 - (1.0 \times 10^{-14})}{1.6 \times 10^{-7}}\right)}{5.0 \times 10^{-7} - \dfrac{(1.6 \times 10^{-7})^2 - (1.0 \times 10^{-14})}{1.6 \times 10^{-7}}}, \ [H^+] = 1.1 \times 10^{-7}$$

We continue the process using $1.1 \times 10^{-7}$ as our estimate for $[H^+]$.  This gives $[H^+]$ = $1.5 \times 10^{-7}$.  We continue the process until we get a self consistent answer.  After three more iterations, we converge on $[H^+]$ = $1.3 \times 10^{-7}$ M.  Solving for the pH:

pH = $-\log(1.3 \times 10^{-7})$ = 6.89

Note that if we were to solve this problem exactly (using the quadratic formula) while ignoring the $H^+$ contribution from water, the answer comes out to $[H^+]$ = $1.0 \times 10^{-7}$ M.  We get a significantly different answer when we consider the $H^+$ contribution from $H_2O$.

## Acid-Base Titrations

47.

B + H$^+$ → BH$^+$; added H$^+$ from the strong acid converts the weak base B into its conjugate acid BH$^+$. Initially, before any H$^+$ is added (point d), B is the dominant species present. After H$^+$ is added, both B and BH$^+$ are present, and a buffered solution results (region b). At the equivalence point (points a and e), exactly enough H$^+$ has been added to convert all the weak base present initially into its conjugate acid BH$^+$. Past the equivalence point (region f), excess H$^+$ is present. For the answer to b, we included almost the entire buffer region. The maximum buffer region is around the halfway point to equivalence (point c), where [B] = [BH$^+$]. Here, pH = pK$_a$, which is a characteristic of a best buffer.

49.     Titration i is a strong acid titrated by a strong base. The pH is very acidic until just before the equivalence point; at the equivalence point, pH = 7.00; and past the equivalence the pH is very basic. Titration ii is a strong base titrated by a strong acid. Here the pH is very basic until just before the equivalence point; at the equivalence point, pH = 7.00; and past the equivalence point, the pH is very acidic. Titration iii is a weak base titrated by a strong acid. The pH starts out basic because a weak base is present. However, the pH will not be as basic as in titration ii, where a strong base is titrated. The pH drops as HCl is added; then at the halfway point to equivalence, pH = pK$_a$. Because K$_b$ = 4.4 ×10$^{-4}$ for CH$_3$NH$_2$, CH$_3$NH$_3^+$ has K$_a$ = K$_w$/K$_b$ = 2.3 ×10$^{-11}$ and pK$_a$ = 10.64. So, at the halfway point to equivalence for this weak base-strong acid titration, pH = 10.64. The pH continues to drop as HCl is added; then at the equivalence point the pH is acidic (pH < 7.00) because the only important major species present is a weak acid (the conjugate acid of the weak base). Past the equivalence point the pH becomes more acidic as excess HCl is added. Titration iv is a weak acid titrated by a strong base. The pH starts off acidic, but not nearly as acidic as the strong acid titration (i). The pH increases as NaOH is added; then, at the halfway point to equivalence, pH = pK$_a$ for HF = –log(7.2 ×10$^{-4}$) = 3.14. The pH continues to increase past the halfway point; then at the equivalence point, the pH is basic (pH > 7.0) because the only important major species present is a weak base (the conjugate base of the weak acid). Past the equivalence point, the pH becomes more basic as excess NaOH is added.

a.   All require the same volume of titrant to reach the equivalence point. At the equivalence point for all these titrations, moles acid = moles base ($M_AV_A = M_BV_B$). Because all the molarities and volumes are the same in the titrations, the volume of titrant will be the same (50.0 mL titrant added to reach equivalence point).

b.   Increasing initial pH: i < iv < iii < ii;  the strong acid titration has the lowest pH, the weak acid titration is next, followed by the weak base titration, with the strong base titration having the highest pH.

c.   i < iv < iii < ii; the strong acid titration has the lowest pH at the halfway point to equivalence, and the strong base titration has the highest halfway point pH. For the weak acid titration, pH = $pK_a$ = 3.14, and for the weak base titration, pH = $pK_a$ = 10.64.

d.   Equivalence point pH: iii < ii = i < iv;  the strong-by-strong titrations have pH = 7.00 at the equivalence point. The weak base titration has an acidic pH at the equivalence point, and a weak acid titration has a basic equivalence point pH.

The only different answer when the weak acid and weak base are changed would be for part c.  This is for  the halfway point to equivalence, where pH = $pK_a$.

$$HOC_6H_5; \; K_a = 1.6 \times 10^{-10}, \; pK_a = -\log(1.6 \times 10^{-10}) = 9.80$$

$$C_5H_5NH^+, \; K_a = \frac{K_w}{K_{b, C_5H_5N}} = \frac{1.0 \times 10^{-14}}{1.7 \times 10^{-9}} = 5.9 \times 10^{-6}, \; pK_a = 5.23$$

From the $pK_a$ values, the correct ordering at the halfway point to equivalence would be  i < iii < iv < ii. Note that for the weak base-strong acid titration using $C_5H_5N$, the pH is acidic at the halfway point to equivalence, whereas the weak acid-strong base titration using $HOC_6H_5$ is basic at the halfway point to equivalence. This is fine; this will always happen when the weak base titrated has a $K_b < 1 \times 10^{-7}$ (so $K_a$ of the conjugate acid is greater than $1 \times 10^{-7}$) and when the weak acid titrated has a $K_a < 1 \times 10^{-7}$ (so $K_b$ of the conjugate base is greater than $1 \times 10^{-7}$).

51.   a.   Because all acids are the same initial concentration, the pH curve with the highest pH at 0 mL of NaOH added will correspond to the titration of the weakest acid.  This is curve f.

b.   The pH curve with the lowest pH at 0 mL of  NaOH added will correspond to the titration of the strongest acid.  This is pH curve a.

The best point to look at to differentiate a strong acid from a weak acid titration (if initial concentrations are not known) is the equivalence point pH.  If the pH = 7.00, the acid titrated is a strong acid; if the pH is greater than 7.00, the acid titrated is a weak acid.

c.   For a weak acid-strong base titration, the pH at the halfway point to equivalence is equal to the $pK_a$ value.  The pH curve, which represents the titration of an acid with $K_a$ = 1.0 × $10^{-6}$, will have a pH = $-\log(1 \times 10^{-6})$ = 6.0 at the halfway point.  The equivalence point, from the plots, occurs at 50 mL NaOH added, so the halfway point is 25 mL.  Plot d has a pH ≈ 6.0 at 25 mL of NaOH added, so the acid titrated in this pH curve (plot d) has $K_a \approx 1 \times 10^{-6}$.

53.    This is a strong acid ($HClO_4$) titrated by a strong base (KOH).  Added $OH^-$ from the strong base will react completely with the $H^+$ present from the strong acid to produce $H_2O$.

a.    Only strong acid present.  $[H^+] = 0.200 \ M$;  pH = 0.699

b.    mmol $OH^-$ added = $10.0 \ mL \times \dfrac{0.100 \ mmol \ OH^-}{mL} = 1.00 \ mmol \ OH^-$

mmol $H^+$ present = $40.0 \ mL \times \dfrac{0.200 \ mmol \ H^+}{mL} = 8.0 \ mmol \ H^+$

*Note*: The units millimoles are usually easier numbers to work with.  The units for molarity are moles per liter but are also equal to millimoles per milliliter.

$$H^+ \quad\quad + \quad\quad OH^- \quad\quad \rightarrow \quad H_2O$$

| | | | |
|---|---|---|---|
| Before | 8.00 mmol | 1.00 mmol | |
| Change | −1.00 mmol | −1.00 mmol | Reacts completely |
| After | 7.00 mmol | 0 | |

The excess $H^+$ determines the pH.  $[H^+]_{excess} = \dfrac{7.00 \ mmol \ H^+}{40.0 \ mL + 10.0 \ mL} = 0.140 \ M$

pH = −log(0.140) = 0.854

c.    mmol $OH^-$ added = $40.0 \ mL \times 0.100 \ M = 4.00 \ mmol \ OH^-$

$$H^+ \quad\quad + \quad\quad OH^- \quad\quad \rightarrow \quad H_2O$$

| | | | |
|---|---|---|---|
| Before | 8.00 mmol | 4.00 mmol | |
| After | 4.00 mmol | 0 | |

$[H^+]_{excess} = \dfrac{4.00 \ mmol}{(40.0 + 40.0) \ mL} = 0.0500 \ M$;  pH = 1.301

d.    mmol $OH^-$ added = $80.0 \ mL \times 0.100 \ M = 8.00 \ mmol \ OH^-$;  this is the equivalence point because we have added just enough $OH^-$ to react with all the acid present.  For a strong acid-strong base titration, pH = 7.00 at the equivalence point because only neutral species are present ($K^+$, $ClO_4^-$, $H_2O$).

e.    mmol $OH^-$ added = $100.0 \ mL \times 0.100 \ M = 10.0 \ mmol \ OH^-$

$$H^+ \quad\quad + \quad\quad OH^- \quad\quad \rightarrow \quad H_2O$$

| | | | |
|---|---|---|---|
| Before | 8.00 mmol | 10.0 mmol | |
| After | 0 | 2.0 mmol | |

Past the equivalence point, the pH is determined by the excess $OH^-$ present.

$[OH^-]_{excess} = \dfrac{2.0 \ mmol}{(40.0 + 100.0) \ mL} = 0.014 \ M$;  pOH = 1.85;  pH = 12.15

55.     This is a weak acid ($HC_2H_3O_2$) titrated by a strong base (KOH).

   a.   Only weak acid is present.  Solving the weak acid problem:

$$HC_2H_3O_2 \quad \rightleftharpoons \quad H^+ \quad + \quad C_2H_3O_2^-$$

| | | | |
|---|---|---|---|
| Initial | 0.200 $M$ | ~0 | 0 |

$x$ mol/L $HC_2H_3O_2$ dissociates to reach equilibrium

| | | | |
|---|---|---|---|
| Change | $-x$ | $\rightarrow$ $+x$ | $+x$ |
| Equil. | $0.200 - x$ | $x$ | $x$ |

$$K_a = 1.8 \times 10^{-5} = \frac{x^2}{0.200 - x} \approx \frac{x^2}{0.200}, \ x = [H^+] = 1.9 \times 10^{-3} \ M$$

pH = 2.72;  assumptions good.

   b.   The added $OH^-$ will react completely with the best acid present, $HC_2H_3O_2$.

$$\text{mmol } HC_2H_3O_2 \text{ present} = 100.0 \text{ mL} \times \frac{0.200 \text{ mmol } HC_2H_3O_2}{\text{mL}} = 20.0 \text{ mmol } HC_2H_3O_2$$

$$\text{mmol } OH^- \text{ added} = 50.0 \text{ mL} \times \frac{0.100 \text{ mmol } OH^-}{\text{mL}} = 5.00 \text{ mmol } OH^-$$

$$HC_2H_3O_2 \quad + \quad OH^- \quad \rightarrow \quad C_2H_3O_2^- \quad + \quad H_2O$$

| | | | | |
|---|---|---|---|---|
| Before | 20.0 mmol | 5.00 mmol | 0 | |
| Change | $-5.00$ mmol | $-5.00$ mmol $\rightarrow$ | $+5.00$ mmol | Reacts completely |
| After | 15.0 mmol | 0 | 5.00 mmol | |

After reaction of all the strong base, we have a buffer solution containing a weak acid ($HC_2H_3O_2$) and its conjugate base ($C_2H_3O_2^-$).  We will use the Henderson-Hasselbalch equation to solve for the pH.

$$pH = pK_a + \log \frac{[C_2H_3O_2^-]}{[HC_2H_3O_2]} = -\log (1.8 \times 10^{-5}) + \log \left( \frac{5.00 \text{ mmol}/V_T}{15.0 \text{ mmol}/V_T} \right), \text{ where } V_T = \text{total volume}$$

$$pH = 4.74 + \log \left( \frac{5.00}{15.0} \right) = 4.74 + (-0.477) = 4.26$$

Note that the total volume cancels in the Henderson-Hasselbalch equation.  For the [base]/[acid] term, the mole ratio equals the concentration ratio because the components of the buffer are always in the same volume of solution.

   c.   mmol $OH^-$ added = 100.0 mL $\times$ (0.100 mmol $OH^-$/mL) = 10.0 mmol $OH^-$;  the same amount (20.0 mmol) of $HC_2H_3O_2$ is present as before (it doesn't change).  As before, let the $OH^-$ react to completion, then see what is remaining in solution after this reaction.

$$HC_2H_3O_2 \quad + \quad OH^- \quad \rightarrow \quad C_2H_3O_2^- \; + \; H_2O$$

| | | | |
|---|---|---|---|
| Before | 20.0 mmol | 10.0 mmol | 0 |
| After | 10.0 mmol | 0 | 10.0 mmol |

A buffer solution results after reaction.   Because $[C_2H_3O_2^-] = [HC_2H_3O_2] = 10.0$ mmol/total volume, pH = $pK_a$.  This is always true at the halfway point to equivalence for a weak acid-strong base titration, pH = $pK_a$.

$$pH = -\log(1.8 \times 10^{-5}) = 4.74$$

d.  mmol $OH^-$ added = 150.0 mL $\times$ 0.100 $M$ = 15.0 mmol $OH^-$.  Added $OH^-$ reacts completely with the weak acid.

$$HC_2H_3O_2 \quad + \quad OH^- \quad \rightarrow \quad C_2H_3O_2^- \; + \; H_2O$$

| | | | |
|---|---|---|---|
| Before | 20.0 mmol | 15.0 mmol | 0 |
| After | 5.0 mmol | 0 | 15.0 mmol |

We have a buffer solution after all the $OH^-$ reacts to completion.  Using the Henderson-Hasselbalch equation:

$$pH = 4.74 + \log \frac{[C_2H_3O_2^-]}{[HC_2H_3O_2]} = 4.74 + \log\left(\frac{15.0\ \text{mmol}}{5.0\ \text{mmol}}\right)$$

$$pH = 4.74 + 0.48 = 5.22$$

e.  mmol $OH^-$ added = 200.00 mL $\times$ 0.100 $M$ = 20.0 mmol $OH^-$; as before, let the added $OH^-$ react to completion with the weak acid;  then see what is in solution after this reaction.

$$HC_2H_3O_2 \quad + \quad OH^- \quad \rightarrow \quad C_2H_3O_2^- \; + \; H_2O$$

| | | | |
|---|---|---|---|
| Before | 20.0 mmol | 20.0 mmol | 0 |
| After | 0 | 0 | 20.0 mmol |

This is the equivalence point.  Enough $OH^-$ has been added to exactly neutralize all the weak acid present initially.  All that remains that affects the pH at the equivalence point is the conjugate base of the weak acid ($C_2H_3O_2^-$).  This is a weak base equilibrium problem.

$$C_2H_3O_2^- + H_2O \; \rightleftharpoons \; HC_2H_3O_2 + OH^- \qquad K_b = \frac{K_w}{K_b} = \frac{1.0 \times 10^{-14}}{1.8 \times 10^{-5}}$$

| | | | | |
|---|---|---|---|---|
| Initial | 20.0 mmol/300.0 mL | 0 | 0 | $K_b = 5.6 \times 10^{-9}$ |

$x$ mol/L $C_2H_3O_2^-$ reacts with $H_2O$ to reach equilibrium

| | | | | |
|---|---|---|---|---|
| Change | $-x$ | $\rightarrow$ | $+x$ | $+x$ |
| Equil. | $0.0667 - x$ | | $x$ | $x$ |

$$K_b = 5.6 \times 10^{-10} = \frac{x^2}{0.0667 - x} \approx \frac{x^2}{0.0667} , \; x = [OH^-] = 6.1 \times 10^{-6}\ M$$

pOH = 5.21;  pH = 8.79;  assumptions good.

f. mmol $OH^-$ added = 250.0 mL × 0.100 $M$ = 25.0 mmol $OH^-$

$$HC_2H_3O_2 \quad + \quad OH^- \quad \rightarrow \quad C_2H_3O_2^- \quad + \quad H_2O$$

| | | | |
|---|---|---|---|
| Before | 20.0 mmol | 25.0 mmol | 0 |
| After | 0 | 5.0 mmol | 20.0 mmol |

After the titration reaction, we have a solution containing excess $OH^-$ and a weak base $C_2H_3O_2^-$. When a strong base and a weak base are both present, assume that the amount of $OH^-$ added from the weak base will be minimal; that is, the pH past the equivalence point is determined by the amount of excess strong base.

$$[OH^-]_{excess} = \frac{5.0 \text{ mmol}}{100.0 \text{ mL} + 250.0 \text{ mL}} = 0.014 \ M; \quad pOH = 1.85; \quad pH = 12.15$$

57. We will do sample calculations for the various parts of the titration. All results are summarized in Table 8.1 at the end of Exercise 59.

At the beginning of the titration, only the weak acid $HC_3H_5O_3$ is present. Let HLac = $HC_3H_5O_3$ and $Lac^- = C_3H_5O_3^-$.

$$HLac \quad \rightleftharpoons \quad H^+ \quad + \quad Lac^- \quad K_a = 10^{-3.86} = 1.4 \times 10^{-4}$$

| | | | |
|---|---|---|---|
| Initial | 0.100 $M$ | ~0 | 0 |
| | $x$ mol/L HLac dissociates to reach equilibrium | | |
| Change | $-x$ | $\rightarrow$ $+x$ | $+x$ |
| Equil. | $0.100 - x$ | $x$ | $x$ |

$$1.4 \times 10^{-4} = \frac{x^2}{0.100 - x} \approx \frac{x^2}{0.100}, \ x = [H^+] = 3.7 \times 10^{-3} \ M; \quad pH = 2.43; \quad \text{assumptions good.}$$

Up to the stoichiometric point, we calculate the pH using the Henderson-Hasselbalch equation. This is the buffer region. For example, at 4.0 mL of NaOH added:

$$\text{initial mmol HLac present} = 25.0 \text{ mL} \times \frac{0.100 \text{ mmol}}{\text{mL}} = 2.50 \text{ mmol HLac}$$

$$\text{mmol } OH^- \text{ added} = 4.0 \text{ mL} \times \frac{0.100 \text{ mmol}}{\text{mL}} = 0.40 \text{ mmol } OH^-$$

*Note*: The units millimoles are usually easier numbers to work with. The units for molarity are moles per liter but are also equal to millimoles per milliliter.

The 0.40 mmol of added $OH^-$ converts 0.40 mmol HLac to 0.40 mmol $Lac^-$ according to the equation:

$$HLac + OH^- \rightarrow Lac^- + H_2O \qquad \text{Reacts completely.}$$

mmol HLac remaining = 2.50 − 0.40 = 2.10 mmol; mmol $Lac^-$ produced = 0.40 mmol

We have a buffer solution.  Using the Henderson-Hasselbalch equation where $pK_a = 3.86$:

$$pH = pK_a + \log \frac{[Lac^-]}{[HLac]} = 3.86 + \log \frac{(0.40)}{(2.10)} \qquad \text{(Total volume cancels, so we can use}$$
$$\text{use the ratio of moles or millimoles.)}$$

$$pH = 3.86 - 0.72 = 3.14$$

Other points in the buffer region are calculated in a similar fashion.  Perform a stoichiometry problem first, followed by a buffer problem.  The buffer region includes all points up to 24.9 mL $OH^-$ added.

At the stoichiometric point (25.0 mL $OH^-$ added), we have added enough $OH^-$ to convert all of the HLac (2.50 mmol) into its conjugate base ($Lac^-$).  All that is present is a weak base. To determine the pH, we perform a weak base calculation.

$$[Lac^-]_0 = \frac{2.50 \text{ mmol}}{25.0 \text{ mL } + 25.0 \text{ mL}} = 0.0500 \; M$$

$$Lac^- + H_2O \quad \rightleftharpoons \quad HLac \quad + \quad OH^- \qquad K_b = \frac{1.0 \times 10^{-14}}{1.4 \times 10^{-4}} = 7.1 \times 10^{-11}$$

| | | | |
|---|---|---|---|
| Initial | 0.0500 $M$ | 0 | 0 |
| | $x$ mol/L $Lac^-$ reacts with $H_2O$ to reach equilibrium | | |
| Change | $-x$ $\rightarrow$ | $+x$ | $+x$ |
| Equil. | $0.0500 - x$ | $x$ | $x$ |

$$K_b = \frac{x^2}{0.0500 - x} \approx \frac{x^2}{0.0500} = 7.1 \times 10^{-11}$$

$x = [OH^-] = 1.9 \times 10^{-6} \; M$;   pOH = 5.72;   pH = 8.28;   assumptions good.

Past the stoichiometric point, we have added more than 2.50 mmol of NaOH.  The pH will be determined by the excess $OH^-$ ion present.  An example of this calculation follows.

At 25.1 mL:  $OH^-$ added = 25.1 mL $\times \dfrac{0.100 \text{ mmol}}{\text{mL}} = 2.51$ mmol $OH^-$

2.50 mmol $OH^-$ neutralizes all the weak acid present.  The remainder is excess $OH^-$.

Excess $OH^- = 2.51 - 2.50 = 0.01$ mmol $OH^-$

$$[OH^-]_{excess} = \frac{0.01 \text{ mmol}}{(25.0 + 25.1) \text{ mL}} = 2 \times 10^{-4} \; M; \quad pOH = 3.7; \quad pH = 10.3$$

All results are listed in Table 8.1 at the end of the solution to Exercise 59.

59.     At beginning of the titration, only the weak base $NH_3$ is present.  As always, solve for the pH using the $K_b$ reaction for $NH_3$.

$$NH_3 + H_2O \rightleftharpoons NH_4^+ + OH^- \qquad K_b = 1.8 \times 10^{-5}$$

Initial     0.100 $M$                    0          ~0
Equil.      0.100 $-x$                   $x$          $x$

$$K_b = \frac{x^2}{0.100 - x} \approx \frac{x^2}{0.100} = 1.8 \times 10^{-5}$$

$x = [OH^-] = 1.3 \times 10^{-3} \ M$;  pOH = 2.89;  pH = 11.11;  assumptions good.

In the buffer region (4.0 – 24.9 mL),  we can use the Henderson-Hasselbalch equation:

$$K_a = \frac{1.0 \times 10^{-14}}{1.8 \times 10^{-5}} = 5.6 \times 10^{-10}; \ \ pK_a = 9.25; \ \ pH = 9.25 + \log \frac{[NH_3]}{[NH_4^+]}$$

We must determine the amounts of $NH_3$ and $NH_4^+$ present after the added $H^+$ reacts completely with the $NH_3$.  For example, after 8.0 mL HCl added:

$$\text{initial mmol } NH_3 \text{ present} = 25.0 \text{ mL} \times \frac{0.100 \text{ mmol}}{\text{mL}} = 2.50 \text{ mmol } NH_3$$

$$\text{mmol } H^+ \text{ added} = 8.0 \text{ mL} \times \frac{0.100 \text{ mmol}}{\text{mL}} = 0.80 \text{ mmol } H^+$$

Added $H^+$ reacts with $NH_3$ to completion:  $NH_3 + H^+ \rightarrow NH_4^+$

mmol $NH_3$ remaining = 2.50 – 0.80 = 1.70 mmol;  mmol $NH_4^+$ produced = 0.80 mmol

$$pH = 9.25 + \log \frac{1.70}{0.80} = 9.58 \qquad \text{(Mole ratios can be used since the total volume cancels.)}$$

Other points in the buffer region are calculated in similar fashion.  Results are summarized in Table 8.1 at the end of this problem.

At the stoichiometric point (25.0 mL $H^+$ added), just enough HCl has been added to convert all the weak base ($NH_3$) into its conjugate acid ($NH_4^+$).  Perform a weak acid calculation.  $[NH_4^+]_0 = 2.50 \text{ mmol}/50.0 \text{ mL} = 0.0500 \ M$

$$NH_4^+ \rightleftharpoons H^+ + NH_3 \qquad K_a = 5.6 \times 10^{-10}$$

Initial     0.0500 $M$                  0          0
Equil.      0.0500 $-x$                 $x$          $x$

$$5.6 \times 10^{-10} = \frac{x^2}{0.0500 - x} \approx \frac{x^2}{0.0500}, \ \ x = [H^+] = 5.3 \times 10^{-6} \ M; \ \ pH = 5.28; \ \ \text{assumptions good.}$$

Beyond the stoichiometric point, the pH is determined by the excess $H^+$.  For example, at 28.0 mL of $H^+$ added:

$$H^+ \text{ added} = 28.0 \text{ mL} \times \frac{0.100 \text{ mmol}}{\text{mL}} = 2.80 \text{ mmol } H^+$$

$$\text{Excess } H^+ = 2.80 \text{ mmol} - 2.50 \text{ mmol} = 0.30 \text{ mmol excess } H^+$$

$$[H^+]_{excess} = \frac{0.30 \text{ mmol}}{(25.0 + 28.0) \text{ mL}} = 5.7 \times 10^{-3} \text{ M}; \text{ pH} = 2.24$$

Table 8.1:  Summary of pH Results for Exercises 57 and 59 (Graph follows)

| Titrant mL | Exercise 57 | Exercise 59 |
|---|---|---|
| 0.0 | 2.43 | 11.11 |
| 4.0 | 3.14 | 9.97 |
| 8.0 | 3.53 | 9.58 |
| 12.5 | 3.86 | 9.25 |
| 20.0 | 4.46 | 8.65 |
| 24.0 | 5.24 | 7.87 |
| 24.5 | 5.6 | 7.6 |
| 24.9 | 6.3 | 6.9 |
| 25.0 | 8.28 | 5.28 |
| 25.1 | 10.3 | 3.7 |
| 26.0 | 11.29 | 2.71 |
| 28.0 | 11.75 | 2.24 |
| 30.0 | 11.96 | 2.04 |

61.    a.   This is a weak acid-strong base titration.  At the halfway point to equivalence, [weak acid] = [conjugate base], so pH = $pK_a$ (always for a weak acid-strong base titration).

pH = $-\log(6.4 \times 10^{-5})$ = 4.19

mmol $HC_7H_5O_2$ present = 100.0 mL × 0.10 $M$ = 10. mmol $HC_7H_5O_2$.  For the equivalence point, 10. mmol of $OH^-$ must be added.  The volume of $OH^-$ added to reach the equivalence point is:

$$10. \text{ mmol } OH^- \times \frac{1 \text{ mL}}{0.10 \text{ mmol } OH^-} = 1.0 \times 10^2 \text{ mL } OH^-$$

At the equivalence point, 10. mmol of $HC_7H_5O_2$ is neutralized by 10. mmol of $OH^-$ to produce 10. mmol of $C_7H_5O_2^-$.  This is a weak base.  The total volume of the solution is 100.0 mL + $1.0 \times 10^2$ mL = $2.0 \times 10^2$ mL.  Solving the weak base equilibrium problem:

$$C_7H_5O_2^- + H_2O \rightleftharpoons HC_7H_5O_2 + OH^- \quad K_b = \frac{1.0 \times 10^{-14}}{6.4 \times 10^{-5}} = 1.6 \times 10^{-10}$$

Initial    10. mmol/2.0 × $10^2$ mL          0          0
Equil.    0.050 − $x$                                    $x$         $x$

$$K_b = 1.6 \times 10^{-10} = \frac{x^2}{0.050 - x} \approx \frac{x^2}{0.050}, \quad x = [OH^-] = 2.8 \times 10^{-6} \, M$$

pOH = 5.55;  pH = 8.45;  assumptions good.

   b.   At the halfway point to equivalence for a weak base-strong acid titration, pH = $pK_a$ because [weak base] = [conjugate acid].

$$K_a = \frac{K_w}{K_b} = \frac{1.0 \times 10^{-14}}{5.6 \times 10^{-4}} = 1.8 \times 10^{-11}; \quad pH = pK_a = -\log(1.8 \times 10^{-11}) = 10.74$$

For the equivalence point (mmol acid added = mmol base present):

mmol $C_2H_5NH_2$ present = 100.0 mL × 0.10 $M$ = 10. mmol $C_2H_5NH_2$

$$\text{mL } H^+ \text{ added} = 10. \text{ mmol } H^+ \times \frac{1 \text{ mL}}{0.20 \text{ mmol } H^+} = 50. \text{ mL } H^+$$

The strong acid added completely converts the weak base into its conjugate acid. Therefore, at the equivalence point, $[C_2H_5NH_3^+]_0$ = 10. mmol/(100.0 + 50.) mL = 0.067 $M$.  Solving the weak acid equilibrium problem:

$$C_2H_5NH_3^+ \quad \rightleftharpoons \quad H^+ + \quad C_2H_5NH_2$$

Initial   $0.067\ M$           0           0
Equil.   $0.067 - x$           $x$          $x$

$$K_a = 1.8 \times 10^{-11} = \frac{x^2}{0.067 - x} \approx \frac{x^2}{0.067}, \quad x = [H^+] = 1.1 \times 10^{-6}\ M$$

pH = 5.96;  assumptions good.

c.  In a strong acid-strong base titration, the halfway point has no special significance other than that exactly one-half of the original amount of acid present has been neutralized.

mmol $H^+$ present = 100.0 mL $\times$ 0.50 $M$ = 50. mmol $H^+$

mL $OH^-$ added = 25 mmol $OH^-$ $\times \dfrac{1\ mL}{0.25\ mmol}$ = $1.0 \times 10^2$ mL $OH^-$

$$H^+ \qquad + \qquad OH^- \quad \rightarrow \quad H_2O$$

Before    50. mmol         25 mmol
After      25 mmol         0

$$[H^+]_{excess} = \frac{25\ mmol}{(100.0 + 1.0 \times 10^2)\ mL} = 0.13\ M; \ pH = 0.89$$

At the equivalence point of a strong acid-strong base titration, only neutral species are present ($Na^+$, $Cl^-$, and $H_2O$), so the pH = 7.00.

63.  a.  1.00 L $\times$ 0.100 mol/L = 0.100 mol HCl added to reach stoichiometric point.

The 10.00-g sample must have contained 0.100 mol of NaA.  $\dfrac{10.00\ g}{0.100\ mol}$ = 100. g/mol

b.  500.0 mL of HCl added represents the halfway point to equivalence.  Thus pH = $pK_a$ = 5.00 and $K_a = 1.0 \times 10^{-5}$.  At the equivalence point, enough $H^+$ has been added to convert all the $A^-$ present initially into HA.  The concentration of HA at the equivalence point is:

$$[HA]_0 = \frac{0.100\ mol}{1.10\ L} = 0.0909\ M$$

$$HA \quad \rightleftharpoons \quad H^+ + \quad A^- \qquad K_a = 1.0 \times 10^{-5}$$

Initial   $0.0909\ M$         0         0
Equil.   $0.0909 - x$        $x$        $x$

$$K_a = 1.0 \times 10^{-5} = \frac{x^2}{0.0909 - x} \approx \frac{x^2}{0.0909}$$

$x = 9.5 \times 10^{-4}\ M = [H^+];\ pH = 3.02;\ \text{assumptions good.}$

65.     $HIn \rightleftharpoons In^- + H^+ \quad K_a = \dfrac{[In^-][H^+]}{[HIn]} = 1.0 \times 10^{-9}$

a.   In a very acid solution, the HIn form dominates, so the solution will be yellow.

b.   The color change occurs when the concentration of the more dominant form is approximately ten times as great as the less dominant form of the indicator.

$\dfrac{[HIn]}{[In^-]} = \dfrac{10}{1};\ K_a = 1.0 \times 10^{-9} = \left(\dfrac{1}{10}\right)[H^+],\ [H^+] = 1 \times 10^{-8}\ M;\ pH = 8.0$ at color change

c.   This is way past the equivalence point (100.0 mL OH$^-$ added), so the solution is very basic and the In$^-$ form of the indicator dominates. The solution will be blue.

67.     pH > 5 for bromcresol green to be blue. pH < 8 for thymol blue to be yellow. The pH is between 5 and 8.

69.     When choosing an indicator, we want the color change of the indicator to occur approximately at the pH of the equivalence point. Since the pH generally changes very rapidly at the equivalence point, we don't have to be exact. This is especially true for strong acid-strong base titrations. The following are some indicators where the color change occurs at about the pH of the equivalence point.

| Exercise | pH at Eq. Pt. | Indicator |
| --- | --- | --- |
| 53 | 7.00 | bromthymol blue or phenol red |
| 55 | 8.79 | o-cresolphthalein or phenolphthalein |

71.

| Exercise | pH at Eq. Pt. | Indicator |
| --- | --- | --- |
| 57 | 8.28 | o-cresolphthalein or phenolphthalein |
| 59 | 5.28 | bromcresol green |

73.     The color of the indicator will change over the approximate range of pH = p$K_a$ ± 1 = 5.3 ± 1. Therefore, the useful pH range of methyl red where it changes color would be about 4.3 (red) to 6.3 (yellow). Note that at pH < 4.3, the HIn form of the indicator dominates, and the color of the solution is the color of HIn (red). At pH > 6.3, the In$^-$ form of the indicator dominates, and the color of the solution is the color of In$^-$ (yellow). In titrating a weak acid with base, we start off with an acidic solution with pH < 4.3, so the color would change from red to reddish orange at pH ≈ 4.3. In titrating a weak base with acid, the color change would be from yellow to yellowish orange at pH ≈ 6.3. Only a weak base-strong acid titration would have an acidic pH at the equivalence point, so only in this type of titration would the color change of methyl red indicate the approximate endpoint.

75.       100.0 mL $\times$ 0.0500 $M$ = 5.00 mmol $H_3X$ initially

a.  Because $K_{a_1} \gg K_{a_2} \gg K_{a_3}$,  pH initially is determined by $H_3X$ equilibrium reaction.

$$H_3X \quad \rightleftharpoons \quad H^+ \quad + \quad H_2X^-$$

| | | | |
|---|---|---|---|
| Initial | 0.0500 $M$ | ~0 | 0 |
| Equil. | 0.0500 $-$ $x$ | $x$ | $x$ |

$K_{a_1} = 1.0 \times 10^{-3} = \dfrac{x^2}{0.0500 - x} \approx \dfrac{x^2}{0.0500}$ , $x = 7.1 \times 10^{-3}$;  assumption poor.

Using the quadratic formula:

$$x^2 + (1.0 \times 10^{-3})x - 5.0 \times 10^{-5} = 0, \quad x = 6.6 \times 10^{-3} \ M = [H^+]; \ pH = 2.18$$

b.  1.00 mmol $OH^-$ added converts $H_3X$ into $H_2X^-$.  After this reaction goes to completion, 4.00 mmol $H_3X$ and 1.00 mmol $H_2X^-$ are in a total volume of 110.0 mL.  Solving the buffer problem:

$$H_3X \quad \rightleftharpoons \quad H^+ \quad + \quad H_2X^-$$

| | | | |
|---|---|---|---|
| Initial | 0.0364 $M$ | ~0 | 0.00909 $M$ |
| Equil. | 0.0364 $-$ $x$ | $x$ | 0.00909 $+$ $x$ |

$K_{a_1} = 1.0 \times 10^{-3} = \dfrac{x(0.00909 + x)}{0.0364 - x}$ ;  assumption that $x$ is small does not work here.

Using the quadratic formula and carrying extra sig. figs.
$$x^2 + (1.01 \times 10^{-2})x - 3.64 \times 10^{-5} = 0, \quad x = 2.8 \times 10^{-3} \ M = [H^+]; \ pH = 2.55$$

c.  2.50 mmol $OH^-$ added results in 2.50 mmol $H_3X$ and 2.50 mmol $H_2X^-$ after $OH^-$ reacts completely with $H_3X$.  This is the first halfway point to equivalence.  pH = p$K_{a_1}$ = 3.00; assumptions good (5% error).

d.  5.00 mmol $OH^-$ added results in 5.00 mmol $H_2X^-$ after $OH^-$ reacts completely with $H_3X$. This is the first stoichiometric point.

$$pH = \frac{pK_{a_1} + pK_{a_2}}{2} = \frac{3.00 + 7.00}{2} = 5.00$$

e.  6.00 mmol $OH^-$ added results in 4.00 mmol $H_2X^-$ and 1.00 mmol $HX^{2-}$ after $OH^-$ reacts completely with $H_3X$ and then reacts completely with $H_2X^-$.

Using the $H_2X^- \rightleftharpoons H^+ + HX^{2-}$ reaction:

$$pH = pK_{a_2} + \log \frac{[HX^{2-}]}{[H_2X^-]} = 7.00 - \log(1.00/4.00) = 6.40; \ \text{assumptions good.}$$

f.   7.50 mmol KOH added results in 2.50 mmol $H_2X^-$ and 2.50 mmol $HX^{2-}$ after $OH^-$ reacts completely. This is the second halfway point to equivalence.

pH = $pK_{a_2}$ = 7.00;   assumptions good.

g.   10.0 mmol $OH^-$ added results in 5.0 mmol $HX^{2-}$ after $OH^-$ reacts completely. This is the second stoichiometric point.

$$pH = \frac{pK_{a_2} + pK_{a_3}}{2} = \frac{7.00 + 12.00}{2} = 9.50$$

h.   12.5 mmol $OH^-$ added results in 2.5 mmol $HX^{2-}$ and 2.5 mmol $X^{3-}$ after $OH^-$ reacts completely with $H_3X$ first, then $H_2X^-$, and finally $HX^{2-}$. This is the third halfway point to equivalence. Usually pH = $pK_{a_3}$ but normal assumptions don't hold. We must solve for the pH exactly.

$[X^{3-}]$ = $[HX^{2-}]$ = 2.5 mmol/225.0 mL = $1.1 \times 10^{-2}$ $M$

| | $X^{3-}$ | + | $H_2O$ | $\rightleftharpoons$ | $HX^{2-}$ | + | $OH^-$ | $K_b = \dfrac{K_w}{K_{a_3}} = 1.0 \times 10^{-2}$ |
|---|---|---|---|---|---|---|---|---|
| Initial | 0.011 $M$ | | | | 0.011 $M$ | | 0 | |
| Equil. | 0.011 − x | | | | 0.011 + x | | x | |

$K_b = 1.0 \times 10^{-2} = \dfrac{x(0.011 + x)}{0.011 - x}$ ; using the quadratic formula:

$x^2 + (2.1 \times 10^{-2})x - 1.1 \times 10^{-4} = 0$,   $x = 4.3 \times 10^{-3}$ $M = OH^-$;   pH = 11.63

i.   15.0 mmol $OH^-$ added results in 5.0 mmol $X^{3-}$ after $OH^-$ reacts completely. This is the third stoichiometric point.

| | $X^{3-}$ | + | $H_2O$ | $\rightleftharpoons$ | $HX^{2-}$ | + | $OH^-$ | $K_b = \dfrac{K_w}{K_{a_3}} = 1.0 \times 10^{-2}$ |
|---|---|---|---|---|---|---|---|---|
| Initial | $\dfrac{5.0 \text{ mmol}}{250.0 \text{ mL}} = 0.020$ $M$ | | | | 0 | | 0 | |
| Equil. | 0.020 − x | | | | x | | x | |

$K_b = \dfrac{x^2}{0.020 - x}$,   $1.0 \times 10^{-2} \approx \dfrac{x^2}{0.020}$,   $x = 1.4 \times 10^{-2}$; assumption poor.

Using the quadratic formula: $x^2 + (1.0 \times 10^{-2})x - 2.0 \times 10^{-4} = 0$

$x = [OH^-] = 1.0 \times 10^{-2}$ $M$; pH = 12.00

j.  20.0 mmol $OH^-$ added results in 5.0 mmol $X^{3-}$ and 5.0 mmol $OH^-$ excess after $OH^-$ reacts completely.  Because $K_b$ for $X^{3-}$ is fairly large for a weak base, we have to worry about the $OH^-$ contribution from $X^{3-}$.

$$[X^{3-}] = [OH^-] = \frac{5.0 \text{ mmol}}{300.0 \text{ mL}} = 1.7 \times 10^{-2} \ M$$

$$X^{3-} \quad + \quad H_2O \quad \rightleftharpoons \quad\quad OH^- \quad + \quad HX^{2-}$$

| | | | |
|---|---|---|---|
| Initial | $1.7 \times 10^{-2} \ M$ | $1.7 \times 10^{-2} \ M$ | 0 |
| Equil. | $1.7 \times 10^{-2} - x$ | $1.7 \times 10^{-2} + x$ | $x$ |

$$K_b = 1.0 \times 10^{-2} = \frac{(1.7 \times 10^{-2} + x)x}{(1.7 \times 10^{-2} - x)}$$

Using the quadratic formula:  $x^2 + (2.7 \times 10^{-2})x - 1.7 \times 10^{-4} = 0$, $x = 5.3 \times 10^{-3} \ M$

$$[OH^-] = (1.7 \times 10^{-2}) + x = (1.7 \times 10^{-2}) + (5.3 \times 10^{-3}) = 2.2 \times 10^{-2} \ M; \ \text{pH} = 12.34$$

77.  $\dfrac{0.200 \text{ g}}{165.0 \text{ g/mol}} = 1.212 \times 10^{-3}$ mol = 1.212 mmol $H_3A$  (carrying extra sig. figs.)

a.  (10.50 mL)(0.0500 $M$) = 0.525 mmol $OH^-$ added;  $H_3A + OH^- \rightarrow H_2A^- + H_2O$; 1.212 − 0.525 = 0.687 mmol $H_3A$ remains after $OH^-$ reacts completely and 0.525 mmol $H_2A^-$ formed.  Solving the buffer problem using the $K_{a_1}$ reaction gives:

$$K_{a_1} = \frac{(10^{-3.73})\left(\dfrac{0.525}{60.50} + 10^{-3.73}\right)}{\dfrac{0.687}{60.50} - 10^{-3.73}} = 1.5 \times 10^{-4} \ (\text{p}K_{a_1} = 3.82)$$

First stoichiometric point:  $\text{pH} = \dfrac{\text{p}K_{a_1} + \text{p}K_{a_2}}{2} = 5.19 = \dfrac{3.82 + \text{p}K_{a_2}}{2}$

$\text{p}K_{a_2} = 6.56 \ (K_{a_2} = 2.8 \times 10^{-7})$

Second stoichiometric point:    $\text{pH} = \dfrac{\text{p}K_{a_2} + \text{p}K_{a_3}}{2}$, $\ 8.00 = \dfrac{6.56 + \text{p}K_{a_3}}{2}$

$\text{p}K_{a_3} = 9.44 \ (K_{a_3} = 3.6 \times 10^{-10})$

b.  1.212 mmol $H_3A$ = $(0.0500 \ M \ OH^-)(V_{OH^-})$, $V_{OH^-}$ = 24.2 mL; 24.2 mL of $OH^-$ are necessary to reach the first stoichiometric point.  It will require 60.5 mL to reach the third halfway point to equivalence, where pH = $\text{p}K_{a_3}$ = 9.44.  The pH at 59.0 mL of NaOH added should be a little lower than 9.44.

c.   59.0 mL of 0.0500 $M$ OH$^-$ = 2.95 mmol OH$^-$ added

$$H_3A \quad + \quad OH^- \quad \rightarrow \quad H_2A^- + H_2O$$

Before  1.212 mmol  2.95 mmol     0

After      0              1.74           1.212

$$H_2A^- \quad + \quad OH^- \quad \rightarrow \quad HA^{2-} + H_2O$$

Before  1.212          1.74            0

After      0              0.53           1.212

$$HA^{2-} \quad + \quad OH^- \quad \rightarrow \quad A^{3-} \quad + \quad H_2O$$

Before  1.212          0.53            0

After   0.68 mmol      0           0.53 mmol

Use the $K_{a_3}$ reaction to solve for the [H$^+$], making the normal assumptions.

$$K_{a_3} = 3.6 \times 10^{-10} = \frac{\left(\dfrac{0.53 \text{ mmol}}{109 \text{ mL}}\right)[H^+]}{\left(\dfrac{0.68 \text{ mmol}}{109 \text{ mL}}\right)}, \quad [H^+] = 4.6 \times 10^{-10} \ M; \ pH = 9.34$$

Assumptions good.

79.   a.   Na$^+$ is present in all solutions.  The added H$^+$ from HCl reacts completely with CO$_3^{2-}$ to convert it into HCO$_3^-$.  After all CO$_3^{2-}$ is reacted (after point C, the first equivalence point), H$^+$ then reacts completely with the next best base present, HCO$_3^-$.  Point E represents the second equivalence point.  The major species present at the various points after H$^+$ reacts completely follow.

A.  CO$_3^{2-}$, H$_2$O, Na$^+$                     B.  CO$_3^{2-}$, HCO$_3^-$, H$_2$O, Cl$^-$, Na$^+$

C.  HCO$_3^-$, H$_2$O, Cl$^-$, Na$^+$               D.  HCO$_3^-$, CO$_2$ (H$_2$CO$_3$), H$_2$O, Cl$^-$, Na$^+$

E.  CO$_2$ (H$_2$CO$_3$), H$_2$O, Cl$^-$, Na$^+$    F.  H$^+$ (excess), CO$_2$ (H$_2$CO$_3$), H$_2$O, Cl$^-$, Na$^+$

b.   <u>Point A</u> (initially):

$$CO_3^{2-} + H_2O \ \rightleftharpoons \ HCO_3^- + OH^- \qquad K_{b,CO_3^{2-}} = \frac{K_w}{K_{a_2}} = \frac{1.0 \times 10^{-14}}{4.8 \times 10^{-11}}$$

Initial   0.100 $M$                     0              ~0                    $K_b = 2.1 \times 10^{-4}$

Equil.   0.100 − $x$                    $x$             $x$

$$K_b = 2.1 \times 10^{-4} = \frac{[HCO_3^-][OH^-]}{[CO_3^{2-}]} = \frac{x^2}{0.100 - x} \approx \frac{x^2}{0.100}$$

$x = 4.6 \times 10^{-3} \ M = [OH^-]; \ pH = 11.66; \ \text{assumptions good.}$

Point B:  The first halfway point where $[CO_3^{2-}] = [HCO_3^-]$.

$pH = pK_{a_2} = -\log(4.8 \times 10^{-11}) = 10.32$;   assumptions good.

Point C:     First equivalence point (25.00 mL of 0.100 $M$ HCl added).  The amphoteric $HCO_3^-$ is the major acid-base species present.

$$pH = \frac{pK_{a_1} + pK_{a_2}}{2};\ \ pK_{a_1} = -\log(4.3 \times 10^{-7}) = 6.37$$

$$pH = \frac{6.37 + 10.32}{2} = 8.35$$

Point D:  The second halfway point where $[HCO_3^-] = [H_2CO_3]$.

$pH = pK_{a_1} = 6.37$;  assumptions good.

Point E:     This is the second equivalence point, where all of the $CO_3^{2-}$ present initially has been converted into $H_2CO_3$ by the added strong acid.  50.0 mL HCl added.

$[H_2CO_3] = 2.50$ mmol/75.0 mL $= 0.0333$ $M$

$$H_2CO_3 \quad \rightleftharpoons \quad H^+ \ + \ HCO_3^- \qquad K_{a_1} = 4.3 \times 10^{-7}$$

Initial   0.0333 $M$                0        0
Equil.    0.0333 $- x$              $x$        $x$

$$K_{a_1} = 4.3 \times 10^{-7} = \frac{x^2}{0.0333 - x} \approx \frac{x^2}{0.0333}$$

$x = [H^+] = 1.2 \times 10^{-4}$ $M$;  $pH = 3.92$;  assumptions good.

## Solubility Equilibria

81.     In our setups, $s$ = solubility in mol/L.  Because solids do not appear in the $K_{sp}$ expression, we do not need to worry about their initial or equilibrium amounts.

a.             $Ag_3PO_4(s) \quad \rightleftharpoons \quad 3\ Ag^+(aq) \ + \ PO_4^{3-}(aq)$

Initial                                      0                0
               $s$ mol/L of $Ag_3PO_4(s)$ dissolves to reach equilibrium
Change          $-s$         $\rightarrow$      $+3s$            $+s$
Equil.                                      $3s$              $s$

$$K_{sp} = 1.8 \times 10^{-18} = [Ag^+]^3[PO_4^{3-}] = (3s)^3(s) = 27s^4$$

$$27s^4 = 1.8 \times 10^{-18}, \quad s = (6.7 \times 10^{-20})^{1/4} = 1.6 \times 10^{-5} \text{ mol/L} = \text{molar solubility}$$

$$\frac{1.6 \times 10^{-5} \text{ mol Ag}_3\text{PO}_4}{L} \times \frac{418.7 \text{ g Ag}_3\text{PO}_4}{\text{mol Ag}_3\text{PO}_4} = 6.7 \times 10^{-3} \text{ g/L}$$

b.
$$CaCO_3(s) \rightleftharpoons Ca^{2+}(aq) + CO_3^{2-}(aq)$$

| | | | |
|---|---|---|---|
| Initial | $s$ = solubility (mol/L) | 0 | 0 |
| Equil. | | $s$ | $s$ |

$$K_{sp} = 8.7 \times 10^{-9} = [Ca^{2+}][CO_3^{2-}] = s^2, \quad s = 9.3 \times 10^{-5} \text{ mol/L}$$

$$\frac{9.3 \times 10^{-5} \text{ mol}}{L} \times \frac{100.1 \text{ g}}{\text{mol}} = 9.3 \times 10^{-3} \text{ g/L}$$

c.
$$Hg_2Cl_2(s) \rightleftharpoons Hg_2^{2+}(aq) + 2 Cl^-(aq)$$

| | | | |
|---|---|---|---|
| Initial | $s$ = solubility (mol/L) | 0 | 0 |
| Equil. | | $s$ | $2s$ |

$$K_{sp} = 1.1 \times 10^{-18} = [Hg_2^{2+}][Cl^-]^2 = (s)(2s)^2 = 4s^3, \quad s = 6.5 \times 10^{-7} \text{ mol/L}$$

$$\frac{6.5 \times 10^{-7} \text{ mol}}{L} \times \frac{472.1 \text{ g}}{\text{mol}} = 3.1 \times 10^{-4} \text{ g/L}$$

83.  In our setup, $s$ = solubility of the ionic solid in mol/L.  This is defined as the maximum amount of a salt that can dissolve.  Because solids do not appear in the $K_{sp}$ expression, we do not need to worry about their initial and equilibrium amounts.

a.
$$CaC_2O_4(s) \rightleftharpoons Ca^{2+}(aq) + C_2O_4^{2-}(aq)$$

| | | | |
|---|---|---|---|
| Initial | | 0 | 0 |
| | $s$ mol/L of CaC$_2$O$_4$(s) dissolves to reach equilibrium | | |
| Change | $-s$ $\rightarrow$ | $+s$ | $+s$ |
| Equil. | | $s$ | $s$ |

From the problem, $s = \dfrac{6.1 \times 10^{-3} \text{ g}}{L} \times \dfrac{1 \text{ mol CaC}_2\text{O}_4}{128.10 \text{ g}} = 4.8 \times 10^{-5} \text{ mol/L}.$

$$K_{sp} = [Ca^{2+}][C_2O_4^{2-}] = (s)(s) = s^2, \quad K_{sp} = (4.8 \times 10^{-5})^2 = 2.3 \times 10^{-9}$$

b.                   $BiI_3(s)$    $\rightleftharpoons$         $Bi^{3+}(aq)$    +    $3\ I^-(aq)$

|          |        |               |      |
|----------|--------|---------------|------|
| Initial  |        | 0             | 0    |

*s* mol/L of $BiI_3(s)$ dissolves to reach equilibrium

| Change | $-s$ | $\rightarrow$ | $+s$ | $+3s$ |
|--------|------|---------------|------|-------|
| Equil. |      |               | $s$  | $3s$  |

$K_{sp} = [Bi^{3+}][I^-]^3 = (s)(3s)^3 = 27s^4$,  $K_{sp} = 27(1.32 \times 10^{-5})^4 = 8.20 \times 10^{-19}$

85.                   $Ag_2C_2O_4(s)$   $\rightleftharpoons$   $2\ Ag^+(aq)$   +   $C_2O_4^{2-}(aq)$

| Initial | $s$ = solubility (mol/L) | 0   | 0   |
|---------|--------------------------|-----|-----|
| Equil.  |                          | $2s$ | $s$ |

From problem, $[Ag^+] = 2s = 2.2 \times 10^{-4}\ M$,  $s = 1.1 \times 10^{-4}\ M$

$K_{sp} = [Ag^+]^2[C_2O_4^{2-}] = (2s)^2(s) = 4s^3 = 4(1.1 \times 10^{-4})^3 = 5.3 \times 10^{-12}$

87.  a.  Because both solids dissolve to produce three ions in solution, we can compare values of $K_{sp}$ to determine relative solubility.  Because the $K_{sp}$ for $CaF_2$ is the smallest, $CaF_2(s)$ has the smallest molar solubility.

b.  We must calculate molar solubilities because each salt yields a different number of ions when it dissolves.

$Ca_3(PO_4)_2(s)$   $\rightleftharpoons$   $3\ Ca^{2+}(aq)$  +  $2\ PO_4^{3-}(aq)$      $K_{sp} = 1.3 \times 10^{-32}$

| Initial | $s$ = solubility (mol/L) | 0    | 0    |
|---------|--------------------------|------|------|
| Equil.  |                          | $3s$ | $2s$ |

$K_{sp} = [Ca^{2+}]^3[PO_4^{3-}]^2 = (3s)^3(2s)^2 = 108s^5$,  $s = (1.3 \times 10^{-32}/108)^{1/5} = 1.6 \times 10^{-7}$ mol/L

$FePO_4(s)$        $\rightleftharpoons$      $Fe^{3+}(aq)$   +   $PO_4^{3-}(aq)$    $K_{sp} = 1.0 \times 10^{-22}$

| Initial | $s$ = solubility (mol/L) | 0   | 0   |
|---------|--------------------------|-----|-----|
| Equil.  |                          | $s$ | $s$ |

$K_{sp} = [Fe^{3+}][PO_4^{3-}] = s^2$,  $s = \sqrt{1.0 \times 10^{-22}} = 1.0 \times 10^{-11}$ mol/L

$FePO_4$ has the smallest molar solubility.

89.  a.        $Fe(OH)_3(s)$    $\rightleftharpoons$     $Fe^{3+}$   +   $3\ OH^-$     pH = 7.0,  $[OH^-] = 1 \times 10^{-7}\ M$

| Initial |        | 0    | $1 \times 10^{-7}\ M$ |
|---------|--------|------|-----------------------|

*s* mol/L of $Fe(OH)_3(s)$ dissolves to reach equilibrium = molar solubility

| Change | $-s$ | $\rightarrow$ | $+s$ | $+3s$                   |
|--------|------|---------------|------|-------------------------|
| Equil. |      |               | $s$  | $1 \times 10^{-7} + 3s$ |

$$K_{sp} = 4 \times 10^{-38} = [Fe^{3+}][OH^-]^3 = (s)(1 \times 10^{-7} + 3s)^3 \approx s(1 \times 10^{-7})^3$$

$s = 4 \times 10^{-17}$ mol/L;  assumption good $(3s \ll 1 \times 10^{-7})$.

b.                  $Fe(OH)_3(s) \;\rightleftharpoons\; Fe^{3+} + 3\,OH^-$   pH = 5.0,  $OH^- = 1 \times 10^{-9}\ M$

| | | | |
|---|---|---|---|
| Initial | | 0 | $1 \times 10^{-9}\ M$ (Buffered) |

$s$ mol/L dissolves to reach equilibrium

| | | | | |
|---|---|---|---|---|
| Change | $-s$ | $\rightarrow$ | $+s$ | — (Assume no pH change in buffer) |
| Equil. | | | $s$ | $1 \times 10^{-9}$ |

$$K_{sp} = 4 \times 10^{-38} = [Fe^{3+}][OH^-]^3 = (s)(1 \times 10^{-9})^3,\ s = 4 \times 10^{-11}\ \text{mol/L} = \text{molar solubility}$$

c.                  $Fe(OH)_3(s) \;\rightleftharpoons\; Fe^{3+} + 3\,OH^-$   pH = 11.0,  $[OH^-] = 1 \times 10^{-3}\ M$

| | | | |
|---|---|---|---|
| Initial | | 0 | 0.001 $M$   (Buffered) |

$s$ mol/L dissolves to reach equilibrium

| | | | | |
|---|---|---|---|---|
| Change | $-s$ | $\rightarrow$ | $+s$ | – (Assume no pH change) |
| Equil. | | | $s$ | 0.001 |

$$K_{sp} = 4 \times 10^{-38} = [Fe^{3+}][OH^-]^3 = (s)(0.001)^3,\ s = 4 \times 10^{-29}\ \text{mol/L} = \text{molar solubility}$$

*Note*:  As $[OH^-]$ increases, solubility decreases.  This is the common ion effect.

91.                  $ZnS(s) \;\rightleftharpoons\; Zn^{2+} + S^{2-}$          $K_{sp} = [Zn^{2+}][S^{2-}]$

| | | | |
|---|---|---|---|
| Initial | $s$ = solubility (mol/L) | 0.050 $M$ | 0 |
| Equil. | | 0.050 + $s$ | $s$ |

$$K_{sp} = 2.5 \times 10^{-22} = (0.050 + s)(s) \approx (0.050)s,\ s = 5.0 \times 10^{-21}\ \text{mol/L};\ \text{assumption good.}$$

$$\text{Mass ZnS that dissolves} = 0.3000\ \text{L} \times \frac{5.0 \times 10^{-21}\ \text{mol ZnS}}{\text{L}} \times \frac{97.45\ \text{g ZnS}}{\text{mol}} = 1.5 \times 10^{-19}\ \text{g}$$

93.    If the anion in the salt can act as a base in water, then the solubility of the salt will increase as the solution becomes more acidic.  Added $H^+$ will react with the base, forming the conjugate acid.  As the basic anion is removed, more of the salt will dissolve to replenish the basic anion.  The salts with basic anions are $Ag_3PO_4$, $CaCO_3$, $CdCO_3$ and $Sr_3(PO_4)_2$. $Hg_2Cl_2$ and $PbI_2$ do not have any pH dependence because $Cl^-$ and $I^-$ are terrible bases (the conjugate bases of a strong acids).

$$Ag_3PO_4(s) + H^+(aq) \rightarrow 3\,Ag^+(aq) + HPO_4^{2-}(aq) \xrightarrow{\text{excess } H^+} 3\,Ag^+(aq) + H_3PO_4(aq)$$

$$CaCO_3(s) + H^+ \rightarrow Ca^{2+} + HCO_3^- \xrightarrow{\text{excess } H^+} Ca^{2+} + H_2CO_3\ [H_2O(l) + CO_2(g)]$$

$$CdCO_3(s) + H^+ \rightarrow Cd^{2+} + HCO_3^- \xrightarrow{\text{excess } H^+} Cd^{2+} + H_2CO_3 \ [H_2O(l) + CO_2(g)]$$

$$Sr_3(PO_4)_2(s) + 2\,H^+ \rightarrow 3\,Sr^{2+} + 2\,HPO_4^{2-} \xrightarrow{\text{excess } H^+} 3\,Sr^{2+} + 2\,H_3PO_4$$

95.    a.  AgF          b.  $Pb(OH)_2$          c.  $Sr(NO_2)_2$          d.  $Ni(CN)_2$

All these salts have anions that are bases.  The anions of the other choices are conjugate bases of strong acids.  They have no basic properties in water and, therefore, do not have solubilities that depend on pH.

97.    $[BaBr_2]_0 = \dfrac{0.150 \text{ L}(1.0 \times 10^{-4} \text{ mol}/\text{L})}{0.250 \text{ L}} = 6.0 \times 10^{-5}\,M$

$[K_2C_2O_4]_0 = \dfrac{0.100 \text{ L}(6.0 \times 10^{-4} \text{ mol}/\text{L})}{0.250 \text{ L}} = 2.4 \times 10^{-4}\,M$

$Q = [Ba^{2+}]_0[C_2O_4^{2-}]_0 = (6.0 \times 10^{-5})(2.4 \times 10^{-4}) = 1.5 \times 10^{-8}\,M$

Because $Q < K_{sp}$, $BaC_2O_4(s)$ will not precipitate.  The final concentration of ions will be:

$[Ba^{2+}] = 6.0 \times 10^{-5}\,M, \ [Br^-] = 1.2 \times 10^{-4}\,M$

$[K^+] = 4.8 \times 10^{-4}\,M, \ [C_2O_4^{2-}] = 2.4 \times 10^{-4}\,M$

99.    $Ag_3PO_4(s) \rightleftharpoons 3\,Ag^+(aq) + PO_4^{3-}(aq);$  when $Q$ is greater than $K_{sp}$, precipitation will occur. We will calculate the $[Ag^+]_0$ necessary for $Q = K_{sp}$.  Any $[Ag^+]_0$ greater than this calculated number will cause precipitation of $Ag_3PO_4(s)$.  In this problem, $[PO_4^{3-}]_0 = [Na_3PO_4]_0 = 1.0 \times 10^{-5}\,M$.

$K_{sp} = 1.8 \times 10^{-18};$  $Q = 1.8 \times 10^{-18} = [Ag^+]_0^3[PO_4^{3-}]_0 = [Ag^+]_0^3(1.0 \times 10^{-5}\,M)$

$[Ag^+]_0 = \left( \dfrac{1.8 \times 10^{-18}}{1.0 \times 10^{-5}} \right)^{1/3}, \ \ [Ag^+]_0 = 5.6 \times 10^{-5}\,M$

When $[Ag^+]_0 = [AgNO_3]_0$ is greater than $5.6 \times 10^{-5}\,M$, $Ag_3PO_4(s)$ will precipitate.

101.     a.

b.

c.

d.

## Complex Ion Equilibria

103.   $Hg^{2+}(aq) + 2 I^-(aq) \rightarrow HgI_2(s)$, orange ppt;  $HgI_2(s) + 2 I^-(aq) \rightarrow HgI_4^{2-}(aq)$

Soluble complex ion

105.   $\dfrac{65 \text{ g KI}}{0.500 \text{ L}} \times \dfrac{1 \text{ mol KI}}{166.0 \text{ g KI}} = 0.78 \ M \text{ KI}$

The formation constant for $HgI_4^{2-}$ is an extremely large number.  Because of this, we will let the $Hg^{2+}$ and $I^-$ ions present initially react to completion and then solve an equilibrium problem to determine the $Hg^{2+}$ concentration.

$$Hg^{2+} \quad + \quad 4 I^- \quad \rightleftharpoons \quad HgI_4^{2-} \qquad K = 1.0 \times 10^{30}$$

| | | | | |
|---|---|---|---|---|
| Before | 0.010 $M$ | 0.78 $M$ | 0 | |
| Change | −0.010 | −0.040  $\rightarrow$ | +0.010 | Reacts completely (K large) |
| After | 0 | 0.74 | 0.010 | New initial |

$x$ mol/L $HgI_4^{2-}$ dissociates to reach equilibrium

| | | | | |
|---|---|---|---|---|
| Change | +$x$ | +4$x$  $\leftarrow$ | −$x$ | |
| Equil. | $x$ | 0.74 + 4$x$ | 0.010 − $x$ | |

$K = 1.0 \times 10^{30} = \dfrac{[HgI_4^{2-}]}{[Hg^{2+}][I^-]^4} = \dfrac{(0.010 - x)}{(x)(0.74 + 4x)^4}$ ;  making usual assumptions:

$1.0 \times 10^{30} \approx \dfrac{(0.010)}{(x)(0.74)^4}$ ,  $x = [Hg^{2+}] = 3.3 \times 10^{-32} \ M$ ;  assumptions good.

*Note*: $3.3 \times 10^{-32}$ mol/L corresponds to one $Hg^{2+}$ ion per $5 \times 10^7$ L.  It is very reasonable to approach the equilibrium in two steps.  The reaction does essentially go to completion.

107.

$$Ag^+ + NH_3 \rightleftharpoons AgNH_3^+ \qquad K_1 = 2.1 \times 10^3$$
$$AgNH_3^+ + NH_3 \rightleftharpoons Ag(NH_3)_2^+ \qquad K_2 = 8.2 \times 10^3$$

$$\overline{Ag^+ + 2 NH_3 \rightleftharpoons Ag(NH_3)_2^+ \qquad K = K_1K_2 = 1.7 \times 10^7}$$

The initial concentrations are halved because equal volumes of the two solutions are mixed. Let the reaction go to completion since K is large; then solve an equilibrium problem.

$$Ag^+ \quad + \quad 2 NH_3 \quad \rightleftharpoons \quad Ag(NH_3)_2^+$$

| | | | |
|---|---|---|---|
| Before | 0.20 $M$ | 2.0 $M$ | 0 |
| After | 0 | 1.6 | 0.20 |
| Equil. | $x$ | $1.6 + 2x$ | $0.20 - x$ |

$$K = 1.7 \times 10^7 = \frac{[Ag(NH_3)_2^+]}{[Ag^+][NH_3]^2} = \frac{0.20 - x}{x(1.6 + 2x)^2} \approx \frac{0.20}{x(1.6)^2}, \ x = 4.6 \times 10^{-9} \ M;$$

assumptions good.

$$[Ag^+] = x = 4.6 \times 10^{-9} \ M; \quad [NH_3] = 1.6 \ M; \quad [Ag(NH_3)_2^+] = 0.20 \ M$$

Use either the $K_1$ or $K_2$ equilibrium expression to calculate $[AgNH_3^+]$.

$$AgNH_3^+ + NH_3 \rightleftharpoons Ag(NH_3)_2^+ \qquad K_2 = 8.2 \times 10^3$$

$$8.2 \times 10^3 = \frac{[Ag(NH_3)_2^+]}{[AgNH_3^+][NH_3]} = \frac{0.20}{[AgNH_3^+](1.6)}, \ [AgNH_3^+] = 1.5 \times 10^{-5} \ M$$

109.    a.

$$Cu(OH)_2 \rightleftharpoons Cu^{2+} + 2 OH^- \qquad\qquad K_{sp} = 1.6 \times 10^{-19}$$
$$Cu^{2+} + 4 NH_3 \rightleftharpoons Cu(NH_3)_4^{2+} \qquad\qquad K_f = 1.0 \times 10^{13}$$

$$\overline{Cu(OH)_2(s) + 4 NH_3(aq) \rightleftharpoons Cu(NH_3)_4^{2+}(aq) + 2 OH^-(aq) \qquad K = K_{sp}K_f = 1.6 \times 10^{-6}}$$

b.

$$Cu(OH)_2(s) \ + \ 4 NH_3 \rightleftharpoons Cu(NH_3)_4^{2+} + \ 2 OH^- \qquad K = 1.6 \times 10^{-6}$$

| | | | |
|---|---|---|---|
| Initial | | 5.0 $M$ | 0 | 0.0095 $M$ |
| | $s$ mol/L Cu(OH)$_2$ dissolves to reach equilibrium | | |
| Equil. | | 5.0 − 4$s$ | $s$ | 0.0095 + 2$s$ |

$$K = 1.6 \times 10^{-6} = \frac{[Cu(NH_3)_4^{2+}][OH^-]^2}{[NH_3]^4} = \frac{s(0.0095 + 2s)^2}{(5.0 - 4s)^4}$$

If $s$ is small: $1.6 \times 10^{-6} = \dfrac{s(0.0095)^2}{(5.0)^4}, \ s = 11. \ \text{mol/L}$

Assumptions are not good. We will solve the problem by successive approximations.

$$s_{calc} = \frac{1.6 \times 10^{-6}(5.0 - 4s_{guess})^4}{(0.0095 + 2s_{guess})^2} ; \quad \text{the results from six trials are:}$$

$$s_{guess}: \qquad 0.10, 0.050, 0.060, 0.055, 0.056$$

$$s_{calc}: \quad 1.6 \times 10^{-2}, 0.071, 0.049, 0.058, 0.056$$

Thus the solubility of $Cu(OH)_2$ is 0.056 mol/L in 5.0 $M\,NH_3$.

111.

$$AgBr(s) \rightleftharpoons Ag^+ + Br^- \qquad K_{sp} = 5.0 \times 10^{-13}$$
$$Ag^+ + 2\, S_2O_3^{2-} \rightleftharpoons Ag(S_2O_3)_2^{3-} \qquad K_f = 2.9 \times 10^{13}$$

$$AgBr(s) + 2\, S_2O_3^{2-} \rightleftharpoons Ag(S_2O_3)_2^{3-} + Br^- \qquad K = K_{sp} \times K_f = 14.5 \qquad \text{(Carry extra sig. figs.)}$$

|  | $AgBr(s)$ | $+$ | $2\, S_2O_3^{2-}$ | $\rightleftharpoons$ | $Ag(S_2O_3)_2^{3-}$ | $+$ | $Br^-$ |
|---|---|---|---|---|---|---|---|
| Initial |  |  | 0.500 $M$ |  | 0 |  | 0 |

$s$ mol/L $AgBr(s)$ dissolves to reach equilibrium

|  |  |  |  |  |  |  |  |
|---|---|---|---|---|---|---|---|
| Change | $-s$ |  | $-2s$ | $\rightarrow$ | $+s$ |  | $+s$ |
| Equil. |  |  | $0.500 - 2s$ |  | $s$ |  | $s$ |

$$K = \frac{s^2}{(0.500 - 2s)^2} = 14.5; \quad \text{taking the square root of both sides:}$$

$$\frac{s}{0.500 - 2s} = 3.81, \quad s = 1.91 - (7.62)s, \quad s = 0.222 \text{ mol/L}$$

$$1.00 \text{ L} \times \frac{0.222 \text{ mol AgBr}}{L} \times \frac{187.8 \text{ g AgBr}}{\text{mol AgBr}} = 41.7 \text{ g AgBr} = 42 \text{ g AgBr}$$

113.    Test tube 1: Added $Cl^-$ reacts with $Ag^+$ to form a silver chloride precipitate.  The net ionic equation is $Ag^+(aq) + Cl^-(aq) \rightarrow AgCl(s)$.  Test tube 2: Added $NH_3$ reacts with $Ag^+$ ions to form a soluble complex ion, $Ag(NH_3)_2^+$.  As this complex ion forms, $Ag^+$ is removed from the solution, which causes the $AgCl(s)$ to dissolve.  When enough $NH_3$ is added, all the silver chloride precipitate will dissolve.  The equation is $AgCl(s) + 2\, NH_3(aq) \rightarrow Ag(NH_3)_2^+(aq) + Cl^-(aq)$.  Test tube 3:  Added $H^+$ reacts with the weak base, $NH_3$, to form $NH_4^+$.  As $NH_3$ is removed from the $Ag(NH_3)_2^+$ complex ion, $Ag^+$ ions are released to solution and can then react with $Cl^-$ to re-form $AgCl(s)$.   The equations are $Ag(NH_3)_2^+(aq) + 2\, H^+(aq) \rightarrow Ag^+(aq) + 2\, NH_4^+(aq)$, and $Ag^+(aq) + Cl^-(aq) \rightarrow AgCl(s)$.

## Additional Exercises

115.  a.  The optimum pH for a buffer is when pH = $pK_a$. At this pH a buffer will have equal neutralization capacity for both added acid and base. As shown next, because the $pK_a$ for TRISH$^+$ is 8.1, the optimal buffer pH is about 8.1.

$$K_b = 1.19 \times 10^{-6}; \quad K_a = K_w/K_b = 8.40 \times 10^{-9}; \quad pK_a = -\log(8.40 \times 10^{-9}) = 8.076$$

b.  $pH = pK_a + \log \dfrac{[TRIS]}{[TRISH^+]}, \quad 7.00 = 8.076 + \log \dfrac{[TRIS]}{[TRISH^+]}$

$$\frac{[TRIS]}{[TRISH^+]} = 10^{-1.08} = 0.083 \quad (\text{at pH} = 7.00)$$

$$9.00 = 8.076 + \log \frac{[TRIS]}{[TRISH^+]}, \quad \frac{[TRIS]}{[TRISH^+]} = 10^{0.92} = 8.3 \quad (\text{at pH} = 9.00)$$

c.  $\dfrac{50.0 \text{ g TRIS}}{2.0 \text{ L}} \times \dfrac{1 \text{ mol}}{121.14 \text{ g}} = 0.206\, M = 0.21\, M = [TRIS]$

$\dfrac{65.0 \text{ g TRISHCl}}{2.0 \text{ L}} \times \dfrac{1 \text{ mol}}{157.60 \text{ g}} = 0.206\, M = 0.21\, M = [TRISHCl] = [TRISH^+]$

$pH = pK_a + \log \dfrac{[TRIS]}{[TRISH^+]} = 8.076 + \log \dfrac{(0.21)}{(0.21)} = 8.08$

The amount of H$^+$ added from HCl is: $(0.50 \times 10^{-3} \text{ L}) \times 12 \text{ mol/L} = 6.0 \times 10^{-3} \text{ mol H}^+$

The H$^+$ from HCl will convert TRIS into TRISH$^+$. The reaction is:

|          | TRIS       | H$^+$                                    |               | TRISH$^+$ |                    |
|----------|-----------|------------------------------------------|---------------|-----------|--------------------|
|          | TRIS       | + H$^+$                                  | $\rightarrow$ | TRISH$^+$ |                    |
| Before   | 0.21 $M$   | $\dfrac{6.0 \times 10^{-3}}{0.2005} = 0.030\, M$ |        | 0.21 $M$  |                    |
| Change   | −0.030     | −0.030                                   | $\rightarrow$ | +0.030    | Reacts completely  |
| After    | 0.18       | 0                                        |               | 0.24      |                    |

Now use the Henderson-Hasselbalch equation to solve this buffer problem.

$$pH = 8.076 + \log\left(\frac{0.18}{0.24}\right) = 7.95$$

117.  A best buffer is when pH $\approx$ $pK_a$; these solutions have about equal concentrations of weak acid and conjugate base. Therefore, choose combinations that yield a buffer where pH $\approx$ $pK_a$; that is, look for acids whose $pK_a$ is closest to the pH.

a.  Potassium fluoride + HCl will yield a buffer consisting of HF ($pK_a = 3.14$) and $F^-$.

b.  Benzoic acid + NaOH will yield a buffer consisting of benzoic acid ($pK_a = 4.19$) and benzoate anion.

c.  Sodium acetate + acetic acid ($pK_a = 4.74$) is the best choice for pH = 5.0 buffer since acetic acid has a $pK_a$ value closest to 5.0.

d.  HOCl and NaOH: This is the best choice to produce a conjugate acid-base pair with pH = 7.0.  This mixture would yield a buffer consisting of HOCl ($pK_a = 7.46$) and $OCl^-$. Actually, the best choice for a pH = 7.0 buffer is an equimolar mixture of ammonium chloride and sodium acetate.  $NH_4^+$ is a weak acid ($K_a = 5.6 \times 10^{-10}$), and $C_2H_3O_2^-$ is a weak base ($K_b = 5.6 \times 10^{-10}$).  A mixture of the two will give a buffer at pH = 7.0 because the weak acid and weak base are the same strengths ($K_a$ for $NH_4^+ = K_b$ for $C_2H_3O_2^-$). $NH_4C_2H_3O_2$ is commercially available, and its solutions are used for pH = 7.0 buffers.

e.  Ammonium chloride + NaOH will yield a buffer consisting of $NH_4^+$ ($pK_a = 9.26$) and $NH_3$.

119.  NaOH added = $50.0 \text{ mL} \times \dfrac{0.500 \text{ mmol}}{\text{mL}} = 25.0$ mmol NaOH

NaOH left unreacted = $31.92 \text{ mL HCl} \times \dfrac{0.289 \text{ mmol}}{\text{mL}} \times \dfrac{1 \text{ mmol NaOH}}{\text{mmol HCl}} = 9.22$ mmol NaOH

NaOH reacted with aspirin = $25.0 - 9.22 = 15.8$ mmol NaOH

$15.8 \text{ mmol NaOH} \times \dfrac{1 \text{ mmol aspirin}}{2 \text{ mmol NaOH}} \times \dfrac{180.2 \text{ mg}}{\text{mmol}} = 1420 \text{ mg} = 1.42$ g aspirin

Purity = $\dfrac{1.42 \text{ g}}{1.427 \text{ g}} \times 100 = 99.5\%$

Here, a strong base is titrated by a strong acid.  The equivalence point will be at pH = 7.0. Bromthymol blue would be the best indicator since it changes color at pH $\approx$ 7 (from base color to acid color).  See Fig. 8.8 of the text.

121.  At the equivalence point, $P^{2-}$ is the major species.  $P^{2-}$ is a weak base in water because it is the conjugate base of a weak acid.

$$P^{2-} \quad + \quad H_2O \quad \rightleftharpoons \quad HP^- \quad + \quad OH^-$$

| | | | | |
|---|---|---|---|---|
| Initial | $\dfrac{0.5 \text{ g}}{0.1 \text{ L}} \times \dfrac{1 \text{ mol}}{204.2 \text{ g}} = 0.024\ M$ | | 0 | ~0 | (carry extra sig. fig.) |
| Equil. | $0.024 - x$ | | $x$ | $x$ |

$$K_b = \frac{[HP^-][OH^-]}{P^{2-}} = \frac{K_w}{K_a} = \frac{1.0 \times 10^{-14}}{10^{-5.51}}, \ 3.2 \times 10^{-9} = \frac{x^2}{0.024 - x} \approx \frac{x^2}{0.024}$$

$x = [OH^-] = 8.8 \times 10^{-6} \ M; \ pOH = 5.1; \ pH = 8.9; \ $ assumptions good.

Phenolphthalein would be the best indicator for this titration because it changes color at $pH \approx 9$ (from acid color to base color).

123.    a.                 $Pb(OH)_2(s) \ \rightleftharpoons \ Pb^{2+} \ + \ 2 \ OH^-$

| Initial | $s$ = solubility (mol/L) | 0 | $1.0 \times 10^{-7} \ M$ from water |
|---|---|---|---|
| Equil. | | $s$ | $1.0 \times 10^{-7} + 2s$ |

$K_{sp} = 1.2 \times 10^{-15} = [Pb^{2+}][OH^-]^2 = s(1.0 \times 10^{-7} + 2s)^2 \approx s(2s^2) = 4s^3$

$s = [Pb^{2+}] = 6.7 \times 10^{-6} \ M; \ $ assumption is good by the 5% rule.

b.                 $Pb(OH)_2(s) \ \rightleftharpoons \ Pb^{2+} \ + \ 2 \ OH^-$

| Initial | | 0 | $0.10 \ M$    $pH = 13.00, \ [OH^-] = 0.10 \ M$ |
|---|---|---|---|
| | $s$ mol/L $Pb(OH)_2(s)$ dissolves to reach equilibrium | | |
| Equil. | | $s$ | $0.10$    (Buffered solution) |

$1.2 \times 10^{-15} = (s)(0.10)^2, \ s = [Pb^{2+}] = 1.2 \times 10^{-13} \ M$

c.  We need to calculate the $Pb^{2+}$ concentration in equilibrium with $EDTA^{4-}$. Since K is large for the formation of $PbEDTA^{2-}$, let the reaction go to completion, and then solve an equilibrium problem to get the $Pb^{2+}$ concentration.

$$Pb^{2+} \ + \ EDTA^{4-} \ \rightleftharpoons \ PbEDTA^{2-} \quad K = 1.1 \times 10^{18}$$

| Before | $0.010 \ M$ | $0.050 \ M$ | 0 | |
|---|---|---|---|---|
| | $0.010$ mol/L $Pb^{2+}$ reacts completely (large K) | | | |
| Change | $-0.010$ | $-0.010$    $\rightarrow$ | $+0.010$ | Reacts completely |
| After | 0 | $0.040$ | $0.010$ | New initial |
| | $x$ mol/L $PbEDTA^{2-}$ dissociates to reach equilibrium | | | |
| Equil. | $x$ | $0.040 + x$ | $0.010 - x$ | |

$$1.1 \times 10^{18} = \frac{(0.010 - x)}{(x)(0.040 + x)} \approx \frac{0.010}{x(0.040)}, \ x = [Pb^{2+}] = 2.3 \times 10^{-19} \ M; \ \text{assumptions good.}$$

Now calculate the solubility quotient for $Pb(OH)_2$ to see if precipitation occurs. The concentration of $OH^-$ is $0.10 \ M$ since we have a solution buffered at $pH = 13.00$.

$Q = [Pb^{2+}]_0[OH^-]_0^2 = (2.3 \times 10^{-19})(0.10)^2 = 2.3 \times 10^{-21} < K_{sp} \ (1.2 \times 10^{-15})$

$Pb(OH)_2(s)$ will not form since Q is less than $K_{sp}$.

125.   $HC_2H_3O_2 \rightleftharpoons H^+ + C_2H_3O_2^-$;  let $C_0$ = initial concentration of $HC_2H_3O_2$

From normal weak acid setup:  $K_a = 1.8 \times 10^{-5} = \dfrac{[H^+][C_2H_3O_2^-]}{[HC_2H_3O_2]} = \dfrac{[H^+]^2}{C_0 - [H^+]}$

$[H^+] = 10^{-2.68} = 2.1 \times 10^{-3}\ M;\ 1.8 \times 10^{-5} = \dfrac{(2.1 \times 10^{-3})^2}{C_0 - (2.1 \times 10^{-3})},\ C_0 = 0.25\ M$

$25.0\ mL \times 0.25\ mmol/mL = 6.3\ mmol\ HC_2H_3O_2$

Need 6.3 mmol KOH = $V_{KOH} \times 0.0975$ mmol/mL, $V_{KOH} = 65$ mL

127.   $0.400\ mol/L \times V_{NH_3} = mol\ NH_3 = mol\ NH_4^+$ after reaction with HCl at the equivalence point.

At the equivalence point:  $[NH_4^+]_0 = \dfrac{mol\ NH_4^+}{total\ volume} = \dfrac{0.400 \times V_{NH_3}}{1.50 \times V_{NH_3}} = 0.267\ M$

|  | $NH_4^+$ | $\rightleftharpoons$ | $H^+$ | $+$ | $NH_3$ |
|---|---|---|---|---|---|
| Initial | 0.267 $M$ | | 0 | | 0 |
| Equil. | $0.267 - x$ | | $x$ | | $x$ |

$K_a = \dfrac{K_w}{K_b} = \dfrac{1.0 \times 10^{-14}}{1.8 \times 10^{-5}},\ 5.6 \times 10^{-10} = \dfrac{x^2}{0.267 - x} \approx \dfrac{x^2}{0.267}$

$x = [H^+] = 1.2 \times 10^{-5}\ M;\ pH = 4.92;$  assumptions good.

129.   $HA + OH^- \rightarrow A^- + H_2O$, where HA = acetylsalicylic acid

mmol HA present = $27.36\ mL\ OH^- \times \dfrac{0.5106\ mmol\ OH^-}{mL\ OH^-} \times \dfrac{1\ mmol\ HA}{mmol\ OH^-} = 13.97\ mmol\ HA$

Molar mass of HA = $\dfrac{2.51\ g\ HA}{13.97 \times 10^{-3}\ mol\ HA} = 180.\ g/mol$

To determine the $K_a$ value, use the pH data. After complete neutralization of acetylsalicylic acid by $OH^-$, we have 13.97 mmol of $A^-$ produced from the neutralization reaction. $A^-$ will react completely with the added $H^+$ and re-form acetylsalicylic acid HA.

mmol $H^+$ added = $15.44\ mL \times \dfrac{0.4524\ mmol\ H^+}{mL} = 6.985\ mmol\ H^+$

|  | $A^-$ | $+$ | $H^+$ | $\rightarrow$ | HA |  |
|---|---|---|---|---|---|---|
| Before | 13.97 mmol | | 6.985 mmol | | 0 | |
| Change | $-6.985$ | | $-6.985$ | $\rightarrow$ | $+6.985$ | Reacts completely |
| After | 6.985 mmol | | 0 | | 6.985 mmol | |

We have back titrated this solution to the halfway point to equivalence, where pH = pK$_a$ (assuming HA is a weak acid). This is true because after H$^+$ reacts completely, equal milliliters of HA and A$^-$ are present, which only occurs at the halfway point to equivalence. Assuming acetylsalicylic acid is a weak acid, then pH = pK$_a$ = 3.48. K$_a$ = $10^{-3.48}$ = 3.3 × $10^{-4}$.

131.    K$_{a_3}$ is so small (4.8 × $10^{-13}$) that a break is not seen at the third stoichiometric point.

## Challenge Problems

133.    mmol HC$_3$H$_5$O$_2$ present initially = 45.0 mL × $\dfrac{0.750 \text{ mmol}}{\text{mL}}$ = 33.8 mmol HC$_3$H$_5$O$_2$

mmol C$_3$H$_5$O$_2^-$ present initially = 55.0 mL × $\dfrac{0.700 \text{ mmol}}{\text{mL}}$ = 38.5 mmol C$_3$H$_5$O$_2^-$

The initial pH of the buffer is:

$$\text{pH} = \text{pK}_a + \log \frac{[\text{C}_3\text{H}_5\text{O}_2^-]}{[\text{HC}_3\text{H}_5\text{O}_2]} = -\log(1.3 \times 10^{-5}) + \log \frac{\dfrac{38.5 \text{ mmol}}{100.0 \text{ mL}}}{\dfrac{33.8 \text{ mmol}}{100.0 \text{ mL}}} = 4.89 + \log \frac{38.5}{33.8} = 4.95$$

*Note*: Because the buffer components are in the same volume of solution, we can use the mole (or millimole) ratio in the Henderson-Hasselbalch equation to solve for pH instead of using the concentration ratio of [C$_3$H$_5$O$_2^-$]/[HC$_3$H$_5$O$_2$]. The total volume always cancels for buffer solutions.

When NaOH is added, the pH will increase, and the added OH$^-$ will convert HC$_3$H$_5$O$_2$ into C$_3$H$_5$O$_2^-$. The pH after addition of OH$^-$ increases by 2.5%, so the resulting pH is:

4.95 + 0.025(4.95) = 5.07

At this pH, a buffer solution still exists, and the millimole ratio between C$_3$H$_5$O$_2^-$ and HC$_3$H$_5$O$_2$ is:

$$\text{pH} = \text{pK}_a + \log \frac{\text{mmol C}_3\text{H}_5\text{O}_2^-}{\text{mmol HC}_3\text{H}_5\text{O}_2}, \quad 5.07 = 4.89 + \log \frac{\text{mmol C}_3\text{H}_5\text{O}_2^-}{\text{mmol HC}_3\text{H}_5\text{O}_2}$$

$$\frac{\text{mmol C}_3\text{H}_5\text{O}_2^-}{\text{mmol HC}_3\text{H}_5\text{O}_2} = 10^{0.18} = 1.5$$

Let $x$ = mmol OH$^-$ added to increase pH to 5.07. Because OH$^-$ will essentially react to completion with HC$_3$H$_5$O$_2$, the setup to the problem using millimoles is:

|            | HC$_3$H$_5$O$_2$ | + | OH$^-$   | → | C$_3$H$_5$O$_2^-$ |                   |
| ---------- | ---------------- | - | -------- | - | ----------------- | ----------------- |
| Before     | 33.8 mmol        |   | $x$ mmol |   | 38.5 mmol         |                   |
| Change     | $-x$             |   | $-x$     | → | $+x$              | Reacts completely |
| After      | $33.8 - x$       |   | 0        |   | $38.5 + x$        |                   |

$$\frac{\text{mmol } C_3H_5O_2^-}{\text{mmol } HC_3H_5O_2} = 1.5 = \frac{38.5 + x}{33.8 - x}, \quad 1.5(33.8 - x) = 38.5 + x, \quad x = 4.9 \text{ mmol } OH^- \text{ added}$$

The volume of NaOH necessary to raise the pH by 2.5% is:

$$4.9 \text{ mmol NaOH} \times \frac{1 \text{ mL}}{0.10 \text{ mmol NaOH}} = 49 \text{ mL}$$

49 mL of 0.10 $M$ NaOH must be added to increase the pH by 2.5%.

135.   a.   Best acid will react with the best base present, so the dominate equilibrium is:

$$NH_4^+ + X^- \rightleftharpoons NH_3 + HX \qquad K_{eq} = \frac{[NH_3][HX]}{[NH_4^+][X^-]} = \frac{K_{a, NH_4^+}}{K_{a, HX}}$$

Because initially $[NH_4^+]_0 = [X^-]_0$ and $[NH_3]_0 = [HX]_0 = 0$, at equilibrium $[NH_4^+] = [X^-]$ and $[NH_3] = [HX]$.   Therefore:

$$K_{eq} = \frac{K_{a, NH_4^+}}{K_{a, HX}} = \frac{[HX]^2}{[X^-]^2}$$

The $K_a$ expression for HX is:   $K_{a, HX} = \frac{[H^+][X^-]}{[HX]}, \quad \frac{[HX]}{[X^-]} = \frac{[H^+]}{K_{a, HX}}$

Substituting into the $K_{eq}$ expression:   $K_{eq} = \frac{K_{a, NH_4^+}}{K_{a, HX}} = \frac{[HX]^2}{[X^-]^2} = \left(\frac{[H^+]}{K_{a, HX}}\right)^2$

Rearranging:   $[H^+]^2 = K_{a, NH_4^+} \times K_{a, HX}$,   or taking the $-\log$ of both sides:

$$pH = \frac{pK_{a, NH_4^+} + pK_{a, HX}}{2}$$

b.   Ammonium formate = $NH_4(HCO_2)$

$$K_{a, NH_4^+} = \frac{1.0 \times 10^{-14}}{1.8 \times 10^{-5}} = 5.6 \times 10^{-10}, \text{ pK}_a = 9.25; \quad K_{a, HCO_2H} = 1.8 \times 10^{-4}, \quad pK_a = 3.74$$

$$pH = \frac{pK_{a, NH_4^+} + pK_{a, HCO_2H}}{2} = \frac{9.25 + 3.74}{2} = 6.50$$

Ammonium acetate = $NH_4(C_2H_3O_2)$; $K_{a, HC_2H_3O_2} = 1.8 \times 10^{-5}$; $pK_a = 4.74$

$$pH = \frac{9.25 + 4.74}{2} = 7.00$$

Ammonium bicarbonate = $NH_4(HCO_3)$; $K_{a, H_2CO_3} = 4.3 \times 10^{-7}$; $pK_a = 6.37$

$$pH = \frac{9.25 + 6.37}{2} = 7.81$$

c.  $NH_4^+(aq) + OH^-(aq) \rightarrow NH_3(aq) + H_2O(l)$; $C_2H_3O_2^-(aq) + H^+(aq) \rightarrow HC_2H_3O_2(aq)$

137.  a.   $\begin{array}{lll} CuBr(s) & \rightleftharpoons \ Cu^+ + Br^- & K_{sp} = 1.0 \times 10^{-5} \\ Cu^+ + 3\ CN^- & \rightleftharpoons \ Cu(CN)_3{}^{2-} & K_f = 1.0 \times 10^{11} \end{array}$

$CuBr(s) + 3\ CN \rightleftharpoons Cu(CN)_3{}^{2-} + Br^-$   $K = 1.0 \times 10^6$

Because K is large, assume that enough CuBr(s) dissolves to completely use up the 1.0 $M$ $CN^-$; then solve the back equilibrium problem to determine the equilibrium concentrations.

$$CuBr(s) + 3\ CN^- \rightleftharpoons Cu(CN)_3{}^{2-} + Br^-$$

| Before | $x$ | 1.0 $M$ | 0 | 0 |
|---|---|---|---|---|

$x$ mol/L of CuBr(s) dissolves to react completely with 1.0 $M$ $CN^-$

| Change | $-x$ | $-3x$ $\rightarrow$ | $+x$ | $+x$ |
|---|---|---|---|---|
| After | 0 | $1.0 - 3x$ | $x$ | $x$ |

For reaction to go to completion, $1.0 - 3x = 0$ and $x = 0.33$ mol/L.  Now solve the back equilibrium problem.

$$CuBr(s) + 3\ CN^- \rightleftharpoons Cu(CN)_3{}^{2-} + Br^-$$

| Initial | | 0 | 0.33 $M$ | 0.33 $M$ |
|---|---|---|---|---|

Let $y$ mol/L of $Cu(CN)_3{}^{2-}$ react to reach equilibrium.

| Change | | $+3y$ $\leftarrow$ | $-y$ | $-y$ |
|---|---|---|---|---|
| Equil. | | 3y | $0.33 - y$ | $0.33 - y$ |

$$K = 1.0 \times 10^6 = \frac{(0.33 - y)^2}{(3y)^3} \approx \frac{(0.33)^2}{27y^3},\ y = 1.6 \times 10^{-3}\ M;\ \text{assumptions good.}$$

Of the initial 1.0 $M$ $CN^-$, only $3(1.6 \times 10^{-3}) = 4.8 \times 10^{-3}\ M$ is present at equilibrium. Indeed, enough CuBr(s) did dissolve to essentially remove the initial 1.0 $M$ $CN^-$. This amount, 0.33 mol/L, is the solubility of CuBr(s) in 1.0 $M$ NaCN.

b.   $[Br^-] = 0.33 - y = 0.33 - 1.6 \times 10^{-3} = 0.33 \; M$

c.   $[CN^-] = 3y = 3(1.6 \times 10^{-3}) = 4.8 \times 10^{-3} \; M$

139.   a.                                  $SrF_2(s) \quad \rightleftharpoons \quad Sr^{2+}(aq) \; + \; 2 \, F^-(aq)$

Initial                                          0               0

$s$ mol/L SrF$_2$ dissolves to reach equilibrium

Equil.                                           $s$              $2s$

$[Sr^{2+}][F^-]^2 = K_{sp} = 7.9 \times 10^{-10} = 4s^3$,  $s = 5.8 \times 10^{-4}$ mol/L in pure water

b.   Greater, because some of the F$^-$ would react with water:

$$F^- + H_2O \rightleftharpoons HF + OH^- \quad K_b = \frac{K_w}{K_{a,\,HF}} = 1.4 \times 10^{-11}$$

This lowers the concentration of F$^-$,  forcing more SrF$_2$ to dissolve.

c.   $SrF_2(s) \rightleftharpoons Sr^{2+} + 2 \, F^- \quad K_{sp} = 7.9 \times 10^{-10} = [Sr^{2+}][F^-]^2$

Let $s$ = solubility = $[Sr^{2+}]$;  then $2s$ = total F$^-$ concentration.

Since F$^-$ is a weak base, some of the F$^-$ is converted into HF.  Therefore:

total F$^-$ concentration = $2s = [F^-] + [HF]$

$$HF \rightleftharpoons H^+ + F^- \quad K_a = 7.2 \times 10^{-4} = \frac{[H^+][F^-]}{[HF]} = \frac{1.0 \times 10^{-2}[F^-]}{[HF]} \text{ (since pH = 2.00 buffer)}$$

$$7.2 \times 10^{-2} = \frac{[F^-]}{[HF]}, \quad [HF] = 14[F^-]; \text{ Solving:}$$

$[Sr^{2+}] = s$;  $2s = [F^-] + [HF] = [F^-] + 14[F^-]$,  $2s = 15[F^-]$,  $[F^-] = 2s/15$

$$K_{sp} = 7.9 \times 10^{-10} = [Sr^{2+}][F^-]^2 = (s)\left(\frac{2s}{15}\right)^2, \quad s = 3.5 \times 10^{-3} \text{ mol/L in pH = 2.00 solution}$$

141.   a.   $Al(OH)_3(s) \rightleftharpoons Al^{3+} + 3 \, OH^-$;  $Al(OH)_3(s) + OH^- \rightleftharpoons Al(OH)_4^-$

$S$ = solubility = total Al$^{3+}$ concentration = $[Al^{3+}] + [Al(OH)_4^-]$

$$[Al^{3+}] = \frac{K_{sp}}{[OH^-]^3} = K_{sp} \times \frac{[H^+]^3}{K_w^{\,3}}, \text{ because } [OH^-]^3 = (K_w/[H^+])^3$$

$$\frac{[\text{Al(OH)}_4^-]}{[\text{OH}^-]} = \text{K}; \quad [\text{OH}^-] = \frac{\text{K}_\text{w}}{[\text{H}^+]}; \quad [\text{Al(OH)}_4^-] = \text{K}[\text{OH}^-] = \frac{\text{KK}_\text{w}}{[\text{H}^+]}$$

$$\text{S} = [\text{Al}^{3+}] + [\text{Al(OH)}_4^-] = [\text{H}^+]^3 \text{K}_\text{sp}/\text{K}_\text{w}^3 + \text{KK}_\text{w}/[\text{H}^+]$$

b.   $\text{K}_\text{sp} = 2 \times 10^{-32}$; $\text{K}_\text{w} = 1.0 \times 10^{-14}$; $\text{K} = 40.0$

$$\text{S} = \frac{[\text{H}^+]^3 (2 \times 10^{-32})}{(1.0 \times 10^{-14})^3} + \frac{40.0(1.0 \times 10^{-14})}{[\text{H}^+]} = [\text{H}^+]^3(2 \times 10^{10}) + \frac{4.0 \times 10^{-13}}{[\text{H}^+]}$$

| pH | solubility (S, mol/L) | log S |
|-----|-----|-----|
| 4.0 | $2 \times 10^{-2}$ | −1.7 |
| 5.0 | $2 \times 10^{-5}$ | −4.7 |
| 6.0 | $4.2 \times 10^{-7}$ | −6.38 |
| 7.0 | $4.0 \times 10^{-6}$ | −5.40 |
| 8.0 | $4.0 \times 10^{-5}$ | −4.40 |
| 9.0 | $4.0 \times 10^{-4}$ | −3.40 |
| 10.0 | $4.0 \times 10^{-3}$ | −2.40 |
| 11.0 | $4.0 \times 10^{-2}$ | −1.40 |
| 12.0 | $4.0 \times 10^{-1}$ | −0.40 |

As expected, the solubility of $\text{Al(OH)}_3$(s) is increased by very acidic solutions and by very basic solutions.

143.    For HOCl, $\text{K}_\text{a} = 3.5 \times 10^{-8}$ and $\text{pK}_\text{a} = -\log(3.5 \times 10^{-8}) = 7.46$.  This will be a buffer solution because the pH is close to the $\text{pK}_\text{a}$ value.

$$\text{pH} = \text{pK}_\text{a} + \log\frac{[\text{OCl}^-]}{[\text{HOCl}]}, \quad 8.00 = 7.46 + \log\frac{[\text{OCl}^-]}{[\text{HOCl}]}, \quad \frac{[\text{OCl}^-]}{[\text{HOCl}]} = 10^{0.54} = 3.5$$

1.00 L × 0.0500 $M$ = 0.0500 mol HOCl initially.  Added $\text{OH}^-$ converts HOCl into $\text{OCl}^-$.  The total moles of $\text{OCl}^-$ and HOCl must equal 0.0500 mol.  Solving where $n$ = moles:

$$n_{\text{OCl}^-} + n_{\text{HOCl}} = 0.0500 \text{ and } n_{\text{OCl}^-} = (3.5)n_{\text{HOCl}}$$

$$(4.5)n_{\text{HOCl}} = 0.0500, \quad n_{\text{HOCl}} = 0.011 \text{ mol}; \quad n_{\text{OCl}^-} = 0.039 \text{ mol}$$

Need to add 0.039 mol NaOH to produce 0.039 mol $\text{OCl}^-$.

0.039 mol = V × 0.0100 $M$, V = 3.9 L NaOH

*Note:*  Normal buffer assumptions hold.

145.    a.    200.0 mL $\times$ 0.250 mmol $Na_3PO_4$/mL = 50.0 mmol $Na_3PO_4$

135.0 mL $\times$ 1.000 mmol HCl/mL = 135.0 mmol HCl

100.0 mL $\times$ 0.100 mmol NaCN/mL = 10.0 mmol NaCN

Let $H^+$ from the HCl react to completion with the bases in solution.  In general, react the strongest base first and so on.  Here, 110.0 mmol of HCl reacts to convert all $CN^-$ to HCN and all $PO_4^{3-}$ to $H_2PO_4^-$.  At this point 10.0 mmol HCN, 50.0 mmol $H_2PO_4^-$, and 25.0 mmol HCl are in solution.  The remaining HCl reacts completely with $H_2PO_4^-$, converting 25.0 mmol to $H_3PO_4$.  The final solution contains  25.0 mmol $H_3PO_4$, (50.0 − 25.0 =) 25.0 mmol $H_2PO_4^-$, and 10.0 mmol HCN.  HCN ($K_a = 6.2 \times 10^{-10}$) is a much weaker acid than either $H_3PO_4$ ( $K_{a_1} = 7.5 \times 10^{-3}$) or $H_2PO_4^-$ ( $K_{a_2} = 6.2 \times 10^{-8}$), so ignore it.  We have a buffer solution.  Principal equilibrium reaction is:

$$H_3PO_4 \quad \rightleftharpoons \quad H^+ \quad + \quad H_2PO_4^- \quad K_{a_1} = 7.5 \times 10^{-3}$$

| | | | |
|---|---|---|---|
| Initial | 25.0 mmol/435.0 mL | 0 | 25.0/435.0 |
| Equil. | 0.0575 − x | x | 0.0575 + x |

$K_{a_1} = 7.5 \times 10^{-3} = \dfrac{x(0.0575 + x)}{0.0575 - x}$ ;  normal assumptions don't hold here.

Using the quadratic formula and carrying extra sig. figs.:

$$x^2 + (0.0650)x - 4.31 \times 10^{-4} = 0, \ x = 0.0061 \ M = [H^+]; \ pH = 2.21$$

b.    $[HCN] = \dfrac{10.0 \ mmol}{435.0 \ mL} = 2.30 \times 10^{-2} \ M$;  HCN dissociation will be minimal.

147.    Major species:  $PO_4^{3-}$,  $H^+$,  $HSO_4^-$,  $H_2O$, and $Na^+$;  let the best base ($PO_4^{3-}$) react with the best acid ($H^+$). Assume the reaction goes to completion because $H^+$ is reacting.  Note that the concentrations are halved when equal volumes of the two reagents are mixed.

$$PO_4^{3-} + \ H^+ \ \rightarrow \quad HPO_4^{2-}$$

| | | | |
|---|---|---|---|
| Before | 0.25 M | 0.050 M | 0 |
| After | 0.20 M | 0 | 0.050 M |

Major species:  $PO_4^{3-}$, $HPO_4^{2-}$, $HSO_4^-$, $H_2O$, and $Na^+$;  react the best base ($PO_4^{3-}$) with the best acid ($HSO_4^-$). Because K for this reaction is very large, assume the reaction goes to completion.

$$PO_4^{3-} + \quad HSO_4^- \quad \rightarrow \quad HPO_4^{2-} \ + \ SO_4^{2-} \quad K = \frac{K_{a, \ HSO_4^-}}{K_{a, \ HPO_4^{2-}}} = 2.5 \times 10^{10}$$

| | | | |
|---|---|---|---|
| Before | 0.20 M | 0.050 M | 0.050 M | 0 |
| After | 0.15 M | 0 | 0.100 M | 0.050 M |

Major species: $PO_4^{3-}$, $HPO_4^{2-}$, $SO_4^{2-}$ (a very weak base with $K_b = 8.3 \times 10^{-13}$), $H_2O$, and $Na^+$; because the best base present ($PO_4^{3-}$) and best acid present ($HPO_4^{2-}$) are conjugate acid-base pairs, a buffer solution exists. Because $K_b$ for $PO_4^{3-}$ is a relatively large value ($K_b = K_w/K_{a, HPO_4^{2-}} = 0.021$), the usual assumptions that the amount of base that reacts to each equilibrium is negligible compared with the initial concentration of base will not hold. Solving using the $K_b$ reaction for $PO_4^{3-}$:

$$PO_4^{3-} + H_2O \rightleftharpoons HPO_4^{2-} + OH^- \qquad K_b = 0.021$$

| | | | | |
|---|---|---|---|---|
| Initial | 0.15 $M$ | | 0.100 $M$ | 0 |
| Change | $-x$ | $\rightarrow$ | $+x$ | $+x$ |
| Equil. | $0.15 - x$ | | $0.100 + x$ | $x$ |

$K_b = 0.021 = \dfrac{(0.100 + x)(x)}{0.15 - x}$; using quadratic equation:

$x = [OH^-] = 0.022\ M$; pOH = 1.66; pH = 12.34

149. $H_3A$: $pK_{a_1} = 3.00$, $pK_{a_2} = 7.30$, $pK_{a_3} = 11.70$

The pH at the second stoichiometric point is:

$$pH = \frac{pK_{a_2} + pK_{a_3}}{2} = \frac{7.30 + 11.70}{2} = 9.50$$

Thus to reach a pH of 9.50, we must go to the second stoichiometric point. 100.0 mL × 0.0500 $M$ = 5.00 mmol $H_3A$ initially. To reach the second stoichiometric point, we need 10.0 mmol $OH^- = 1.00$ mmol/mL × $V_{NaOH}$. Solving for $V_{NaOH}$:

$V_{NaOH} = 10.0$ mL  (to reach pH = 9.50)

pH = 4.00 is between the first halfway point to equivalence (pH = $pK_{a_1}$ = 3.00) and the first stoichiometric point (pH = $\dfrac{pK_{a_1} + pK_{a_2}}{2}$ = 5.15).

This is the buffer region controlled by $H_3A \rightleftharpoons H_2A^- + H^+$.

$pH = pK_{a_1} + \log\dfrac{[H_2A^-]}{[H_3A]}$,  $4.00 = 3.00 + \log\dfrac{[H_2A^-]}{[H_3A]}$,  $\dfrac{[H_2A^-]}{[H_3A]} = 10.$

Because both species are in the same volume, the mole ratio also equals 10. Let $n$ = mmol:

$\dfrac{n_{H_2A^-}}{n_{H_3A}} = 10.$ and $n_{H_2A^-} + n_{H_3A} = 5.00$ mmol  (mole balance)

$11 n_{H_3A} = 5.00$,  $n_{H_3A} = 0.45$ mmol;  $n_{H_2A^-} = 4.55$ mmol

We need to add 4.55 mmol $OH^-$ to get 4.55 mmol $H_2A^-$ from the original $H_3A$ present.

4.55 mmol = 1.00 mmol/mL $\times$ $V_{NaOH}$,   $V_{NaOH}$ = 4.55 mL of NaOH (to reach pH = 4.00)

*Note*:  Normal buffer assumptions are good.

151.  a.  $V_1$ corresponds to the titration reaction of $CO_3^{2-} + H^+ \rightarrow HCO_3^-$;   $V_2$ corresponds to the titration reaction of $HCO_3^- + H^+ \rightarrow H_2CO_3$.

   Here, there are two sources of $HCO_3^-$:  $NaHCO_3$ and the titration of $Na_2CO_3$, so $V_2 > V_1$.

   b.  $V_1$ corresponds to two titration reactions:  $OH^- + H^+ \rightarrow H_2O$ and $CO_3^{2-} + H^+ \rightarrow HCO_3^-$. $V_2$ corresponds to just one titration reaction:  $HCO_3^- + H^+ \rightarrow H_2CO_3$.

   Here, $V_1 > V_2$ due to the presence of $OH^-$ , which is titrated in the $V_1$ region.

   c.  0.100 mmol HCl/mL $\times$ 18.9 mL = 1.89 mmol $H^+$;  Because the first stoichiometric point only involves the titration of $Na_2CO_3$ by $H^+$, 1.89 mmol of $CO_3^{2-}$ has been converted into $HCO_3^-$.  The sample contains 1.89 mmol $Na_2CO_3$ $\times$ 105.99 mg/mmol = $2.00 \times 10^2$ mg = 0.200 g $Na_2CO_3$.

   The second stoichiometric point involves the titration of $HCO_3^-$ by $H^+$.

$$\frac{0.100 \text{ mmol } H^+}{mL} \times 36.7 \text{ mL} = 3.67 \text{ mmol } H^+ = 3.67 \text{ mmol } HCO_3^-$$

   1.89 mmol $NaHCO_3$ came from the first stoichiometric point of the $Na_2CO_3$ titration.

   3.67 − 1.89 = 1.78 mmol $HCO_3^-$ came from $NaHCO_3$ in the original mixture.

   1.78 mmol $NaHCO_3$ $\times$ 84.01 mg $NaHCO_3$/mmol = $1.50 \times 10^2$ mg $NaHCO_3$ = 0.150 g $NaHCO_3$

$$\text{Mass \% } Na_2CO_3 = \frac{0.200 \text{ g}}{(0.200 + 0.150) \text{ g}} \times 100 = 57.1\% \text{ } Na_2CO_3$$

$$\text{Mass \% } NaHCO_3 = \frac{0.150 \text{ g}}{0.350 \text{ g}} \times 100 = 42.9\% \text{ } NaHCO_3$$

# CHAPTER 9

# ENERGY, ENTHALPY, AND THERMOCHEMISTRY

## The Nature of Energy

15.    Ball A:  $PE = mgz = 2.00 \text{ kg} \times \dfrac{9.81 \text{ m}}{s^2} \times 10.0 \text{ m} = \dfrac{196 \text{ kg m}^2}{s^2} = 196 \text{ J}$

At point I:    All this energy is transferred to ball B.  All of B's energy is kinetic energy at this point.  $E_{total} = KE = 196 \text{ J}$.  At point II, the sum of the total energy will equal 196 J.

At point II:  $PE = mgz = 4.00 \text{ kg} \times \dfrac{9.81 \text{ m}}{s^2} \times 3.00 \text{ m} = 118 \text{ J}$

$KE = E_{total} - PE = 196 \text{ J} - 118 \text{ J} = 78 \text{ J}$

17.    Path-dependent functions for a trip from Chicago to Denver are those quantities that depend on the route taken. One can fly directly from Chicago to Denver or one could fly from Chicago to Atlanta to Los Angeles and then to Denver. Some path-dependent quantities are miles traveled, fuel consumption of the airplane, time traveling, airplane snacks eaten, etc. State functions are path independent; they only depend on the initial and final states. Some state functions for an airplane trip from Chicago to Denver would be longitude change, latitude change, elevation change, and overall time zone change.

19.    Step 1:  $\Delta E_1 = q + w = 72 \text{ J} + 35 \text{ J} = 107 \text{ J}$;   step 2:  $\Delta E_2 = 35 \text{ J} - 72 \text{ J} = -37 \text{ J}$

$\Delta E_{overall} = \Delta E_1 + \Delta E_2 = 107 \text{ J} - 37 \text{ J} = 70. \text{ J}$

21.    $q = \text{molar heat capacity} \times \text{mol} \times \Delta T = \dfrac{20.8 \text{ J}}{^\circ \text{C mol}} \times 39.1 \text{ mol} \times (38.0 - 0.0)^\circ \text{C} = 30,900 \text{ J}$

$= 30.9 \text{ kJ}$

$w = -P\Delta V = -1.00 \text{ atm} \times (998 \text{ L} - 876 \text{ L}) = -122 \text{ L atm} \times \dfrac{101.3 \text{ J}}{\text{L atm}} = -12,400 \text{ J} = -12.4 \text{ kJ}$

$\Delta E = q + w = 30.9 \text{ kJ} + (-12.4 \text{ kJ}) = 18.5 \text{ kJ}$

23.   $H_2O(g) \rightarrow H_2O(l)$;   $\Delta E = q + w$;   $q = -40.66$ kJ;   $w = -P\Delta V$

Volume of 1 mol $H_2O(l)$ = 1.000 mol $H_2O(l) \times \dfrac{18.02 \text{ g}}{\text{mol}} \times \dfrac{1 \text{ cm}^3}{0.996 \text{ g}} = 18.1 \text{ cm}^3 = 18.1 \text{ mL}$

$w = -P\Delta V = -1.00 \text{ atm} \times (0.0181 \text{ L} - 30.6 \text{ L}) = 30.6 \text{ L atm} \times \dfrac{101.3 \text{ J}}{\text{L atm}} = 3.10 \times 10^3 \text{ J}$

$= 3.10 \text{ kJ}$

$\Delta E = q + w = -40.66 \text{ kJ} + 3.10 \text{ kJ} = -37.56 \text{ kJ}$

## Properties of Enthalpy

25.   $\Delta H = \Delta E + P\Delta V$ at constant P; from the definition of enthalpy, the difference between $\Delta H$ and $\Delta E$ at constant P is the quantity $P\Delta V$. Thus when a system at constant P can do pressure-volume work, then $\Delta H \neq \Delta E$. When the system cannot do PV work, then $\Delta H = \Delta E$ at constant pressure. An important way to differentiate $\Delta H$ from $\Delta E$ is to concentrate on q, the heat flow; the heat flow by a system at constant pressure equals $\Delta H$, and the heat flow by a system at constant volume equals $\Delta E$.

27.   One should try to cool the reaction mixture or provide some means of removing heat since the reaction is very exothermic (heat is released).  The $H_2SO_4(aq)$ will get very hot and possibly boil unless cooling is provided.

29.   $4 \text{ Fe}(s) + 3 \text{ O}_2(g) \rightarrow 2 \text{ Fe}_2O_3(s)$   $\Delta H = -1652$ kJ;  note that 1652 kJ of heat is released when 4 mol Fe reacts with 3 mol $O_2$ to produce 2 mol $Fe_2O_3$.

a.   $4.00 \text{ mol Fe} \times \dfrac{-1652 \text{ kJ}}{4 \text{ mol Fe}} = -1650 \text{ kJ}$; 1650 kJ of  heat released

b.   $1.00 \text{ mol Fe}_2O_3 \times \dfrac{-1652 \text{ kJ}}{2 \text{ mol Fe}_2O_3} = -826 \text{ kJ}$;  826 kJ of heat released

c.   $1.00 \text{ g Fe} \times \dfrac{1 \text{ mol Fe}}{55.85 \text{ g}} \times \dfrac{-1652 \text{ kJ}}{4 \text{ mol Fe}} = -7.39 \text{ kJ}$; 7.39 kJ of heat released

d.   $10.0 \text{ g Fe} \times \dfrac{1 \text{ mol Fe}}{55.85 \text{ g}} = 0.179 \text{ mol Fe}$;  $2.00 \text{ g O}_2 \times \dfrac{1 \text{ mol O}_2}{32.00 \text{ g}} = 0.0625 \text{ mol O}_2$

0.179 mol Fe/0.0625 mol $O_2$ = 2.86;  the balanced equation requires a 4 mol Fe/3 mol $O_2$ = 1.33 mole ratio. $O_2$ is limiting since the actual mole Fe/mole $O_2$ ratio is greater than the required mole ratio.

$0.0625 \text{ mol O}_2 \times \dfrac{-1652 \text{ kJ}}{3 \text{ mol O}_2} = -34.4 \text{ kJ}$;  34.4 kJ of  heat released

31.   When a liquid is converted into gas, there is an increase in volume. The 2.5 kJ/mol quantity is the work done by the vaporization process in pushing back the atmosphere.

## The Thermodynamics of Ideal Gases

33.    Consider the constant volume process first.

$$n = 1.00 \times 10^3 \text{ g} \times \frac{1 \text{ mol}}{30.07 \text{ g}} = 33.3 \text{ mol } C_2H_6; \quad C_v = \frac{44.60 \text{ J}}{\text{K mol}} = \frac{44.60 \text{ J}}{\text{°C mol}}$$

$$\Delta E = nC_v\Delta T = (33.3 \text{ mol})(44.60 \text{ J °C}^{-1} \text{ mol}^{-1})(75.0 - 25.0\text{°C}) = 74{,}300 \text{ J} = 74.3 \text{ kJ}$$

$$\Delta E = q + w; \text{ since } \Delta V = 0, w = 0; \quad \Delta E = q_v = 74.3 \text{ kJ}$$

$$\Delta H = \Delta E + \Delta PV = \Delta E + nR\Delta T$$

$$\Delta H = 74.3 \text{ kJ} + (33.3 \text{ mol})(8.3145 \text{ J K}^{-1} \text{ mol}^{-1})(50.0 \text{ K})(1 \text{ kJ}/1000 \text{ J})$$

$$\Delta H = 74.3 \text{ kJ} + 13.8 \text{ kJ} = 88.1 \text{ kJ}$$

Now consider the constant pressure process.

$$q_p = \Delta H = nC_p\Delta T = (33.3 \text{ mol})(52.92 \text{ J K}^{-1} \text{ mol}^{-1})(50.0 \text{ K})$$

$$q_p = 88{,}100 \text{ J} = 88.1 \text{ kJ} = \Delta H$$

$$w = -P\Delta V = -nR\Delta T = -(33.3 \text{ mol})(8.3145 \text{ J K}^{-1} \text{ mol}^{-1})(50.0 \text{ K}) = -13{,}800 \text{ J} = -13.8 \text{ kJ}$$

$$\Delta E = q + w = 88.1 \text{ kJ} - 13.8 \text{ kJ} = 74.3 \text{ kJ}$$

| Summary | Constant V | Constant P |
|---------|-----------|-----------|
| q | 74.3 kJ | 88.1 kJ |
| $\Delta E$ | 74.3 kJ | 74.3 kJ |
| $\Delta H$ | 88.1 kJ | 88.1 kJ |
| w | 0 | −13.8 kJ |

35.    Pathway I:

Step 1:  (5.00 mol, 3.00 atm,  15.0 L) $\rightarrow$ (5.00 mol,  3.00 atm,  55.0 L)

$$w = -P\Delta V = -(3.00 \text{ atm})(55.0 - 15.0 \text{ L}) = -120. \text{ L atm}$$

$$w = -120. \text{ L atm} \times \frac{101.3 \text{ J}}{\text{L atm}} \times \frac{1 \text{ kJ}}{1000 \text{ J}} = -12.2 \text{ kJ}$$

$$\Delta H = q_p = nC_p\Delta T = nC_p \times \frac{\Delta(PV)}{nR} = \frac{C_p\Delta(PV)}{R}; \quad \Delta(PV) = (P_2V_2 - P_1V_1)$$

For an ideal monatomic gas: $C_p = \frac{5}{2}R; \quad \Delta H = \left(\frac{5}{2}R\right)\frac{\Delta(PV)}{R} = \frac{5}{2}\Delta(PV)$

$$\Delta H = q_p = \frac{5}{2}\Delta(PV) = \frac{5}{2}(165 - 45.0) \text{ L atm} = 300. \text{ L atm}$$

$$\Delta H = q_p = 300. \text{ L atm} \times \frac{101.3 \text{ J}}{\text{L atm}} \times \frac{1 \text{ kJ}}{1000 \text{ J}} = 30.4 \text{ kJ}$$

$$\Delta E = q + w = 30.4 \text{ kJ} - 12.2 \text{ kJ} = 18.2 \text{ kJ}$$

Step 2: (5.00 mol, 3.00 atm, 55.0 L) $\rightarrow$ (5.00 mol, 6.00 atm, 20.0 L)

$$\Delta E = nC_v\Delta T = n\left(\frac{3}{2}R\right)\left(\frac{\Delta(PV)}{nR}\right) = \frac{3}{2}\Delta PV$$

$$\Delta E = \frac{3}{2}(120. - 165) \text{ L atm} = -67.5 \text{ L atm} \quad \text{(Carry an extra significant figure.)}$$

$$\Delta E = -67.5 \text{ L atm} \times \frac{101.3 \text{ J}}{\text{L atm}} \times \frac{1 \text{ kJ}}{1000 \text{ J}} = -6.8 \text{ kJ}$$

$$\Delta H = nC_p\Delta T = n\left(\frac{5}{2}R\right)\left(\frac{\Delta(PV)}{nR}\right) = \frac{5}{2}\Delta PV$$

$$\Delta H = \frac{5}{2}(-45 \text{ L atm}) = -113 \text{ L atm} \quad \text{(Carry an extra significant figure.)}$$

$$\Delta H = -113 \text{ L atm} \times \frac{101.3 \text{ J}}{\text{L atm}} \times \frac{1 \text{ kJ}}{1000 \text{ J}} = -11.4 = -11 \text{ kJ}$$

$$w = -P_{ext}\Delta V = -(6.00 \text{ atm})(20.0 - 55.0) \text{ L} = 210. \text{ L atm}$$

$$w = 210. \text{ L atm} \times \frac{101.3 \text{ J}}{\text{L atm}} \times \frac{1 \text{ kJ}}{1000 \text{ J}} = 21.3 \text{ kJ}$$

$$\Delta E = q + w, \quad -6.8 \text{ kJ} = q + 21.3 \text{ kJ}, \quad q = -28.1 \text{ kJ}$$

| Summary: | Path I | Step 1 | Step 2 | Total |
|---|---|---|---|---|
| | q | 30.4 kJ | -28.1 kJ | 2.3 kJ |
| | w | -12.2 kJ | 21.3 kJ | 9.1 kJ |
| | $\Delta E$ | 18.2 kJ | -6.8 kJ | 11.4 kJ |
| | $\Delta H$ | 30.4 kJ | -11 kJ | 19 kJ |

Pathway II:

Step 3:  (5.00 mol,  3.00 atm,  15.0 L) $\rightarrow$ (5.00 mol,  6.00 atm,  15.0 L)

$$\Delta E = q_v = \frac{3}{2} \Delta(PV) = \frac{5}{2} (90.0 - 45.0) \text{ L atm} = 67.5 \text{ L atm}$$

$$\Delta E = q_v = 67.5 \text{ L atm} \times \frac{101.3 \text{ J}}{\text{L atm}} \times \frac{1 \text{ kJ}}{1000 \text{ J}} = 6.84 \text{ kJ}$$

$$w = -P\Delta V = 0 \text{ because } \Delta V = 0$$

$$\Delta H = \Delta E + \Delta(PV) = 67.5 \text{ L atm} + 45.0 \text{ L atm} = 112.5 \text{ L atm} = 11.40 \text{ kJ}$$

Step 4:  (5.00 mol,  6.00 atm,  15.0 L) $\rightarrow$ (5.00 mol,  6.00 atm,  20.0 L)

$$\Delta H = q_p = nC_p\Delta T = \left(\frac{5}{2}R\right)\left(\frac{\Delta(PV)}{nR}\right) = \frac{5}{2}\Delta PV$$

$$\Delta H = \frac{5}{2}(120. - 90.0) \text{ L atm} = 75 \text{ L atm}$$

$$\Delta H = q_p = 75 \text{ L atm} \times \frac{101.3 \text{ J}}{\text{L atm}} \times \frac{1 \text{ kJ}}{1000 \text{ J}} = 7.6 \text{ kJ}$$

$$w = -P\Delta V = -(6.00 \text{ atm})(20.0 - 15.0) \text{ L} = -30. \text{ L atm}$$

$$w = -30. \text{ L atm} \times \frac{101.3 \text{ J}}{\text{L atm}} \times \frac{1 \text{ kJ}}{1000 \text{ J}} = -3.0 \text{ kJ}$$

$$\Delta E = q + w = 7.6 \text{ kJ} - 3.0 \text{ kJ} = 4.6 \text{ kJ}$$

| Summary: | Path II | Step 3 | Step 4 | Total |
|---|---|---|---|---|
| q | | 6.84 kJ | 7.6 kJ | 14.4 kJ |
| w | | 0 | –3.0 kJ | –3.0 kJ |
| $\Delta E$ | | 6.84 kJ | 4.6 kJ | 11.4 kJ |
| $\Delta H$ | | 11.40 kJ | 7.6 kJ | 19.0 kJ |

State functions are independent of the particular pathway taken between two states; path functions are dependent on the particular pathway.  In this problem, the overall values of $\Delta H$ and $\Delta E$ for the two pathways are the same; hence $\Delta H$ and $\Delta E$ are state functions.  The overall values of q and w for the two pathways are different; hence q and w are path functions.

## Calorimetry and Heat Capacity

37.    In calorimetry, heat flow is determined into or out of the surroundings. Because $\Delta E_{univ} = 0$ by the first law of thermodynamics, $\Delta E_{sys} = -\Delta E_{surr}$; what happens to the surroundings is the exact opposite of what happens to the system. To determine heat flow, we need to know the heat capacity of the surroundings, the mass of the surroundings that accepts/donates the heat, and the change in temperature. If we know these quantities, $q_{surr}$ can be calculated and then equated to $q_{sys}$ ($-q_{surr} = q_{sys}$). For an endothermic reaction, the surroundings (the calorimeter contents) donates heat to the system. This is accompanied by a decrease in temperature of the surroundings. For an exothermic reaction, the system donates heat to the surroundings (the calorimeter), so temperature increases.

       $q_P = \Delta H$;   $q_V = \Delta E$; a coffee-cup calorimeter is at constant (atmospheric) pressure. The heat released or gained at constant pressure is $\Delta H$. A bomb calorimeter is at constant volume. The heat released or gained at constant volume is $\Delta E$.

39.    Specific heat capacity is defined as the amount of heat necessary to raise the temperature of one gram of substance by one degree Celsius. Therefore, $H_2O(l)$ with the largest heat capacity value requires the largest amount of heat for this process. The amount of heat for $H_2O(l)$ is:

$$\text{energy} = s \times m \times \Delta T = \frac{4.18 \text{ J}}{\text{°C g}} \times 25.0 \text{ g} \times (37.0\text{°C} - 15.0\text{°C}) = 2.30 \times 10^3 \text{ J}$$

       The largest temperature change when a certain amount of energy is added to a certain mass of substance will occur for the substance with the smallest specific heat capacity. This is Hg(l), and the temperature change for this process is:

$$\Delta T = \frac{\text{energy}}{s \times m} = \frac{10.7 \text{ kJ} \times \dfrac{1000 \text{ J}}{\text{kJ}}}{\dfrac{0.14 \text{ J}}{\text{°C g}} \times 550. \text{ g}} = 140\text{°C}$$

41.    Heat loss by hot water = heat gain by cold water; keeping all quantities positive to avoid sign errors:

$$\frac{4.18 \text{ J}}{\text{°C g}} \times m_{hot} \times (55.0\text{°C} - 37.0\text{°C}) = \frac{4.18 \text{ J}}{\text{°C g}} \times 90.0 \text{ g} \times (37.0 \text{ °C} - 22.0\text{°C})$$

$$m_{hot} = \frac{90.0 \text{ g} \times 15.0\text{°C}}{18.0\text{°C}} = 75.0 \text{ g hot water needed}$$

43.    $50.0 \times 10^{-3}$ L $\times 0.100$ mol/L $= 5.00 \times 10^{-3}$ mol of both $AgNO_3$ and HCl are reacted. Thus $5.00 \times 10^{-3}$ mol of AgCl will be produced because there is a 1 : 1 mole ratio between reactants.

       Heat lost by chemicals = heat gained by solution

Heat gain $= \dfrac{4.18 \, J}{°C \, g} \times 100.0 \, g \times (23.40 - 22.60)°C = 330 \, J$

Heat loss $= 330 \, J$; this is the heat evolved (exothermic reaction) when $5.00 \times 10^{-3}$ mol of AgCl is produced. So $q = -330 \, J$ and $\Delta H$ (heat per mol AgCl formed) is negative with a value of:

$$\Delta H = \dfrac{-330 \, J}{5.00 \times 10^{-3} \, mol} \times \dfrac{1 \, kJ}{1000 \, J} = -66 \, kJ/mol$$

*Note*: Sign errors are common with calorimetry problems. However, the correct sign for $\Delta H$ can be determined easily from the $\Delta T$ data; i.e., if $\Delta T$ of the solution increases, then the reaction is exothermic because heat was released, and if $\Delta T$ of the solution decreases, then the reaction is endothermic because the reaction absorbed heat from the water. For calorimetry problems, keep all quantities positive until the end of the calculation and then decide the sign for $\Delta H$. This will help eliminate sign errors.

45.     Because $\Delta H$ is exothermic, the temperature of the solution will increase as $CaCl_2(s)$ dissolves.   Keeping all quantities positive:

$$\text{heat loss as } CaCl_2 \text{ dissolves} = 11.0 \, g \, CaCl_2 \times \dfrac{1 \, mol \, CaCl_2}{110.98 \, g \, CaCl_2} \times \dfrac{81.5 \, kJ}{mol \, CaCl_2} = 8.08 \, kJ$$

$$\text{heat gained by solution} = 8.08 \times 10^3 \, J = \dfrac{4.18 \, J}{°C \, g} \times (125 + 11.0) \, g \times (T_f - 25.0°C)$$

$$T_f - 25.0°C = \dfrac{8.08 \times 10^3}{4.18 \times 136} = 14.2°C, \ T_f = 14.2°C + 25.0°C = 39.2°C$$

47.     Heat gain by calorimeter $= \dfrac{1.56 \, kJ}{°C} \times 3.2°C = 5.0 \, kJ = $ heat loss by quinone

Heat loss $= 5.0 \, kJ$, which is the heat evolved (exothermic reaction) by the combustion of 0.1964 g of quinone. Because we are at constant volume, $q_v = \Delta E$.

$$\Delta E_{comb} = \dfrac{-5.0 \, kJ}{0.1964 \, g} = -25 \, kJ/g; \qquad \Delta E_{comb} = \dfrac{-25 \, kJ}{g} \times \dfrac{108.09 \, g}{mol} = -2700 \, kJ/mol$$

49.     a.   $C_{12}H_{22}O_{11}(s) + 12 \, O_2(g) \rightarrow 12 \, CO_2(g) + 11 \, H_2O(l)$

        b.   A bomb calorimeter is at constant volume, so heat released $= q_v = \Delta E$:

$$\Delta E = \dfrac{-24.00 \, kJ}{1.46 \, g} \times \dfrac{342.30 \, g}{mol} = -5630 \, kJ/mol \ C_{12}H_{22}O_{11}$$

        c.   $\Delta H = \Delta E + \Delta(PV) = \Delta E + \Delta(nRT) = \Delta E + \Delta nRT$, where $\Delta n =$ moles of gaseous products – moles of gaseous reactants.

        For this reaction, $\Delta n = 12 - 12 = 0$, so $\Delta H = \Delta E = -5630 \, kJ/mol$.

## Hess's Law

51.
$$2\,C + 2\,O_2 \rightarrow 2\,CO_2 \qquad\qquad \Delta H = 2(-394\text{ kJ})$$
$$H_2 + 1/2\,O_2 \rightarrow H_2O \qquad\qquad \Delta H = -286\text{ kJ}$$
$$2\,CO_2 + H_2O \rightarrow C_2H_2 + 5/2\,O_2 \qquad \Delta H = -(-1300.\text{kJ})$$

$$\overline{2\,C(s) + H_2(g) \rightarrow C_2H_2(g) \qquad\qquad \Delta H = 226\text{ kJ}}$$

*Note*:  The enthalpy change for a reaction that is reversed is the negative quantity of the enthalpy change for the original reaction.  If the coefficients in a balanced reaction are multiplied by an integer, then the value of $\Delta H$ is multiplied by the same integer.

53.
$$CaC_2 \rightarrow Ca + 2\,C \qquad\qquad \Delta H = -(-62.8\text{ kJ})$$
$$CaO + H_2O \rightarrow Ca(OH)_2 \qquad\qquad \Delta H = -653.1\text{ kJ}$$
$$2\,CO_2 + H_2O \rightarrow C_2H_2 + 5/2\,O_2 \qquad \Delta H = -(-1300.\text{ kJ})$$
$$Ca + 1/2\,O_2 \rightarrow CaO \qquad\qquad \Delta H = -635.5\text{ kJ}$$
$$2\,C + 2\,O_2 \rightarrow 2\,CO_2 \qquad\qquad \Delta H = 2(-393.5\text{ kJ})$$

$$\overline{CaC_2(s) + 2\,H_2O(l) \rightarrow Ca(OH)_2(aq) + C_2H_2(g) \qquad \Delta H = -713\text{ kJ}}$$

55.
$$C_4H_4(g) + 5\,O_2(g) \rightarrow 4\,CO_2(g) + 2\,H_2O(l) \qquad \Delta H_{comb} = -2341\text{ kJ}$$

$$C_4H_8(g) + 6\,O_2(g) \rightarrow 4\,CO_2(g) + 4\,H_2O(l) \qquad \Delta H_{comb} = -2755\text{ kJ}$$

$$H_2(g) + 1/2\,O_2(g) \rightarrow H_2O(l) \qquad\qquad \Delta H_{comb} = -286\text{ kJ}$$

By convention, $H_2O(l)$ is produced when enthalpies of combustion are given, and because per mole quantities are given, the combustion reaction refers to 1 mole of that quantity reacting with $O_2(g)$.

Using Hess's Law to solve:

$$C_4H_4(g) + 5\,O_2(g) \rightarrow 4\,CO_2(g) + 2\,H_2O(l) \qquad \Delta H_1 = -2341\text{ kJ}$$
$$4\,CO_2(g) + 4\,H_2O(l) \rightarrow C_4H_8(g) + 6\,O_2(g) \qquad \Delta H_2 = -(-2755\text{ kJ})$$
$$2\,H_2(g) + O_2(g) \rightarrow 2\,H_2O(l) \qquad\qquad \Delta H_3 = 2(-286\text{ kJ})$$

$$\overline{C_4H_4(g) + 2\,H_2(g) \rightarrow C_4H_8(g) \qquad \Delta H = \Delta H_1 + \Delta H_2 + \Delta H_3 = -158\text{ kJ}}$$

57.
$$C_6H_4(OH)_2 \rightarrow C_6H_4O_2 + H_2 \qquad\qquad \Delta H = 177.4\text{ kJ}$$
$$H_2O_2 \rightarrow H_2 + O_2 \qquad\qquad \Delta H = -(-191.2\text{ kJ})$$
$$2\,H_2 + O_2 \rightarrow 2\,H_2O(g) \qquad\qquad \Delta H = 2(-241.8\text{ kJ})$$
$$2\,H_2O(g) \rightarrow 2\,H_2O(l) \qquad\qquad \Delta H = 2(-43.8\text{ kJ})$$

$$\overline{C_6H_4(OH)_2(aq) + H_2O_2(aq) \rightarrow C_6H_4O_2(aq) + 2\,H_2O(l) \qquad \Delta H = -202.6\text{ kJ}}$$

59.
$$2\,N_2(g) + 6\,H_2(g) \rightarrow 4\,NH_3(g) \qquad\qquad \Delta H = -4(46\text{ kJ})$$
$$6\,H_2O(g) \rightarrow 6\,H_2(g) + 3\,O_2(g) \qquad \Delta H = -3(-484\text{ kJ})$$

$$\overline{2\,N_2(g) + 6\,H_2O(g) \rightarrow 3\,O_2(g) + 4\,NH_3(g) \qquad \Delta H = 1268\text{ kJ}}$$

No, because the reaction is very endothermic (requires a lot of heat to react), it would not be a practical way of making ammonia because of the high energy costs required.

## Standard Enthalpies of Formation

61.    In general: $\Delta H° = \sum n_p \Delta H°_{f,\ products} - \sum n_r \Delta H°_{f,\ reactants}$, and all elements in their standard state have $\Delta H°_f = 0$ by definition.

a.    The balanced equation is:  $2\ NH_3(g) + 3\ O_2(g) + 2\ CH_4(g) \rightarrow 2\ HCN(g) + 6\ H_2O(g)$

$$\Delta H° = (2\ mol\ HCN \times \Delta H°_{f,\ HCN} + 6\ mol\ H_2O(g) \times \Delta H°_{f,\ H_2O})$$
$$- (2\ mol\ NH_3 \times \Delta H°_{f,\ NH_3} + 2\ mol\ CH_4 \times \Delta H°_{f,\ CH_4})$$

$$\Delta H° = [2(135.1) + 6(-242)] - [2(-46) + 2(-75)] = -940.\ kJ$$

b.    $Ca_3(PO_4)_2(s) + 3\ H_2SO_4(l) \rightarrow 3\ CaSO_4(s) + 2\ H_3PO_4(l)$

$$\Delta H° = \left[ 3\ mol\ CaSO_4(s)\left(\frac{-1433\ kJ}{mol}\right) + 2\ mol\ H_3PO_4(l)\left(\frac{-1267\ kJ}{mol}\right)\right]$$
$$- \left[1\ mol\ Ca_3(PO_4)_2(s)\left(\frac{-4126\ kJ}{mol}\right) + 3\ mol\ H_2SO_4(l)\left(\frac{-814\ kJ}{mol}\right)\right]$$

$$\Delta H° = -6833\ kJ - (-6568\ kJ) = -265\ kJ$$

c.    $NH_3(g) + HCl(g) \rightarrow NH_4Cl(s)$

$$\Delta H° = (1\ mol\ NH_4Cl \times \Delta H°_{f,\ NH_4Cl}) - (1\ mol\ NH_3 \times \Delta H°_{f,\ NH_3} + 1\ mol\ HCl \times \Delta H°_{f,\ HCl})$$

$$\Delta H° = \left[1\ mol\left(\frac{-314\ kJ}{mol}\right)\right] - \left[1\ mol\left(\frac{-46\ kJ}{mol}\right) + 1\ mol\left(\frac{-92\ kJ}{mol}\right)\right]$$

$$\Delta H° = -314\ kJ + 138\ kJ = -176\ kJ$$

d.    The balanced equation is:  $C_2H_5OH(l) + 3\ O_2(g) \rightarrow 2\ CO_2(g) + 3\ H_2O(g)$

$$\Delta H° = \left[2\ mol\left(\frac{-393.5\ kJ}{mol}\right) + 3\ mol\left(\frac{-242\ kJ}{mol}\right)\right] - \left[1\ mol\left(\frac{-278\ kJ}{mol}\right)\right]$$

$$\Delta H° = -1513\ kJ - (-278\ kJ) = -1235\ kJ$$

e.   $SiCl_4(l) + 2\ H_2O(l) \rightarrow SiO_2(s) + 4\ HCl(aq)$

Because $HCl(aq)$ is $H^+(aq) + Cl^-(aq)$, $\Delta H_f^\circ = 0 - 167 = -167$ kJ/mol.

$$\Delta H^\circ = \left[ 4\ mol\left(\frac{-167\ kJ}{mol}\right) + 1\ mol\left(\frac{-911\ kJ}{mol}\right) \right]$$

$$- \left[ 1\ mol\left(\frac{-687\ kJ}{mol}\right) + 2\ mol\left(\frac{-286\ kJ}{mol}\right) \right]$$

$$\Delta H^\circ = -1579\ kJ - (-1259\ kJ) = -320.\ kJ$$

f.   $MgO(s) + H_2O(l) \rightarrow Mg(OH)_2(s)$

$$\Delta H^\circ = \left[ 1\ mol\left(\frac{-925\ kJ}{mol}\right) \right] - \left[ 1\ mol\left(\frac{-602\ kJ}{mol}\right) + 1\ mol\left(\frac{-286\ kJ}{mol}\right) \right]$$

$$\Delta H^\circ = -925\ kJ - (-888\ kJ) = -37\ kJ$$

63.   $4\ Na(s) + O_2(g) \rightarrow 2\ Na_2O(s)$,   $\Delta H^\circ = 2\ mol\left(\frac{-416\ kJ}{mol}\right) = -832\ kJ$

$2\ Na(s) + 2\ H_2O(l) \rightarrow 2\ NaOH(aq) + H_2(g)$

$$\Delta H^\circ = \left[ 2\ mol\left(\frac{-470.\ kJ}{mol}\right) \right] - \left[ 2\ mol\left(\frac{-286\ kJ}{mol}\right) \right] = -368\ kJ$$

$2Na(s) + CO_2(g) \rightarrow Na_2O(s) + CO(g)$

$$\Delta H^\circ = \left[ 1\ mol\left(\frac{-416\ kJ}{mol}\right) + 1\ mol\left(\frac{-110.5\ kJ}{mol}\right) \right] - \left[ 1\ mol\left(\frac{-393.5\ kJ}{mol}\right) \right] = -133\ kJ$$

In reactions 2 and 3, sodium metal reacts with the "extinguishing agent."  Both reactions are exothermic and each reaction produces a flammable gas, $H_2$ and CO, respectively.

65.   $5\ N_2O_4(l) + 4\ N_2H_3CH_3(l) \rightarrow 12\ H_2O(g) + 9\ N_2(g) + 4\ CO_2(g)$

$$\Delta H^\circ = \left[ 12\ mol\left(\frac{-242\ kJ}{mol}\right) + 4\ mol\left(\frac{-393.5\ kJ}{mol}\right) \right]$$

$$- \left[ 5\ mol\left(\frac{-20.\ kJ}{mol}\right) + 4\ mol\left(\frac{54\ kJ}{mol}\right) \right] = -4594\ kJ$$

67.     a.    $\Delta H° = 3 \text{ mol}(227 \text{ kJ/mol}) - 1 \text{ mol}(49 \text{ kJ/mol}) = 632 \text{ kJ}$

      b.    Because $3 C_2H_2(g)$ is higher in energy than $C_6H_6(l)$, acetylene will release more energy per gram when burned in air.

69.     $2 ClF_3(g) + 2 NH_3(g) \rightarrow N_2(g) + 6 HF(g) + Cl_2(g) \quad \Delta H° = -1196 \text{ kJ}$

$$\Delta H° = (6 \Delta H°_{f, \text{ HF}}) - (2 \Delta H°_{f, \text{ ClF}_3} + 2\Delta H°_{f, \text{ NH}_3})$$

$$-1196 \text{ kJ} = 6 \text{ mol}\left(\frac{-271 \text{ kJ}}{\text{mol}}\right) - 2 \Delta H°_{f, \text{ ClF}_3} - 2 \text{ mol}\left(\frac{-46 \text{ kJ}}{\text{mol}}\right)$$

$$-1196 \text{ kJ} = -1626 \text{ kJ} - 2 \Delta H°_{f, \text{ ClF}_3} + 92 \text{ kJ}, \quad \Delta H°_{f, \text{ ClF}_3} = \frac{(-1626 + 92 + 1196) \text{ kJ}}{2 \text{ mol}} = \frac{-169 \text{ kJ}}{\text{mol}}$$

## Energy Consumption and Sources

71.     Mass of $H_2O = 1.00 \text{ gal} \times \dfrac{3.785 \text{ L}}{\text{gal}} \times \dfrac{1000 \text{ mL}}{\text{L}} \times \dfrac{1.00 \text{ g}}{\text{mL}} = 3790 \text{ g } H_2O$

Energy required (theoretical) $= s \times m \times \Delta T = \dfrac{4.18 \text{ J}}{°C \text{ g}} \times 3790 \text{ g} \times 10.0 \text{ °C} = 1.58 \times 10^5 \text{ J}$

For an actual (80.0% efficient) process, more than this quantity of energy is needed since heat is always lost in any transfer of energy.  The energy required is:

$$1.58 \times 10^5 \text{ J} \times \frac{100. \text{ J}}{80.0 \text{ J}} = 1.98 \times 10^5 \text{ J}$$

Mass of $C_2H_2 = 1.98 \times 10^5 \text{ J} \times \dfrac{1 \text{ mol } C_2H_2}{1300. \times 10^3 \text{ J}} \times \dfrac{26.04 \text{ g } C_2H_2}{\text{mol } C_2H_2} = 3.97 \text{ g } C_2H_2$

73.     $CH_3OH(l) + 3/2 O_2(g) \rightarrow CO_2(g) + 2 H_2O(l)$

$\Delta H° = [-393.5 \text{ kJ} + 2(-286 \text{ kJ})] - (-239 \text{ kJ}) = -727 \text{ kJ/mol } CH_3OH$

$$\frac{-727 \text{ kJ}}{\text{mol}} \times \frac{1 \text{ mol}}{32.04 \text{ g}} = -22.7 \text{ kJ/g versus } -29.67 \text{ kJ/g for ethanol}$$

Ethanol has a higher fuel value than methanol.

## Additional Exercises

75.     $\Delta E_{\text{overall}} = \Delta E_{\text{step 1}} + \Delta E_{\text{step 2}}$;  this is a cyclic process, which means that the overall initial state and final state are the same.  Because $\Delta E$ is a state function, $\Delta E_{\text{overall}} = 0$ and $\Delta E_{\text{step 1}} = -\Delta E_{\text{step 2}}$.

$\Delta E_{step\ 1} = q + w = 45\ J + (-10.\ J) = 35\ J$

$\Delta E_{step\ 2} = -\Delta E_{step\ 1} = -35\ J = q + w,\ -35\ J = -60.\ J + w,\ w = 25\ J$

77.    $H_2(g) + 1/2\ O_2(g) \rightarrow H_2O(l)\quad \Delta H° = \Delta H°_{f,\ H_2O(l)} = -285.8\ kJ$

$H_2O(l) \rightarrow H_2(g) + 1/2\ O_2(g)\quad \Delta H° = 285.8\ kJ$

$\Delta E° = \Delta H° - P\Delta V = \Delta H° - \Delta nRT$

$\Delta E° = 285.8\ kJ - (1.50 - 0\ mol)(8.3145\ J\ K^{-1}\ mol^{-1})(298\ K)\left(\dfrac{1\ kJ}{1000\ J}\right)$

$\Delta E° = 285.8\ kJ - 3.72\ kJ = 282.1\ kJ$

79.    The specific heat of water is $4.18\ J\ °C^{-1}\ g^{-1}$, which is equal to $4.18\ kJ\ °C^{-1}\ kg^{-1}$

We have 1.00 kg of $H_2O$, so:   $1.00\ kg \times \dfrac{4.18\ kJ}{°C\ kg} = 4.18\ kJ/°C$

This is the portion of the heat capacity that can be attributed to $H_2O$.

Total heat capacity $= C_{cal} + C_{H_2O},\quad C_{cal} = 10.84 - 4.18 = 6.66\ kJ/°C$

81.    $Na_2SO_4(aq) + Ba(NO_3)_2(aq) \rightarrow BaSO_4(s) + 2\ NaNO_3(aq)\qquad \Delta H = ?$

$1.00\ L \times \dfrac{2.00\ mol}{L} = 2.00\ mol\ Na_2SO_4;\ 2.00\ L \times \dfrac{0.750\ mol}{L} = 1.50\ mol\ Ba(NO_3)_2$

The balanced equation requires a 1 : 1 mole ratio between $Na_2SO_4$ and $Ba(NO_3)_2$. Because we have fewer moles of $Ba(NO_3)_2$ present, it is limiting and 1.50 mol $BaSO_4$ will be produced [there is a 1 : 1 mole ratio between $Ba(NO_3)_2$ and $BaSO_4$].

Heat gain by solution = heat loss by reaction

Mass of solution $= 3.00\ L\ \times \dfrac{1000\ mL}{1\ L} \times \dfrac{2.00\ g}{mL} = 6.00 \times 10^3\ g$

Heat gain by solution $= \dfrac{6.37\ J}{°C\ g} \times 6.00 \times 10^3\ g \times (42.0 - 30.0)°C = 4.59 \times 10^5\ J$

Because the solution gained heat, the reaction is exothermic;   $q = -4.59 \times 10^5\ J$ for the reaction.

$\Delta H = \dfrac{-4.59 \times 10^5\ J}{1.50\ mol\ BaSO_4} = -3.06 \times 10^5\ J/mol = -306\ kJ/mol$

83.    $HCl(aq) + NaOH(aq) \rightarrow H_2O(l) + NaCl(aq)$    $\Delta H = -56$ kJ

$$0.2000 \text{ L} \times \frac{0.400 \text{ mol HCl}}{\text{L}} = 8.00 \times 10^{-2} \text{ mol HCl}$$

$$0.1500 \text{ L} \times \frac{0.500 \text{ mol NaOH}}{\text{L}} = 7.50 \times 10^{-2} \text{ mol NaOH}$$

Because the balanced reaction requires a 1 : 1 mole ratio between HCl and NaOH, and because fewer moles of NaOH are actually present as compared with HCl, NaOH is the limiting reagent.

$$7.50 \times 10^{-2} \text{ mol NaOH} \times \frac{-56 \text{ kJ}}{\text{mol NaOH}} = -4.2 \text{ kJ}; \quad 4.2 \text{ kJ of heat is released.}$$

85.    $w = -P\Delta V$;  $\Delta n$ = moles of gaseous products – moles of gaseous reactants.  Only gases can do PV work (we ignore solids and liquids).  When a balanced reaction has more moles of product gases than moles of reactant gases ($\Delta n$ positive), the reaction will expand in volume ($\Delta V$ positive), and the system will do work on the surroundings.  For example, in reaction c, $\Delta n = 2 - 0 = 2$ moles, and this reaction would do expansion work against the surroundings.  When a balanced reaction has a decrease in the moles of gas from reactants to products ($\Delta n$ negative), the reaction will contract in volume ($\Delta V$ negative), and the surroundings will do compression work on the system, e.g., reaction a, where $\Delta n = 0 - 1 = -1$.  When there is no change in the moles of gas from reactants to products, $\Delta V = 0$ and $w = 0$, e.g., reaction b, where $\Delta n = 2 - 2 = 0$.

When $\Delta V > 0$ ($\Delta n > 0$), then $w < 0$, and the system does work on the surroundings (c and e).

When $\Delta V < 0$ ($\Delta n < 0$), then $w > 0$, and the surroundings do work on the system (a and d).

When $\Delta V = 0$ ($\Delta n = 0$), then $w = 0$ (b).

87.    a.    aluminum oxide = $Al_2O_3$;  $2 \text{ Al}(s) + 3/2 \text{ O}_2(g) \rightarrow Al_2O_3(s)$

b.    $C_2H_5OH(l) + 3 \text{ O}_2(g) \rightarrow 2 \text{ CO}_2(g) + 3 \text{ H}_2O(l)$

c.    $Ba(OH)_2(aq) + 2 \text{ HCl}(aq) \rightarrow 2 \text{ H}_2O(l) + BaCl_2(aq)$

d.    $2 \text{ C(graphite, s)} + 3/2 \text{ H}_2(g) + 1/2 \text{ Cl}_2(g) \rightarrow C_2H_3Cl(g)$

e.    $C_6H_6(l) + 15/2 \text{ O}_2(g) \rightarrow 6 \text{ CO}_2(g) + 3 \text{ H}_2O(l)$

Note:  $\Delta H_{comb}$ values assume 1 mole of compound combusted.

f.    $NH_4Br(s) \rightarrow NH_4^+(aq) + Br^-(aq)$

## Challenge Problems

89.   $H_2O(s) \rightarrow H_2O(l)$   $\Delta H = \Delta H_{fus}$;  for 1 mol of supercooled water at $-15.0°C$ (or 258.2 K), $\Delta H_{fus, 258.2 K} = 10.9$ kJ/2.00 mol $= 5.45$ kJ/mol.  Using Hess's law and the equation $\Delta H = nC_p\Delta T$:

$H_2O(s, 273.2 \text{ K}) \rightarrow H_2O(s, 258.2 \text{ K})$      $\Delta H_1 = 1 \text{ mol}(37.5 \text{ J K}^{-1} \text{ mol}^{-1})(-15.0 \text{ K})$
$= -563 \text{ J} = -0.563 \text{ kJ}$

$H_2O(s, 258.2 \text{ K}) \rightarrow H_2O(l, 258.2 \text{ K})$      $\Delta H_2 = 1 \text{ mol}(5.45 \text{ kJ/mol}) = 5.45 \text{ kJ}$

$H_2O(l, 258.2 \text{ K}) \rightarrow H_2O(l, 273.2 \text{ K})$      $\Delta H_3 = 1 \text{ mol}(75.3 \text{ J K}^{-1} \text{ mol}^{-1})(15.0 \text{ K})$
$= 1130 \text{ J} = 1.13 \text{ kJ}$

$H_2O(s, 273.2 \text{ K}) \rightarrow H_2O(l, 273.2 \text{ K})$      $\Delta H_{fus, 273.2} = \Delta H_1 + \Delta H_2 + \Delta H_3$

$\Delta H_{fus, 273.2} = -0.563 \text{ kJ} + 5.45 \text{ kJ} + 1.13 \text{ kJ},$   $\Delta H_{fus, 273.2} = 6.02 \text{ kJ/mol}$

91.   Molar heat capacity of $H_2O(l) = 4.184$ J K$^{-1}$ g$^{-1}$(18.015 g/mol) $= 75.37$ J K$^{-1}$ mol$^{-1}$

Molar heat capacity of $H_2O(g) = 2.02$ J K$^{-1}$ g$^{-1}$(18.015 g/mol) $= 36.4$ J K$^{-1}$ mol$^{-1}$

Using Hess's law and the equation $\Delta H = nC_p\Delta T$:

$H_2O(l, 298.2 \text{ K}) \rightarrow H_2O(l, 373.2 \text{ K})$   $\Delta H_1 = 1 \text{ mol}(75.37 \text{ J K}^{-1} \text{ mol}^{-1})(75.0 \text{ K})(1 \text{ kJ}/1000 \text{ J})$
$= 5.65 \text{ kJ}$

$H_2O(l, 373.2 \text{ K}) \rightarrow H_2O(g, 373.2 \text{ K})$   $\Delta H_2 = 1 \text{ mol}(40.66 \text{ kJ/mol}) = 40.66 \text{ kJ}$

$H_2O(g, 373.2 \text{ K}) \rightarrow H_2O(g, 298.2 \text{ K})$   $\Delta H_3 = 1 \text{ mol}(36.4 \text{ J K}^{-1} \text{ mol}^{-1})(-75.0 \text{ K})(1 \text{ kJ}/1000 \text{ J})$
$= -2.73 \text{ kJ}$

$H_2O(l, 298.2 \text{ K}) \rightarrow H_2O(g, 298.2 \text{ K})$   $\Delta H_{vap, 298.2 K} = \Delta H_1 + \Delta H_2 + \Delta H_3 = 43.58 \text{ kJ/mol}$

Using $\Delta H_f^o$ values in Appendix 4 (which are determined at 25° C):

$\Delta H°_{vap} = -242 \text{ kJ} - (-286 \text{ kJ}) = 44 \text{ kJ}$

To two significant figures, the two calculated $\Delta H_{vap}$ values agree (as they should).

93.   Energy used in 8.0 hours $= 40.$ kWh $= \dfrac{40. \text{ kJ h}}{s} \times \dfrac{3600 \text{ s}}{h} = 1.4 \times 10^5$ kJ

Energy from the sun in 8.0 hours $= \dfrac{10. \text{ kJ}}{s \text{ m}^2} \times \dfrac{60 \text{ s}}{\min} \times \dfrac{60 \min}{h} \times 8.0 \text{ h} = 2.9 \times 10^4 \text{ kJ/m}^2$

Only 13% of the sunlight is converted into electricity:

$0.13 \times (2.9 \times 10^4 \text{ kJ/m}^2) \times \text{area} = 1.4 \times 10^5 \text{ kJ},$   area $= 37 \text{ m}^2$

95.   For an isothermal (constant T) process involving the expansion or compression of a gas, $\Delta E = nC_v\Delta T = 0$ ($\Delta H$ is also zero).  Because $\Delta E = q + w = 0$, $q = -w = -(-P_{ex}\Delta V)$. So if the expansion or compression occurs against some nonzero external pressure, $w \neq 0$ and $q \neq 0$. Instead, $q = -w = P_{ex}\Delta V$ (if the gas expands or contracts against some constant pressure).

# CHAPTER 10

# SPONTANEITY, ENTROPY, AND FREE ENERGY

## Spontaneity and Entropy

13.    We draw all the possible arrangements of the two particles in the three levels.

| 2 kJ | __ | | x | __ | x | xx |
|------|----|---|---|----|---|----|
| 1 kJ | __ | x | __ | xx | x | __ |
| 0 kJ | xx | x | x | __ | __ | __ |

Total E =    0 kJ    1 kJ    2 kJ    2 kJ    3 kJ    4 kJ

The most likely total energy is 2 kJ.

15.    Processes a, b, d, and g are spontaneous.  Processes c, e, and f require an external source of energy in order to occur since they are nonspontaneous.

17.    a.    Positional probability increases; there is a greater volume accessible to the randomly moving gas molecules, which increases disorder.

   b.    The positional probability doesn't change.  There is no change in volume and thus no change in the numbers of positions of the molecules.

   c.    Positional probability decreases because the volume decreases (P and V are inversely related).

19.    There are more ways to roll a seven.  We can consider all the possible throws by constructing a table.

| One die | 1 | 2 | 3 | 4 | 5 | 6 | |
|---------|---|---|---|---|---|---|---|
| 1 | 2 | 3 | 4 | 5 | 6 | 7 | |
| 2 | 3 | 4 | 5 | 6 | 7 | 8 | |
| 3 | 4 | 5 | 6 | 7 | 8 | 9 | Sum of the two dice |
| 4 | 5 | 6 | 7 | 8 | 9 | 10 | |
| 5 | 6 | 7 | 8 | 9 | 10 | 11 | |
| 6 | 7 | 8 | 9 | 10 | 11 | 12 | |

There are six ways to get a seven, more than any other number.  The seven is not favored by energy; rather, it is favored by probability.  To change the probability, we would have to expend energy (do work).

## Energy, Enthalpy, and Entropy Changes Involving Ideal Gases and Physical Changes

21.    $1.00 \times 10^3$ g $C_2H_6 \times \dfrac{1 \text{ mol}}{30.07 \text{ g}} = 33.3$ mol

$q_v = \Delta E = nC_v\Delta T = 33.3$ mol$(44.60$ J K$^{-1}$ mol$^{-1})(48.4$ K$) = 7.19 \times 10^4$ J $= 71.9$ kJ

At constant volume, 71.9 kJ of energy is required, and $\Delta E = 71.9$ kJ.

At constant pressure (assuming ethane acts as an ideal gas):

$C_p = C_v + R = 44.60 + 8.31 = 52.91$ J K$^{-1}$ mol$^{-1}$

Energy required $= q_p = \Delta H = nC_p\Delta T = 33.3$ mol$(52.91$ J K$^{-1}$ mol$^{-1})(48.4$ K$) = 8.53 \times 10^4$ J
$= 85.3$ kJ

For the constant pressure process, $\Delta E = 71.9$ kJ, as calculated previously ($\Delta E$ is unchanged).

23.    Heat gain by He = heat loss by $N_2$;  because $\Delta T$ in °C = $\Delta T$ in K, the units on the heat capacities also could be J °C$^{-1}$ mol$^{-1}$.

$(0.400$ mol$)(12.5$ J °C$^{-1}$ mol$^{-1})(T_f - 20.0$ °C$) = (0.600$ mol$)(20.7$ J °C$^{-1}$ mol$^{-1})(100.0$ °C $- T_f)$

$(5.00)T_f - 100. = 1240 - (12.4)T_f,$  $T_f = \dfrac{1340}{17.4} = 77.0$ °C

25.    $P_1V_1 = P_2V_2$ since n and T are constant;  $P_2 = \dfrac{P_1V_1}{V_2} = \dfrac{(5.0)(1.0)}{(2.0)} = 2.5$ atm

Gas expands isothermally against no pressure, so $\Delta E = 0$, $w = 0$, and $q = 0$.

$\Delta E = 0$, so $q_{rev} = -w_{rev} = nRT \ln(V_2/V_1)$;  $T = \dfrac{PV}{nR} = 61$ K

$q_{rev} = (1.0$ mol$)(8.3145$ J K$^{-1}$ mol$^{-1})(61$ K$) \ln(2.0/1.0) = 350$ J

27.    a.   $q_v = \Delta E = nC_v\Delta T = (1.000$ mol$)(28.95$ J K$^{-1}$ mol$^{-1})(350.0 - 298.0$ K$)$

$q_v = 1.51 \times 10^3$ J $= 1.51$ kJ

$q_p = \Delta H = nC_p\Delta T = (1.000)(37.27)(350.0 - 298.0) = 1.94 \times 10^3$ J $= 1.94$ kJ

b.   $\Delta S = S_{350} - S_{298} = nC_p \ln(T_2/T_1)$

$S_{350} - 213.64$  J/K $= (1.000 \text{ mol})(37.27 \text{ J K}^{-1} \text{ mol}^{-1}) \ln(350.0/298.0)$

$S_{350} = 213.64$ J/K $+ 5.994$ J/K $= 219.63$ J/K $=$ molar entropy at 350.0 K and 1.000 atm

c.   $\Delta S = nR \ln(V_2/V_1)$,   $V = nRT/P$,   $\Delta S = nR \ln(P_1/P_2) = S_{(350, 1.174)} - S_{(350, 1.000)}$

$\Delta S = S_{(350, 1.174)} - 219.63$ J/K $= (1.000 \text{ mol})(8.3145 \text{ J K}^{-1} \text{ mol}^{-1}) \ln(1.000 \text{ atm}/1.174 \text{ atm})$

$\Delta S = -1.334$ J/K $= S_{(350, 1.174)} - 219.63$,   $S_{(350, 1.174)} = 218.30 \text{ J K}^{-1} \text{ mol}^{-1}$

29.     For A(l, 125°C) → A(l, 75°C):

$\Delta S = nC_p \ln(T_2/T_1) = 1.00 \text{ mol}(75.0 \text{ J K}^{-1} \text{ mol}^{-1}) \ln(348 \text{ K}/398 \text{ K}) = -10.1 \text{ J/K}$

For A(l, 75°C) → A(g, 155°C):   $\Delta S = 75.0 \text{ J K}^{-1} \text{ mol}^{-1}$

For A(g, 155°C) → A(g, 125°C):

$\Delta S = nC_p \ln(T_2/T_1) = 1.00 \text{ mol}(29.0 \text{ J K}^{-1} \text{ mol}^{-1}) \ln(398 \text{ K}/428 \text{ K}) = -2.11 \text{ J/K}$

The sum of the three step gives A(l, 125°C) → A(g, 125°C).  $\Delta S$ for this process is the sum of $\Delta S$ for each of the three steps.

$\Delta S = -10.1 + 75.0 - 2.11 = 62.8 \text{ J/K}$

For a phase change, $\Delta S = \Delta H/T$.  At 125°C:  $\Delta H_{vap} = T\Delta S = 398 \text{ K}(62.8 \text{ J/K}) = 2.50 \times 10^4 \text{ J}$

31.     Calculate the final temperature by equating heat loss to heat gain.  Keep all quantities positive to avoid sign errors.

$(3.00 \text{ mol})(75.3 \text{ J °C}^{-1} \text{ mol}^{-1})(T_f - 0°C) = (1.00 \text{ mol})(75.3 \text{ J °C}^{-1} \text{ mol}^{-1})(100.°C - T_f)$

Solving:  $T_f = 25°C = 298 \text{ K}$

Now we can calculate $\Delta S$ for the various changes using $\Delta S = nC_p \ln(T_2/T_1)$.

Heat 3 mol $H_2O$:  $\Delta S_1 = (3.00 \text{ mol})(75.3 \text{ J K}^{-1} \text{ mol}^{-1}) \ln(298 \text{ K}/273 \text{ K}) = 19.8 \text{ J/K}$

Cool 1 mol $H_2O$:  $\Delta S_2 = (1.00 \text{ mol})(75.3 \text{ J K}^{-1} \text{ mol}^{-1}) \ln(298/373) = -16.9 \text{ J/K}$

$\Delta S_{total} = \Delta S_{heat} + \Delta S_{cool} = 19.8 - 16.9 = 2.9 \text{ J/K}$

## Entropy and the Second Law of Thermodynamics:  Free Energy

33.     Living organisms need an external source of energy to carry out these processes.  Green plants use the energy from sunlight to produce glucose from carbon dioxide and water by

photosynthesis.  In the human body the energy released from the metabolism of glucose helps drive the synthesis of proteins.  For all processes combined, $\Delta S_{univ}$ must be greater than zero (second law).

35.   $\Delta S_{surr}$ is primarily determined by heat flow. This heat flow into or out of the surroundings comes from the heat flow out of or into the system. In an exothermic process ($\Delta H < 0$), heat flows into the surroundings from the system. The heat flow into the surroundings increases the random motions in the surroundings and increases the entropy of the surroundings ($\Delta S_{surr} > 0$). This is a favorable driving force for spontaneity. In an endothermic reaction ($\Delta H > 0$), heat is transferred from the surroundings into the system. This heat flow out of the surroundings decreases the random motions in the surroundings and decreases the entropy of the surroundings ($\Delta S_{surr} < 0$). This is unfavorable. The magnitude of $\Delta S_{surr}$ also depends on the temperature. The relationship is inverse; at low temperatures, a specific amount of heat exchange makes a larger percent change in the surroundings than the same amount of heat flow at a higher temperature. The negative sign in the $\Delta S_{surr} = -\Delta H/T$ equation is necessary to get the signs correct. For an exothermic reaction where $\Delta H$ is negative, this increases $\Delta S_{surr}$, so the negative sign converts the negative $\Delta H$ value into a positive quantity. For an endothermic process where $\Delta H$ is positive, the sign of $\Delta S_{surr}$ is negative, and the negative sign converts the positive $\Delta H$ value into a negative quantity.

37.   a.   $\Delta S_{surr} = \dfrac{-\Delta H}{T} = \dfrac{-(-2221\,kJ)}{298\,K} = 7.45\,kJ/K = 7.45 \times 10^3\,J/K$

      b.   $\Delta S_{surr} = \dfrac{-\Delta H}{T} = \dfrac{-112\,kJ}{298\,K} = -0.376\,kJ/K = -376\,J/K$

39.   a.   Decrease in positional probability; $\Delta S°(-)$

      b.   Increase in positional probability; $\Delta S°(+)$

      c.   Decrease in positional probability ($\Delta n < 0$); $\Delta S°(-)$

      d.   Decrease in positional probability ($\Delta n < 0$);  $\Delta S°(-)$

      e.   HCl(g) has a greater positional probability due to the huge volume of gas;  $\Delta S°(-)$.

      f.   Increase in positional probability;  $\Delta S°(+)$

      For c, d, and e, concentrate on the gaseous products and reactants.  When there are more gaseous product molecules than gaseous reactant molecules ($\Delta n > 0$), then $\Delta S°$ will be positive.  When $\Delta n$ is negative, then $\Delta S°$ is negative.

41.   $C_2H_2(g) + 4\,F_2(g) \rightarrow 2\,CF_4(g) + H_2(g)$;  $\Delta S° = 2S°_{CF_4} + S°_{H_2} - [S°_{C_2H_2} + 4S°_{F_2}]$

      $-358\,J/K = (2\,mol)S°_{CF_4} + 131\,J/K - [201\,J/K + 4(203\,J/K)]$,  $S°_{CF_4} = 262\,J\,K^{-1}\,mol^{-1}$

43.   $-144\,J/K = (2\,mol)\,S°_{AlBr_3} - [2(28\,J/K) + 3(152\,J/K)]$,  $S°_{AlBr_3} = 184\,J\,K^{-1}\,mol^{-1}$

45.   At the boiling point, $\Delta G = 0$ so $\Delta H = T\Delta S$.   $T = \dfrac{\Delta H}{\Delta S} = \dfrac{58.51 \times 10^3\,J/mol}{92.92\,J\,K^{-1}\,mol^{-1}} = 629.7\,K$

47.    a.    $NH_3(s) \rightarrow NH_3(l)$;  $\Delta G = \Delta H - T\Delta S = 5650 \text{ J/mol} - 200. \text{ K}(28.9 \text{ J K}^{-1} \text{ mol}^{-1})$

$\Delta G = 5650 \text{ J/mol} - 5780 \text{ J/mol} = -130 \text{ J/mol}$

Yes, $NH_3$ will melt because $\Delta G < 0$ at this temperature.

b.    At the melting point, $\Delta G = 0$, so $T = \dfrac{\Delta H}{\Delta S} = \dfrac{5650 \text{ J/mol}}{28.9 \text{ J K}^{-1} \text{ mol}^{-1}} = 196 \text{ K}$.

49.    Solid I $\rightarrow$ solid II;  equilibrium occurs when $\Delta G = 0$.

$\Delta G = \Delta H - T\Delta S$,  $\Delta H = T\Delta S$,  $T = \Delta H/\Delta S = \dfrac{-743.1 \text{ J/mol}}{-17.0 \text{ J K}^{-1} \text{ mol}^{-1}} = 43.7 \text{ K} = -229.5°C$

## Free Energy and Chemical Reactions

51.    $-5490. \text{ kJ} = 8(-394 \text{ kJ}) + 10(-237 \text{ kJ}) - 2 \Delta G^{\circ}_{f, C_4H_{10}}$,  $\Delta G^{\circ}_{f, C_4H_{10}} = -16 \text{ kJ/mol}$

53.    $\Delta G° = -58.03 \text{ kJ} - (298 \text{ K})(-0.1766 \text{ kJ/K}) = -5.40 \text{ kJ}$

$\Delta G° = 0 = \Delta H° - T\Delta S°$,  $T = \dfrac{\Delta H°}{\Delta S°} = \dfrac{-58.03 \text{ kJ}}{-0.1766 \text{ kJ/K}} = 328.6 \text{ K}$

$\Delta G°$ is negative below 328.6 K, where the favorable $\Delta H°$ term dominates.

55.    a.    $\Delta G° = 2(-270. \text{ kJ}) - 2(-502 \text{ kJ}) = 464 \text{ kJ}$

b.    Because $\Delta G°$ is positive, this reaction is not spontaneous at standard conditions at 298 K.

c.    $\Delta G° = \Delta H° - T\Delta S°$,  $\Delta H° = \Delta G° + T\Delta S° = 464 \text{ kJ} + 298 \text{ K}(0.179 \text{ kJ/K}) = 517 \text{ kJ}$

We need to solve for the temperature when $\Delta G° = 0$:

$\Delta G° = 0 = \Delta H° - T\Delta S°$,  $T = \dfrac{\Delta H°}{\Delta S°} = \dfrac{517 \text{ kJ}}{0.179 \text{ kJ/K}} = 2890 \text{ K}$

This reaction will be spontaneous at standard conditions ($\Delta G° < 0$) when T > 2890 K.  At these temperatures, the favorable entropy term will dominate.

57.    $CH_4(g) + CO_2(g) \rightarrow CH_3CO_2H(l)$

$\Delta H° = -484 - [-75 + (-393.5)] = -16 \text{ kJ}$;  $\Delta S° = 160. - (186 + 214) = -240. \text{ J/K}$

$\Delta G° = \Delta H° - T\Delta S° = -16 \text{ kJ} - (298 \text{ K})(-0.240 \text{ kJ/K}) = 56 \text{ kJ}$

At standard concentrations, where $\Delta G = \Delta G°$, this reaction is spontaneous only at temperatures below $T = \Delta H°/\Delta S° = 67$ K (where the favorable $\Delta H°$ term will dominate, giving a negative $\Delta G°$ value). This is not practical. Substances will be in condensed phases and rates will be very slow at this extremely low temperature.

$CH_3OH(g) + CO(g) \rightarrow CH_3CO_2H(l)$

$\Delta H° = -484 - [-110.5 + (-201)] = -173$ kJ;  $\Delta S° = 160. - (198 + 240.) = -278$ J/K

$\Delta G° = -173$ kJ $- (298$ K$)(-0.278$ kJ/K$) = -90.$ kJ

This reaction also has a favorable enthalpy and an unfavorable entropy term. This reaction is spontaneous at temperatures below $T = \Delta H°/\Delta S° = 622$ K (assuming standard concentrations). The reaction of $CH_3OH$ and CO will be preferred at standard conditions. It is spontaneous at high enough temperatures that the rates of reaction should be reasonable.

59.    Enthalpy is not favorable, so $\Delta S$ must provide the driving force for the change. Thus $\Delta S$ is positive. There is an increase in positional probability, so the original enzyme has the more ordered structure (has the smaller positional probability).

61.    Because there are more product gas molecules than reactant gas molecules ($\Delta n > 0$), $\Delta S$ will be positive. From the signs of $\Delta H$ and $\Delta S$, this reaction is spontaneous at all temperatures. It will cost money to heat the reaction mixture. Because there is no thermodynamic reason to do this, the purpose of the elevated temperature must be to increase the rate of the reaction, i.e., kinetic reasons.

## Free Energy:  Pressure Dependence and Equilibrium

63.    At constant temperature and pressure, the sign of $\Delta G$ (positive or negative) tells us which reaction is spontaneous (the forward or reverse reaction). If $\Delta G < 0$, then the forward reaction is spontaneous and if $\Delta G > 0$, then the reverse reaction is spontaneous. If $\Delta G = 0$, then the reaction is at equilibrium (neither the forward nor reverse reaction is spontaneous). $\Delta G°$ gives the equilibrium position by deter-mining K for a reaction utilizing the equation $\Delta G° = -RT \ln$ K. $\Delta G°$ can only be used to predict spontaneity when all reactants and products are present at standard pressures of 1 atm and/or standard concentrations of 1 $M$.

65.    $\Delta G = \Delta G° + RT \ln Q = \Delta G° + RT \ln \dfrac{P_{N_2O_4}}{P_{NO_2}^2}$

$\Delta G° = 1$ mol$(98$ kJ/mol$) - 2$ mol$(52$ kJ/mol$) = -6$ kJ

a.    These are standard conditions, so $\Delta G = \Delta G°$ since $Q = 1$ and $\ln Q = 0$. Because $\Delta G°$ is negative, the forward reaction is spontaneous. The reaction shifts right to reach equilibrium.

b.    $\Delta G = -6 \times 10^3$ J $+ 8.3145$ J K$^{-1}$ mol$^{-1}(298$ K$) \ln \dfrac{0.50}{(0.21)^2}$

$\Delta G = -6 \times 10^3$ J $+ 6.0 \times 10^3$ J $= 0$

Since $\Delta G = 0$, this reaction is at equilibrium (no shift).

c.  $\Delta G = -6 \times 10^3 \text{ J} + 8.3145 \text{ J K}^{-1} \text{ mol}^{-1}(298 \text{ K}) \ln \dfrac{1.6}{(0.29)^2}$

$\Delta G = -6 \times 10^3 \text{ J} + 7.3 \times 10^3 \text{ J} = 1.3 \times 10^3 \text{ J} = 1 \times 10^3 \text{ J}$

Since $\Delta G$ is positive, the reverse reaction is spontaneous, so the reaction shifts to the left to reach equilibrium.

67.   $\Delta H^\circ = 2 \Delta H^\circ_{f, NH_3} = 2(-46) = -92 \text{ kJ}; \ \Delta G^\circ = 2 \Delta G^\circ_{f, NH_3} = 2(-17) = -34 \text{ kJ}$

$\Delta S^\circ = 2(193 \text{ J/K}) - [192 \text{ J/K} + 3(131 \text{ J/K})] = -199 \text{ J/K}; \ \Delta G^\circ = -RT \ln K$

$K = \dfrac{-\Delta G^\circ}{RT} = \exp\left( \dfrac{-(-34,000 \text{ J})}{(8.3145 \text{ J K}^{-1} \text{ mol}^{-1})(298 \text{ K})} \right) = e^{13.72} = 9.1 \times 10^5$

*Note*: When determining exponents, we will round off after the calculation is complete. This helps eliminate excessive round off error.

a.   $\Delta G = \Delta G^\circ + RT \ln \dfrac{P^2_{NH_3}}{P_{N_2} \times P^3_{H_2}} = -34 \text{ kJ} + \dfrac{(8.3145 \text{ J K}^{-1} \text{ mol}^{-1})(298 \text{ K})}{1000 \text{ J} / \text{kJ}} \ln \dfrac{(50.)^2}{(200.)(200.)^3}$

$\Delta G = -34 \text{ kJ} - 33 \text{ kJ} = -67 \text{ kJ}$

b.   $\Delta G = -34 \text{ kJ} + \dfrac{(8.3145 \text{ J K}^{-1} \text{ mol}^{-1})(298 \text{ K})}{1000 \text{ J} / \text{kJ}} \ln \dfrac{(200.)^2}{(200.)(600.)^3}$

$\Delta G = -34 \text{ kJ} - 34.4 \text{ kJ} = -68 \text{ kJ}$

c.   Assume $\Delta H^\circ$ and $\Delta S^\circ$ are temperature independent.

$\Delta G^\circ_{100} = \Delta H^\circ - T \Delta S^\circ, \ \ \Delta G^\circ_{100} = -92 \text{ kJ} - (100. \text{ K})(-0.199 \text{ kJ/K}) = -72 \text{ kJ}$

$\Delta G_{100} = \ \Delta G^\circ_{100} + RT \ln Q = -72 \text{ kJ} + \dfrac{(8.3145 \text{ J K}^{-1} \text{ mol}^{-1})(100. \text{ K})}{1000 \text{ J} / \text{kJ}} \ln \dfrac{(10.)^2}{(50.)(200.)^3}$

$\Delta G_{100} = -72 \text{ kJ} - 13 \text{ kJ} = -85 \text{ kJ}$

d.   $\Delta G^\circ_{700} = -92 \text{ kJ} - (700. \text{ K})(-0.199 \text{ kJ/K}) = 47 \text{ kJ}$

$\Delta G_{700} = 47 \text{ kJ} + \dfrac{(8.3145 \text{ J K}^{-1} \text{ mol}^{-1})(700. \text{ K})}{1000 \text{ J} / \text{kJ}} \ln \dfrac{(10.)^2}{(50.)(200.)^3}$

$\Delta G_{700} = 47 \text{ kJ} - 88 \text{ kJ} = -41 \text{ kJ}$

69.    $\Delta G° = -RT \ln K$.  To determine K at a temperature other than 25°C, one needs to know $\Delta G°$ at that temperature. We assume $\Delta H°$ and $\Delta S°$ are temperature-independent and use the equation $\Delta G° = \Delta H° - T\Delta S°$ to estimate $\Delta G°$ at the different temperature. For K = 1, we want $\Delta G° = 0$, which occurs when $\Delta H° = T\Delta S°$. Again, assume $\Delta H°$ and $\Delta S°$ are temperature independent, and then solve for T $(= \Delta H°/\Delta S°)$. At this temperature, K = 1 because $\Delta G° = 0$. This only works for reactions where the signs of $\Delta H°$ and $\Delta S°$ are the same (either both positive or both negative). When the signs are opposite, K will always be greater than one (when $\Delta H°$ is negative and $\Delta S°$ is positive) or K will always be less than one (when $\Delta H°$ is positive and $\Delta S°$ is negative). When the signs of $\Delta H°$ and $\Delta S°$ are opposite, K can never equal one.

71.    $K = \dfrac{P_{NF_3}^2}{P_{N_2}^2 \times P_{F_2}^3} = \dfrac{(0.48)^2}{0.021(0.063)^3} = 4.4 \times 10^4$

$\Delta G_{800}° = -RT \ln K = -8.3145 \text{ J K}^{-1} \text{ mol}^{-1}(800. \text{ K}) \ln(4.4 \times 10^4) = -7.1 \times 10^4 \text{ J/mol}$

$= -71 \text{ kJ/mol}$

73.    Because the partial pressure of C(g) decreased, the net change that occurs for this reaction to reach equilibrium is for products to convert to reactants.

|          | A(g)        | + | 2 B(g)      | ⇌ | C(g)        |
|----------|-------------|---|-------------|---|-------------|
| Initial  | 0.100 atm   |   | 0.100 atm   |   | 0.100 atm   |
| Change   | $+x$        |   | $+2x$       | ← | $-x$        |
| Equil.   | $0.100 + x$ |   | $0.100 + 2x$|   | $0.100 - x$ |

From the problem, $P_C = 0.040$ atm $= 0.100 - x$,  $x = 0.060$ atm.

The equilibrium partial pressures are:  $P_A = 0.100 + x = 0.100 + 0.060 = 0.160$ atm,

$P_B = 0.100 + 2(0.60) = 0.220$ atm, and $P_C = 0.040$ atm.

$K = \dfrac{0.040}{0.160(0.220)^2} = 5.2$

$\Delta G° = -RT \ln K = -8.3145 \text{ J K}^{-1} \text{ mol}^{-1}(298 \text{ K}) \ln(5.2) = -4.1 \times 10^3 \text{ J/mol} = -4.1 \text{ kJ/mol}$

75.         $\text{HgbO}_2 \rightarrow \text{Hgb} + \text{O}_2 \qquad \Delta G° = -(-70 \text{ kJ})$
$\text{Hgb} + \text{CO} \rightarrow \text{HgbCO} \qquad \Delta G° = -80 \text{ kJ}$

$\overline{\text{HgbO}_2 + \text{CO} \rightarrow \text{HgbCO} + \text{O}_2 \quad \Delta G° = -10 \text{ kJ}}$

$\Delta G° = -RT \ln K, \quad K = \exp\left(\dfrac{-\Delta G°}{RT}\right) = \exp\left(\dfrac{-(-10 \times 10^3 \text{ J})}{(8.3145 \text{ J K}^{-1} \text{ mol}^{-1})(298 \text{ K})}\right) = 60$

77.    At 25.0°C:  $\Delta G° = \Delta H° - T\Delta S° = -58.03 \times 10^3 \text{ J/mol} - (298.2 \text{ K})(-176.6 \text{ J K}^{-1} \text{ mol}^{-1})$

$= -5.37 \times 10^3 \text{ J/mol}$

$$\Delta G^\circ = -RT \ln K, \ln K = \frac{-\Delta G^\circ}{RT} = \frac{-(-5.37 \times 10^3 \, J/mol)}{(8.3145 \, J \, K^{-1} \, mol^{-1})(298.2 \, K)} = 2.166; \quad K = e^{2.166} = 8.72$$

At 100.0°C: $\Delta G^\circ = -58.03 \times 10^3 \, J/mol - (373.2 \, K)(-176.6 \, J \, K^{-1} \, mol^{-1}) = 7.88 \times 10^3 \, J/mol$

$$\ln K = \frac{-(7.88 \times 10^3 \, J/mol)}{(8.3145 \, J \, K^{-1} \, mol^{-1})(373.2 \, K)} = -2.540, \quad K = e^{-2.540} = 0.0789$$

79.    A graph of ln K vs. 1/T will yield a straight line with slope equal to -$\Delta H^\circ$/R and $y$ intercept equal to $\Delta S^\circ$/R.

a.

| Temp (°C) | T(K) | 1000/T (K$^{-1}$) | $K_w$ | ln $K_w$ |
|---|---|---|---|---|
| 0 | 273 | 3.66 | $1.14 \times 10^{-15}$ | −34.408 |
| 25 | 298 | 3.36 | $1.00 \times 10^{-14}$ | −32.236 |
| 35 | 308 | 3.25 | $2.09 \times 10^{-14}$ | −31.499 |
| 40. | 313 | 3.19 | $2.92 \times 10^{-14}$ | −31.165 |
| 50. | 323 | 3.10 | $5.47 \times 10^{-14}$ | −30.537 |

The straight-line equation (from a calculator) is: $\ln K = -6.91 \times 10^3 \left(\dfrac{1}{T}\right) - 9.09$

$$\text{Slope} = -6.91 \times 10^3 \, K = \frac{-\Delta H^\circ}{R}$$

$$\Delta H^\circ = -(-6.91 \times 10^3 \, K \times 8.3145 \, J \, K^{-1} \, mol^{-1}) = 5.75 \times 10^4 \, J/mol = 57.5 \, kJ/mol$$

$$y \text{ intercept} = -9.09 = \frac{\Delta S^\circ}{R}, \quad \Delta S^\circ = -9.09 \times 8.3145 \, J \, K^{-1} \, mol^{-1} = -75.6 \, J \, K^{-1} \, mol^{-1}$$

b.   From part a,  $\Delta H° = 57.5$ kJ/mol and $\Delta S° = -75.6$ J $K^{-1}$ $mol^{-1}$.  Assuming that $\Delta H°$ and $\Delta S°$ are temperature independent:

$$\Delta G° = 57,500 \text{ J/mol} - 647 \text{ K}(-75.6 \text{ J K}^{-1} \text{ mol}^{-1}) = 106,400 \text{ J/mol} = 106.4 \text{ kJ/mol}$$

81.   From Exercise 10.78, $\ln K = \dfrac{-\Delta H°}{RT} + \dfrac{\Delta S°}{R}$,  R = 8.3145 J $K^{-1}$ $mol^{-1}$

For two sets of K and T:

$$\ln K_1 = \frac{-\Delta H°}{R}\left(\frac{1}{T_1}\right) + \frac{\Delta S°}{R}; \quad \ln K_2 = \frac{-\Delta H°}{R}\left(\frac{1}{T_2}\right) + \frac{\Delta S°}{R}$$

Subtracting the first expression from the second:

$$\ln K_2 - \ln K_1 = \frac{\Delta H°}{R}\left(\frac{1}{T_1} - \frac{1}{T_2}\right) \text{ or } \ln \frac{K_2}{K_1} = \frac{\Delta H°}{R}\left(\frac{1}{T_1} - \frac{1}{T_2}\right)$$

$$\ln\left(\frac{3.25 \times 10^{-2}}{8.84}\right) = \frac{\Delta H°}{8.3145 \text{ J K}^{-1}\text{mol}^{-1}}\left(\frac{1}{298 \text{ K}} - \frac{1}{348 \text{ K}}\right)$$

$-5.61 = (5.8 \times 10^{-5} \text{ mol/J})(\Delta H°)$,  $\Delta H° = -9.7 \times 10^4$ J/mol

For K = 8.84 at T = 25°C:

$$\ln(8.84) = \frac{-(-9.7 \times 10^4 \text{ J/mol})}{(8.3145 \text{ J K}^{-1}\text{mol}^{-1})(298 \text{ K})} + \frac{\Delta S°}{8.3145 \text{ J K}^{-1}\text{mol}^{-1}}$$

$$\frac{\Delta S°}{8.3145} = -37, \ \Delta S° = -310 \text{ J K}^{-1}\text{mol}^{-1}$$

We get the same value for $\Delta S°$ using K = $3.25 \times 10^{-2}$ at T = 348 K data.  $\Delta G° = -RT \ln K$. When K = 1.00 then $\Delta G° = 0$ since $\ln(1.00) = 0$.  $\Delta G° = 0 = \Delta H° - T\Delta S°$.  Assuming that $\Delta H°$ and $\Delta S°$ do not depend on temperature:

$$\Delta H° = T\Delta S°, \ T = \frac{\Delta H°}{\Delta S°} = \frac{-9.7 \times 10^4 \text{ J/mol}}{-310 \text{ J K}^{-1}\text{mol}^{-1}} = 310 \text{ K}$$

## Adiabatic Processes

83.   For reversible adaiabatic process, $P_1 V_1^{\gamma} = P_2 V_2^{\gamma}$, where $\gamma = C_p/C_v$. For an ideal gas, $C_p = C_v + R$.

$C_p = C_v + R = 20.5$ J $K^{-1}$ $mol^{-1} + 8.3145$ J $K^{-1}$ $mol^{-1} = 28.8$ J $K^{-1}$ $mol^{-1}$

$$V_2^\gamma = \frac{P_1 V_1^\gamma}{P_2}, \text{ where } \gamma = \frac{C_p}{C_v} = \frac{28.8}{20.5} = 1.40$$

$$V_1 = \frac{nRT_1}{P_1} = \frac{1.75 \text{ mol}(0.08206 \text{ L atm K}^{-1} \text{ mol}^{-1})(294 \text{ K})}{1.50 \text{ atm}} = 28.1 \text{ L}$$

$$V_2^{1.40} = \frac{1.50 \text{ atm}(28.1 \text{ L})^{1.40}}{4.50 \text{ atm}} = 35.6, \ V_2 = (35.6)^{1/1.40} = 12.8 \text{ L}$$

To calculate $w = \Delta E = nC_V \Delta T$, we need $T_2$ in order to calculate $\Delta T$.

$$T_2 = \frac{P_2 V_2}{nR} = \frac{4.50 \text{ atm}(12.8 \text{ L})}{1.75 \text{ mol}(0.08206 \text{ L atm K}^{-1} \text{ mol}^{-1})} = 401 \text{ K}$$

Note: we also could have used the formula $T_2/T_1 = (V_1/V_2)^{\gamma-1}$ to calculate $T_2$.

$$w = \Delta E = nC_V \Delta T = 1.75 \text{ mol}(20.5 \text{ J K}^{-1} \text{ mol}^{-1})(401 \text{ K} - 294 \text{ K}) = 3.84 \times 10^3 \text{ J}$$

Note that for the adiabatic reversible compression, T increases, so $\Delta E$ increases. As work is added during the compression, the internal energy of the system increases ($\Delta E = w$).

85.    a.    Isothermal reversible expansion:  $\Delta T = 0$ so $\Delta E = 0$

$$w = -nRT \ln(V_2/V_1) = -nRT \ln(P_1/P_2)$$

$$w = -1.00 \text{ mol}(8.3145 \text{ J K}^{-1} \text{ mol}^{-1})(300. \text{ K}) \ln(5.00/1.00) = -4.01 \times 10^3 \text{ J}$$

Since $\Delta E = 0$,  $q = -w = 4.01 \times 10^3$ J.

b.    Isothermal irreversible expansion against a constant external pressure of 1.00 atm, so $\Delta E = 0$,  and $q = -w$.

$$w = -P_{ex}\Delta V; \ V_1 = \frac{1.00 \text{ mol}(0.08206 \text{ L atm K}^{-1} \text{ mol}^{-1})300. \text{ K}}{5.00 \text{ atm}} = 4.92 \text{ L}$$

$$V_f = \frac{1.00 \text{ mol}(0.08206)(300. \text{ K})}{1.00 \text{ atm}} = 24.6 \text{ L}$$

$$w = -(1.00 \text{ atm})(24.6 \text{ L} - 4.92 \text{ L}) \times 101.3 \text{ J L}^{-1} \text{ atm}^{-1} = -1.99 \times 10^3 \text{ J}$$

$$q = -w = 1.99 \times 10^3 \text{ J}$$

As expected, we get more work from the reversible process.

c.  Adiabatic reversible expansion:  $q = 0$ so $\Delta E = w$, $\Delta E = nC_v\Delta T$;  we need $C_v$ and $\Delta T$ to calculate $\Delta E$ and, in turn, $w$. In order to calculate $T_2$, we need to determine $V_2$.

$P_1V_1^{\gamma} = P_2V_2^{\gamma}$, where $\gamma = C_p/C_v$.  From part b, $V_1 = 4.92$ L.

For a gas behaving ideally:

$$C_p = C_v + R,\ C_v = C_p - R = 37.1\ \text{J K}^{-1}\ \text{mol}^{-1} - 8.3145\ \text{J K}^{-1}\ \text{mol}^{-1} = 28.8\ \text{J K}^{-1}\ \text{mol}^{-1}$$

$$\frac{C_p}{C_v} = \frac{37.1}{28.8} = 1.29;\ V_2^{1.29} = \frac{P_1V_1^{1.29}}{P_2}$$

$$V_2^{1.29} = \frac{5.00\ \text{atm}(4.92\ \text{L})^{1.29}}{1.00\ \text{atm}} = 39.0,\ V_2 = 17.1\ \text{L}$$

$$T_2 = \frac{P_2V_2}{nR} = \frac{1.00\ \text{atm}(17.1\ \text{L})}{1.00\ \text{mol}(0.08206\ \text{L atm K}^{-1}\ \text{mol}^{-1})} = 208\ \text{K}$$

$$\Delta E = w = nC_v\Delta T = 1.00\ \text{mol}(28.8\ \text{J K}^{-1}\ \text{mol}^{-1})(208\ \text{K} - 300.\ \text{K}) = -2.65 \times 10^3\ \text{J}$$

As expected, we do not get as much work in the adiabatic reversible process as when the gas expands isothermally and reversibly.

## Additional Exercises

87.  When an ionic solid dissolves, positional probability increases, so $\Delta S_{sys}$ is positive.  Since temperature increased as the solid dissolved, this is an exothermic process, and $\Delta S_{surr}$ is positive ($\Delta S_{surr} = -\Delta H/T$).  Since the solid did dissolve, the dissolving process is spontaneous, so $\Delta S_{univ}$ is positive (as it must be when $\Delta S_{sys}$ and $\Delta S_{surr}$ are both positive).

89.  $w_{max} = \Delta G$;  when $\Delta G$ is negative, the magnitude of $\Delta G$ is equal to the maximum possible useful work obtainable from the process (at constant T and P).  When $\Delta G$ is positive, the magnitude of $\Delta G$ is equal to the minimum amount of work that must be expended to make the process spontaneous.  Due to waste energy (heat) in any real process, the amount of useful work obtainable from a spontaneous process is always less than $w_{max}$, and for a nonspontaneous reaction, an amount of work greater than $w_{max}$ must be applied to make the process spontaneous.

91.  $HF(aq) \rightleftharpoons H^+(aq) + F^-(aq)$;  $\Delta G = \Delta G^\circ + RT \ln \dfrac{[H^+][F^-]}{[HF]}$

$\Delta G^\circ = -RT \ln K = -(8.3145\ \text{J K}^{-1}\ \text{mol}^{-1})(298\ \text{K}) \ln(7.2 \times 10^{-4}) = 1.8 \times 10^4\ \text{J/mol}$

a.  The concentrations are all at standard conditions, so $\Delta G = \Delta G = 1.8 \times 10^4$ J/mol (since $Q = 1.0$ and $\ln Q = 0$).  Because $\Delta G^\circ$ is positive, the reaction shifts left to reach equilibrium.

b.   $\Delta G = 1.8 \times 10^4 \text{ J/mol} + (8.3145 \text{ J K}^{-1} \text{ mol}^{-1})(298 \text{ K}) \ln \dfrac{(2.7 \times 10^{-2})^2}{0.98}$

$\Delta G = 1.8 \times 10^4 \text{ J/mol} - 1.8 \times 10^4 \text{ J/mol} = 0$

Because $\Delta G = 0$, the reaction is at equilibrium (no shift).

c.   $\Delta G = 1.8 \times 10^4 + 8.3145(298) \ln \dfrac{(1.0 \times 10^{-5})^2}{1.0 \times 10^{-5}} = -1.1 \times 10^4 \text{ J/mol};$  shifts right

d.   $\Delta G = 1.8 \times 10^4 + 8.3145(298) \ln \dfrac{7.2 \times 10^{-4}(0.27)}{0.27} = 1.8 \times 10^4 - 1.8 \times 10^4 = 0;$
at equilibrium

e.   $\Delta G = 1.8 \times 10^4 + 8.3145(298) \ln \dfrac{1.0 \times 10^{-3}(0.67)}{0.52} = 2 \times 10^3 \text{ J/mol};$  shifts left

93.   $\Delta S°$ will be negative because 2 moles of gaseous reactants forms 1 mole of gaseous product. For $\Delta G°$ to be negative, $\Delta H$ must be negative (exothermic). For this sign combination of $\Delta H°$ and $\Delta S°$, K decreases as T increases because $\Delta G°$ becomes more positive ($\Delta G° = -RT \ln K$). Therefore, the ratio of the partial pressure of $PCl_5$ (a product) to the partial pressure of $PCl_3$ (a reactant) will decrease when T is raised.

95.   Using Le Chatelier's principle:  A decrease in pressure (volume increases) will favor the side with the greater number of particles.  Thus 2 I(g) will be favored at low pressure.

Looking at $\Delta G$:  $\Delta G = \Delta G° + RT \ln(P_I^2/P_{I_2})$;   $\ln(P_I^2/P_{I_2}) > 0$ when $P_I = P_{I_2} = 10$ atm, and $\Delta G$ is positive (not spontaneous).  But at $P_I = P_{I_2} = 0.10$ atm, the logarithm term is negative. If $|RT \ln Q| > \Delta G°$, then $\Delta G$ becomes negative, and the reaction is spontaneous.

97.   $NaCl(s) \rightleftharpoons Na^+(aq) + Cl^-(aq)$      $K = K_{sp} = [Na^+][Cl^-]$

$\Delta G° = [(-262 \text{ kJ}) + (-131 \text{ kJ})] - (-384 \text{ kJ}) = -9 \text{ kJ} = -9000 \text{ J}$

$\Delta G° = -RT \ln K_{sp}, \ K_{sp} = \exp\left[\dfrac{-(-9000 \text{ J})}{8.3145 \text{ J K}^{-1} \text{ mol}^{-1} \times 298 \text{ K}}\right] = 38 = 40$

$$NaCl(s) \rightleftharpoons Na^+(aq) + Cl^-(aq) \qquad K_{sp} = 40$$

Initial   $s$ = solubility (mol/L)   0           0
Equil.                              $s$          $s$

$K_{sp} = 40 = s(s), \ s = (40)^{1/2} = 6.3 = 6 \ M = [Cl^-]$

99.    $\Delta S$ is more favorable (less negative) for reaction 2 than for reaction 1, resulting in $K_2 > K_1$. In reaction 1, seven particles in solution are forming one particle in solution. In reaction 2, four particles are forming one, which results in a smaller decrease in positional probability than for reaction 1.

101.    Note that these substances are not in the solid state but are in the aqueous state; water molecules are also present. There is an apparent increase in ordering (decrease in positional probability) when these ions are placed in water as compared to the separated state. The hydrating water molecules must be in a highly ordered arrangement when surrounding these anions.

103.    $S = k \ln \Omega$; $S$ has units of J $K^{-1}$ mol$^{-1}$, and k has units of J/K (k = $1.38 \times 10^{-23}$ J/K).

To make units match:  $S$ (J $K^{-1}$ mol$^{-1}$) = $N_A k \ln \Omega$ when $N_A$ = Avogadro's number

$$189 \text{ J K}^{-1} \text{ mol}^{-1} = 8.31 \text{ J K}^{-1} \text{ mol}^{-1} \ln \Omega_g$$
$$-(70. \text{ J K}^{-1} \text{ mol}^{-1} = 8.31 \text{ J K}^{-1} \text{ mol}^{-1} \ln \Omega_l)$$

Subtracting:  $119 \text{ J K}^{-1} \text{ mol}^{-1} = 8.31 \text{ J K}^{-1} \text{ mol}^{-1}(\ln \Omega_g - \ln \Omega_l)$

$14.3 = \ln(\Omega_g/\Omega_l)$,  $\dfrac{\Omega_g}{\Omega_l} = e^{14.3} = 1.6 \times 10^6$

105.    Isothermal:  $\Delta H = 0$ (assume ideal gas)

$$\Delta S = nR \ln\left(\frac{V_2}{V_1}\right) = (1.00 \text{ mol})(8.3145 \text{ J K}^{-1} \text{ mol}^{-1}) \ln\left(\frac{1.00 \text{ L}}{100.0 \text{ L}}\right) = -38.3 \text{ J/K}$$

$\Delta G = \Delta H - T\Delta S = 0 - (300. \text{ K})(-38.3 \text{ J/K}) = +11,500 \text{ J} = 11.5 \text{ kJ}$

107.    a.    Isothermal:  $\Delta E = 0$ and $\Delta H = 0$ if gas is ideal.

$\Delta S = nR \ln(P_1/P_2) = (1.00 \text{ mol})(8.3145 \text{ J K}^{-1} \text{ mol}^{-1}) \ln(5.00 \text{ atm}/2.00 \text{ atm}) = 7.62 \text{ J/K}$

$$T = \frac{PV}{nR} = \frac{5.00 \text{ atm} \times 5.00 \text{ L}}{1.00 \text{ mol} \times 0.08206 \text{ L atm K}^{-1} \text{ mol}^{-1}} = 305 \text{ K}$$

$\Delta G = \Delta H - T\Delta S = 0 - (305 \text{ K})(7.62 \text{ J/K}) = -2320 \text{ J}$

$w = -P\Delta V = -(2.00 \text{ atm})\Delta V$, where $V_f = \dfrac{nRT}{2.00 \text{ atm}} = 12.5 \text{ L}$

$w = -2.00 \text{ atm} (12.5 - 5.00 \text{ L}) \times (101.3 \text{ J L}^{-1} \text{ atm}^{-1}) = -1500 \text{ J}$

$\Delta E = 0 = q + w$,  $q = 1500 \text{ J}$

b.   Second law, $\Delta S_{univ} > 0$ for spontaneous processes:

$$\Delta S_{univ} = \Delta S_{sys} + \Delta S_{surr} = \Delta S_{sys} - \frac{q_{actual}}{T}$$

$$\Delta S_{univ} = 7.62 \text{ J/K} - \frac{1500 \text{ J}}{305 \text{ K}} = 7.62 - 4.9 = 2.7 \text{ J/K}; \text{ thus the process is spontaneous.}$$

## Challenge Problems

109.   a.   $V_1 = \dfrac{nRT_1}{P_1} = \dfrac{2.00 \text{ mol} \times 0.08206 \text{ L atm K}^{-1} \text{ mol}^{-1} \times 298 \text{ K}}{2.00 \text{ atm}} = 24.5 \text{ L}$

For an adiabatic reversible process, $P_1V_1^{\gamma} = P_2V_2^{\gamma}$ and $T_1V_1^{\gamma-1} = T_2V_2^{\gamma-1}$, where $\gamma = C_p/C_v$.  Because argon is a monoatomic gas, $C_p = (5/2)R$ and $C_v = (3/2)R$, so $\gamma = 5/3$.

$$V_2^{\gamma} = \frac{P_1V_1^{\gamma}}{P_2} = \frac{2.00 \text{ atm}(24.5 \text{ L})^{5/3}}{1.00 \text{ atm}} = 413, \ V_2 = (413)^{3/5} = 37.1 \text{ L}$$

We can either use the ideal gas law or the $T_1V^{\gamma-1} = T_2V_2^{\gamma-1}$ equation to calculate the final temperature.  Using the ideal gas law:

$$T_2 = \frac{P_2V_2}{nR} = \frac{1.00 \text{ atm} \times 37.1 \text{ L}}{2.00 \text{ mol} \times 0.08206 \text{ L atm K}^{-1} \text{ mol}^{-1}} = 226 \text{ K}$$

b.   For an adiabatic process ($q = 0$), $\Delta E = w = nC_v\Delta T$.  For an expansion against a fixed external pressure, $w = -P\Delta V$.  From the ideal gas equation (see part a), $V_1 = 24.5$ L.

$$w = -P\Delta V = nC_v\Delta T$$

$$1.00 \text{ atm}(V_2 - 24.5 \text{ L})\left(\frac{101.3 \text{ J}}{\text{L atm}}\right) = 2.00 \text{ mol}(3/2)\left(\frac{8.3145 \text{ J}}{\text{mol K}}\right)(T_2 - 298 \text{ K})$$

We will ignore units from here.  Note that both sides of the equation are in units of J.

$$(-101)V_2 + 2480 = (24.9)T_2 - 7430$$

To solve for $T_2$, we need to find an expression for $V_2$.  Using the ideal gas equation:

$$V_2 = \frac{nRT_2}{P_2} = \frac{2.00 \text{ mol}(0.08206)T_2}{1.00} = (0.164)T_2; \text{ substituting:}$$

$$-101(0.164 \ T_2) + 2480 = (24.9)T_2 - 7430, \ (41.5)T_2 = 9910, \ T_2 = 239 \text{ K}$$

111.    a.    $\Delta S = \dfrac{q_{rev}}{T}$ ;  isothermal:  $\Delta T = 0$ so $\Delta E = 0$ and $q = -w$.

Reversible expansion:  $q_{rev} = 855$ J;  $\Delta S = \dfrac{q_{rev}}{T} = \dfrac{855\,\text{J}}{298\,\text{K}} = 2.87$ J/K

For the compression, we go back to the initial state for the overall process (expansion then compression). Because the initial and overall final state are the same, all state functions like $\Delta S$ must equal zero.

$\Delta S_{overall} = 0 = \Delta S_{exp} + \Delta S_{comp}$,  $\Delta S_{comp} = -\Delta S_{exp} = -2.87$ J/K

*Note*: Even though the compression step is not a reversible process, it still must have $q_{rev}$ = $-855$ J.

$\Delta S_{comp} = \dfrac{q_{rev}}{T} = \dfrac{-855\,\text{J}}{298\,\text{K}} = -2.87$ J/K

b.    $\Delta S_{univ,\ overall} = \Delta S_{sys,\ overall} + \Delta S_{surr,\ overall}$

$\Delta S_{sys,\ overall} = 2.87\ \text{J/K} - 2.87\ \text{J/K} = 0$

$\Delta S_{surr,\ exp} = \dfrac{-q_{actual}}{T} = \dfrac{-855\,\text{J}}{298\,\text{K}} = -\,2.87$ J/K

From the problem, the compression step is isothermal, so $q = -w$. The work done on the system is 2(855) J, so $q = -2(855)$ J.

$\Delta S_{surr,\ comp} = \dfrac{-q_{actual}}{T} = \dfrac{-[-2(855)\,\text{J}]}{298\,\text{K}} = 5.74$ J/K

$\Delta S_{surr,\ overall} = -2.87\ \text{J/K} + 5.74\ \text{J/K} = 2.87$ J/K

$\Delta S_{univ,\ overall} = 0 + 2.87\ \text{J/K} = 2.87$ J/K

113.    The system is the 1.00-L sample of water. The process is the cooling of the water from 90.°C to 25°C. The surroundings are the room (at 25°C) and everything else.

$1.00 \times 10^3\ \text{mL} \times \dfrac{1.00\ \text{g}}{\text{mL}} \times \dfrac{1\ \text{mol}}{18.02\ \text{g}} = 55.5\ \text{mol H}_2\text{O}$

$\Delta S = nC_p\ \ln(T_2/T_1) = 55.5\ \text{mol}(75.3\ \text{J K}^{-1}\ \text{mol}^{-1})\ \ln(298/363) = -825$ J/K

$\Delta S_{surr} = \dfrac{-q_{actual}}{T}$ ;  $q_{actual} = nC_p\Delta T = 55.5\ \text{mol}(75.3\ \text{J K}^{-1}\ \text{mol}^{-1})(-65\ \text{K}) = -2.72 \times 10^5$ J

$$\Delta S_{surr} = \frac{-q_{actual}}{T} = \frac{-(-2.72 \times 10^5 \text{ J})}{298 \text{ K}} = 913 \text{ J/K}$$

$$\Delta S_{univ} = \Delta S + \Delta S_{surr} = -825 \text{ J/K} + 913 \text{ J/K} = 88 \text{ J/K}$$

Not too surprising, this is a spontaneous process due to the favorable $\Delta S_{surr}$ term.

115.    $K_p = P_{CO_2}$; to prevent $Ag_2CO_3$ from decomposing, $P_{CO_2}$ should be greater than $K_p$.

From Exercise 10.78, $\ln K = \frac{-\Delta H^\circ}{RT} + \frac{\Delta S^\circ}{R}$. For two conditions of K and T, the equation is:

$$\ln \frac{K_2}{K_1} = \frac{\Delta H^\circ}{R} \left( \frac{1}{T_1} - \frac{1}{T_2} \right)$$

Let $T_1 = 25°C = 298$ K, $K_1 = 6.23 \times 10^{-3}$ torr; $T_2 = 110.°C = 383$ K, $K_2 = ?$

$$\ln \frac{K_2}{6.23 \times 10^{-3} \text{ torr}} = \frac{79.14 \times 10^3 \text{ J/mol}}{8.3145 \text{ J K}^{-1} \text{ mol}^{-1}} \left( \frac{1}{298 \text{ K}} - \frac{1}{383 \text{ K}} \right)$$

$$\ln \frac{K_2}{6.23 \times 10^{-3}} = 7.1, \quad \frac{K_2}{6.23 \times 10^{-3}} = e^{7.1} = 1.2 \times 10^3, \ K_2 = 7.5 \text{ torr}$$

To prevent decomposition of $Ag_2CO_3$, the partial pressure of $CO_2$ should be greater than 7.5 torr.

117.    $3 O_2(g) \rightleftharpoons 2 O_3(g)$; $\Delta H^\circ = 2(143) = 286$ kJ; $\Delta G^\circ = 2(163) = 326$ kJ

$$\ln K = \frac{-\Delta G^\circ}{RT} = \frac{-326 \times 10^3 \text{ J}}{(8.3145 \text{ J K}^{-1} \text{ mol}^{-1})(298 \text{ K})} = -131.573, \ K = e^{-131.573} = 7.22 \times 10^{-58}$$

We need the value of K at 230. K. From Exercise 10.78: $\ln K = \frac{-\Delta H^\circ}{RT} + \frac{\Delta S^\circ}{R}$

For two sets of K and T:

$$\ln \frac{K_2}{K_1} = \frac{\Delta H^\circ}{R} \left( \frac{1}{T_1} - \frac{1}{T_2} \right)$$

Let $K_2 = 7.22 \times 10^{-58}$, $T_2 = 298$; $K_1 = K_{230}$, $T_1 = 230.$ K; $\Delta H^\circ = 286 \times 10^3$ J

$$\ln \frac{7.22 \times 10^{-58}}{K_{230}} = \frac{286 \times 10^3}{8.3145} \left( \frac{1}{230.} - \frac{1}{298} \right) = 34.13$$

$$\frac{7.22 \times 10^{-58}}{K_{230}} = e^{34.13} = 6.6 \times 10^{14}, \quad K_{230} = 1.1 \times 10^{-72}$$

$$K_{230} = 1.1 \times 10^{-72} = \frac{P_{O_3}^2}{P_{O_2}^3} = \frac{P_{O_3}^2}{(1.0 \times 10^{-3} \text{ atm})^3}, \quad P_{O_3} = 3.3 \times 10^{-41} \text{ atm}$$

The volume occupied by one molecule of ozone is:

$$V = \frac{nRT}{P} = \frac{(1/6.022 \times 10^{23} \text{ mol})(0.8206 \text{ L atm K}^{-1} \text{ mol}^{-1})(230.\,K)}{(3.3 \times 10^{-41} \text{ atm})}, \quad V = 9.5 \times 10^{17} \text{ L}$$

Equilibrium is probably not maintained under these conditions. When only two ozone molecules are in a volume of $9.5 \times 10^{17}$ L, the reaction is not at equilibrium. Under these conditions, $Q > K$ and the reaction shifts left. But with only two ozone molecules in this huge volume, it is extremely unlikely that they will collide with each other. At these conditions, the concentration of ozone is not large enough to maintain equilibrium.

119.   a.   $\Delta G° = G_B^° - G_A^° = 11{,}718 - 8996 = 2722$ J

$$K = \exp\left(\frac{-\Delta G°}{RT}\right) = \exp\left[\frac{-2722 \text{ J}}{(8.3145 \text{ J K}^{-1} \text{ mol}^{-1})(298 \text{ K})}\right] = 0.333$$

b.   Since $Q = 1.00 > K$, reaction shifts left. Let $x =$ atm of B(g) that reacts to reach equilibrium.

|          | A(g)        | $\rightleftharpoons$ | B(g)      | $K = P_B/P_A$ |
|----------|-------------|----------------------|-----------|---------------|
| Initial  | 1.00 atm    |                      | 1.00 atm  |               |
| Equil.   | $1.00 + x$  |                      | $1.00 - x$|               |

$$K = \frac{1.00 - x}{1.00 + x} = 0.333, \quad 1.00 - x = 0.333 + (0.333)x, \quad x = 0.50 \text{ atm}$$

$P_B = 1.00 - 0.50 = 0.50$ atm;   $P_A = 1.00 + 0.50 = 1.50$ atm

c.   $\Delta G = \Delta G° + RT \ln Q = \Delta G° + RT \ln(P_B/P_A)$

$\Delta G = 2722 \text{ J} + (8.3145)(298) \ln(0.50/1.50) = 2722 \text{ J} - 2722 \text{ J} = 0$

(carrying extra sig. figs.)

121.   Step 1: $\Delta E = 0$ and $\Delta H = 0$ since $\Delta T = 0$

$$w = -P\Delta V = -(9.87 \times 10^{-3} \text{ atm})\Delta V; \quad V = \frac{nRT}{P}, \quad R = 0.08206 \text{ L atm K}^{-1} \text{ mol}^{-1}$$

$$\Delta V = V_f - V_i = nRT\left(\frac{1}{P_f} - \frac{1}{P_i}\right)$$

$$\Delta V = 1.00 \text{ mol}(0.08206)(298 \text{ K})\left(\frac{1}{9.87 \times 10^{-3} \text{ atm}} - \frac{1}{2.45 \times 10^{-2} \text{ atm}}\right)$$

$\Delta V = 1480$ L (we will carry all values to three sig. figs.)

$$w = -(9.87 \times 10^{-3} \text{ atm})(1480 \text{ L}) = -14.6 \text{ L atm}(101.3 \text{ J L}^{-1} \text{ atm}^{-1}) = -1480 \text{ J}$$

$$\Delta E = q + w = 0, \quad q = -w = +1480 \text{ J}; \quad \Delta S = nR \ln\left(\frac{P_1}{P_2}\right)$$

$$\Delta S = 1.00 \text{ mol}(8.3145 \text{ J K}^{-1} \text{ mol}^{-1}) \ln\left(\frac{2.45 \times 10^{-2} \text{ atm}}{9.87 \times 10^{-3} \text{ atm}}\right), \quad \Delta S = 7.56 \text{ J/K}$$

$$\Delta G = \Delta H - T\Delta S = 0 - 298 \text{ K}(7.56 \text{ J/K}) = -2250 \text{ J}$$

Step 2:  $\Delta E = 0, \quad \Delta H = 0$

$$w = -(4.93 \times 10^{-3} \text{ atm})\left(\frac{nRT}{4.93 \times 10^{-3}} - \frac{nRT}{9.87 \times 10^{-3} \text{ atm}}\right) L \times \frac{101.3 \text{ J}}{\text{L atm}} = -1240 \text{ J}$$

$$q = -w = 1240 \text{ J}; \quad \Delta S = nR \ln\left(\frac{9.87 \times 10^{-3} \text{ atm}}{4.93 \times 10^{-3} \text{ atm}}\right) = 5.77 \text{ J/K}$$

$$\Delta G = 0 - 298 \text{K}(5.77 \text{ J/K}) = -1720 \text{ J}$$

Step 3:  $\Delta E = 0, \quad \Delta H = 0$

$$w = -(2.45 \times 10^{-3} \text{ atm})\left(\frac{nRT}{2.45 \times 10^{-3}} - \frac{nRT}{4.93 \times 10^{-3}}\right) L \times \frac{101.3 \text{ J}}{\text{L atm}} = -1250 \text{ J}$$

$$q = -w = 1250 \text{ J}; \quad \Delta S = nR \ln\left(\frac{4.93 \times 10^{-3} \text{ atm}}{2.45 \times 10^{-3} \text{ atm}}\right) = 5.81 \text{ J/K}$$

$$\Delta G = 0 - 298 \text{ K}(5.81 \text{ J/K}) = -1730 \text{ J}$$

| | q | w | $\Delta E$ | $\Delta S$ | $\Delta H$ | $\Delta G$ |
|---|---|---|---|---|---|---|
| Step 1 | 1480 J | −1480 J | 0 | 7.56 J/K | 0 | −2250 J |
| Step 2 | 1240 J | −1240 J | 0 | 5.77 J/K | 0 | −1720 J |
| Step 3 | 1250 J | −1250 J | 0 | 5.81 J/K | 0 | −1730 J |
| Total | 3970 J | −3970 J | 0 | 19.14 J/K | 0 | $-5.70 \times 10^3$ J |

123.   a.   $\Delta G° = 2 \text{ mol}(-394 \text{ kJ/mol}) - 2 \text{ mol}(-137 \text{ kJ/mol}) = -514 \text{ kJ}$

$$K = \exp\left(\frac{-\Delta G°}{RT}\right) = \exp\left(\frac{-(-514{,}000 \text{ J})}{(8.3145 \text{ J K}^{-1}\text{ mol}^{-1})(298 \text{ K})}\right) = 1.24 \times 10^{90}$$

b.   $\Delta S° = 2(214 \text{ J/K}) - [2(198 \text{ J/K}) + 205 \text{ J/K}] = -173 \text{ J/K}$

| | |
|---|---|
| $2 \text{ CO}(1.00 \text{ atm}) + O_2(1.00 \text{ atm}) \rightarrow 2 \text{ CO}_2(1.00 \text{ atm})$ | $\Delta S° = -173 \text{ J/K}$ |
| $2 \text{ CO}_2(1.00 \text{ atm}) \rightarrow 2 \text{ CO}_2(10.0 \text{ atm})$ | $\Delta S = nR \ln(P_1/P_2) = -38.3 \text{ J/K}$ |
| $2 \text{ CO}(10.0 \text{ atm}) \rightarrow 2 \text{ CO}(1.00 \text{ atm})$ | $\Delta S = nR \ln(P_1/P_2) = 38.3 \text{ J/K}$ |
| $O_2(10.0 \text{ atm}) \rightarrow O_2(1.00 \text{ atm})$ | $\Delta S = nR \ln(P_1/P_2) = 19.1 \text{ J/K}$ |

| | |
|---|---|
| $2 \text{ CO}(10.0 \text{ atm}) + O_2(10.0 \text{ atm}) \rightarrow \text{CO}_2(10.0 \text{ atm})$ | $\Delta S = -173 + 19.1 = -154 \text{ J/K}$ |

125.

| T(°C) | T(K) | $C_p$ (J K$^{-1}$ mol$^{-1}$) | $C_p/T$ (J K$^{-2}$ mol$^{-1}$) |
|---|---|---|---|
| −200. | 73 | 12 | 0.16 |
| -180. | 93 | 15 | 0.16 |
| -160. | 113 | 17 | 0.15 |
| -140. | 133 | 19 | 0.14 |
| -100. | 173 | 24 | 0.14 |
| -60. | 213 | 29 | 0.14 |
| -30. | 243 | 33 | 0.14 |
| -10. | 263 | 36 | 0.14 |
| 0 | 273 | 37 | 0.14 |

Total area of $C_p/T$ versus T plot = $\Delta S$ = I + II + III   (See following plot.)

$\Delta S = (0.16 \text{ J K}^{-2} \text{ mol}^{-1})(20. \text{ K}) + (0.14 \text{ J K}^{-2} \text{ mol}^{-1})(180. \text{ K}) + 1/2 (0.02 \text{ J K}^{-2} \text{ mol}^{-1})(40. \text{ K})$

$\Delta S = 3.2 + 25 + 0.4 = 29 \text{ J K}^{-1} \text{ mol}^{-1}$

127.  To calculate $\Delta S_{sys}$ at 10.0°C, we need a place to start.  From the data in the problem, we can calculate $\Delta S_{sys}$ at the melting point (5.5°C).  For a phase change, $\Delta S_{sys} = q_{rev}/T = \Delta H/T$, where $\Delta H$ is determined at the melting point (5.5°C).  Solving for $\Delta H$ at 5.5°C (using a thermochemical cycle):

$C_6H_6(l, 25.0°C) \rightarrow C_6H_6(s, 25.0°C) \qquad \Delta H = -10.04 \text{ kJ}$
$C_6H_6(l, 5.5°C) \rightarrow C_6H_6(l, 25.0°C) \qquad \Delta H = nC_p\Delta T/1000 = 2.59 \text{ kJ}$
$C_6H_6(s, 25.0°C) \rightarrow C_6H_6(s, 5.5°C) \qquad \Delta H = nC_p\Delta T/1000 = -1.96 \text{ kJ}$

---

$C_6H_6(l, 5.5°C) \rightarrow C_6H_6(s, 5.5°C) \qquad \Delta H = -9.41 \text{ kJ}$

At the melting point, $\Delta S_{sys} = \dfrac{\Delta H}{T} = \dfrac{-9.41 \times 10^3 \text{ J}}{278.7 \text{ K}} = -33.8 \text{ J/K}.$

For the phase change at 10.0°C (283.2 K):

$C_6H_6(l, 278.7 \text{ K}) \rightarrow C_6H_6(s, 278.7 \text{ K}) \qquad \Delta S = -33.8 \text{ J/K}$
$C_6H_6(l, 283.2 \text{ K}) \rightarrow C_6H_6(l, 278.7 \text{ K}) \qquad \Delta S = nC_p \ln(T_2/T_1) = -2.130 \text{ J/K}$
$C_6H_6(s, 278.7 \text{ K}) \rightarrow C_6H_6(s, 283.2 \text{ K}) \qquad \Delta S = nC_p \ln(T_2/T_1) = 1.608 \text{ J/K}$

---

$C_6H_6(l, 283.2 \text{ K}) \rightarrow C_6H_6(s, 283.2 \text{ K}) \qquad \Delta S_{sys} = -34.3 \text{ J/K}$

To calculate $\Delta S_{surr}$, we need $\Delta H$ at 10.0°C ($\Delta S_{surr} = \dfrac{-\Delta H}{T}$ ).

$C_6H_6(l, 25.0°C) \rightarrow C_6H_6(s, 25.0°C) \qquad \Delta H = -10.04 \text{ kJ}$
$C_6H_6(l, 10.0°C) \rightarrow C_6H_6(l, 25.0°C) \qquad \Delta H = nC_p\Delta T/1000 = 2.00 \text{ kJ}$
$C_6H_6(s, 25.0°C) \rightarrow C_6H_6(s, 10.0°C) \qquad \Delta H = nC_p\Delta T/1000 = -1.51 \text{ kJ}$

---

$C_6H_6(l, 10.0°C) \rightarrow C_6H_6(s, 10.0°C) \qquad \Delta H = -9.55 \text{ kJ}$

$\Delta S_{surr} = \dfrac{-\Delta H}{T} = \dfrac{-(-9.55 \times 10^3 \text{ J})}{283.2 \text{ K}} = 33.7 \text{ J/K}$

# CHAPTER 11

# ELECTROCHEMISTRY

## Galvanic Cells, Cell Potentials, and Standard Reduction Potentials

15. Electrochemistry is the study of the interchange of chemical and electrical energy. A redox (oxidation-reduction) reaction is a reaction in which one or more electrons are transferred. In a galvanic cell, a spontaneous redox reaction occurs that produces an electric current. In an electrolytic cell, electricity is used to force a nonspontaneous redox reaction to occur.

17. A typical galvanic cell diagram is:

The diagram for all cells will look like this. The contents of each half-cell will be identified for each reaction, with all concentrations at 1.0 $M$ and partial pressures at 1.0 atm. Note that cations always flow into the cathode compartment and anions always flow into the anode compartment. This is required to keep each compartment electrically neutral.

a. Reference Table 11.1 for standard reduction potentials. Remember that $E^{\circ}_{cell} = E^{\circ}(cathode) - E^{\circ}(anode)$; in the *Solutions Guide*, we will represent $E^{\circ}(cathode)$ as $E^{\circ}_{c}$ and represent $-E^{\circ}(anode)$ as $-E^{\circ}_{a}$. Also remember that standard potentials are *not* multiplied by the integer used to obtain the overall balanced equation.

$$(Cl_2 + 2\ e^- \rightarrow 2\ Cl^-) \times 3 \qquad\qquad E_c^o = 1.36\ V$$

$$7\ H_2O + 2\ Cr^{3+} \rightarrow Cr_2O_7^{2-} + 14\ H^+ + 6\ e^- \qquad -E_a^o = -1.33\ V$$

---

$$7\ H_2O(l) + 2\ Cr^{3+}(aq) + 3\ Cl_2(g) \rightarrow Cr_2O_7^{2-}(aq) + 6\ Cl^-(aq) + 14\ H^+(aq) \quad E_{cell}^o = 0.03\ V$$

The contents of each compartment are:

Cathode: Pt electrode; $Cl_2$ bubbled into solution, $Cl^-$ in solution

Anode: Pt electrode; $Cr^{3+}$, $H^+$, and $Cr_2O_7^{2-}$ in solution

We need a nonreactive metal to use as the electrode in each case since all the reactants and products are in solution. Pt is the most common choice. Another possibility is graphite.

b. 
$$Cu^{2+} + 2\ e^- \rightarrow Cu \qquad\qquad E_c^o = 0.34\ V$$

$$Mg \rightarrow Mg^{2+} + 2\ e^- \qquad\qquad -E_a^o = 2.37\ V$$

---

$$Cu^{2+}(aq) + Mg(s) \rightarrow Cu(s) + Mg^{2+}(aq) \qquad E_{cell}^o = 2.71\ V$$

Cathode: Cu electrode; $Cu^{2+}$ in solution

Anode: Mg electrode; $Mg^{2+}$ in solution

c. 
$$5\ e^- + 6\ H^+ + IO_3^- \rightarrow 1/2\ I_2 + 3\ H_2O \qquad\qquad E_c^o = 1.20\ V$$

$$(Fe^{2+} \rightarrow Fe^{3+} + e^-) \times 5 \qquad\qquad -E_a^o = -0.77\ V$$

---

$$6\ H^+ + IO_3^- + 5\ Fe^{2+} \rightarrow 5\ Fe^{3+} + 1/2\ I_2 + 3\ H_2O \qquad E_{cell}^o = 0.43\ V$$

or $12\ H^+(aq) + 2\ IO_3^-(aq) + 10\ Fe^{2+}(aq) \rightarrow 10\ Fe^{3+}(aq) + I_2(s) + 6\ H_2O(l) \quad E_{cell}^o = 0.43\ V$

Cathode: Pt electrode; $IO_3^-$, $I_2$, and $H_2SO_4$ ($H^+$ source) in solution

Anode: Pt electrode; $Fe^{2+}$ and $Fe^{3+}$ in solution

*Note*: $I_2(s)$ would make a poor electrode since it sublimes.

d. 
$$(Ag^+ + e^- \rightarrow Ag) \times 2 \qquad\qquad E_c^o = 0.80\ V$$

$$Zn \rightarrow Zn^{2+} + 2\ e^- \qquad\qquad -E_a^o = 0.76\ V$$

---

$$Zn(s) + 2\ Ag^+(aq) \rightarrow 2\ Ag(s) + Zn^{2+}(aq) \qquad E_{cell}^o = 1.56\ V$$

Cathode: Ag electrode; $Ag^+$ in solution

Anode: Zn electrode; $Zn^{2+}$ in solution

19.   Reference Exercise 11.17 for a typical galvanic cell design.  The contents of each half-cell compartment are identified below, with all solute concentrations at 1.0 $M$ and all gases at 1.0 atm.  For each pair of half-reactions, the half-reaction with the largest standard reduction potential will be the cathode reaction, and the half-reaction with the smallest reduction potential will be reversed to become the anode reaction.  Only this combination gives a spontaneous overall reaction, i.e., a reaction with a positive overall standard cell potential.

a.          $Cl_2 + 2\,e^- \rightarrow 2\,Cl^-$                            $E_c^o = 1.36\ V$

             $2\,Br^- \rightarrow Br_2 + 2\,e^-$                           $-E_a^o = -1.09\ V$

$$\overline{Cl_2(g) + 2\,Br^-(aq) \rightarrow Br_2(aq) + 2\,Cl^-(aq) \qquad E_{cell}^o = 0.27\ V}$$

The contents of each compartment are:

   Cathode:  Pt electrode;  $Cl_2(g)$ bubbled in, $Cl^-$ in solution

   Anode:  Pt electrode;  $Br_2$ and $Br^-$ in solution

b.          $(2\,e^- + 2\,H^+ + IO_4^- \rightarrow IO_3^- + H_2O) \times 5$          $E_c^o = 1.60\ V$

             $(4\,H_2O + Mn^{2+} \rightarrow MnO_4^- + 8\,H^+ + 5\,e^-) \times 2$          $-E_a^o = -1.51\ V$

$$\overline{10\,H^+ + 5\,IO_4^- + 8\,H_2O + 2\,Mn^{2+} \rightarrow 5\,IO_3^- + 5\,H_2O + 2\,MnO_4^- + 16\,H^+ \qquad E_{cell}^o = 0.09\ V}$$

This simplifies to:

   $3\,H_2O(l) + 5\,IO_4^-(aq) + 2\,Mn^{2+}(aq) \rightarrow 5\,IO_3^-(aq) + 2\,MnO_4^-(aq) + 6\,H^+(aq)$

                                                                                    $E_{cell}^o = 0.09\ V$

   Cathode:  Pt electrode;  $IO_4^-$, $IO_3^-$, and $H_2SO_4$ (as a source of $H^+$) in solution

   Anode:  Pt electrode;  $Mn^{2+}$,  $MnO_4^-$, and $H_2SO_4$ in solution

c.   $H_2O_2 + 2\,H^+ + 2\,e^- \rightarrow 2\,H_2O$               $E_c^o = 1.78\ V$

       $H_2O_2 \rightarrow O_2 + 2\,H^+ + 2\,e^-$                        $-E_a^o = -0.68\ V$

$$\overline{2\,H_2O_2(aq) \rightarrow 2\,H_2O(l) + O_2(g) \qquad E_{cell}^o = 1.10\ V}$$

Cathode:  Pt electrode; $H_2O_2$ and $H^+$ in solution

Anode:  Pt electrode; $O_2(g)$ bubbled in, $H_2O_2$ and $H^+$ in solution

d.          $(Fe^{3+} + 3\,e^- \rightarrow Fe) \times 2$                        $E_c^o = -0.036\ V$

             $(Mn \rightarrow Mn^{2+} + 2\,e^-) \times 3$                     $-E_a^o = 1.18\ V$

$$\overline{2\,Fe^{3+}(aq) + 3\,Mn(s) \rightarrow 2\,Fe(s) + 3\,Mn^{2+}(aq) \qquad E_{cell}^o = 1.14\ V}$$

Cathode:  Fe electrode;  $Fe^{3+}$ in solution;   Anode:  Mn electrode;  $Mn^{2+}$ in solution

21.    In standard line notation, the anode is listed first and the cathode is listed last.  A double line separates the two compartments.  By convention, the electrodes are on the ends, with all solutes and gases toward the middle.  A single line is used to indicate a phase change.  We also included all concentrations.

19a.    $Pt \mid Br^- (1.0\ M), Br_2 (1.0\ M) \parallel Cl_2 (1.0\ atm) \mid Cl^- (1.0\ M) \mid Pt$

19b.    $Pt \mid Mn^{2+} (1.0\ M), MnO_4^- (1.0\ M), H^+ (1.0\ M) \mid\mid IO_4^- (1.0\ M), IO_3^- (1.0\ M),$

$H^+ (1.0\ M) \mid Pt$

19c.    $Pt \mid H_2O_2 (1.0\ M), H^+ (1.0\ M) \mid O_2 (1.0\ atm) \mid\mid H_2O_2 (1.0\ M), H^+ (1.0\ M) \mid Pt$

19d.    $Mn \mid Mn^{2+} (1.0\ M) \mid\mid Fe^{3+} (1.0\ M) \mid Fe$

23.    a.    $2\ H^+ + 2\ e^- \rightarrow H_2$        $E° = 0.00\ V;\ Cu \rightarrow Cu^{2+} + 2\ e^-$    $-E° = -0.34\ V$

$E^\circ_{cell} = -0.34\ V$;  no, $H^+$ cannot oxidize Cu to $Cu^{2+}$ at standard conditions ($E^\circ_{cell} < 0$).

b.    $Fe^{3+} + e^- \rightarrow Fe^{2+}$    $E° = 0.77\ V;\ 2\ I^- \rightarrow I_2 + 2\ e^-$    $-E° = -0.54\ V$

$E^\circ_{cell} = 0.77 - 0.54 = 0.23\ V$; yes, $Fe^{3+}$ can oxidize $I^-$ to $I_2$.

c.    $H_2 \rightarrow 2\ H^+ + 2\ e^-$    $-E° = 0.00\ V;\ Ag^+ + e^- \rightarrow Ag$    $E° = 0.80\ V$

$E^\circ_{cell} = 0.80\ V$;  yes, $H_2$ can reduce $Ag^+$ to Ag at standard conditions ($E^\circ_{cell} > 0$).

d.    $Fe^{2+} \rightarrow Fe^{3+} + e^-$    $-E° = -0.77\ V;\ Cr^{3+} + e^- \rightarrow Cr^{2+}$    $E° = -0.50\ V$

$E^\circ_{cell} = -0.50 - 0.77 = -1.27\ V$;  no, $Fe^{2+}$ cannot reduce $Cr^{3+}$ to $Cr^{2+}$ at standard conditions.

25.    Good reducing agents are easily oxidized.  The reducing agents are on the right side of the reduction half-reactions listed in Table 11.1.  The best reducing agents have the most negative standard reduction potentials ($E°$); i.e., the best reducing agents have the most positive $-E°$ value. The ordering from worst to best reducing agents is:

|  |  | $F^-$ | $<$ | $H_2O$ | $<$ | $I_2$ | $<$ | $Cu^+$ | $<$ | $H^-$ | $<$ | $K$ |
|---|---|---|---|---|---|---|---|---|---|---|---|---|
| $-E°(V)$ |  | $-2.92$ |  | $-1.23$ |  | $-1.20$ |  | $-0.16$ |  | $2.23$ |  | $2.92$ |

27.    a.    $2\ Br^- \rightarrow Br_2 + 2\ e^-\ -E^\circ_a = -1.09\ V;\ 2\ Cl^- \rightarrow Cl_2 + 2\ e^-\ -E^\circ_a = -1.36\ V;\ E^\circ_c > 1.09\ V$ to oxidize $Br^-$; $E^\circ_c < 1.36\ V$ to not oxidize $Cl^-$;  $Cr_2O_7^{2-}$, $O_2$, $MnO_2$, and $IO_3^-$ are all possible because when all these oxidizing agents are coupled with $Br^-$, they give $E^\circ_{cell} > 0$, and when coupled with $Cl^-$, they give $E^\circ_{cell} < 0$ (assuming standard conditions).

b.  $Mn \rightarrow Mn^{2+} + 2 e^- \quad -E_a^\circ = 1.18 \text{ V}; \quad Ni \rightarrow Ni^{2+} + 2 e^- \quad -E_a^\circ = 0.23 \text{ V};$  any oxidizing
agent with $-0.23 \text{ V} > E_c^\circ > -1.18 \text{ V}$ will work. $PbSO_4, Cd^{2+}, Fe^{2+}, Cr^{3+}, Zn^{2+}$, and $H_2O$
will be able to oxidize Mn but not oxidize Ni (assuming standard conditions).

29.     $\begin{array}{lr} ClO^- + H_2O + 2 e^- \rightarrow 2 OH^- + Cl^- & E_c^\circ = 0.90 \text{ V} \\ 2 NH_3 + 2 OH^- \rightarrow N_2H_4 + 2 H_2O + 2 e^- & -E_a^\circ = 0.10 \text{ V} \end{array}$

$$\overline{ClO^-(aq) + 2 NH_3(aq) \rightarrow Cl^-(aq) + N_2H_4(aq) + H_2O(l) \quad E_{cell}^\circ = 1.00 \text{ V}}$$

Because $E_{cell}^\circ$ is positive for this reaction, at standard conditions $ClO^-$ can spontaneously
oxidize $NH_3$ to the somewhat toxic $N_2H_4$.

## Cell Potential, Free Energy, and Equilibrium

31.     An extensive property is one that depends directly on the amount of substance. The free
energy change for a reaction depends on whether 1 mole of product is produced or 2 moles of
product are produced or 1 million moles of product are produced. This is not the case for cell
potentials, which do not depend on the amount of substance. The equation that relates $\Delta G$ to
E is $\Delta G = -nFE$. It is the n term that converts the intensive property E into the extensive
property $\Delta G$. n is the number of moles of electrons transferred in the balanced reaction that
$\Delta G$ is associated with.

33.     Because the cells are at standard conditions, $w_{max} = \Delta G = \Delta G^\circ = -nFE_{cell}^\circ$. See Exercise
11.20 for the balanced overall equations and $E_{cell}^\circ$.

20a.    $w_{max} = -(3 \text{ mol } e^-)(96{,}485 \text{ C/mol } e^-)(1.34 \text{ J/C}) = -3.88 \times 10^5 \text{ J} = -388 \text{ kJ}$

20b.    $w_{max} = -(2 \text{ mol } e^-)(96{,}485 \text{ C/mol } e^-)(1.40 \text{ J/C}) = -2.70 \times 10^5 \text{ J} = -270. \text{ kJ}$

35.     Reference Exercise 11.19 for the balanced reactions and standard cell potentials.  The
balanced reactions are necessary to determine n, the moles of electrons transferred.

19a.    $Cl_2(aq) + 2 Br^-(aq) \rightarrow Br_2(aq) + 2 Cl^-(aq) \quad E_{cell}^\circ = 0.27 \text{ V} = 0.27 \text{ J/C}, n = 2 \text{ mol } e^-$

$\Delta G^\circ = -nFE_{cell}^\circ = -(2 \text{ mol } e^-)(96{,}485 \text{ C/mol } e^-)(0.27 \text{ J/C}) = -5.2 \times 10^4 \text{ J} = -52 \text{ kJ}$

$E_{cell}^\circ = \dfrac{0.0591}{n} \log K, \quad \log K = \dfrac{nE^\circ}{0.0591} = \dfrac{2(0.27)}{0.0591} = 9.14, \quad K = 10^{9.14} = 1.4 \times 10^9$

*Note*:  When determining exponents, we will round off to the correct number of
significant figures after the calculation is complete in order to help eliminate
excessive round-off errors.

19b.    $\Delta G° = -(10 \text{ mol e}^-)(96{,}485 \text{ C/mol e}^-)(0.09 \text{ J/C}) = -9 \times 10^4 \text{ J} = -90 \text{ kJ}$

$$\log K = \frac{10(0.09)}{0.0591} = 15.2, \ K = 10^{15.2} = 2 \times 10^{15}$$

19c.    $\Delta G° = -(2 \text{ mol e}^-)(96{,}485 \text{ C/mol e}^-)(1.10 \text{ J/C}) = -2.12 \times 10^5 \text{ J} = -212 \text{ kJ}$

$$\log K = \frac{2(1.10)}{0.0591} = 37.225, \ K = 1.68 \times 10^{37}$$

19d.    $\Delta G° = -(6 \text{ mol e}^-)(96{,}485 \text{ C/mol e}^-)(1.14 \text{ J/C}) = -6.60 \times 10^5 \text{ J} = -660. \text{ kJ}$

$$\log K = \frac{6(1.14)}{0.0591} = 115.736, \ K = 5.45 \times 10^{115}$$

37.    a.

$$(4 \text{ H}^+ + \text{NO}_3^- + 3 \text{ e}^- \rightarrow \text{NO} + 2 \text{ H}_2\text{O}) \times 2 \qquad E_c° = 0.96 \text{ V}$$
$$(\text{Mn} \rightarrow \text{Mn}^{2+} + 2 \text{ e}^-) \times 3 \qquad -E_a° = 1.18 \text{ V}$$

$$3 \text{ Mn(s)} + 8 \text{ H}^+(aq) + 2 \text{ NO}_3^-(aq) \rightarrow 2 \text{ NO(g)} + 4 \text{ H}_2\text{O(l)} + 3 \text{ Mn}^{2+}(aq) \qquad E_{cell}° = 2.14 \text{ V}$$

$$5 \times (2 \text{ e}^- + 2 \text{ H}^+ + \text{IO}_4^- \rightarrow \text{IO}_3^- + \text{H}_2\text{O}) \qquad E_c° = 1.60 \text{ V}$$
$$2 \times (\text{Mn}^{2+} + 4 \text{ H}_2\text{O} \rightarrow \text{MnO}_4^- + 8 \text{ H}^+ + 5 \text{ e}^-) \qquad -E_a° = -1.51 \text{ V}$$

$$5 \text{ IO}_4^-(aq) + 2 \text{ Mn}^{2+}(aq) + 3 \text{ H}_2\text{O(l)} \rightarrow 5 \text{ IO}_3^-(aq) + 2 \text{ MnO}_4^-(aq) + 6 \text{ H}^+(aq)$$
$$E_{cell}° = 0.09 \text{ V}$$

b.    Nitric acid oxidation (see part a for $E_{cell}°$):

$$\Delta G° = -n F E_{cell}° = -(6 \text{ mol e}^-)(96{,}485 \text{ C/mol e}^-)(2.14 \text{ J/C}) = -1.24 \times 10^6 \text{ J} =$$
$$-1240 \text{ kJ}$$

$$\log K = \frac{n E°}{0.0591} = \frac{6(2.14)}{0.0591} = 217, \ K \approx 10^{217}$$

Periodate oxidation (see part a for $E_{cell}°$):

$$\Delta G° = -(10 \text{ mol e}^-)(96{,}485 \text{ C/mol e}^-)(0.09 \text{ J/C})(1 \text{ kJ}/1000 \text{ J}) = -90 \text{ kJ}$$

$$\log K = \frac{10(0.09)}{0.0591} = 15.2, \ K = 10^{15.2} = 2 \times 10^{15}$$

39.    $\Delta G° = -n F E° = \Delta H° - T\Delta S°, \ E° = \dfrac{T\Delta S°}{nF} - \dfrac{\Delta H°}{nF}$

If we graph E° versus T, we should get a straight line ($y = mx + b$). The slope ($m$) of the line is equal to $\Delta S°/nF$, and the $y$ intercept is equal to $-\Delta H°/nF$. From the equation above, E° will have a small temperature dependence when $\Delta S°$ is close to zero.

41.   a.   $Cu^+ + e^- \rightarrow Cu$                                    $E_c^o = 0.52$ V

            $Cu^+ \rightarrow Cu^{2+} + e^-$                              $-E_a^o = -0.16$ V

---

            $2\ Cu^+(aq) \rightarrow Cu^{2+}(aq) + Cu(s)$        $E_{cell}^o = 0.36$ V;  spontaneous

$\Delta G° = -nFE_{cell}^o = -(1\ mol\ e^-)(96{,}485\ C/mol\ e^-)(0.36\ J/C) = -34{,}700\ J = -35$ kJ

$E_{cell}^o = \dfrac{0.0591}{n}\ \log K,\ \ \log K = \dfrac{nE°}{0.0591} = \dfrac{1(0.36)}{0.0591} = 6.09,\ \ K = 10^{6.09} = 1.2 \times 10^6$

      b.   $Fe^{2+} + 2\ e^- \rightarrow Fe$                             $E_c^o = -0.44$ V

            $(Fe^{2+} \rightarrow Fe^{3+} + e^-) \times 2$              $-E_a^o = -0.77$ V

---

            $3\ Fe^{2+}(aq) \rightarrow 2\ Fe^{3+}(aq) + Fe(s)$        $E_{cell}^o = -1.21$ V;  not spontaneous

      c.   $HClO_2 + 2\ H^+ + 2\ e^- \rightarrow HClO + H_2O$                    $E_c^o = 1.65$ V

            $HClO_2 + H_2O \rightarrow ClO_3^- + 3\ H^+ + 2\ e^-$               $-E_a^o = -1.21$ V

---

            $2\ HClO_2(aq) \rightarrow ClO_3^-(aq) + H^+(aq) + HClO(aq)$     $E_{cell}^o = 0.44$ V;  spontaneous

$\Delta G° = -nFE_{cell}^o = -(2\ mol\ e^-)(96{,}485\ C/mol\ e^-)(0.44\ J/C) = -84{,}900\ J = -85$ kJ

$\log K = \dfrac{nE°}{0.0591} = \dfrac{2(0.44)}{0.0591} = 14.89,\ \ K = 7.8 \times 10^{14}$

43.            $Al^{3+} + 3\ e^- \rightarrow Al$                       $E_c^o = -1.66$ V

               $Al + 6\ F^- \rightarrow AlF_6^{3-} + 3\ e^-$          $-E_a^o = 2.07$ V

---

      $Al^{3+}(aq) + 6\ F^-(aq) \rightarrow AlF_6^{3-}(aq)$            $E_{cell}^o = 0.41$ V     $K = ?$

$\log K = \dfrac{nE°}{0.0591} = \dfrac{3(0.41)}{0.0591} = 20.81,\ \ K = 10^{20.81} = 6.5 \times 10^{20}$

45.   $CuI + e^- \rightarrow Cu + I^-$                           $E_{CuI}^o = ?$

         $Cu \rightarrow Cu^+ + e^-$                             $-E_a^o = -0.52$ V

---

      $CuI(s) \rightarrow Cu^+(aq) + I^-(aq)$              $E_{cell}^o = E_{CuI}^o - 0.52$ V

For this overall reaction,  $K = K_{sp} = 1.1 \times 10^{-12}$:

$$E^{\circ}_{cell} = \frac{0.0591}{n} \log K_{sp} = \frac{0.0591}{1} \log(1.1 \times 10^{-12}) = -0.71 \text{ V}$$

$$E^{\circ}_{cell} = -0.71 \text{ V} = E^{\circ}_{CuI} - 0.52, \quad E^{\circ}_{CuI} = -0.19 \text{ V}$$

## Galvanic Cells:  Concentration Dependence

47.   a.

$$Au^{3+} + 3 \text{ e}^- \rightarrow Au \qquad\qquad E^{\circ}_c = 1.50 \text{ V}$$

$$(Tl \rightarrow Tl^+ + e^-) \times 3 \qquad -E^{\circ}_a = 0.34 \text{ V}$$

$$\overline{Au^{3+}(aq) + 3 \text{ Tl}(s) \rightarrow Au(s) + 3 \text{ Tl}^+(aq) \qquad E^{\circ}_{cell} = 1.84 \text{ V}}$$

b.   $\Delta G^{\circ} = -nFE^{\circ}_{cell} = -(3 \text{ mol e}^-)(96,485 \text{ C/mol e}^-)(1.84 \text{ J/C}) = -5.33 \times 10^5 \text{ J} = -533 \text{ kJ}$

$$\log K = \frac{nE^{\circ}}{0.0591} = \frac{3(1.84)}{0.0591} = 93.401, \quad K = 10^{93.401} = 2.52 \times 10^{93}$$

c.   At 25°C, $E_{cell} = E^{\circ}_{cell} - \frac{0.0591}{n} \log Q$, where $Q = \frac{[Tl^+]^3}{[Au^{3+}]}$.

$$E_{cell} = 1.84 \text{ V} - \frac{0.0591}{3} \log \frac{[Tl^+]^3}{[Au^{3+}]} = 1.84 - \frac{0.0591}{3} \log \frac{(1.0 \times 10^{-4})^3}{1.0 \times 10^{-2}}$$

$$E_{cell} = 1.84 - (-0.20) = 2.04 \text{ V}$$

49.

$$(Pb^{2+} + 2 \text{ e}^- \rightarrow Pb) \times 3 \qquad\qquad E^{\circ}_c = -0.13 \text{ V}$$

$$(Al \rightarrow Al^{3+} + 3 \text{ e}^-) \times 2 \qquad -E^{\circ}_a = 1.66 \text{ V}$$

$$\overline{3 \text{ Pb}^{2+}(aq) + 2 \text{ Al}(s) \rightarrow 3 \text{ Pb}(s) + 2 \text{ Al}^{3+}(aq) \qquad E^{\circ}_{cell} = 1.53 \text{ V}}$$

From the balanced reaction, when the $[Al^{3+}]$ has increased by 0.60 mol/L ($Al^{3+}$ is a product in the spontaneous reaction), then the $Pb^{2+}$ concentration has decreased by 3/2 (0.60 mol/L) = 0.90 *M*.

$$E_{cell} = 1.53 \text{ V} - \frac{0.0591}{6} \log \frac{[Al^{3+}]^2}{[Pb^{2+}]^3} = 1.53 - \frac{0.0591}{6} \log \frac{(1.60)^2}{(0.10)^3}$$

$$E_{cell} = 1.53 \text{ V} - 0.034 \text{ V} = 1.50 \text{ V}$$

51.   a.   n = 2 for this reaction (lead goes from Pb $\rightarrow$ Pb$^{2+}$ in PbSO$_4$).

$$E = E^{\circ}_{cell} - \frac{0.0591}{2} \log \left( \frac{1}{[H^+]^2 [HSO_4^-]^2} \right) = 2.04 \text{ V} - \frac{0.0591}{2} \log \frac{1}{(4.5)^2 (4.5)^2}$$

$$2.04 \text{ V} - (-0.077 \text{ V}) = 2.12 \text{ V}$$

b.  We can calculate $\Delta G°$ from $\Delta G° = \Delta H° - T\Delta S°$ and then $E°$ from $\Delta G° = -nFE°$, or we can use the equation derived in Exercise 11.39.

$$E°_{-20} = \frac{T\Delta S° - \Delta H°}{nF} = \frac{(253\ K)(263.5\ J/K) + 315.9 \times 10^3\ J}{(2\ mol\ e^-)(96{,}485\ C/mol\ e^-)} = 1.98\ J/C = 1.98\ V$$

c.  $E_{-20} = E°_{-20} - \dfrac{RT}{nF}\ \ln Q = 1.98\ V - \dfrac{RT}{nF}\ \ln\dfrac{1}{[H^+]^2[HSO_4^-]^2}$

$$E_{-20} = 1.98\ V - \frac{(8.3145\ J\ K^{-1}\ mol^{-1})(253\ K)}{(2\ mol\ e^-)(96{,}485\ C/mol\ e^-)}\ \ln\frac{1}{(4.5)^2(4.5)^2} = 1.98\ V - (-0.066\ V)$$

$$= 2.05\ V$$

d.  As the temperature decreases, the cell potential decreases.  Also, oil becomes more viscous at lower temperatures, which adds to the difficulty of starting an engine on a cold day.  The combination of these two factors results in batteries failing more often on cold days than on warm days.

53.  Concentration cell: a galvanic cell in which both compartments contain the same components but at different concentrations. All concentration cells have $E°_{cell} = 0$ because both compartments contain the same contents. The driving force for the cell is the different ion concentrations at the anode and cathode. The cell produces a voltage as long as the ion concentrations are different. Equilibrium for a concentration cell is reached ($E = 0$) when the ion concentrations in the two compartments are equal.

The net reaction in a concentration cell is:

$$M^{a+}(\text{cathode},\ x\ M) \rightarrow M^{a+}(\text{anode},\ y\ M)\quad E°_{cell} = 0$$

and the Nernst equation is:

$$E = E° - \frac{0.0591}{n}\ \log Q = -\frac{0.0591}{a}\ \log\frac{[M^{a+}(\text{anode})]}{[M^{a+}(\text{cathode})]},\quad \text{where } a \text{ is the number of electrons transferred.}$$

To register a potential ($E > 0$), the log Q term must be a negative value. This occurs when $M^{a+}(\text{cathode}) > M^{a+}(\text{anode})$. The higher ion concentration is always at the cathode, and the lower ion concentration is always at the anode. The magnitude of the cell potential depends on the magnitude of the differences in ion concentrations between the anode and cathode. The larger the difference in ion concentrations, the more negative is the log Q term, and the more positive is the cell potential. Thus, as the difference in ion concentrations between the anode and cathode compartments increases, the cell potential increases. This can be accomplished by decreasing the ion concentration at the anode and/or by increasing the ion concentration at the cathode.

55.  As is the case for all concentration cells, $E^o_{cell} = 0$, and the smaller ion concentration is always in the anode compartment.  The general Nernst equation for the Ni | Ni$^{2+}$ ($x$ $M$) || Ni$^{2+}$($y$ $M$) | Ni concentration cell is:

$$E_{cell} = E^o_{cell} - \frac{0.0591}{n} \log Q = \frac{0.0591}{2} \log \frac{[Ni^{2+}]_{anode}}{[Ni^{2+}]_{cathode}}$$

a.  Because both compartments are at standard conditions ([Ni$^{2+}$] = 1.0 $M$), $E_{cell} = E^o_{cell} = 0$ V.  No electron flow occurs.

b.  Cathode = 2.0 $M$ Ni$^{2+}$; anode = 1.0 $M$ Ni$^{2+}$; electron flow is always from the anode to the cathode, so electrons flow to the right in the diagram.

$$E_{cell} = \frac{-0.0591}{2} \log \frac{[Ni^{2+}]_{anode}}{[Ni^{2+}]_{cathode}} = \frac{-0.0591}{2} \log \frac{1.0}{2.0} = 8.9 \times 10^{-3} \text{ V}$$

c.  Cathode = 1.0 $M$ Ni$^{2+}$; anode = 0.10 $M$ Ni$^{2+}$; electrons flow to the left in the diagram.

$$E_{cell} = \frac{-0.0591}{2} \log \frac{0.10}{1.0} = 0.030 \text{ V}$$

d.  Cathode = 1.0 $M$ Ni$^{2+}$; anode = 4.0 × 10$^{-5}$ $M$ Ni$^{2+}$; electrons flow to the left in the diagram.

$$E_{cell} = \frac{-0.0591}{2} \log \frac{4.0 \times 10^{-5}}{1.0} = 0.13 \text{ V}$$

e.  Since both concentrations are equal, log(2.5/2.5) = log(1.0) = 0, and $E_{cell}$ = 0.  No electron flow occurs.

57.  Cu$^{2+}$(aq) + H$_2$(g) → 2 H$^+$(aq) + Cu(s)    $E^o_{cell}$ = 0.34 V – 0.00V = 0.34 V and n = 2

Since $P_{H_2}$ = 1.0 atm and [H$^+$] = 1.0 $M$:  $E_{cell} = E^o_{cell} - \frac{0.0591}{2} \log \frac{1}{[Cu^{2+}]}$

a.  $E_{cell} = 0.34 \text{ V} - \frac{0.0591}{2} \log \frac{1}{2.5 \times 10^{-4}} = 0.34 \text{ V} - 0.11 \text{ V} = 0.23 \text{ V}$

b.  Use the K$_{sp}$ expression to calculate the Cu$^{2+}$ concentration in the cell.

Cu(OH)$_2$(s) ⇌ Cu$^{2+}$(aq) + 2 OH$^-$(aq)    K$_{sp}$ = 1.6 × 10$^{-19}$ = [Cu$^{2+}$][OH$^-$]$^2$

From the problem, [OH$^-$] = 0.10 $M$, so:  $[Cu^{2+}] = \frac{1.6 \times 10^{-19}}{(0.10)^2} = 1.6 \times 10^{-17} M$

$$E_{cell} = E°_{cell} - \frac{0.0591}{2} \log\frac{1}{[Cu^{2+}]} = 0.34 - \frac{0.0591}{2} \log\frac{1}{1.6 \times 10^{-17}} = 0.34 - 0.50$$
$$= -0.16 \text{ V}$$

Because $E_{cell} < 0$, the forward reaction is not spontaneous, but the reverse reaction is spontaneous.  The Cu electrode becomes the anode, and $E_{cell} = 0.16$ V for the reverse reaction.  The cell reaction is $2 \text{ H}^+(aq) + Cu(s) \rightarrow Cu^{2+}(aq) + H_2(g)$.

c.   $0.195 \text{ V} = 0.34 \text{ V} - \dfrac{0.0591}{2} \log\dfrac{1}{[Cu^{2+}]}$,   $\log\dfrac{1}{[Cu^{2+}]} = 4.91$,   $[Cu^{2+}] = 10^{-4.91}$
$$= 1.2 \times 10^{-5} M$$

*Note*: When determining exponents, we will carry extra significant figures.

d.   $E_{cell} = E°_{cell} - (0.0591/2) \log (1/[Cu^{2+}]) = E°_{cell} + 0.0296 \log [Cu^{2+}]$;  this equation is in the form of a straight-line equation, $y = mx + b$.  A graph of $E_{cell}$ versus $\log[Cu^{2+}]$ will yield a straight line with slope equal to 0.0296 V or 29.6 mV.

59.   a.   $Ag^+(x \, M, \text{anode}) \rightarrow Ag^+(0.10 \, M, \text{cathode})$;  for the silver concentration cell, $E° = 0.00$ (as is always the case for concentration cells) and n = 1.

$$E = 0.76 \text{ V} = 0.00 - \frac{0.0591}{1} \log\frac{[Ag^+]_{anode}}{[Ag^+]_{cathode}}$$

$$0.76 = -0.0591 \log\frac{[Ag^+]_{anode}}{0.10}, \quad \frac{[Ag^+]_{anode}}{0.10} = 10^{-12.86}, \quad [Ag^+]_{anode} = 1.4 \times 10^{-14} M$$

b.   $Ag^+(aq) + 2 \, S_2O_3^{2-}(aq) \rightleftharpoons Ag(S_2O_3)_2^{3-}(aq)$

$$K = \frac{[Ag(S_2O_3)_2^{3-}]}{[Ag^+][S_2O_3^{2-}]^2} = \frac{1.0 \times 10^{-3}}{1.4 \times 10^{-14}(0.050)^2} = 2.9 \times 10^{13}$$

## Electrolysis

61.   $15 \text{ A} = \dfrac{15 \text{ C}}{s} \times \dfrac{60 \text{ s}}{min} \times \dfrac{60 \text{ min}}{mol} = 5.4 \times 10^4 \text{ C of charge passed in 1 hour}$

a.   $5.4 \times 10^4 \text{ C} \times \dfrac{1 \text{ mol e}^-}{96,485 \text{ C}} \times \dfrac{1 \text{ mol Co}}{2 \text{ mol e}^-} \times \dfrac{58.9 \text{ g}}{mol} = 16 \text{ g Co}$

b.   $5.4 \times 10^4 \text{ C} \times \dfrac{1 \text{ mol e}^-}{96,485 \text{ C}} \times \dfrac{1 \text{ mol Hf}}{4 \text{ mol e}^-} \times \dfrac{178.5 \text{ g}}{mol} = 25 \text{ g Hf}$

c.   $2 \, I^- \rightarrow I_2 + 2e^-$;  $5.4 \times 10^4 \text{ C} \times \dfrac{1 \text{ mol e}^-}{96,485 \text{ C}} \times \dfrac{1 \text{ mol I}_2}{2 \text{ mol e}^-} \times \dfrac{253.8 \text{ g I}_2}{\text{mol I}_2} = 71 \text{ g I}_2$

d.  Cr is in the +6 oxidation state in $CrO_3$.  Six moles of $e^-$ are needed to produce 1 mol Cr from molten $CrO_3$.

$$5.4 \times 10^4 \text{ C} \times \frac{1 \text{ mol } e^-}{96,485 \text{ C}} \times \frac{1 \text{ mol Cr}}{6 \text{ mol } e^-} \times \frac{52.0 \text{ g Cr}}{\text{mol Cr}} = 4.9 \text{ g Cr}$$

63.    The oxidation state of bismuth in $BiO^+$ is +3 because oxygen has a $-2$ oxidation state in this ion.  Therefore, 3 moles of electrons are required to reduce the bismuth in $BiO^+$ to Bi(s).

$$10.0 \text{ g Bi} \times \frac{1 \text{ mol Bi}}{209.0 \text{ g Bi}} \times \frac{3 \text{ mol } e^-}{\text{mol Bi}} \times \frac{96,485 \text{ C}}{\text{mol } e^-} \times \frac{1 \text{ s}}{25.0 \text{ C}} = 554 \text{ s} = 9.23 \text{ min}$$

65.    First determine the species present, and then reference Table 11.1 to help you identify each species as a possible oxidizing agent (species reduced) or as a possible reducing agent (species oxidized).  Of all the possible oxidizing agents, the species that will be reduced at the cathode will have the most positive $E_c^o$ value; the species that will be oxidized at the anode will be the reducing agent with the most positive $-E_a^o$ value.

a.  Species present:  $Ni^{2+}$ and $Br^-$;  $Ni^{2+}$ can be reduced to Ni, and $Br^-$ can be oxidized to $Br_2$ (from Table 11.1).  The reactions are:

Cathode:  $Ni^{2+} + 2e^- \rightarrow Ni$          $E_c^o = -0.23$ V

Anode:    $2 Br^- \rightarrow Br_2 + 2 e^-$      $-E_a^o = -1.09$ V

b.  Species present:  $Al^{3+}$ and $F^-$;  $Al^{3+}$ can be reduced, and $F^-$ can be oxidized.  The reactions are:

Cathode:  $Al^{3+} + 3 e^- \rightarrow Al$          $E_c^o = -1.66$ V

Anode:    $2 F^- \rightarrow F_2 + 2 e^-$      $-E_a^o = -2.87$ V

c.  Species present:  $Mn^{2+}$ and $I^-$;  $Mn^{2+}$ can be reduced, and $I^-$ can be oxidized.  The reactions are:

Cathode:  $Mn^{2+} + 2 e^- \rightarrow Mn$          $E_c^o = -1.18$ V

Anode:    $2 I^- \rightarrow I_2 + 2 e^-$          $-E_a^o = -0.54$ V

d.  For aqueous solutions, we must now consider $H_2O$ as a possible oxidizing agent and a possible reducing agent.  Species present: $Ni^{2+}$, $Br^-$, and $H_2O$.  Possible cathode reactions are:

$Ni^{2+} + 2e^- \rightarrow Ni$                          $E_c^o = -0.23$ V

$2 H_2O + 2 e^- \rightarrow H_2 + 2 OH^-$                $E_c^o = -0.83$ V

Because it is easier to reduce $Ni^{2+}$ than $H_2O$ (assuming standard conditions), $Ni^{2+}$ will be reduced by the above cathode reaction.

Possible anode reactions are:

$$2\ Br^- \rightarrow Br_2 + 2\ e^- \qquad\qquad -E_a^\circ = -1.09\ V$$

$$2\ H_2O \rightarrow O_2 + 4\ H^+ + 4\ e^- \qquad -E_a^\circ = -1.23\ V$$

Because $Br^-$ is easier to oxidize than $H_2O$ (assuming standard conditions), $Br^-$ will be oxidized by the above anode reaction.

e.  Species present: $Al^{3+}$, $F^-$, and $H_2O$; $Al^{3+}$ and $H_2O$ can be reduced.  The reduction potentials are $E_c^\circ = -1.66\ V$ for $Al^{3+}$ and $E_c^\circ = -0.83\ V$ for $H_2O$ (assuming standard conditions).  $H_2O$ should be reduced at the cathode ($2\ H_2O + 2\ e^- \rightarrow H_2 + 2\ OH^-$).

$F^-$ and $H_2O$ can be oxidized.  The oxidation potentials are $-E_a^\circ = -2.87\ V$ for $F^-$ and $-E_a^\circ = -1.23\ V$ for $H_2O$ (assuming standard conditions).  From the potentials, we would predict $H_2O$ to be oxidized at the anode ($2\ H_2O \rightarrow O_2 + 4\ H^+ + 4\ e^-$).

f.  Species present: $Mn^{2+}$, $I^-$, and $H_2O$; $Mn^{2+}$ and $H_2O$ can be reduced.  The possible cathode reactions are:

$$Mn^{2+} + 2\ e^- \rightarrow Mn \qquad\qquad E_c^\circ = -1.18\ V$$

$$2\ H_2O + 2\ e^- \rightarrow H_2 + 2\ OH^- \qquad E_c^\circ = -0.83\ V$$

Reduction of $H_2O$ should occur at the cathode because $E_c^\circ$ for $H_2O$ is most positive.

$I^-$ and $H_2O$ can be oxidized.  The possible anode reactions are:

$$2\ I^- \rightarrow I_2 + 2\ e^- \qquad\qquad -E_a^\circ = -0.54\ V$$

$$2\ H_2O \rightarrow O_2 + 4\ H^+ + 4\ e^- \qquad -E_a^\circ = -1.23\ V$$

Oxidation of $I^-$ will occur at the anode because $-E_a^\circ$ for $I^-$ is most positive.

67.  $Au^{3+} + 3\ e^- \rightarrow Au \qquad E^\circ = 1.50\ V \qquad\qquad Ni^{2+} + 2\ e^- \rightarrow Ni \qquad E^\circ = -0.23\ V$

$Ag^+ + e^- \rightarrow Ag \qquad E^\circ = 0.80\ V \qquad\qquad Cd^{2+} + 2\ e^- \rightarrow Cd \qquad E^\circ = -0.40\ V$

$2\ H_2O + 2e^- \rightarrow H_2 + 2\ OH^- \quad E^\circ = -0.83\ V$

Au(s) will plate out first since it has the most positive reduction potential, followed by Ag(s), which is followed by Ni(s), and finally, Cd(s) will plate out last since it has the most negative reduction potential of the metals listed.

69.  To begin plating out Pd:  $E_c = 0.62\ V - \dfrac{0.0591}{2} \log \dfrac{[Cl^-]^4}{[PdCl_4{}^{2-}]} = 0.62 - \dfrac{0.0591}{2} \log \dfrac{(1.0)^4}{0.020}$

$E_c = 0.62\ V - 0.050\ V = 0.57\ V$

When 99% of Pd has plated out, $[PdCl_4^-] = \dfrac{1}{100}(0.020) = 0.00020\ M$.

$$E_c = 0.62 - \frac{0.0591}{2}\log\frac{(1.0)^4}{2.0\times10^{-4}} = 0.62\text{ V} - 0.11\text{V} = 0.51\text{ V}$$

To begin Pt plating:  $E_c = 0.73\text{ V} - \dfrac{0.0591}{2}\log\dfrac{(1.0)^4}{0.020} = 0.73 - 0.050 = 0.68\text{ V}$

When 99% of Pt is plated:  $E_c = 0.73 - \dfrac{0.0591}{2}\log\dfrac{(1.0)^4}{2.0\times10^{-4}} = 0.73 - 0.11 = 0.62\text{ V}$

To begin Ir plating:  $E_c = 0.77\text{ V} - \dfrac{0.0591}{3}\log\dfrac{(1.0)^4}{0.020} = 0.77 - 0.033 = 0.74\text{ V}$

When 99% of Ir is plated:  $E_c = 0.77 - \dfrac{0.0591}{3}\log\dfrac{(1.0)^4}{2.0\times10^{-4}} = 0.77 - 0.073 = 0.70\text{ V}$

Yes, since the range of potentials for plating out each metal do not overlap, we should be able to separate the three metals.  The exact potential to apply depends on the oxidation reaction. The order of plating will be Ir(s) first, followed by Pt(s), and finally, Pd(s) as the potential is gradually increased.

71.    Alkaline earth metals form +2 ions, so 2 mol of $e^-$ are transferred to form the metal M.

$$\text{Mol M} = 748\text{ s} \times \frac{5.00\text{ C}}{\text{s}} \times \frac{1\text{ mol }e^-}{96{,}485\text{ C}} \times \frac{1\text{ mol M}}{2\text{ mol }e^-} = 1.94\times10^{-2}\text{ mol M}$$

$$\text{Molar mass of M} = \frac{0.471\text{ g M}}{1.94\times10^{-2}\text{ mol M}} = 24.3\text{ g/mol};\ \ MgCl_2\text{ was electrolyzed.}$$

73.    $F_2$ is produced at the anode:  $2\ F^- \rightarrow F_2 + 2\ e^-$

$$2.00\text{ h} \times \frac{60\text{ min}}{\text{h}} \times \frac{60\text{ s}}{\text{min}} \times \frac{10.0\text{ C}}{\text{s}} \times \frac{1\text{ mol }e^-}{96{,}485\text{ C}} = 0.746\text{ mol }e^-$$

$$0.746\text{ mol }e^- \times \frac{1\text{ mol }F_2}{2\text{ mol }e^-} = 0.373\text{ mol }F_2;\ \ PV = nRT,\ \ V = \frac{nRT}{P}$$

$$V = \frac{(0.373\text{ mol})(0.08206\text{ L atm K}^{-1}\text{ mol}^{-1})(298\text{ K})}{1.00\text{ atm}} = 9.12\text{ L }F_2$$

K is produced at the cathode:  $K^+ + e^- \rightarrow K$

$$0.746\text{ mol }e^- \times \frac{1\text{ mol K}}{\text{mol }e^-} \times \frac{39.10\text{ g K}}{\text{mol K}} = 29.2\text{ g K}$$

75.     In the electrolysis of aqueous sodium chloride, $H_2O$ is reduced in preference to $Na^+$, and $Cl^-$ is oxidized in preference to $H_2O$. The anode reaction is $2\ Cl^- \rightarrow Cl_2 + 2\ e^-$, and the cathode reaction is $2\ H_2O + 2\ e^- \rightarrow H_2 + 2\ OH^-$. The overall reaction is $2\ H_2O(l) + 2\ Cl^-\ (aq) \rightarrow Cl_2(g) + H_2(g) + 2\ OH^-\ (aq)$.

From the 1 : 1 mole ratio between $Cl_2$ and $H_2$ in the overall balanced reaction, if 6.00 L of $H_2(g)$ is produced, then 6.00 L of $Cl_2(g)$ also will be produced since moles and volume of gas are directly proportional at constant T and P (see Chapter 5 of text).

## Additional Exercises

77.     $(CO + O^{2-} \rightarrow CO_2 + 2\ e^-) \times 2$
        $O_2 + 4\ e^- \rightarrow 2\ O^{2-}$

$$\overline{2\ CO + O_2 \rightarrow 2\ CO_2}$$

$$\Delta G = -nFE, \quad E = \frac{-\Delta G}{nF} = \frac{-(-380 \times 10^{-3}\ J)}{(4\ mol\ e^-)(96{,}485\ C / mol\ e^-)} = 0.98\ V$$

79.     For $C_2H_5OH$, H has a +1 oxidation state, and O has a –2 oxidiation state. This dictates a –2 oxidation state for C. For $CO_2$, O has a –2 oxidiation state, so carbon has a +4 oxidiation state. Six moles of electrons are transferred per mole of carbon oxidized (C goes from –2 $\rightarrow$ +4). Two moles of carbon are in the balanced reaction, so n = 12.

$w_{max} = -1320\ kJ = \Delta G = -nFE, \quad -1320 \times 10^3\ J = -nFE = -(12\ mol\ e^-)(96{,}485\ C/mol\ e^-)E$

$E = 1.14\ J/C = 1.14\ V$

81.     Cadmium goes from the zero oxidation state to the +2 oxidation state in $Cd(OH)_2$. Because one mole of Cd appears in the balanced reaction, n = 2 mol electrons transferred. At standard conditions:

$$w_{max} = \Delta G° = -nFE°, \quad w_{max} = -(2\ mol\ e^-)(96{,}485\ C/mol\ e^-)(1.10\ J/C) = -2.12 \times 10^5\ J$$
$$= -212\ kJ$$

83.     a.   Paint: covers the metal surface so that no contact occurs between the metal and air. This only works as long as the painted surface is not scratched.

        b.   Durable oxide coatings: covers the metal surface so that no contact occurs between the metal and air.

        c.   Galvanizing:  coating steel with zinc;  Zn forms an effective oxide coating over steel; also, zinc is more easily oxidized than the iron in the steel.

        d.   Sacrificial metal:  attaching a more easily oxidized metal to an iron surface; the more active metal is preferentially oxidized instead of iron.

        e.   Alloying:  adding chromium and nickel to steel; the added Cr and Ni form oxide coatings on the steel surface.

f.  Cathodic protection:  a more easily oxidized metal is placed in electrical contact with the metal we are trying to protect. It is oxidized in preference to the protected metal. The protected metal becomes the cathode electrode, thus cathodic protection.

85.    $Zn \rightarrow Zn^{2+} + 2\ e^- \quad -E_a^o = 0.76\ V;\ Fe \rightarrow Fe^{2+} + 2\ e^- \quad -E_a^o = 0.44\ V$

It is easier to oxidize Zn than Fe, so the Zn will be oxidized, protecting the iron of the *Monitor's* hull.

87.    The potential oxidizing agents are $NO_3^-$ and $H^+$.  Hydrogen ion cannot oxidize Pt under either condition.  Nitrate cannot oxidize Pt unless there is $Cl^-$ in the solution.  Aqua regia has both $Cl^-$ and $NO_3^-$.  The overall reaction is:

$$(NO_3^- + 4\ H^+ + 3\ e^- \rightarrow NO + 2\ H_2O) \times 2 \qquad\qquad E_c^o = \ \ 0.96\ V$$
$$(4\ Cl^- + \ Pt \rightarrow PtCl_4^{2-} + 2\ e^-) \times 3 \qquad\qquad -E_a^o = -0.755\ V$$

---

$$12\ Cl^-(aq) + 3\ Pt(s) + 2\ NO_3^-(aq) + 8\ H^+(aq) \rightarrow 3\ PtCl_4^{2-}(aq) + 2\ NO(g) + 4\ H_2O(l) \quad E_{cell}^o = 0.21\ V$$

89.    a.  $E_{cell} = E_{ref} + 0.05916\ pH,\ \ 0.480\ V = 0.250\ V + 0.05916\ pH$

$$pH = \frac{0.480 - 0.250}{0.05916} = 3.888;\ \ \text{uncertainty} = \pm1\ mV = \pm\ 0.001\ V$$

$$pH_{max} = \frac{0.481 - 0.250}{0.05916} = 3.905;\quad pH_{min} = \frac{0.479 - 0.250}{0.05916} = 3.871$$

Thus if the uncertainty in potential is $\pm0.001$ V, then the uncertainty in pH is $\pm0.017$, or about $\pm0.02$ pH units.  For this measurement, $[H^+] = 10^{-3.888} = 1.29 \times 10^{-4}$ *M*.  For an error of $+1$ mV, $[H^+] = 10^{-3.905} = 1.24 \times 10^{-4}$ *M*.  For an error of $-1$ mV, $[H^+] = 10^{-3.871} = 1.35 \times 10^{-4}$ *M*.  So the uncertainty in $[H^+]$ is $\pm0.06 \times 10^{-4}$ *M* $= \pm6 \times 10^{-6}$ *M*.

b.  From part a, we will be within $\pm0.02$ pH units if we measure the potential to the nearest $\pm0.001$ V ($\pm1$ mV).

91.    $2\ Ag^+(aq) + Cu(s) \rightarrow Cu^{2+}(aq) + 2\ Ag(s) \quad E_{cell}^o = 0.80 - 0.34\ V = 0.46\ V;$  A galvanic cell produces a voltage as the forward reaction occurs.  Any stress that increases the tendency of the forward reaction to occur will increase the cell potential, whereas a stress that decreases the tendency of the forward reaction to occur will decrease the cell potential.

a.  Added $Cu^{2+}$ (a product ion) will decrease the tendency of the forward reaction to occur, which will decrease the cell potential.

b.  Added $NH_3$ removes $Cu^{2+}$ in the form of $Cu(NH_3)_4^{2+}$.  Because a product ion is removed, this will increase the tendency of the forward reaction to occur, which will increase the cell potential.

c.   Added $Cl^-$ removes $Ag^+$ in the form of $AgCl(s)$.  Because a reactant ion is removed, this will decrease the tendency of the forward reaction to occur, which will decrease the cell potential.

d.   $Q_1 = \dfrac{[Cu^{2+}]_0}{[Ag^+]_0^2}$ ;  as the volume of solution is doubled, each concentration is halved.

$$Q_2 = \frac{1/2 \, [Cu^{2+}]_0}{(1/2 \, [Ag^+]_0)^2} = \frac{2[Cu^{2+}]_0}{[Ag^+]_0^2} = 2Q_1$$

The reaction quotient is doubled because the concentrations are halved.  Because reactions are spontaneous when $Q < K$, and because Q increases when the solution volume doubles, the reaction is closer to equilibrium, which will decrease the cell potential.

e.   Because $Ag(s)$ is not a reactant in this spontaneous reaction, and because solids do not appear in the reaction quotient expressions, replacing the silver electrode with a platinum electrode will have no effect on the cell potential.

## Challenge Problems

93.   Chromium(III) nitrate $[Cr(NO_3)_3]$ has chromium in the +3 oxidation state.

$$1.15 \text{ g Cr} \times \frac{1 \text{ mol Cr}}{52.00 \text{ g}} \times \frac{3 \text{ mol e}^-}{\text{mol Cr}} \times \frac{96{,}485 \text{ C}}{\text{mol e}^-} = 6.40 \times 10^3 \text{ C of charge}$$

For the Os cell, $6.40 \times 10^3$ C of charge also was passed.

$$3.15 \text{ g Os} \times \frac{1 \text{ mol Os}}{190.2 \text{ g}} = 0.0166 \text{ mol Os}; \quad 6.40 \times 10^3 \text{ C} \times \frac{1 \text{ mol e}^-}{96{,}485 \text{ C}} = 0.0663 \text{ mol e}^-$$

$$\frac{\text{Mol e}^-}{\text{Mol Os}} = \frac{0.0663}{0.0166} = 3.99 \approx 4$$

This salt is composed of $Os^{4+}$ and $NO_3^-$ ions. The compound is $Os(NO_3)_4$, osmium(IV) nitrate.

For the third cell, identify X by determining its molar mass. Two moles of electrons are transferred when $X^{2+}$ is reduced to X.

$$\text{Molar mass} = \frac{2.11 \text{ g X}}{6.40 \times 10^3 \text{ C} \times \dfrac{1 \text{ mol e}^-}{96{,}485 \text{ C}} \times \dfrac{1 \text{ mol X}}{2 \text{ mol e}^-}} = 63.6 \text{ g/mol}$$

This is copper (Cu).

95.    a.    $Zn(s) + Cu^{2+}(aq) \rightarrow Zn^{2+}(aq) + Cu(s)$    $E^{\circ}_{cell} = 1.10\ V$

$$E_{cell} = 1.10\ V - \frac{0.0591}{2}\ \log\frac{[Zn^{2+}]}{[Cu^{2+}]}$$

$$E_{cell} = 1.10\ V - \frac{0.0591}{2}\ \log\frac{0.10}{2.50} = 1.10\ V - (-0.041\ V) = 1.14\ V$$

b.    $10.0\ h \times \dfrac{60\ min}{h} \times \dfrac{60\ s}{min} \times \dfrac{10.0\ C}{s} \times \dfrac{1\ mol\ e^{-}}{96,485\ C} \times \dfrac{1\ mol\ Cu}{2\ mol\ e^{-}} = 1.87\ mol\ Cu$ produced

The $Cu^{2+}$ concentration decreases by 1.87 mol/L, and the $Zn^{2+}$ concentration will increase by 1.87 mol/L.

$[Cu^{2+}] = 2.50 - 1.87 = 0.63\ M;\ [Zn^{2+}] = 0.10 + 1.87 = 1.97\ M$

$$E_{cell} = 1.10\ V - \frac{0.0591}{2}\ \log\frac{1.97}{0.63} = 1.10\ V - 0.015\ V = 1.09\ V$$

c.    $1.87\ mol\ Zn$ consumed $\times \dfrac{65.38\ g\ Zn}{mol\ Zn} = 122\ g\ Zn$

Mass of electrode = 200. − 122 = 78 g Zn

$1.87\ mol\ Cu$ formed $\times \dfrac{63.55\ g\ Cu}{mol\ Cu} = 119\ g\ Cu$

Mass of electrode = 200. + 119 = 319 g Cu

d.    Three things could possibly cause this battery to go dead:

     1.    All the Zn is consumed.
     2.    All the $Cu^{2+}$ is consumed.
     3.    Equilibrium is reached ($E_{cell} = 0$).

We began with 2.50 mol $Cu^{2+}$ and 200. g Zn × 1 mol Zn/65.38 g Zn = 3.06 mol Zn. $Cu^{2+}$ is the limiting reagent and will run out first. To react all the $Cu^{2+}$ requires:

$$2.50\ mol\ Cu^{2+} \times \frac{2\ mol\ e^{-}}{mol\ Cu^{2+}} \times \frac{96,485\ C}{mol\ e^{-}} \times \frac{1\ s}{10.0\ C} \times \frac{1\ h}{3600\ s} = 13.4\ h$$

For equilibrium to be reached:    $E = 0 = 1.10\ V - \dfrac{0.0591}{2}\ \log\dfrac{[Zn^{2+}]}{[Cu^{2+}]}$

$$\frac{[Zn^{2+}]}{[Cu^{2+}]} = K = 10^{2(1.10)/0.0591} = 1.68 \times 10^{37}$$

This is such a large equilibrium constant that virtually all the $Cu^{2+}$ must react to reach equilibrium. So the battery will go dead in 13.4 hours.

97. a.

$$3 \times (e^- + 2\,H^+ + NO_3^- \rightarrow NO_2 + H_2O) \qquad E_c^o = 0.775\ V$$

$$2\,H_2O + NO \rightarrow NO_3^- + 4\,H^+ + 3\,e^- \qquad -E_a^o = -0.957\ V$$

---

$$2\,H^+(aq) + 2\,NO_3^-(aq) + NO(g) \rightarrow 3\,NO_2(g) + H_2O(l) \qquad E_{cell}^o = -0.182\ V \quad K = ?$$

$$\log K = \frac{E^o}{0.0591} = \frac{3(-0.182)}{0.0591} = -9.239,\ K = 10^{-9.239} = 5.77 \times 10^{-10}$$

b. Let $C$ = concentration of $HNO_3 = [H^+] = [NO_3^-]$.

$$5.77 \times 10^{-10} = \frac{P_{NO_2}^3}{P_{NO} \times [H^+]^2 \times [NO_3^-]^2} = \frac{P_{NO_2}^3}{P_{NO} \times C^4}$$

If 0.20% $NO_2$ by moles and $P_{total} = 1.00$ atm:

$$P_{NO_2} = \frac{0.20\ mol\ NO_2}{100.\ mol\ total} \times 1.00\ atm = 2.0 \times 10^{-3}\ atm;\ P_{NO} = 1.00 - 0.0020 = 1.00\ atm$$

$$5.77 \times 10^{-10} = \frac{(2.0 \times 10^{-3})^3}{(1.00)C^4},\ C = 1.9\ M\ HNO_3$$

99.

$$2\,H^+ + 2\,e^- \rightarrow H_2 \qquad E_c^o = 0.000\ V$$

$$Fe \rightarrow Fe^{2+} + 2\,e^- \qquad -E_a^o = -(-0.440V)$$

---

$$2\,H^+(aq) + Fe(s) \rightarrow H_2(g) + Fe^{3+}(aq) \qquad E_{cell}^o = 0.440\ V$$

$$E_{cell} = E_{cell}^o - \frac{0.0591}{n}\ \log Q,\ where\ n = 2\ and\ Q = \frac{P_{H_2} \times [Fe^{3+}]}{[H^+]^2}$$

To determine $K_a$ for the weak acid, first use the electrochemical data to determine the $H^+$ concentration in the half-cell containing the weak acid.

$$0.333\ V = 0.440\ V - \frac{0.0591}{2}\ \log \frac{1.00\ atm(1.00 \times 10^{-3}\ M)}{[H^+]^2}$$

$$\frac{0.107(2)}{0.0591} = \log \frac{1.0 \times 10^{-3}}{[H^+]^2},\ \frac{1.0 \times 10^{-3}}{[H^+]^2} = 10^{3.621} = 4.18 \times 10^3,\ [H^+] = 4.89 \times 10^{-4}\ M$$

Now we can solve for the $K_a$ value of the weak acid HA through the normal setup for a weak acid problem.

$$HA(aq) \rightleftharpoons H^+(aq) + A^-(aq) \qquad K_a = \frac{[H^+][A^-]}{[HA]}$$

|  | HA | H⁺ | A⁻ |
|---|---|---|---|
| Initial | $1.00\ M$ | ~0 | 0 |
| Equil. | $1.00 - x$ | $x$ | $x$ |

$$K_a = \frac{x^2}{1.00 - x}, \text{ where } x = [H^+] = 4.89 \times 10^{-4}\ M,\ K_a = \frac{(4.89 \times 10^{-4})^2}{1.00 - 4.89 \times 10^{-4}} = 2.39 \times 10^{-7}$$

101.    a.

$$(Ag^+ + e^- \rightarrow Ag) \times 2 \qquad\qquad E_c^\circ = 0.80\ V$$
$$Cu \rightarrow Cu^{2+} + 2\ e^- \qquad\qquad -E_a^\circ = -0.34\ V$$

$$\overline{2\ Ag^+(aq) + Cu(s) \rightarrow 2\ Ag(s) + Cu^{2+}(aq) \qquad E_{cell}^\circ = 0.46\ V}$$

$$E_{cell} = E_{cell}^\circ - \frac{0.0591}{n} \log Q, \text{ where } n = 2 \text{ and } Q = \frac{[Cu^{2+}]}{[Ag^+]^2}.$$

To calculate $E_{cell}$, we need to use the $K_{sp}$ data to determine $[Ag^+]$.

$$AgCl(s) \rightleftharpoons Ag^+(aq) + Cl^-(aq) \quad K_{sp} = 1.6 \times 10^{-10} = [Ag^+][Cl^-]$$

|  | AgCl | Ag⁺ | Cl⁻ |
|---|---|---|---|
| Initial | $s$ = solubility (mol/L) | 0 | 0 |
| Equil. |  | $s$ | $s$ |

$$K_{sp} = 1.6 \times 10^{-10} = s^2,\ s = [Ag^+] = 1.3 \times 10^{-5}\ mol/L$$

$$E_{cell} = 0.46\ V - \frac{0.0591}{2} \log \frac{2.0}{(1.3 \times 10^{-5})^2} = 0.46\ V - 0.30 = 0.16\ V$$

b.   $Cu^{2+}(aq) + 4\ NH_3(aq) \rightleftharpoons Cu(NH_4)_4^{2+}(aq) \qquad K = 1.0 \times 10^{13} = \dfrac{[Cu(NH_3)_4^{2+}]}{[Cu^{2+}][NH_3]^4}$

Because K is very large for the formation of $Cu(NH_3)_4^{2+}$, the forward reaction is dominant. At equilibrium, essentially all the 2.0 $M$ $Cu^{2+}$ will react to form 2.0 $M$ $Cu(NH_3)_4^{2+}$. This reaction requires 8.0 $M$ $NH_3$ to react with all the $Cu^{2+}$ in the balanced equation. Therefore, the moles of $NH_3$ added to 1.0-L solution will be larger than 8.0 mol since some $NH_3$ must be present at equilibrium. In order to calculate how much $NH_3$ is present at equilibrium, we need to use the electrochemical data to determine the $Cu^{2+}$ concentration.

$$E_{cell} = E_{cell}^\circ - \frac{0.0591}{n} \log Q,\ \ 0.52\ V = 0.46\ V - \frac{0.0591}{2} \log \frac{[Cu^{2+}]}{(1.3 \times 10^{-5})^2}$$

$$\log \frac{[Cu^{2+}]}{(1.3 \times 10^{-5})^2} = \frac{-0.06(2)}{0.0591} = -2.03,\ \ \frac{[Cu^{2+}]}{(1.3 \times 10^{-5})^2} = 10^{-2.03} = 9.3 \times 10^{-3}$$

$[Cu^{2+}] = 1.6 \times 10^{-12} = 2 \times 10^{-12}\ M$

(We carried extra significant figures in the calculation.)

*Note*: Our assumption that the 2.0 $M$ $Cu^{2+}$ essentially reacts to completion is excellent as only $2 \times 10^{-12}$ $M$ $Cu^{2+}$ remains after this reaction. Now we can solve for the equilibrium $[NH_3]$.

$$K = 1.0 \times 10^{13} = \frac{[Cu(NH_3)_4^{2+}]}{[Cu^{2+}][NH_3]^4} = \frac{(2.0)}{(2 \times 10^{-12})[NH_3]^4}\ ,\quad [NH_3] = 0.6\ M$$

Because 1.0 L of solution is present, 0.6 mol $NH_3$ remains at equilibrium. The total moles of $NH_3$ added is 0.6 mol plus the 8.0 mol $NH_3$ necessary to form 2.0 $M$ $Cu(NH_3)_4^{2+}$. Therefore, $8.0 + 0.6 = 8.6$ mol $NH_3$ was added.

103.  
$$\begin{array}{llll} (Ag^+ + e^- \rightarrow Ag) \times 2 & & E_c^{\circ} & = 0.80\ V \\ Cd \rightarrow Cd^{2+} + 2\,e^- & & -E_a^{\circ} & = 0.40\ V \end{array}$$

$$\overline{2\ Ag^+(aq)\ + Cd(s) \rightarrow\ Cd^{2+}(aq) + 2\ Ag(s)\qquad E_{cell}^{\circ} = 1.20\ V}$$

Overall complex ion reaction:

$$Ag^+(aq)\ +\ 2\ NH_3(aq)\ \rightarrow\ Ag(NH_3)_2^+(aq)\quad K = K_1K_2 = 2.1 \times 10^3(8.2 \times 10^3) = 1.7 \times 10^7$$

Because K is large, we will let the reaction go to completion, and then solve the back-equilibrium problem.

|          | $Ag^+$ | + | $2\,NH_3$ | $\rightleftharpoons$ | $Ag(NH_3)_2^+$ | $K = 1.7 \times 10^7$ |
|----------|--------|---|-----------|------|----------------|----------------------|
| Before   | 1.00 $M$ |   | 15.0 $M$ |      | 0              |                      |
| After    | 0      |   | 13.0      |      | 1.00           | New initial          |
| Change   | $x$    |   | $+2x$     | $\leftarrow$ | $-x$   |                      |
| Equil.   | $x$    |   | $13.0 + 2x$ |    | $1.00 - x$     |                      |

$$K = \frac{[Ag(NH_3)_2^+]}{[Ag^+][NH_3]^2};\quad 1.7 \times 10^7 = \frac{1.00 - x}{x(13.0 + 2x)^2} \approx \frac{1.00}{x(13.0)^2}$$

Solving: $x = 3.5 \times 10^{-10}\ M = [Ag^+]$; assumptions good.

$$E = E^{\circ} - \frac{0.0591}{2}\ \log\frac{[Cd^{2+}]}{[Ag^+]^2} = 1.20\ V - \frac{0.0591}{2}\ \log\left[\frac{1.0}{(3.5 \times 10^{-10})^2}\right]$$

$$E = 1.20 - 0.56 = 0.64\ V$$

105.    a.    From Table 11.1: $2 H_2O + 2 e^- \rightarrow H_2 + 2 OH^-$   $E° = -0.83$ V

$$E°_{cell} = E°_{H_2O} - E°_{Zr} = -0.83 \text{ V} + 2.36 \text{ V} = 1.53 \text{ V}$$

Yes, the reduction of $H_2O$ to $H_2$ by Zr is spontaneous at standard conditions since $E°_{cell} > 0$.

b.       $(2 H_2O + 2 e^- \rightarrow H_2 + 2 OH^-) \times 2$

$\phantom{b}$       $Zr + 4 OH^- \rightarrow ZrO_2 \bullet H_2O + H_2O + 4 e^-$

$\phantom{b}$       _____

$\phantom{b}$       $3 H_2O(l) + Zr(s) \rightarrow 2 H_2(g) + ZrO_2 \bullet H_2O(s)$

c.    $\Delta G° = -nFE° = -(4 \text{ mol } e^-)(96,485 \text{ C/mol } e^-)(1.53 \text{ J/C}) = -5.90 \times 10^5 \text{ J} = -590. \text{ kJ}$

$$E = E° - \frac{0.0591}{n} \log Q; \text{ at equilibrium, } E = 0 \text{ and } Q = K.$$

$$E° = \frac{0.0591}{n} \log K, \; \log K = \frac{4(1.53)}{0.0591} = 104, \; K \approx 10^{104}$$

d.    $1.00 \times 10^3 \text{ kg Zr} \times \dfrac{1000 \text{ g}}{\text{kg}} \times \dfrac{1 \text{ mol Zr}}{91.22 \text{ g Zr}} \times \dfrac{2 \text{ mol } H_2}{\text{mol Zr}} = 2.19 \times 10^4 \text{ mol } H_2$

$2.19 \times 10^4 \text{ mol } H_2 \times \dfrac{2.016 \text{ g } H_2}{\text{mol } H_2} = 4.42 \times 10^4 \text{ g } H_2$

$V = \dfrac{nRT}{P} = \dfrac{(2.19 \times 10^4 \text{ mol})(0.08206 \text{ L atm K}^{-1} \text{ mol}^{-1})(1273 \text{ K})}{1.0 \text{ atm}} = 2.3 \times 10^6 \text{ L } H_2$

e.    Probably yes;  less radioactivity overall was released by venting the $H_2$ than what would have been released if the $H_2$ had exploded inside the reactor (as happened at Chernobyl). Neither alternative is pleasant, but venting the radioactive hydrogen is the less unpleasant of the two alternatives.

# CHAPTER 12

# QUANTUM MECHANICS AND ATOMIC THEORY

## Light and Matter

21. $\nu = \dfrac{c}{\lambda} = \dfrac{3.00 \times 10^8 \text{ m/s}}{1.0 \times 10^{-2} \text{ m}} = 3.0 \times 10^{10} \text{ s}^{-1}$

$E = h\nu = 6.63 \times 10^{-34} \text{ J s} \times 3.0 \times 10^{10} \text{ s}^{-1} = 2.0 \times 10^{-23} \text{ J/photon}$

$\dfrac{2.0 \times 10^{-23} \text{ J}}{\text{photon}} = \dfrac{6.02 \times 10^{23} \text{ photons}}{\text{mol}} = 12 \text{ J/mol}$

23. a. $\lambda = \dfrac{c}{\nu} = \dfrac{3.00 \times 10^8 \text{ m/s}}{6.0 \times 10^{13} \text{ s}^{-1}} = 5.0 \times 10^{-6} \text{ m}$

b. From Figure 12.3, this is infrared EMR.

c. $E = h\nu = 6.63 \times 10^{-34} \text{ J s} \times 6.0 \times 10^{13} \text{ s}^{-1} = 4.0 \times 10^{-20} \text{ J/photon}$

$\dfrac{4.0 \times 10^{-20} \text{ J}}{\text{photon}} \times \dfrac{6.022 \times 10^{23} \text{ photons}}{\text{mol}} = 2.4 \times 10^4 \text{ J/mol}$

d. Frequency and photon energy are directly related ($E = h\nu$). Because $5.4 \times 10^{13} \text{ s}^{-1}$ EMR has a lower frequency than $6.0 \times 10^{13} \text{ s}^{-1}$ EMR, the $5.4 \times 10^{13} \text{ s}^{-1}$ EMR will have less energetic photons.

25. The energy needed to remove a single electron is:

$\dfrac{279.7 \text{ kJ}}{\text{mol}} \times \dfrac{1 \text{ mol}}{6.0221 \times 10^{23}} = 4.645 \times 10^{-22} \text{ kJ} = 4.645 \times 10^{-19} \text{ J}$

$E = \dfrac{hc}{\lambda}, \quad \lambda = \dfrac{hc}{E} = \dfrac{6.6261 \times 10^{-34} \text{ J s} \times 2.9979 \times 10^8 \text{ m/s}}{4.645 \times 10^{-19} \text{ J}} = 4.277 \times 10^{-7} \text{ m} = 427.7 \text{ nm}$

27.     The energy to remove a single electron is:

$$\frac{208.4 \text{ kJ}}{\text{mol}} \times \frac{1 \text{ mol}}{6.022 \times 10^{23}} = 3.461 \times 10^{-22} \text{ kJ} = 3.461 \times 10^{-19} \text{ J} = E_w$$

Energy of 254-nm light is:

$$E = \frac{hc}{\lambda} = \frac{(6.626 \times 10^{-34} \text{ J s})(2.998 \times 10^8 \text{ m/s})}{254 \times 10^{-9} \text{ m}} = 7.82 \times 10^{-19} \text{ J}$$

$$E_{photon} = E_K + E_w, \ E_K = 7.82 \times 10^{-19} \text{ J} - 3.461 \times 10^{-19} \text{ J} = 4.36 \times 10^{-19} \text{ J} = \text{maximum KE}$$

29.     The photoelectric effect refers to the phenomenon in which electrons are emitted from the surface of a metal when light strikes it. The light must have a certain minimum frequency (energy) in order to remove electrons from the surface of a metal.  Light having a frequency below the minimum results in no electrons being emitted, whereas light at or higher than the minimum frequency does cause electrons to be emitted. For light having a frequency higher than the minimum frequency, the excess energy is transferred into kinetic energy for the emitted electron. Albert Einstein explained the photoelectric effect by applying quantum theory.

31.     a.    $\lambda = \dfrac{h}{mv} = \dfrac{6.626 \times 10^{-34} \text{ J s}}{1.675 \times 10^{-27} \text{ kg} \times (0.0100 \times 2.998 \times 10^8 \text{ m/s})} = 1.32 \times 10^{-13} \text{ m}$

        b.    $\lambda = \dfrac{h}{mv}, \ v = \dfrac{h}{\lambda m} = \dfrac{6.626 \times 10^{-34} \text{ J s}}{75 \times 10^{-12} \text{ m} \times 1.675 \times 10^{-27} \text{ kg}} = 5.3 \times 10^3 \text{ m/s}$

33.     $m = \dfrac{h}{\lambda v} = \dfrac{6.626 \times 10^{-34} \text{ kg m}^2/\text{s}}{3.31 \times 10^{-15} \text{ m} \times (0.0100 \times 2.998 \times 10^8 \text{ m/s})} = 6.68 \times 10^{-26} \text{ kg/atom}$

$$\frac{6.68 \times 10^{-26} \text{ kg}}{\text{atom}} \times \frac{6.022 \times 10^{23} \text{ atoms}}{\text{mol}} \times \frac{1000 \text{ g}}{\text{kg}} = 40.2 \text{ g/mol}$$

The element is calcium (Ca).

## Hydrogen Atom:  The Bohr Model

35.     For the H atom (Z = 1): $E_n = -2.178 \times 10^{-18} \text{ J}/n^2$;  for a spectral transition, $\Delta E = E_f - E_i$:

$$\Delta E = -2.178 \times 10^{-18} \text{ J} \left( \frac{1}{n_f^2} - \frac{1}{n_i^2} \right)$$

where $n_i$ and $n_f$ are the levels of the initial and final states, respectively.  A positive value of $\Delta E$ always corresponds to an absorption of light, and a negative value of $\Delta E$ always corresponds to an emission of light.

a.   $\Delta E = -2.178 \times 10^{-18} \text{ J} \left( \dfrac{1}{2^2} - \dfrac{1}{3^2} \right) = -2.178 \times 10^{-18} \text{ J} \left( \dfrac{1}{4} - \dfrac{1}{9} \right)$

$\Delta E = -2.178 \times 10^{-18} \text{ J} \times (0.2500 - 0.1111) = -3.025 \times 10^{-19} \text{ J}$

The photon of light  must have precisely this energy $(3.025 \times 10^{-19} \text{ J})$.

$$|\Delta E| = E_{photon} = h\nu = \frac{hc}{\lambda}, \quad \lambda = \frac{hc}{|\Delta E|} = \frac{6.6261 \times 10^{-34} \text{ J s} \times 2.9979 \times 10^8 \text{ m/s}}{3.025 \times 10^{-19} \text{ J}}$$

$$= 6.567 \times 10^{-7} \text{ m} = 656.7 \text{ nm}$$

From Figure 12.3, this is visible electromagnetic radiation (red light).

b.   $\Delta E = -2.178 \times 10^{-18} \text{ J} \left( \dfrac{1}{2^2} - \dfrac{1}{4^2} \right) = -4.084 \times 10^{-19} \text{ J}$

$$\lambda = \frac{hc}{|\Delta E|} = \frac{6.6261 \times 10^{-34} \text{ J s} \times 2.9979 \times 10^8 \text{ m/s}}{4.084 \times 10^{-19} \text{ J}} = 4.864 \times 10^{-7} \text{ m} = 486.4 \text{ nm}$$

This is visible electromagnetic radiation (green-blue light).

c.   $\Delta E = -2.178 \times 10^{-18} \text{ J} \left( \dfrac{1}{1^2} - \dfrac{1}{2^2} \right) = -1.634 \times 10^{-18} \text{ J}$

$$\lambda = \frac{6.6261 \times 10^{-34} \text{ J s} \times 2.9979 \times 10^8 \text{ m/s}}{1.634 \times 10^{-18} \text{ J}} = 1.216 \times 10^{-7} \text{ m} = 121.6 \text{ nm}$$

This is ultraviolet electromagnetic radiation.

d.   $\Delta E = -2.178 \times 10^{-18} \text{ J} \left( \dfrac{1}{3^2} - \dfrac{1}{4^2} \right) = -1.059 \times 10^{-19} \text{ J}$

$$\lambda = \frac{hc}{|\Delta E|} = \frac{6.6261 \times 10^{-34} \text{ J s} \times 2.9979 \times 10^8 \text{ m/s}}{1.059 \times 10^{-19} \text{ J}} = 1.876 \times 10^{-6} \text{ m or 1876 nm}$$

This is infrared electromagnetic radiation.

37.  a.  False;  it takes less energy to ionize an electron from $n = 3$ than from the ground state.

b.  True

c.  False;  the energy difference between $n = 3$ and $n = 2$ is smaller than the energy difference to $n = 2$ electronic transition than for the $n = 3$ to $n = 1$ transition.  E and $\lambda$ are inversely proportional to each other (E = hc/$\lambda$).

d.  True

e.  False;  the ground state in hydrogen is $n = 1$ and all other allowed energy states are called excited states;  $n = 2$ is the first excited state, and $n = 3$ is the second excited state.

39.  $$|\Delta E| = E_{photon} = \frac{hc}{\lambda} = \frac{6.6261 \times 10^{-34} \text{ J s} \times 2.9979 \times 10^8 \text{ m/s}}{397.2 \times 10^{-9} \text{ m}} = 5.001 \times 10^{-19} \text{ J}$$

$\Delta E = -5.001 \times 10^{-19}$ J because we have an emission.

$$-5.001 \times 10^{-19} \text{ J} = E_2 - E_n = -2.178 \times 10^{-18} \text{ J} \left( \frac{1}{2^2} - \frac{1}{n^2} \right)$$

$$0.2296 = \frac{1}{4} - \frac{1}{n^2}, \quad \frac{1}{n^2} = 0.0204, \quad n = 7$$

41.  $$\Delta E = E_4 ! E_n = -E_n = 2.178 \times 10^{-18} \text{ J} \left( \frac{1}{n^2} \right)$$

$$E_{photon} = \frac{hc}{\lambda} = \frac{6.626 \times 10^{-34} \text{ J s} \times 2.9979 \times 10^8 \text{ m/s}}{1460 \times 10^{-9} \text{ m}} = 1.36 \times 10^{-19} \text{ J}$$

$$E_{photon} = \Delta E = 1.36 \times 10^{-19} \text{ J} = 2.178 \times 10^{-18} \left( \frac{1}{n^2} \right), \quad n^2 = 16.0, \quad n = 4$$

43.  $$E_{photon} = \frac{hc}{\lambda} = \frac{6.6261 \times 10^{-34} \text{ J s} \times 2.9979 \times 10^8 \text{ m/s}}{253.4 \times 10^{-9} \text{ m}} = 7.839 \times 10^{-19} \text{ J}$$

$\Delta E = 7.839 \times 10^{-19}$ J

The general energy equation for one-electron ions is $E_n = !2.178 \times 10^{-18}$ J $(Z^2)/n^2$, where Z = atomic number.

$$\Delta E = !2.178 \times 10^{-18} \text{ J } (Z)^2 \left( \frac{1}{n_f^2} - \frac{1}{n_i^2} \right), \quad Z = 4 \text{ for Be}^{3+}$$

$$\Delta E = !7.839 \times 10^{-19} \text{ J} = !2.178 \times 10^{-18} (4)^2 \left( \frac{1}{n_f^2} - \frac{1}{5^2} \right)$$

$$\frac{7.839 \times 10^{-19}}{2.178 \times 10^{-18} \times 16} + \frac{1}{25} = \frac{1}{n_f^2}, \quad \frac{1}{n_f^2} = 0.06249, \quad n_f = 4$$

This emission line corresponds to the $n = 5 \rightarrow n = 4$ electronic transition.

## Wave Mechanics and Particle in a Box

45.   a.   $\Delta p = m\Delta v = 9.11 \times 10^{-31}$ kg $\times 0.100$ m/s $= \dfrac{9.11 \times 10^{-32} \text{ kg m}}{\text{s}}$

$$\Delta p \Delta x \geq \frac{h}{4\pi}, \quad \Delta x = \frac{h}{4\pi \Delta p} = \frac{6.626 \times 10^{-34} \text{ J s}}{4 \times 3.142 \times (9.11 \times 10^{-32} \text{ kg m/s})} = 5.79 \times 10^{-4} \text{ m}$$

b.   $\Delta x = \dfrac{h}{4\pi \Delta p} = \dfrac{6.626 \times 10^{-34} \text{ J s}}{4 \times 3.142 \times 0.145 \text{ kg} \times 0.100 \text{ m/s}} = 3.64 \times 10^{-33} \text{ m}$

c.   The diameter of an H atom is roughly $1.0 \times 10^{-8}$ cm.  The uncertainty in position is much larger than the size of the atom.

d.   The uncertainty is insignificant compared to the size of a baseball.

47.   $E_n = \dfrac{n^2 h^2}{8mL^2}$;   $\Delta E = E_3 - E_2 = \dfrac{9h^2}{8mL^2} - \dfrac{4h^2}{8mL^2} = \dfrac{5h^2}{8mL^2}$

$$\Delta E = \frac{hc}{\lambda} = \frac{(6.626 \times 10^{-34} \text{ J s})(2.998 \times 10^8 \text{ m/s})}{8080 \times 10^{-9} \text{ m}} = 2.46 \times 10^{-20} \text{ J}$$

$$\Delta E = 2.46 \times 10^{-20} \text{ J} = \frac{5h^2}{8mL^2} = \frac{5(6.626 \times 10^{-34} \text{ J s})^2}{8(9.109 \times 10^{-31} \text{ kg}) L^2}, \quad L = 3.50 \times 10^{-9} \text{ m} = 3.50 \text{ nm}$$

49.   $E_n = \dfrac{n^2 h^2}{8mL^2}$;   as L increases, $E_n$ will decrease, and the spacing between energy levels will also decrease.

51.   $E_n = \dfrac{n^2 h^2}{8mL^2}$, $n = 1$ for ground state;  from equation, as L increases, $E_n$ decreases.

Using numbers: $10^{-6}$ m box: $E_1 = \dfrac{h^2}{8m}(1 \times 10^{12} \text{ m}^{-2})$;  $10^{-10}$ m box: $E_1 = \dfrac{h^2}{8m}(1 \times 10^{20} \text{ m}^{-2})$

As expected, the electron in the $1 \times 10^{-6}$ m box has the lowest ground state energy.

## Orbitals and Quantum Numbers

53.    The 2p orbitals differ from each other in the direction in which they point in space. The 2p and 3p orbitals differ from each other in their size, energy, and number of nodes.  A nodal surface in an atomic orbital is a surface in which the probability of finding an electron is zero.

The 1p, 1d, 2d, 1f, 2f, and 3f orbitals are not allowed solutions to the Schrödinger equation. For $n = 1$, $P \neq 1, 2, 3$, etc., so 1p, 1d, and 1f orbitals are forbidden. For $n = 2$, $P \neq 2, 3, 4$, etc., so 2d and 2f orbitals are forbidden. For $n = 3$, $P \neq 3, 4, 5$, etc., so 3f orbitals are forbidden.

The penetrating term refers to the fact that there is a higher probability of finding a 4s electron closer to the nucleus than a 3d electron. This leads to a lower energy for the 4s orbital relative to the 3d orbitals in polyelectronic atoms and ions.

55.    a.    For $n = 3$, $\ell = 3$ is not possible.

d.    $m_s$ cannot equal !1.

e.    $\ell$ cannot be a negative number.

f.    For $\ell = 1$, $m_\ell$ cannot equal 2.

The quantum numbers in parts b and c are allowed.

57.    1p, 0 electrons ($\ell \neq 1$ when $n = 1$); $6d_{x^2-y^2}$, 2 electrons (specifies one atomic orbital); 4f, 14 electrons (7 orbitals have 4f designation); $7p_y$, 2 electrons (specifies one atomic orbital); 2s,  2 electrons (specifies one atomic orbital); $n = 3$, 18 electrons (3s, 3p, and 3d orbitals are possible;  there are one 3s orbital, three 3p orbitals, and five 3d orbitals).

59.    The diagrams of the orbitals in the text give only 90% probabilities of where the electron may reside. We can never be 100% certain of the location of the electrons due to Heisenburg's uncertainty principle.

61.    For $r = a_o$ and $\theta = 0°$ ($Z = 1$ for H):

$$\psi_{2p_z} = \frac{1}{4(2\pi)^{1/2}} \left( \frac{1}{5.29 \times 10^{-11}} \right)^{3/2} (1)\, e^{-1/2} \cos 0 = 1.57 \times 10^{14}; \quad \psi^2 = 2.46 \times 10^{28}$$

For $r = a_o$ and $\theta = 90°$:  $\psi_{2p_z} = 0$ because $\cos 90° = 0$;  $\psi^2 = 0$;  the xy plane is a node for the $2p_z$ atomic orbital.

## Polyelectronic Atoms

63.    Valence electrons are the electrons in the outermost principal quantum level of an atom (those electrons in the highest $n$ value orbitals). The electrons in the lower $n$ value orbitals are all inner core or just core electrons. The key is that the outermost electrons are the valence electrons. When atoms interact with each other, it will be the outermost electrons that are involved in these interactions. In addition, how tightly the nucleus holds these outermost

electrons determines atomic size, ionization energy, and other properties of atoms. Elements in the same group have similar valence electron configurations and, as a result, have similar chemical properties.

65.    a.  $n = 4$:  $\ell$ can be 0, 1, 2, or 3.  Thus we have s (2 e⁻), p (6 e⁻), d (10 e⁻) and f (14 e⁻) orbitals present.  Total number of electrons to fill these orbitals is 32.

b.  $n = 5$, $m_\ell = +1$: for $n = 5$, $\ell = 0, 1, 2, 3, 4$;  for $\ell = 1, 2, 3, 4$, all can have $m_\ell = +1$.  Four distinct orbitals, thus 8 electrons.

c.  $n = 5$, $m_s = +1/2$: for $n = 5$,  $\ell = 0, 1, 2, 3, 4$.  Number of orbitals = 1, 3, 5, 7, 9 for each value of $\ell$, respectively.  There are 25 orbitals with $n = 5$.  They can hold 50 electrons, and 25 of these electrons can have $m_s = +1/2$.

d.  $n = 3$,  $\ell = 2$:  these quantum numbers define a set of 3d orbitals.  There are 5 degenerate 3d orbitals that can hold a total of 10 electrons.

e.  $n = 2$,  $\ell = 1$:  these define a set of 2p orbitals.  There are 3 degenerate 2p orbitals that can hold a total of 6 electrons.

f.  It is impossible for $n = 0$.  Thus no electrons can have this set of quantum numbers.

g.  The four quantum numbers completely specify a single electron.

h.   $n = 3$:  3s, 3p, and 3d orbitals all have $n = 3$.  These orbitals can hold 18 electrons, and 9 of these electrons can have $m_s = +1/2$.

i.  $n = 2$, $\ell = 2$:  this combination is not possible ($\ell \neq 2$ for $n = 2$).  Zero electrons in an atom can have these quantum numbers.

j.   $n = 1$, $\ell = 0$, $m_\ell = 0$:  these define a 1s orbital that can hold 2 electrons.

67.    The following are complete electron configurations.  Noble gas shorthand notation could also be used.

Sc: $1s^2 2s^2 2p^6 3s^2 3p^6 4s^2 3d^1$;  Fe: $1s^2 2s^2 2p^6 3s^2 3p^6 4s^2 3d^6$

P: $1s^2 2s^2 2p^6 3s^2 3p^3$;  Cs: $1s^2 2s^2 2p^6 3s^2 3p^6 4s^2 3d^{10} 4p^6 5s^2 4d^{10} 5p^6 6s^1$

Eu: $1s^2 2s^2 2p^6 3s^2 3p^6 4s^2 3d^{10} 4p^6 5s^2 4d^{10} 5p^6 6s^2 4f^6 5d^1$*

Pt: $1s^2 2s^2 2p^6 3s^2 3p^6 4s^2 3d^{10} 4p^6 5s^2 4d^{10} 5p^6 6s^2 4f^{14} 5d^8$*

Xe: $1s^2 2s^2 2p^6 3s^2 3p^6 4s^2 3d^{10} 4p^6 5s^2 4d^{10} 5p^6$;  Br: $1s^2 2s^2 2p^6 3s^2 3p^6 4s^2 3d^{10} 4p^5$

*Note:  These electron configurations were written down using only the periodic table.

Actual electron configurations are:  Eu: $[Xe]6s^2 4f^7$ and Pt: $[Xe]6s^1 4f^{14} 5d^9$

69.    Exceptions:  Cr, Cu, Nb, Mo, Tc, Ru, Rh, Pd, Ag, Pt, and Au;  Tc, Ru, Rh, Pd, and Pt do not correspond to the supposed extra stability of half-filled and filled subshells.

71.    The two exceptions are Cr and Cu.

Cr:  $1s^2 2s^2 2p^6 3s^2 3p^6 4s^1 3p^5$;  Cr has 6 unpaired electrons.

Cu:  $1s^2 2s^2 2p^6 3s^2 3p^6 4s^1 3d^{10}$;  Cu has 1 unpaired electron.

73.    Element 115, Uup, is in Group 5A under Bi (bismuth):

Uup:  $1s^2 2s^2 2p^6 3s^2 3p^6 4s^2 3d^{10} 4p^6 5s^2 4d^{10} 5p^6 6s^2 4f^{14} 5d^{10} 6p^6 7s^2 5f^{14} 6d^{10} 7p^3$

a.    $5s^2$, $5p^6$, $5d^{10}$, and $5f^{14}$; 32 electrons have $n = 5$ as one of their quantum numbers

b.    $\ell = 3$ are f orbitals.  $4f^{14}$ and $5f^{14}$ are the f orbitals used. They are all filled so 28 electrons have $\ell = 3$.

c.    p, d, and f orbitals all have one of the degenerate orbitals with $m_\ell = 1$. There are 6 orbitals with $m_\ell = 1$ for the various p orbitals used; there are 4 orbitals with $m_\ell = 1$ for the various d orbitals used; and there are 2 orbitals with $m_\ell = 1$ for the various f orbitals used. We have a total of $6 + 4 + 2 = 12$ orbitals with $m_\ell = 1$. Eleven of these orbitals are filled with 2 electrons, and the 7p orbitals are only half-filled. The number of electrons with $m_\ell = 1$ is $11 \times (2\ e^-) + 1 \times (1\ e^-) = 23$ electrons.

d.    The first 112 electrons are all paired; one-half of these electrons (56 e⁻) will have $m_s =$ !1/2. The 3 electrons in the 7p orbitals singly occupy each of the three degenerate 7p orbitals; the three electrons are spin parallel, so the 7p electrons either have $m_s = +1/2$ or $m_s =$ !1/2. Therefore, either 56 electrons have $m_s =$ !1/2 or 59 electrons have $m_s =$ !1/2.

75.     We get the number of unpaired electrons by examining the incompletely filled subshells.  The paramagnetic substances have unpaired electrons, and the ones with no unpaired electrons are not paramagnetic (they are called diamagnetic).

Li:  $1s^2 2s^1$  ↑ ;  paramagnetic with 1 unpaired electron.
            2s

N:  $1s^2 2s^2 2p^3$  ↑ ↑ ↑ ;  paramagnetic with 3 unpaired electrons.
                2p

Ni:  $[Ar]4s^2 3d^8$  ↑↓ ↑↓ ↑↓ ↑ ↑ ;  paramagnetic with 2 unpaired electrons.
                3d

Te:  $[Kr]5s^2 4d^{10} 5p^4$  ↑↓ ↑ ↑ ;  paramagnetic with 2 unpaired electrons.
                        5p

Ba:  $[Xe]6s^2$  ↑↓ ;  not paramagnetic because no unpaired electrons are present.
            6s

Hg:  $[Xe]6s^2 4f^{14} 5d^{10}$  ↑↓ ↑↓ ↑↓ ↑↓ ↑↓ ;  not paramagnetic because no unpaired electrons.
                    5d

77.     The s block elements with $ns^1$ for a valence electron configuration have one unpaired electrons. These are elements H, Li, Na, and K for the first 36 elements. The p block elements with $ns^2 np^1$ or $ns^2 np^5$ valence electron configurations have one unpaired electron. These are elements B, Al, and Ga ($ns^2 np^1$) and elements F, Cl, and Br ($ns^2 np^5$) for the first 36 elements. In the d block, Sc ($[Ar]4s^2 3d^1$) and Cu ($[Ar]4s^1 3d^{10}$) each have one unpaired electron.  A total of 12 elements from the first 36 elements have one unpaired electron in the ground state.

79.     We get the number of unpaired electrons by examining the incompletely filled subshells.

O:  $[He]2s^2 2p^4$          $2p^4$:  ↑↓ ↑ ↑          Two unpaired e⁻

O⁺:  $[He]2s^2 2p^3$          $2p^3$:  ↑ ↑ ↑          Three unpaired e⁻

O⁻:  $[He]2s^2 2p^5$          $2p^5$:  ↑↓ ↑↓ ↑          One unpaired e⁻

Os:  $[Xe]6s^2 4f^{14} 5d^6$          $5d^6$:  ↑↓ ↑ ↑ ↑ ↑          Four unpaired e⁻

Zr:  $[Kr]5s^2 4d^2$          $4d^2$:  ↑ ↑ _ _ _          Two unpaired e⁻

S:  $[Ne]3s^2 3p^4$          $3p^4$:  ↑↓ ↑ ↑          Two unpaired e⁻

F:  $[He]2s^2 2p^5$          $2p^5$:  ↑↓ ↑↓ ↑          One unpaired e⁻

Ar:  $[Ne]3s^2 3p^6$          $3p^6$:  ↑↓ ↑↓ ↑↓          Zero unpaired e⁻

## The Periodic Table and Periodic Properties

81.    Ionization energy: $P(g) \rightarrow P^+(g) + e^-$; electron affinity: $P(g) + e^- \rightarrow P^-(g)$

83.    As successive electrons are removed, the net positive charge on the resulting ion increases. This increase in positive charge binds the remaining electrons more firmly, and the ionization energy increases.

The electron configuration for Si is $1s^2 2s^2 2p^6 3s^2 3p^2$. There is a large jump in ionization energy when going from the removal of valence electrons to the removal of core electrons. For silicon, this occurs when the fifth electron is removed since we go from the valence electrons in $n = 3$ to the core electrons in $n = 2$. There should be another big jump when the thirteenth electron is removed, i.e., when a 1s electron is removed.

85.    Size (radius) decreases left to right across the periodic table, and size increases from top to bottom of the periodic table.

a.  S < Se < Te          b.  Br < Ni < K          c.  F < Si < Ba

d.  Be < Na < Rb        e.  Ne < Se < Sr        f.  O < P < Fe

All follow the general radius trend.

87.    a.  Ba                              b.  K

c.  O;  in general, Group 6A elements have a lower ionization energy than neighboring Group 5A elements.  This is an exception to the general ionization energy trend across the periodic table.

d.  $S^{2-}$; this ion has the most electrons compared to the other sulfur species present.  $S^{2-}$ has the largest number of electron-electron repulsions, which leads to $S^{2-}$ having the largest size and smallest ionization energy.

e.  Cs;  this follows the general ionization energy trend.

89.    As: $[Ar]4s^2 3d^{10} 4p^3$; Se: $[Ar]4s^2 3d^{10} 4p^4$; the general ionization energy trend predicts that Se should have a higher ionization energy than As.  Se is an exception to the general ionization energy trend.  There are extra electron-electron repulsions in Se because two electrons are in the same 4p orbital, resulting in a lower ionization energy for Se than predicted.

91.    Size also decreases going across a period.  Sc and Ti along with Y and Zr are adjacent elements.  There are 14 elements (the lanthanides) between La and Hf, making Hf considerably smaller.

93.    a.  Uus will have 117 electrons. $[Rn]7s^2 5f^{14} 6d^{10} 7p^5$

b.  It will be in the halogen family and will be most similar to astatine (At).

c.  Like the other halogens: $NaUus$, $Mg(Uus)_2$, $C(Uus)_4$, $O(Uus)_2$

d.   Like the other halogens: $UusO^-$, $UusO_2^-$, $UusO_3^-$, $UusO_4^-$

95.    Electron-electron repulsions become important when we try to add electrons to an atom. From the standpoint of electron-electron repulsions, larger atoms would have more favorable (more exothermic) electron affinities.  Considering only electron-nucleus attractions, smaller atoms would be expected to have the more favorable (more exothermic) EA values.  These two factors are the opposite of each other.  Thus the overall variation in EA is not as great as ionization energy, in which attractions to the nucleus dominate.

97.    Electron-electron repulsions are much greater in $O^-$ than in $S^-$ because the electron goes into a smaller 2p orbital versus the larger 3p orbital in sulfur.  This results in a more favorable (more exothermic) EA for sulfur.

99.    a.   The electron affinity of $Mg^{2+}$ is $\Delta H$ for $Mg^{2+}(g) + e^- \rightarrow Mg^+(g)$; this is just the reverse of the second ionization energy for Mg.   $EA(Mg^{2+}) = -IE_2(Mg) = -1445$ kJ/mol (Table 12.6)

       b.   EA of $Al^+$ is $\Delta H$ for $Al^+(g) + e^- \rightarrow Al(g)$;  $EA(Al^+) = -IE_1(Al) = $
                                                                              $-580$ kJ/mol (Table 12.6)

       c.   IE of $Cl^-$is $\Delta H$ for $Cl^-(g) \rightarrow Cl(g) + e^-$;  $IE(Cl^-) = -EA(Cl) = +348.7$ kJ/mol
                                                                              (Table 12.8)

       d.   $Cl(g) \rightarrow Cl^+(g) + e^-$    $\Delta H = IE_1(Cl) = 1255$ kJ/mol (Table 12.6)

       e.   $Cl^+(g) + e^- \rightarrow Cl(g)$    $\Delta H = -IE_1(Cl) = -1255$ kJ/mol $= EA(Cl^+)$

## The Alkali Metals

101.   Yes; the ionization energy general trend is to decrease down a group, and the atomic radius trend is to increase down a group.  The data in Table 12.9 confirm both of these general trends.

103.   a.   $6 Li(s) + N_2(g) \rightarrow 2 Li_3N(s)$          b.   $2 Rb(s) + S(s) \rightarrow Rb_2S(s)$

       c.   $2 Cs(s) + 2 H_2O(l) \rightarrow 2 CsOH(aq) + H_2(g)$          d.   $2 Na(s) + Cl_2(g) \rightarrow 2 NaCl(s)$

105.   For 589.0 nm: $\nu = \dfrac{c}{\lambda} = \dfrac{2.9979 \times 10^8 \text{ m/s}}{589.0 \times 10^{-9} \text{ m}} = 5.090 \times 10^{14} \text{s}^{-1}$

       $E = h\nu = 6.6261 \times 10^{-34} \text{ J s} \times 5.090 \times 10^{14} \text{s}^{-1} = 3.373 \times 10^{-19} \text{ J}$

       For 589.6 nm: $\nu = c/\lambda = 5.085 \times 10^{14} \text{s}^{-1}$;   $E = h\nu = 3.369 \times 10^{-19} \text{ J}$

The energies in kJ/mol are:

$$3.373 \times 10^{-19} \text{ J} \times \frac{1 \text{ kJ}}{1000 \text{ J}} \times \frac{6.0221 \times 10^{23}}{\text{mol}} = 203.1 \text{ kJ/mol}$$

$$3.369 \times 10^{-19} \text{ J} \times \frac{1 \text{ kJ}}{1000 \text{ J}} \times \frac{6.0221 \times 10^{23}}{\text{mol}} = 202.9 \text{ kJ/mol}$$

107.    It should be element 119 with ground state electron configuration: $[\text{Rn}]7s^2 5f^{14} 6d^{10} 7p^6 8s^1$

## Additional Exercises

109.    a.   $n$              b.   $n$ and $\ell$

111.    Size decreases from left to right and increases going down the periodic table. Thus going one element right and one element down would result in a similar size for the two elements diagonal to each other. The ionization energies will be similar for the diagonal elements since the periodic trends also oppose each other. Electron affinities are harder to predict, but atoms with similar sizes and ionization energies should also have similar electron affinities.

113.    $60 \times 10^6 \text{ km} \times \dfrac{1000 \text{ m}}{\text{km}} \times \dfrac{1 \text{ s}}{3.00 \times 10^8 \text{ m}} = 200 \text{ s}$  (about 3 minutes)

115.    $\lambda = \dfrac{hc}{E} = \dfrac{6.626 \times 10^{-34} \text{ J s} \times 2.998 \times 10^8 \text{ m/s}}{3.59 \times 10^{-19} \text{ J}} = 5.53 \times 10^{-7} \text{ m} \times \dfrac{100 \text{ cm}}{\text{m}}$

$$= 5.53 \times 10^{-5} \text{ cm}$$

From the spectrum, $\lambda = 5.53 \times 10^{-5}$ cm is greenish yellow light.

117.    When the p and d orbital functions are evaluated at various points in space, the results sometimes have positive values and sometimes have negative values. The term phase is often associated with the + and ! signs. For example, a sine wave has alternating positive and negative phases. This is analogous to the positive and negative values (phases) in the p and d orbitals.

119.    a.        $\text{Na(g)} \rightarrow \text{Na}^+\text{(g)} + \text{e}^-$        $\text{IE}_1 = 495$ kJ

$\phantom{a.}$         $\text{Cl(g)} + \text{e}^- \rightarrow \text{Cl}^-\text{(g)}$        $\text{EA} = !348.7$ kJ

$\phantom{a.}$         $\text{Na(g)} + \text{Cl(g)} \rightarrow \text{Na}^+\text{(g)} + \text{Cl}^-\text{(g)}$        $\Delta H = 146$ kJ

$\phantom{a.}$    b.        $\text{Mg(g)} \rightarrow \text{Mg}^+\text{(g)} + \text{e}^-$        $\text{IE}_1 = 735$ kJ

$\phantom{a.}$         $\text{F(g)} + \text{e}^- \rightarrow \text{F}^-\text{(g)}$        $\text{EA} = !327.8$ kJ

$\phantom{a.}$         $\text{Mg(g)} + \text{F(g)} \rightarrow \text{Mg}^+\text{(g)} + \text{F}^-\text{(g)}$        $\Delta H = 407$ kJ

c.
$$Mg^+(g) \rightarrow Mg^{2+}(g) + e^- \qquad IE_2 = 1445 \text{ kJ}$$
$$F(g) + e^- \rightarrow F^-(g) \qquad EA = !327.8 \text{ kJ}$$

$$Mg^+(g) + F(g) \rightarrow Mg^{2+}(g) + F^-(g) \qquad \Delta H = 1117 \text{ kJ}$$

d.  From parts b and c, we get:

$$Mg(g) + F(g) \rightarrow Mg^+(g) + F^-(g) \qquad \Delta H = 407 \text{ kJ}$$
$$Mg^+(g) + F(g) \rightarrow Mg^{2+}(g) + F^-(g) \qquad \Delta H = 1117 \text{ kJ}$$

$$Mg(g) + 2 F(g) \rightarrow Mg^{2+}(g) + 2 F^-(g) \qquad \Delta H = 1524 \text{ kJ}$$

121.    Valence electrons are easier to remove than inner core electrons. The large difference in energy between $I_2$ and $I_3$ indicates that this element has two valence electrons. This element is most likely an alkaline earth metal since alkaline earth metal elements all have two valence electrons.

123.    a.  The 4+ ion contains 20 electrons. Thus the electrically neutral atom will contain 24 electrons. The atomic number is 24 which identifies it as chromium.

b.  The ground state electron configuration of the ion must be: $1s^2 2s^2 2p^6 3s^2 3p^6 4s^0 3d^2$; there are 6 electrons in s orbitals.

c.  12                                d.  2

e.  This is the isotope $^{50}_{24}Cr$. There are 26 neutrons in the nucleus.

f.  $3.01 \times 10^{23}$ atoms $\times \dfrac{1 \text{ mol}}{6.022 \times 10^{23} \text{ atoms}} \times \dfrac{49.9 \text{ g}}{\text{mol}} = 24.9 \text{ g}$

g.  $1s^2 2s^2 2p^6 3s^2 3p^6 4s^1 3d^5$ is the ground state electron configuration for Cr. Cr is an exception to the normal filling order.

125.    a.  Each orbital could hold 4 electrons.

b.  The first period corresponds to $n = 1$, which can only have 1s orbitals. The 1s orbital could hold 4 electrons; hence the first period would have four elements. The second period corresponds to $n = 2$, which has 2s and 2p orbitals. These four orbitals can each hold four electrons. A total of 16 elements would be in the second period.

c.  20                    d.  28

127.    At $x = 0$, the value of the square of the wave function must be zero. The particle must be inside the box. For $\psi = A \cos(Lx)$, at $x = 0$, $\cos(0) = 1$ and $\psi^2 = A^2$. This violates the boundary condition.

129.   a.   Because wavelength is inversely proportional to energy, the spectral line to the right of B (at a larger wavelength) represents the lowest possible energy transition; this is $n = 4$ to $n = 3$.  The B line represents the next lowest energy transition, which is $n = 5$ to $n = 3$, and the A line corresponds to the $n = 6$ to $n = 3$ electronic transition.

b.   Because this spectrum is for a one-electron ion, $E_n = -2.178 \times 10^{-18}$ J $(Z^2/n^2)$.  To determine $\Delta E$ and, in turn, the wavelength of spectral line A, we must determine Z, the atomic number of the one electron species.  Use spectral line B data to determine Z.

$$\Delta E_{5 \rightarrow 3} = -2.178 \times 10^{-18} \text{ J} \left( \frac{Z^2}{3^2} - \frac{Z^2}{5^2} \right) = -2.178 \times 10^{-18} \left( \frac{16Z^2}{9 \times 25} \right)$$

$$E = \frac{hc}{\lambda} = \frac{6.6261 \times 10^{-34} \text{ J s}(2.9979 \times 10^8 \text{ m/s})}{142.5 \times 10^{-9} \text{ m}} = 1.394 \times 10^{-18} \text{ J}$$

Because an emission occurs, $\Delta E_{5 \rightarrow 3} = -1.394 \times 10^{-18}$ J.

$$\Delta E = -1.394 \times 10^{-18} \text{ J} = -2.178 \times 10^{-18} \text{ J} \left( \frac{16\,Z^2}{9 \times 25} \right), \ Z^2 = 9.001, \ Z = 3; \text{ the ion is Li}^{2+}.$$

Solving for the wavelength of line A:

$$\Delta E_{6 \rightarrow 3} = -2.178 \times 10^{-18}(3)^2 \left( \frac{1}{3^2} - \frac{1}{6^2} \right) = -1.634 \times 10^{-18} \text{ J}$$

$$\lambda = \frac{hc}{|\Delta E|} = \frac{6.6261 \times 10^{-34} \text{ J s}(2.9979 \times 10^8 \text{ m/s})}{1.634 \times 10^{-18} \text{ J}} = 1.216 \times 10^{-7} \text{ m} = 121.6 \text{ nm}$$

## Challenge Problems

131.   For one-electron species, $E_n = -R_H Z^2/n^2$.  IE is for the $n = 1 \rightarrow n = \infty$ transition.  So:

$$\text{IE} = E_\infty - E_1 = -E_1 = R_H Z^2/n^2 = R_H Z^2$$

$$\frac{4.72 \times 10^4 \text{ kJ}}{\text{mol}} \times \frac{1 \text{ mol}}{6.022 \times 10^{23}} \times \frac{1000 \text{ J}}{\text{kJ}} = 2.178 \times 10^{-18} \text{ J } (Z^2); \text{ solving: } Z = 6$$

Element 6 is carbon (X = carbon), and the charge for a one-electron carbon ion is 5+ ($m = 5$).  The one-electron ion is $C^{5+}$.

133.   a.   Because the energy levels $E_{xy}$ are inversely proportional to $L^2$, the $n_x = 2$, $n_y = 1$ energy level will be lower in energy than the $n_x = 1$, $n_y = 2$ energy level since $L_x > L_y$.  The first three energy levels $E_{xy}$ in order of increasing energy are:

$$E_{11} < E_{21} < E_{12}$$

The quantum numbers are:

Ground state ($E_{11}$)          $\rightarrow$  $n_x = 1, n_y = 1$
First excited state ($E_{21}$)     $\rightarrow$  $n_x = 2, n_y = 1$
Second excited state ($E_{12}$)   $\rightarrow$  $n_x = 1, n_y = 2$

b.  $E_{21} \rightarrow E_{12}$ is the transition.  $E_{xy} = \dfrac{h^2}{8m}\left(\dfrac{n_x^2}{L_x^2} + \dfrac{n_y^2}{L_y^2}\right)$

$$E_{12} = \frac{h^2}{8m}\left[\frac{1^2}{(8.00 \times 10^{-9}\ m)^2} + \frac{2^2}{(5.00 \times 10^{-9}\ m)^2}\right] = \frac{1.76 \times 10^{17}\ h^2}{8m}$$

$$E_{21} = \frac{h^2}{8m}\left[\frac{2^2}{(8.00 \times 10^{-9}\ m)^2} + \frac{1^2}{(5.00 \times 10^{-9}\ m)^2}\right] = \frac{1.03 \times 10^{17}\ h^2}{8m}$$

$$\Delta E = E_{12} - E_{21} = \frac{1.76 \times 10^{17}\ h^2}{8m} - \frac{1.03 \times 10^{17}\ h^2}{8m} = \frac{7.3 \times 10^{16}\ h^2}{8m}$$

$$\Delta E = \frac{(7.3 \times 10^{16}\ m^{-2})(6.626 \times 10^{-34}\ J\ s)^2}{8(9.11 \times 10^{-31}\ kg)} = 4.4 \times 10^{-21}\ J$$

$$\lambda = \frac{hc}{\Delta E} = \frac{6.626 \times 10^{-34}\ J\ s(2.998 \times 10^8\ m/s)}{4.4 \times 10^{-21}\ J} = 4.5 \times 10^{-5}\ m$$

135.   $\psi_{1s} = \dfrac{1}{\sqrt{\pi}}\left(\dfrac{Z}{a_0}\right)^{3/2} e^{-\sigma}$;  $Z = 1$ for H,  $\sigma = \dfrac{Zr}{a_0} = \dfrac{-r}{a_0}$,  $a_0 = 5.29 \times 10^{-11}\ m$

$$\psi_{1s} = \frac{1}{\sqrt{\pi}}\left(\frac{1}{a_0}\right)^{3/2} \exp\left(\frac{-r}{a_0}\right)$$

Probability is proportional to $\psi^2$:  $\psi_{1s}^2 = \dfrac{1}{\pi}\left(\dfrac{1}{a_0}\right)^3 \exp\left(\dfrac{-2r}{a_0}\right)$   (units of $\psi^2 = m^{-3}$)

a.  $\psi_{1s}^2$ (at nucleus) $= \dfrac{1}{\pi}\left(\dfrac{1}{a_0}\right)^3 \exp\left[\dfrac{-2\,(0)}{a_0}\right] = 2.15 \times 10^{30}\ m^{-3}$

If we assume this probability is constant throughout the $1 \times 10^{-3}\ pm^3$ volume, then the total probability p is $\psi_{1s}^2 \times V$.

$1.0 \times 10^{-3}\ pm^3 = (1.0 \times 10^{-3}\ pm) \times (1 \times 10^{-12}\ m/pm)^3 = 1.0 \times 10^{-39}\ m^3$

Total probability $= p = (2.15 \times 10^{30}\ m^{-3}) \times (1.0 \times 10^{-39}\ m^3) = 2.2 \times 10^{-9}$

b.  For an electron that is $1.0 \times 10^{-11}$ m from the nucleus:

$$\psi_{1s}^2 = \frac{1}{\pi} \left( \frac{1}{5.29 \times 10^{-11}} \right)^3 \exp\left[ \frac{-2(1.0 \times 10^{-11})}{(5.29 \times 10^{-11})} \right] = 1.5 \times 10^{30} \text{ m}^{-3}$$

$V = 1.0 \times 10^{-39} \text{ m}^3; \quad p = \psi_{1s}^2 \times V = 1.5 \times 10^{-9}$

c.  $\psi_{1s}^2 = 2.15 \times 10^{30} \text{ m}^{-3} \exp\left[ \frac{-2(53 \times 10^{-12})}{(5.29 \times 10^{-11})} \right] = 2.9 \times 10^{29}; \quad V = 1.0 \times 10^{-39} \text{ m}^3$

$p = \psi_{1s}^2 \times V = 2.9 \times 10^{-10}$

d.  $V = \frac{4}{3} \pi [(10.05 \times 10^{-12} \text{ m})^3 - (9.95 \times 10^{-12} \text{ m})^3] = 1.3 \times 10^{-34} \text{ m}^3$

We shall evaluate $\psi_{1s}^2$ at the middle of the shell, $r = 10.00$ pm, and assume $\psi_{1s}^2$ is constant from $r = 9.95$ to 10.05 pm.  The concentric spheres are assumed centered about the nucleus.

$$\psi_{1s}^2 = 2.15 \times 10^{30} \text{ m}^{-3} \exp\left[ \frac{-2(10.0 \times 10^{-12} \text{ m})}{(5.29 \times 10^{-11} \text{ m})} \right] = 1.47 \times 10^{30} \text{ m}^{-3}$$

$p = (1.47 \times 10^{30} \text{ m}^{-3}) (1.3 \times 10^{-34} \text{ m}^3) = 1.9 \times 10^{-4}$

e.  $V = \frac{4}{3} \pi [(52.95 \times 10^{-12} \text{ m})^3 - (52.85 \times 10^{-12} \text{ m})^3] = 4 \times 10^{-33} \text{ m}^3$

Evaluate $\psi_{1s}^2$ at $r = 52.90$ pm: $\psi_{1s}^2 = 2.15 \times 10^{30} \text{ m}^{-3} (e^{-2}) = 2.91 \times 10^{29} \text{ m}^{-3}; \quad p = 1 \times 10^{-3}$

137.  $E_{xyz} = \dfrac{h^2(n_x^2 + n_y^2 + n_z^2)}{8mL^2}$, where $L = L_x = L_y = L_z$.

The first four energy levels will be filled with the 8 electrons.  The first four energy levels are:

$$E_{111} = \frac{h^2 + 1^2 + 1^2}{8mL^2} = \frac{3h^2}{8mL^2}$$

$$E_{211} = E_{121} = E_{112} = \frac{6h^2}{8mL^2} \quad \text{(These three energy levels are degenerate.)}$$

The next energy levels correspond to the first excited state.  The energy for these levels are:

$$E_{221} = E_{212} = E_{122} = \frac{9h^2}{8mL^2} \quad \text{(These three energy levels are degenerate.)}$$

The electronic transition in question is from one of the degenerate $E_{211}$, $E_{121}$, or $E_{112}$ levels to one of the degenerate $E_{221}$, $E_{212}$, or $E_{122}$ levels.

$$\Delta E = \frac{9h^2}{8mL^2} - \frac{6h^2}{8mL^2} = \frac{3h^2}{8mL^2}$$

$$\Delta E = \frac{3(6.626 \times 10^{-34}\ \text{J s})^2}{8(9.109 \times 10^{-31}\ \text{kg})(1.50 \times 10^{-9}\ \text{m})^2} = 8.03 \times 10^{-20}\ \text{J}$$

$$\lambda = \frac{hc}{\Delta E} = \frac{(6.626 \times 10^{-34}\ \text{J s})(2.998 \times 10^8\ \text{m/s})}{8.03 \times 10^{-20}\ \text{J}} = 2.47 \times 10^{-6}\ \text{m} = 2470\ \text{nm}$$

139.   a.   Assuming the Bohr model applies to the 1s electron, $E_{1s} = -R_H Z^2/n^2 = -R_H Z^2_{eff}$, where $n = 1$.

   $$IE = E_\infty - E_{1s} = 0 - E_{1s} = R_H Z^2_{eff}$$

   $$\frac{2.462 \times 10^6\ \text{kJ}}{\text{mol}} \times \frac{1\ \text{mol}}{6.0221 \times 10^{23}} \times \frac{1000\ \text{J}}{\text{kJ}} = 2.178 \times 10^{-18}\ \text{J}\ (Z_{eff})^2$$

   Solving:  $Z_{eff} = 43.33$

   b.   Silver is element 47, so $Z = 47$ for silver.  Our calculated $Z_{eff}$ value is slightly less than 47.  Electrons in other orbitals can penetrate the 1s orbital.  Thus a 1s electron can be slightly shielded from the nucleus, giving a $Z_{eff}$ close to but less than $Z$.

# CHAPTER 13

# BONDING: GENERAL CONCEPTS

## Chemical Bonds and Electronegativity

11. Electronegativity is the ability of an atom in a molecule to attract electrons to itself. Electronegativity is a bonding term. Electron affinity is the energy change when an electron is added to a substance. Electron affinity deals with isolated atoms in the gas phase.

A covalent bond is a sharing of electron pair(s) in a bond between two atoms. An ionic bond is a complete transfer of electrons from one atom to another to form ions. The electrostatic attraction of the oppositely charged ions is the ionic bond.

A pure covalent bond is an equal sharing of shared electron pair(s) in a bond. A polar covalent bond is an unequal sharing.

Ionic bonds form when there is a large difference in electronegativity between the two atoms bonding together. This usually occurs when a metal with a small electronegativity is bonded to a nonmetal having a large electronegativity. A pure covalent bond forms between atoms having identical or nearly identical eletronegativities. A polar covalent bond forms when there is an intermediate electronegativity difference. In general, nonmetals bond together by forming covalent bonds, either pure covalent or polar covalent.

Ionic bonds form due to the strong electrostatic attraction between two oppositely charged ions. Covalent bonds form because the shared electrons in the bond are attracted to two different nuclei, unlike the isolated atoms where electrons are only attracted to one nuclei. The attraction to another nuclei overrides the added electron-electron repulsions.

13. Using the periodic table, we expect the general trend for electronegativity to be:

    1. Increase as we go from left to right across a period

    2. Decrease as we go down a group

  a. $C < N < O$      b. $Se < S < Cl$      c. $Sn < Ge < Si$

  d. $Tl < Ge < S$      e. $Rb < K < Na$      f. $Ga < B < O$

15. The general trends in electronegativity used in Exercises 13.13 and 13.14 are only rules of thumb. In this exercise we use experimental values of electronegativities and can begin to see several exceptions. The order of EN using Figure 13.3 is:

a.   C (2.6) < N (3.0) < O (3.4)    same as predicted

b.   Se (2.6) = S (2.6) < Cl (3.2)   different

c.   Si (1.9) < Ge (2.0) = Sn (2.0) different      d.   Tl (2.0) = Ge (2.0) < S (2.6)  different

e.   Rb (0.8) = K (0.8) < Na (0.9) different       f.   Ga (1.8) < B (2.0) < O (3.4)   same

Most polar bonds using actual EN values:

a.   Si–F (Ge–F predicted)              b.   P–Cl (same as predicted)

c.   S–F (same as predicted)            d.   Ti–Cl (same as predicted)

e.   C–H (Sn–H predicted)               f.   Al–Br (Tl–Br predicted)

17.    Ionic character is proportional to the difference in electronegativity values between the two elements forming the bond. Using the trend in electronegativity, the order will be:

Br–Br  <  N–O  <  C–F  <  Ca–O  <  K–F
least                              most
ionic character                   ionic character

Note that Br–Br, N–O and C–F bonds are all covalent bonds since the elements are all non-metals. The Ca–O and K–F bonds are ionic, as is generally the case when a metal forms a bond with a nonmetal.

## Ionic Compounds

19.    Anions are larger than the neutral atom, and cations are smaller than the neutral atom. For anions, the added electrons increase the electron-electron repulsions. To counteract this, the size of the electron cloud increases, placing the electrons further apart from one another. For cations, as electrons are removed, there are fewer electron-electron repulsions, and the electron cloud can be pulled closer to the nucleus.

Isoelectronic: same number of electrons. Two variables, the number of protons and the number of electrons, determine the size of an ion. Keeping the number of electrons constant, we only have to consider the number of protons to predict trends in size. The ion with the most protons attracts the same number of electrons most strongly, resulting in a smaller size.

21.    a.   $Cu > Cu^+ > Cu^{2+}$          b.   $Pt^{2+} > Pd^{2+} > Ni^{2+}$          c.   $O^{2-} > O^- > O$

d.   $La^{3+} > Eu^{3+} > Gd^{3+} > Yb^{3+}$       e.   $Te^{2-} > I^- > Cs^+ > Ba^{2+} > La^{3+}$

For answer a, as electrons are removed from an atom, size decreases.  Answers b and d follow the radius trend.  For answer c, as electrons are added to an atom, size increases. Answer e follows the trend for an isoelectronic series, i.e., the smallest ion has the most protons.

23.    a.  $Cs_2S$ is composed of $Cs^+$ and $S^{2-}$. $Cs^+$ has the same electron configuration as Xe, and $S^{2-}$ has the same configuration as Ar.

       b.  $SrF_2$; $Sr^{2+}$ has the Kr electron configuration, and $F^-$ has the Ne configuration.

       c.  $Ca_3N_2$; $Ca^{2+}$ has the Ar electron configuration, and $N^{3-}$ has the Ne configuration.

       d.  $AlBr_3$; $Al^{3+}$ has the Ne electron configuration, and $Br^-$ has the Kr configuration.

25.    $Se^{2-}$, $Br^-$, $Rb^+$, $Sr^{2+}$, $Y^{3+}$, and $Zr^{4+}$ are some ions that are isoelectronic with Kr (36 electrons). In terms of size, the ion with the most protons will hold the electrons tightest and will be the smallest. The size trend is:

$$Zr^{4+} < Y^{3+} < Sr^{2+} < Rb^+ < Br^- < Se^{2-}$$

smallest                                    largest

27.    a.  $Al^{3+}$ and $S^{2-}$ are the expected ions. The formula of the compound would be $Al_2S_3$ (aluminum sulfide).

       b.  $K^+$ and $N^{3-}$; $K_3N$, potassium nitride

       c.  $Mg^{2+}$ and $Cl^-$; $MgCl_2$, magnesium chloride

       d.  $Cs^+$ and $Br^-$; CsBr, cesium bromide

29.    $$K(s) \rightarrow K(g) \qquad\qquad \Delta H = 64 \text{ kJ (sublimation)}$$
       $$K(g) \rightarrow K^+(g) + e^- \qquad\qquad \Delta H = 419 \text{ kJ (ionization energy)}$$
       $$1/2 \ Cl_2(g) \rightarrow Cl(g) \qquad\qquad \Delta H = 239/2 \text{ kJ (bond energy)}$$
       $$Cl(g) + e^- \rightarrow Cl^-(g) \qquad\qquad \Delta H = -349 \text{ kJ (electron affinity)}$$
       $$K^+(g) + Cl^-(g) \rightarrow KCl(s) \qquad\qquad \Delta H = -690. \text{ kJ (lattice energy)}$$

       $$K(s) + 1/2 \ Cl_2(g) \rightarrow KCl(s) \qquad\qquad \Delta H^{\circ}_f = -437 \text{ kJ/mol}$$

31.    Use Figure 13.11 as a template for this problem.

       $$Li(s) \rightarrow Li(g) \qquad\qquad \Delta H_{sub} = ?$$
       $$Li(g) \rightarrow Li^+(g) + e^- \qquad\qquad \Delta H = 520. \text{ kJ}$$
       $$1/2 \ I_2(g) \rightarrow I(g) \qquad\qquad \Delta H = 151/2 \text{ kJ}$$
       $$I(g) + e^- \rightarrow I^-(g) \qquad\qquad \Delta H = -295 \text{ kJ}$$
       $$Li^+(g) + I^-(g) \rightarrow LiI(s) \qquad\qquad \Delta H = -753 \text{ kJ}$$

       $$Li(s) + 1/2 \ I_2(g) \rightarrow LiI(s) \qquad\qquad \Delta H = -272 \text{ kJ}$$

       $\Delta H_{sub} + 520. + 151/2 - 295 - 753 = -272$,  $\Delta H_{sub} = 181 \text{ kJ}$

33.    a.  From the data given, less energy is required to produce $Mg^+(g) + O^-(g)$ than to produce $Mg^{2+}(g) + O^{2-}(g)$. However, the lattice energy for $Mg^{2+}O^{2-}$ will be much more exothermic than for $Mg^+O^-$ (due to the greater charges in $Mg^{2+}O^{2-}$). The favorable lattice energy term will dominate and $Mg^{2+}O^{2-}$ forms.

b.   $Mg^+$ and $O^-$ both have unpaired electrons.  In $Mg^{2+}$ and $O^{2-}$ there are no unpaired electrons.  Hence $Mg^+O^-$ would be paramagnetic; $Mg^{2+}O^{2-}$ would be diamagnetic. Paramagnetism can be detected by measuring the mass of a sample in the presence and absence of a magnetic field.  The apparent mass of a paramagnetic substance will be larger in a magnetic field because of the force between the unpaired electrons and the field.

35.   $Ca^{2+}$ has a greater charge than $Na^+$, and $Se^{2-}$ is smaller than $Te^{2-}$.  The effect of charge on the lattice energy is greater than the effect of size.  We expect the trend from most exothermic to least exothermic to be:

$$CaSe \;>\; CaTe \;>\; Na_2Se \;>\; Na_2Te$$

$$(-2862) \quad (-2721) \quad (-2130) \quad (-2095 \text{ kJ/mol}) \qquad \text{This is what we observe.}$$

## Bond Energies

37.   a.   $H{-}H \;+\; Cl{-}Cl \;\longrightarrow\; 2\,H{-}Cl$

Bonds broken:                           Bonds formed:

1 H–H (432 kJ/mol)                   2 H–Cl (427 kJ/mol)
1 Cl–Cl (239 kJ/mol)

$\Delta H = \Sigma D_{broken} - \Sigma D_{formed},\;\; \Delta H = 432 \text{ kJ} + 239 \text{ kJ} - 2(427) \text{ kJ} = -183 \text{ kJ}$

b.   $N{\equiv}N + 3\,H{-}H \longrightarrow 2\,H{-}\underset{\underset{\textstyle H}{|}}{N}{-}H$

Bonds broken:                           Bonds formed:

1 N $\equiv$ N (941 kJ/mol)           6 N–H (391 kJ/mol)
3 H–H (432 kJ/mol)

$\Delta H = 941 \text{ kJ} + 3(432) \text{ kJ} - 6(391) \text{ kJ} = -109 \text{ kJ}$

c.  Sometimes some of the bonds remain the same between reactants and products.  To save time, only break and form bonds that are involved in the reaction.

$$H-C\equiv N + 2\ H-H \longrightarrow H-\underset{\underset{H}{|}}{\overset{\overset{H}{|}}{C}}-\underset{\underset{H}{|}}{\overset{\overset{H}{|}}{N}}$$

Bonds broken:

1 C≡N (891 kJ/mol)
2 H–H (432 kJ/mol)

Bonds formed:

1 C–N (305 kJ/mol)
2 C–H (413 kJ/mol)
2 N–H (391 kJ/mol)

$\Delta H = 891\ kJ + 2(432\ kJ) - [305\ kJ + 2(413\ kJ) + 2(391\ kJ)] = -158\ kJ$

d.

$$\underset{\underset{H}{}}{\overset{\overset{H}{}}{N}}-\underset{}{\overset{}{N}}\overset{H}{\underset{H}{}} + 2\ F-F \longrightarrow 4\ H-F + N=N$$

Bonds broken:

1 N–N (160. kJ/mol)
4 N–H (391 kJ/mol)
2 F–F (154 kJ/mol)

Bonds formed:

4 H–F (565 kJ/mol)
1 N≡N (941 kJ/mol)

$\Delta H = 160.\ kJ + 4(391\ kJ) + 2(154\ kJ) - [4(565\ kJ) + 941\ kJ] = -1169\ kJ$

39.

$$H-\underset{\underset{H}{|}}{\overset{\overset{H}{|}}{C}}-N\equiv C \longrightarrow H-\underset{\underset{H}{|}}{\overset{\overset{H}{|}}{C}}-C\equiv N$$

Bonds broken:  1 C–N (305 kJ/mol)     Bonds formed:  1 C–C (347 kJ/mol)

$\Delta H = \Sigma D_{broken} - \Sigma D_{formed},\ \Delta H = 305 - 347 = -42\ kJ$

*Note*:  Sometimes some of the bonds remain the same between reactants and products. To save time, only break and form bonds that are involved in the reaction.

41.  $H-C\equiv C-H + 5/2\ O=O \rightarrow 2\ O=C=O + H-O-H$

Bonds broken:

2 C–H (413 kJ/mol)
1 C≡C (839 kJ/mol)
5/2 O = O (495 kJ/mol)

Bonds formed:

$2 \times 2$ C=O (799 kJ/mol)
2 O–H (467 kJ/mol)

$\Delta H = 2(413\ kJ) + 839\ kJ + 5/2\ (495\ kJ) - [4(799\ kJ) + 2(467\ kJ)] = -1228\ kJ$

43.

Bonds broken:                                          Bonds formed:

9 N–N  (160. kJ/mol)                                   24 O–H  (467 kJ/mol)
4 N–C  (305 kJ/mol)                                    9 N≡N  (941 kJ/mol)
12 C–H  (413 kJ/mol)                                   8 C=O  (799 kJ/mol)
12 N–H  (391 kJ/mol)
10 N=O  (607 kJ/mol)
10 N–O  (201 kJ/mol)

$\Delta H = 9(160.) + 4(305) + 12(413) + 12(391) + 10(607) + 10(201)$

$- [24(467) + 9(941) + 8(799)]$

$\Delta H = 20{,}388 \text{ kJ} - 26{,}069 \text{ kJ} = -5681 \text{ kJ}$

45.    Because both reactions are highly exothermic, the high temperature is not needed to provide energy.  It must be necessary for some other reason.  The reason is to increase the speed of the reaction.  This will be discussed in Chapter 15 on kinetics.

47.    a.    $HF(g) \rightarrow H(g) + F(g)$                    $\Delta H = 565 \text{ kJ}$
              $H(g) \rightarrow H^+(g) + e^-$                    $\Delta H = 1312 \text{ kJ}$
              $F(g) + e^- \rightarrow F^-(g)$                    $\Delta H = -327.8 \text{ kJ}$
              _____
              $HF(g) \rightarrow H^+(g) + F^-(g)$               $\Delta H = 1549 \text{ kJ}$

       b.    $HCl(g) \rightarrow H(g) + Cl(g)$                  $\Delta H = 427 \text{ kJ}$
              $H(g) \rightarrow H^+(g) + e^-$                    $\Delta H = 1312 \text{ kJ}$
              $Cl(g) + e^- \rightarrow Cl^-(g)$                  $\Delta H = -348.7 \text{ kJ}$
              _____
              $HCl(g) \rightarrow H^+(g) + Cl^-(g)$             $\Delta H = 1390. \text{ kJ}$

       c.    $HI(g) \rightarrow H(g) + I(g)$                    $\Delta H = 295 \text{ kJ}$
              $H(g) \rightarrow H^+(g) + e^-$                    $\Delta H = 1312 \text{ kJ}$
              $I(g) + e^- \rightarrow I^-(g)$                    $\Delta H = -295.2 \text{ kJ}$
              _____
              $HI(g) \rightarrow H^+(g) + I^-(g)$               $\Delta H = 1312 \text{ kJ}$

       d.    $H_2O(g) \rightarrow OH(g) + H(g)$                 $\Delta H = 467 \text{ kJ}$
              $H(g) \rightarrow H^+(g) + e^-$                    $\Delta H = 1312 \text{ kJ}$
              $OH(g) + e^- \rightarrow OH^-(g)$                  $\Delta H = -180. \text{ kJ}$
              _____
              $H_2O(g) \rightarrow H^+(g) + OH^-(g)$            $\Delta H = 1599 \text{ kJ}$

49.     $NH_3(g) \rightarrow N(g) + 3 H(g)$

$\Delta H° = 3D_{NH} = 472.7 \text{ kJ} + 3(216.0 \text{ kJ}) - (-46.1 \text{ kJ}) = 1166.8 \text{ kJ}$

$D_{NH} = \dfrac{1166.8 \text{ kJ}}{3 \text{ mol NH bonds}} = 388.93 \text{ kJ/mol}$

$D_{calc}$ = 389 kJ/mol compared with 391 kJ/mol in the table.  There is good agreement.

## Lewis Structures and Resonance

51.     Drawing Lewis structures is mostly trial and error.  However, the first two steps are always the same.  These steps are (1) count the valence electrons available in the molecule/ion, and (2) attach all atoms to each other with single bonds (called the skeletal structure).  Unless noted otherwise, the atom listed first is assumed to be the atom in the middle, called the central atom, and all other atoms in the formula are attached to this atom.  The most notable exceptions to the rule are formulas that begin with H, e.g., $H_2O$, $H_2CO$, etc.  Hydrogen can never be a central atom since this would require H to have more than two electrons. In these compounds, the atom listed second is assumed to be the central atom.

After counting valence electrons and drawing the skeletal structure, the rest is trial and error. We place the remaining electrons around the various atoms in an attempt to satisfy the octet rule (or duet rule for H).  Keep in mind that practice makes perfect.  After practicing, you can (and will) become very adept at drawing Lewis structures.

a.   HCN has 1 + 4 + 5 = 10 valence electrons.

b.   $PH_3$ has 5 + 3(1) = 8 valence electrons.

Skeletal structure uses 4 e⁻;  6 e⁻ remain

Skeletal structures uses 6 e⁻;  2 e⁻ remain

c.   $CHCl_3$ has 4 + 1 + 3(7) = 26 valence electrons.

d.   $NH_4^+$ has 5 + 4(1) − 1 = 8 valence electrons.

*Note*: Subtract valence electrons for positive charged ions.

e.   $H_2CO$ has $2(1) + 4 + 6 = 12$ valence electrons.     f.   $SeF_2$ has $6 + 2(7) = 20$ valence electrons.

g.   $CO_2$ has $4 + 2(6) = 16$ valence electrons.     h.   $O_2$ has $2(6) = 12$ valence electrons.

i.   HBr has $1 + 7 = 8$ valence electrons.

53.     Molecules/ions that have the same number of valence electrons and the same number of atoms will have similar Lewis structures.

55.     Ozone:  $O_3$ has $3(6) = 18$ valence electrons.  Two resonance structures can be drawn.

Sulfur dioxide:  $SO_2$  has $6 + 2(6) = 18$ valence electrons.  Two resonance structures are possible.

Sulfur trioxide:  $SO_3$ has $6 + 3(6) = 24$ valence electrons.  Three resonance structures are possible.

57.     $CH_3NCO$ has $4 + 3(1) + 5 + 4 + 6 = 22$ valence electrons.  The order of the elements in the formula give the skeletal structure.

59.     Benzene has $6(4) + 6(1) = 30$ valence electrons.  Two resonance structures can be drawn for benzene.  The actual structure of benzene is an average of these two resonance structures; i.e., all carbon-carbon bonds are equivalent with a bond length and bond strength somewhere between a single and a double bond.

61.     Borazine ($B_3N_3H_6$) has $3(3) + 3(5) + 6(1) = 30$ valence electrons.  The possible resonance structures are similar to those of benzene in Exercise 13.59.

63.     Statements a and c are true. For statement a, $XeF_2$ has 22 valence electrons and it is impossible to satisfy the octet rule for all atoms with this number of electrons. The best Lewis structure is:

For statement c, $NO^+$ has 10 valence electrons, whereas $NO^-$ has 12 valence electrons. The Lewis structures are:

$$\left[ :N{\equiv}O: \right]^+ \qquad \left[ \ddot{N}{=}\ddot{O} \right]^-$$

Because a triple bond is stronger than a double bond, $NO^+$ has a stronger bond.

For statement b, $SF_4$ has five electron pairs around the sulfur in the best Lewis structure; it is an exception to the octet rule. Because $OF_4$ has the same number of valence electrons as $SF_4$, $OF_4$ would also have to be an exception to the octet rule. However, Row 2 elements such as O never have more than 8 electrons around them, so $OF_4$ does not exist. For statement d, two resonance structures can be drawn for ozone:

When resonance structures can be drawn, the actual bond lengths and strengths are all equal to each other. Even though each Lewis structure implies the two O–O bonds are different, this is not the case in real life. In real life, both of the O–O bonds are equivalent. When resonance structures can be drawn, you can think of the bonding as an average of all of the resonance structures.

65.     $PF_5$, 5 +5(7) = 40 valence electrons          $SF_4$, 6 + 4(7) = 34 $e^-$

$ClF_3$, 7 + 3(7) = 28 $e^-$                    $Br_3^-$, 3(7) + 1 = 22 $e^-$

$$\left[ :\ddot{B}r{-}\ddot{B}r{-}\ddot{B}r: \right]^-$$

Row 3 and heavier nonmetals can have more than 8 electrons around them when they have to. Row 3 and heavier elements have empty d orbitals that are close in energy to valence s and p orbitals. These empty d orbitals can accept extra electrons.

For example, P in $PF_5$ has its five valence electrons in the 3s and 3p orbitals. These s and p orbitals have room for three more electrons, and if it has to, P can use the empty 3d orbitals for any electrons above 8.

67.    $CO_3^{2-}$ has $4 + 3(6) + 2 = 24$ valence electrons.

Three resonance structures can be drawn for $CO_3^{2-}$.  The actual structure for $CO_3^{2-}$ is an average of these three resonance structures.  That is, the three C–O bond lengths are all equivalent, with a length somewhere between a single and a double bond.  The actual bond length of 136 pm is consistent with this resonance view of $CO_3^{2-}$.

69.

   $N_2$ (10 e $^-$):        :N≡N:              Triple bond between N and N.

   $N_2F_4$ (38 e $^-$):                          Single bond between N and N.

   $N_2F_2$ (24 e $^-$):     :F—N=N—F:          Double bond between N and N.

As the number of bonds increase between two atoms, bond strength increases and bond length decreases.  From the Lewis structure, the shortest to longest N-N bonds is $N_2 < N_2F_2 < N_2F_4$.

## Formal Charge

71.    See Exercise 13.52a for the Lewis structures of $POCl_3$, $SO_4^{2-}$, $ClO_4^-$ and $PO_4^{3-}$.  All of these compounds/ions have similar Lewis structures to those of $SO_2Cl_2$ and $XeO_4$ shown below.

   a.   $POCl_3$:  P, FC = $5 - 1/2(8) = +1$            b.   $SO_4^{2-}$:  S, FC = $6 - 1/2(8) = +2$

   c.   $ClO_4^-$:  Cl, FC = $7 - 1/2(8) = +3$            d.   $PO_4^{3-}$:  P, FC = $5 - 1/2(8) = +1$

   e.   $SO_2Cl_2$, $6 + 2(6) + 2(7) = 32$ e$^-$          f.   $XeO_4$, $8 + 4(6) = 32$ e$^-$

      S, FC = $6 - 1/2(8) = +2$                   Xe, FC = $8 - 1/2(8) = +4$

g.  $ClO_3^-$,  $7 + 3(6) + 1 = 26 \ e^-$          h.  $NO_4^{3-}$,  $5 + 4(6) + 3 = 32 \ e^-$

$$\left[ \begin{array}{c} :\ddot{O}-\ddot{Cl}-\ddot{O}: \\ \ddot{\underset{..}{O}}: \end{array} \right]^-  \qquad\qquad \left[ \begin{array}{c} :\ddot{O}: \\ :\ddot{O}-N-\ddot{O}: \\ :\ddot{\underset{..}{O}}: \end{array} \right]^{3-}$$

Cl, FC $= 7 - 2 - 1/2(6) = +2$          N, FC $= 5 - 1/2(8) = +1$

73.    SCl, $6 + 7 = 13$; the formula could be SCl (13 valence electrons), $S_2Cl_2$ (26 valence electrons), $S_3Cl_3$ (39 valence electrons), etc. For a formal charge of zero on S, we will need each sulfur in the Lewis structure to have two bonds to it and two lone pairs [FC $= 6 - 4 - 1/2(4) = 0$]. Cl will need one bond and three lone pairs for a formal charge of zero  [FC $= 7 - 6 - 1/2(2) = 0$]. Since chlorine wants only one bond to it, it will not be a central atom here. With this in mind, only $S_2Cl_2$ can have a Lewis structure with a formal charge of zero on all atoms. The structure is:

$$:\ddot{Cl}\!\!-\!\!\!-\!\!\!-\ddot{\underset{..}{S}}\!\!-\!\!\!-\!\!\!-\ddot{\underset{..}{S}}\!\!-\!\!\!-\!\!\!-\ddot{Cl}:$$

## Molecular Structure and Polarity

75.    The first step always is to draw a valid Lewis structure when predicting molecular structure. When resonance is possible, only one of the possible resonance structures is necessary to predict the correct structure because all resonance structures give the same structure.  The Lewis structures are in Exercises 13.51, 13.52 and 13.54.  The structures and bond angles for each follow.

13.51  a.  HCN: linear, 180°                    b.  $PH_3$: trigonal pyramid, <109.5°

       c.  $CHCl_3$: tetrahedral, 109.5°        d.  $NH_4^+$: tetrahedral, 109.5°

       e.  $H_2CO$: trigonal planar, 120°       f.  $SeF_2$: V-shaped or bent, <109.5°

       g.  $CO_2$: linear, 180°                 h and i. $O_2$ and HBr are both linear, but
                                                         there is no bond angle in either.

*Note*:  $PH_3$ and $SeF_2$ both have lone pairs of electrons on the central atom, which result in bond angles that are something less than predicted from a tetrahedral arrangement (109.5°). However, we cannot predict the exact number.  For these cases we will just insert a less than sign to indicate this phenomenon.

13.52  a.  All are tetrahedral;  109.5°

       b.  All are trigonal pyramid;  <109.5°

       c.  All are V-shaped;  <109.5°

13.54   a.   $NO_2^-$: V-shaped, $\approx 120°$;  $NO_3^-$: trigonal planar, $120°$

$N_2O_4$: trigonal planar, $120°$ about both N atoms

b.   $OCN^-$, $SCN^-$, and $N_3^-$ are all linear with $180°$ bond angles.

77.   From the Lewis structures (see Exercise 13.65), $Br_3^-$ would have a linear molecular structure, $ClF_3$ would have a T-shaped molecular structure, and $SF_4$ would have a see-saw molecular structure.  For example, consider $ClF_3$ (28 valence electrons):

The central Cl atom is surrounded by five electron pairs, which requires a trigonal bipyramid geometry.  Since there are three bonded atoms and two lone pairs of electrons about Cl, we describe the molecular structure of $ClF_3$ as T-shaped with predicted bond angles of about $90°$.  The actual bond angles will be slightly less than $90°$ due to the stronger repulsive effect of the lone pair electrons as compared to the bonding electrons.

79.   a.   $XeCl_2$ has $8 + 2(7) = 22$ valence electrons.

180°

There are five pairs of electrons about the central Xe atom.  The structure will be based on a trigonal bipyramid geometry.  The most stable arrangement of the atoms in $XeCl_2$ is a linear molecular structure with a $180°$ bond angle.

b.   $ICl_3$ has $7 + 3(7) = 28$ valence electrons.

$\approx 90°$

$\approx 90°$

T-shaped;  The ClICl angles are $\approx 90°$.  Since the lone pairs will take up more space, the ClICl  bond angles will probably be slightly less than $90°$.

c.   $TeF_4$ has $6 + 4(7) = 34$ valence electrons.

d.   $PCl_5$ has $5 + 5(7) = 40$ valence electrons.

$\approx 120°$

$\approx 90°$

$90°$

$120°$

See-saw or teeter-totter
or distorted tetrahedron

Trigonal bipyramid

All the species in this exercise have five pairs of electrons around the central atom. All the structures are based on a trigonal bipyramid geometry, but only in $PCl_5$ are all the pairs bonding pairs. Thus $PCl_5$ is the only one we describe the molecular structure as trigonal bipyramid. Still, we had to begin with the trigonal bipyramid geometry to get to the structures (and bond angles) of the others.

81. Let us consider the molecules with three pairs of electrons around the central atom first; these molecules are $SeO_3$ and $SeO_2$, and both have a trigonal planar arrangement of electron pairs. Both these molecules have polar bonds, but only $SeO_2$ has an overall net dipole moment. The net effect of the three bond dipoles from the three polar Se–O bonds in $SeO_3$ will be to cancel each other out when summed together. Hence $SeO_3$ is nonpolar since the overall molecule has no resulting dipole moment. In $SeO_2$, the two Se–O bond dipoles do not cancel when summed together; hence $SeO_2$ has a net dipole moment (is polar). Since O is more electronegative than Se, the negative end of the dipole moment is between the two O atoms, and the positive end is around the Se atom. The arrow in the following illustration represents the overall dipole moment in $SeO_2$. Note that to predict polarity for $SeO_2$, either of the two resonance structures can be used.

The other molecules in Exercise 13.76 ($PCl_3$, $SCl_2$, and $SiF_4$) have a tetrahedral arrangement of electron pairs. All have polar bonds; in $SiF_4$ the individual bond dipoles cancel when summed together, and in $PCl_3$ and $SCl_2$ the individual bond dipoles do not cancel. Therefore, $SiF_4$ has no net dipole moment (is nonpolar), and $PCl_3$ and $SCl_2$ have net dipole moments (are polar). For $PCl_3$, the negative end of the dipole moment is between the more electronegative chlorine atoms, and the positive end is around P. For $SCl_2$, the negative end is between the more electronegative Cl atoms, and the positive end of the dipole moment is around S.

83. The two general requirements for a polar molecule are:

    1.  Polar bonds

    2.  A structure such that the bond dipoles of the polar bonds do not cancel

$CF_4$, $4 + 4(7) = 32$ valence electrons      $XeF_4$, $8 + 4(7) = 36$ e$^-$

Tetrahedral, 109.5°                Square planar, 90°

$SF_4$, $6 + 4(7) = 34$ e$^-$

≈90°

≈120°

See-saw, ≈90°, ≈120°

≈90°

The arrows indicate the individual bond dipoles in the three molecules (the arrows point to the more electronegative atom in the bond, which will be the partial negative end of the bond dipole). All three of these molecules have polar bonds. To determine the polarity of the overall molecule, we sum the effect of all of the individual bond dipoles. In $CF_4$, the fluorines are symmetrically arranged about the central carbon atom. The net result is for all the individual C–F bond dipoles to cancel each other out, giving a nonpolar molecule. In $XeF_4$, the 4 Xe–F bond dipoles are also symmetrically arranged, and $XeF_4$ is also nonpolar. The individual bond dipoles cancel out when summed together. In $SF_4$, we also have four polar bonds. But in $SF_4$ the bond dipoles are not symmetrically arranged, and they do not cancel each other out. $SF_4$ is polar. It is the positioning of the lone pair that disrupts the symmetry in $SF_4$.

$CO_2$, $4 + 2(6) = 16$ e$^-$                    COS, $4 + 6 + 6 = 16$ e$^-$

$CO_2$ and COS both have linear molecular structures with a 180° bond angle. $CO_2$ is nonpolar because the individual bond dipoles cancel each other out, but COS is polar. By replacing an O with a less electronegative S atom, the molecule is not symmetric any more. The individual bond dipoles do not cancel because the C–S bond dipole is smaller than the C–O bond dipole resulting in a polar molecule.

85.    Only statement c is true. The bond dipoles in $CF_4$ and $KrF_4$ are arranged in a manner that they all cancel each other out, making them nonpolar molecules ($CF_4$ has a tetrahedral molecular structure, whereas $KrF_4$ has a square planar molecular structure). In $SeF_4$ the bond dipoles in this see-saw molecule do not cancel each other out, so $SeF_4$ is polar. For statement a, all the molecules have either a trigonal planar geometry or a trigonal bipyramid geometry, both of which have 120° bond angles. However, $XeCl_2$ has three lone pairs and two bonded chlorine atoms around it. $XeCl_2$ has a linear molecular structure with a 180° bond angle. With three lone pairs, we no longer have a 120° bond angle in $XeCl_2$. For statement b, $SO_2$ has a V-shaped molecular structure with a bond angle of about 120°. $CS_2$ is linear with a 180° bond angle and $SCl_2$ is V-shaped but with an approximate 109.5° bond angle. The three compounds do not have the same bond angle. For statement d, central atoms adopt a geometry to minimize electron repulsions, not maximize them.

87.     The formula is $EF_2O^{2-}$, and the Lewis structure has 28 valence electrons.

$$28 = x + 2(7) + 6 + 2, \quad x = 6 \text{ valence electrons for element E}$$

Element E must belong to the Group 6A elements since E has six valence electrons. E must also be a Row 3 or heavier element since this ion has more than eight electrons around the central E atom (Row 2 elements never have more than eight electrons around them). Some possible identities for E are S, Se and Te. The ion has a T-shaped molecular structure (see Exercise 13.77) with bond angles of ≈90°.

89.     Molecules that have an overall dipole moment are called polar molecules, and molecules that do not have an overall dipole moment are called nonpolar molecules.

a.   $OCl_2$, $6 + 2(7) = 20\ e^-$                                 $KrF_2$, $8 + 2(7) = 22\ e^-$

V-shaped, polar; $OCl_2$ is polar because the two O–Cl bond dipoles don't cancel each other. The resulting dipole moment is shown in the drawing.

Linear, nonpolar; The molecule is nonpolar because the two Kr–F bond dipoles cancel each other.

$BeH_2$, $2 + 2(1) = 4\ e^-$                                       $SO_2$, $6 + 2(6) = 18\ e^-$

Linear, nonpolar; Be–H bond dipoles are equal and point in opposite directions. They cancel each other. $BeH_2$ is nonpolar.

V-shaped, polar; The S–O bond dipoles do not cancel, so $SO_2$ is polar (has a net dipole moment). Only one resonance structure is shown.

*Note*: All four species contain three atoms. They have different structures because the number of lone pairs of electrons around the central atom are different in each case.

b.   $SO_3$,  $6 + 3(6) = 24$ e$^-$                     $NF_3$,  $5 + 3(7) = 26$ e$^-$

Trigonal planar, nonpolar;                Trigonal pyramid, polar;
bond dipoles cancel.  Only one            bond dipoles do not cancel.
resonance structure is shown.

$IF_3$ has $7 + 3(7) = 28$ valence electrons.

T-shaped, polar; bond dipoles do not cancel.

*Note*: Each molecule has the same number of atoms but different structures because of
differing numbers of lone pairs around each central atom.

c.   $CF_4$,  $4 + 4(7) = 32$ e$^-$                     $SeF_4$,  $6 + 4(7) = 34$ e$^-$

Tetrahedral, nonpolar;                    See-saw, polar;
bond dipoles cancel.                      bond dipoles do not cancel.

$KrF_4$, $8 + 4(7) = 36$ valence electrons

Square planar, nonpolar;
bond dipoles cancel.

*Note*:  Again, each molecule has the same number of atoms but different structures
because of differing numbers of lone pairs around the central atom.

d.   $IF_5$,  $7 + 5(7) = 42$ e$^-$                              $AsF_5$,  $5 + 5(7) = 40$ e$^-$

Square pyramid, polar;                        Trigonal bipyramid, nonpolar;
bond dipoles do not cancel.                   bond dipoles cancel.

*Note*:  Yet again, the molecules have the same number of atoms but different structures
because of the presence of differing numbers of lone pairs.

91.   All these molecules have polar bonds that are symmetrically arranged about the central
atoms.  In each molecule the individual bond dipoles cancel to give no net overall dipole
moment.  All these molecules are nonpolar even though they all contain polar bonds.

## Additional Exercises

93.   $CO_3{}^{2-}$ has $4 + 3(6) + 2 = 24$ valence electrons.

$HCO_3{}^-$ has $1 + 4 + 3(6) + 1 = 24$ valence electrons.

$H_2CO_3$ has $2(1) + 4 + 3(6) = 24$ valence electrons.

The Lewis structures for the reactants and products are:

Bonds broken:                              Bonds formed:

    2 C–O (358 kJ/mol)                         1 C=O (799 kJ/mol)
    1 O–H (467 kJ/mol)                         1 O–H (467 kJ/mol)

$\Delta H = 2(358) + 467 - (799 + 467) = -83$ kJ;  the carbon-oxygen double bond is stronger than two carbon-oxygen single bonds; hence $CO_2$ and $H_2O$ are more stable than $H_2CO_3$.

95.     As the halogen atoms get larger, it becomes more difficult to fit three halogen atoms around the small nitrogen atom, and the $NX_3$ molecule becomes less stable.

97.     The stable species are:

    a.   NaBr:  In $NaBr_2$, the sodium ion would have a 2+ charge, assuming that each bromine has a 1– charge.  Sodium doesn't form stable $Na^{2+}$ compounds.

    b.   $ClO_4^-$:  $ClO_4$ has 31 valence electrons, so it is impossible to satisfy the octet rule for all atoms in $ClO_4$.  The extra electron from the 1– charge in $ClO_4^-$ allows for complete octets for all atoms.

    c.   $XeO_4$:  We can't draw a Lewis structure that obeys the octet rule for $SO_4$ (30 electrons), unlike with $XeO_4$ (32 electrons).

    d.   $SeF_4$:  Both compounds require the central atom to expand its octet.  O is too small and doesn't have low-energy d orbitals to expand its octet (which is true for all Row 2 elements).

99.     a.   Radius:  $N^+ < N < N^-$;  IE:  $N^- < N < N^+$

        $N^+$ has the fewest electrons held by the seven protons in the nucleus whereas $N^-$ has the most electrons held by the seven protons.  The seven protons in the nucleus will hold the electrons most tightly in $N^+$ and least tightly in $N^-$.  Therefore, $N^+$ has the smallest radius with the largest ionization energy (IE), and $N^-$ is the largest species with the smallest IE.

    b.   Radius:  $Cl^+ < Cl < Se < Se^-$;  IE:  $Se^- < Se < Cl < Cl^+$

        The general trends tell us that Cl has a smaller radius than Se and a larger IE than Se.  $Cl^+$, with fewer electron-electron repulsions than Cl, will be smaller than Cl and have a larger IE.  $Se^-$, with more electron-electron repulsions than Se, will be larger than Se and have a smaller IE.

c.   Radius: $Sr^{2+} < Rb^+ < Br^-$;  IE: $Br^- < Rb^+ < Sr^{2+}$

These ions are isoelectronic.  The species with the most protons ($Sr^{2+}$) will hold the electrons most tightly and will have the smallest radius and largest IE.   The ion with the fewest protons ($Br^-$) will hold the electrons least tightly and will have the largest radius and smallest IE.

101.   Nonmetals, which form covalent bonds to each other, have valence electrons in the s and p orbitals. Since there are four total s and p orbitals, there is room for only eight valence electrons (the octet rule). The valence shell for hydrogen is just the 1s orbital. This orbital can hold two electrons, so hydrogen follows the duet rule.

103.   Assuming 100.00 g of compound:  $42.81 \text{ g F} \times \dfrac{1 \text{ mol F}}{19.00 \text{ g F}} = 2.253 \text{ mol F}$

The number of moles of X in $XF_5$ is:  $2.253 \text{ mol F} \times \dfrac{1 \text{ mol X}}{5 \text{ mol F}} = 0.4506 \text{ mol X}$

This number of moles of X has a mass of 57.19 g (= 100.00 g – 42.81 g). The molar mass of X is:

$$\dfrac{57.19 \text{ g X}}{0.4506 \text{ mol X}} = 126.9 \text{ g/mol};  \text{This is element I.}$$

$IF_5$, $7 + 5(7) = 42 \text{ e}^-$

The molecular structure is square pyramid.

105.   Yes, each structure has the same number of effective pairs around the central atom, giving the same predicted molecular structure for each compound/ion.  (A multiple bond is counted as a single group of electrons.)

## Challenge Problems

107.   $KrF_2$, $8 + 2(7) = 22$ e$^-$;  from the Lewis structure, we have a trigonal bipyramid arrangement of electron pairs with a linear molecular structure.

$$:\ddot{F}—\ddot{Kr}—\ddot{F}:$$

Hyperconjugation assumes that the overall bonding in $KrF_2$ is a combination of covalent and ionic contributions (see Section 13.12 of the text for discussion of hyperconjugation).  Using hyperconjugation, two resonance structures are possible that keep the linear structure.

109.  $$\begin{array}{ll} 2\,Li^+(g) + 2\,Cl^-(g) \rightarrow 2\,LiCl(s) & \Delta H = 2(-829\text{ kJ}) \\ 2\,Li(g) \rightarrow 2\,Li^+(g) + 2\,e^- & \Delta H = 2(520.\text{ kJ}) \\ 2\,Li(s) \rightarrow 2\,Li(g) & \Delta H = 2(166\text{ kJ}) \\ 2\,HCl(g) \rightarrow 2\,H(g) + 2\,Cl(g) & \Delta H = 2(427\text{ kJ}) \\ 2\,Cl(g) + 2\,e^- \rightarrow 2\,Cl^-(g) & \Delta H = 2(-349\text{ kJ}) \\ 2\,H(g) \rightarrow H_2(g) & \Delta H = -(432\text{ kJ}) \end{array}$$

$$2\,Li(s) + 2\,HCl(g) \rightarrow 2\,LiCl(s) + H_2(g) \qquad \Delta H = -562\text{ kJ}$$

111.   a.   $N(NO_2)_2^-$ contains $5 + 2(5) + 4(6) + 1 = 40$ valence electrons.

The most likely structures are:

There are other possible resonance structures, but these are most likely.

b.  The NNN and all ONN and ONO bond angles should be about 120°.

c.  $NH_4N(NO_2)_2 \rightarrow 2 N_2 + 2 H_2O + O_2$;  break and form all bonds.

Bonds broken:                                Bonds formed:

    4 N–H (391 kJ/mol)                       2 N≡N (941 kJ/mol)

    1 N–N (160. kJ/mol)                      4 H–O (467 kJ/mol)

    1 N=N (418 kJ/mol)                       1 O=O (495 kJ/mol)

    3 N–O (201 kJ/mol)                       _____

    1 N=O (607 kJ/mol)                       $\Sigma D_{formed} = 4245$ kJ

    _____

    $\Sigma D_{broken} = 3352$ kJ

    $\Delta H = \Sigma D_{broken} - \Sigma D_{formed} = 3352$ kJ $- 4245$ kJ $= -893$ kJ

d.  To estimate $\Delta H$, we completely ignored the ionic interactions between $NH_4{}^+$ and $N(NO_2)_2{}^-$.  In addition, we assumed the bond energies in Table 13.6 applied to the $N(NO_2)^-$ bonds in any one of the resonance structures above.  This is a bad assumption since molecules that exhibit resonance generally have stronger overall bonds than predicted.  All these assumptions give an estimated $\Delta H$ value which is too negative.

113.    a.    i.    $C_6H_6N_{12}O_{12} \rightarrow 6 CO + 6 N_2 + 3 H_2O + 3/2 O_2$

The $NO_2$ groups have one N–O single bond and one N=O double bond, and each carbon atom has one C–H single bond.  We must break and form all bonds.

Bonds broken:                                Bonds formed:

    3 C–C  (347 kJ/mol)                      6 C≡O  (1072 kJ/mol)

    6 C–H  (413 kJ/mol)                      6 N≡N  (941 kJ/mol)

    12 C–N (305 kJ/mol)                      6 H–O  (467 kJ/mol)

    6 N–N  (160. kJ/mol)                     3/2 O=O (495 kJ/mol)

    6 N–O  (201 kJ/mol)                      _____

    6 N=O  (607 kJ/mol)                      $\Sigma D_{formed} = 15,623$ kJ

    _____

    $\Sigma D_{broken} = 12,987$ kJ

    $\Delta H = \Sigma D_{broken} - \Sigma D_{formed} = 12,987$ kJ $- 15,623$ kJ $= -2636$ kJ

ii.    $C_6H_6N_{12}O_{12} \rightarrow 3 CO + 3 CO_2 + 6 N_2 + 3 H_2O$

*Note*: The bonds broken will be the same for all three reactions.

Bonds formed:

3 C≡O  (1072 kJ/mol)
6 C=O  (799 kJ/mol)
6 N≡N  (941 kJ/mol)
6 H–O  (467 kJ/mol)

$\Sigma D_{formed} = 16,458$ kJ

$\Delta H = 12,987$ kJ $- 16,458$ kJ $= -3471$ kJ

iii.  $C_6H_6N_{12}O_{12} \rightarrow 6\ CO_2 + 6\ N_2 + 3\ H_2$

Bonds formed:

12 C=O  (799 kJ/mol)
6 N≡N  (941 kJ/mol)
3 H–H  (432 kJ/mol)

$\Sigma D_{formed} = 16,530.$ kJ

$\Delta H = 12,987$ kJ $- 16,530.$ kJ $= -3543$ kJ

b.  Reaction iii yields the most energy per mole of CL-20, so it will yield the most energy per kilogram.

$$\frac{-3543\ kJ}{mol} \times \frac{1\ mol}{438.23\ g} \times \frac{1000\ g}{kg} = -8085\ kJ/kg$$

115.    The reaction is:

$1/2\ I_2(s) + 1/2\ Cl_2(g) \rightarrow ICl(g)$      $\Delta H_f^o = ?$

Using Hess's law:

$1/2\ I_2(s) \rightarrow 1/2\ I_2(g)$        $\Delta H = 1/2\ (62$ kJ)     (Appendix 4)
$1/2\ I_2(g) \rightarrow I(g)$        $\Delta H = 1/2\ (149$ kJ)    (Table 13.6)
$1/2\ Cl_2(g) \rightarrow Cl(g)$        $\Delta H = 1/2\ (239$ kJ)    (Table 13.6)
$I(g) + Cl(g) \rightarrow ICl(g)$        $\Delta H = -208$ kJ        (Table 13.6)

$1/2\ I_2(s) + 1/2\ Cl_2(g) \rightarrow ICl(g)$        $\Delta H = 17$ kJ so $\Delta H_f^o = 17$ kJ/mol

# CHAPTER 14

# COVALENT BONDING: ORBITALS

## The Localized Electron Model and Hybrid Orbitals

9.      The valence orbitals of the nonmetals are the s and p orbitals. The lobes of the p orbitals are 90° and 180° apart from each other. If the p orbitals were used to form bonds, then all bond angles shoud be 90° or 180°. This is not the case. In order to explain the observed geometry (bond angles) that molecules exhibit, we need to make up (hybridize) orbitals that point to where the bonded atoms and lone pairs are located. We know the geometry; we hybridize orbitals to explain the geometry.

Sigma bonds have shared electrons in the area centered on a line joining the atoms. The orbitals that overlap to form the sigma bonds must overlap head to head or end to end. The hybrid orbitals about a central atom always are directed at the bonded atoms. Hybrid orbitals will always overlap head to head to form sigma bonds.

11.     We use d orbitals when we have to, i.e., we use d orbitals when the central atom on a molecule has more than eight electrons around it. The d orbitals are necessary to accommodate the electrons over eight. Row 2 elements never have more than eight electrons around them, so they never hybridize d orbitals. We rationalize this by saying there are no d orbitals close in energy to the valence 2s and 2p orbitals (2d orbitals are forbidden energy levels). However, for Row 3 and heavier elements, there are 3d, 4d, 5d, etc. orbitals that will be close in energy to the valence s and p orbitals. It is Row 3 and heavier nonmetals that hybridize d orbitals when they have to.

For phosphorus, the valence electrons are in 3s and 3p orbitals. Therefore, 3d orbitals are closest in energy and are available for hybridization. Arsenic would hybridize 4d orbitals to go with the valence 4s and 4p orbitals, whereas iodine would hybridize 5d orbitals since the valence electrons are in $n = 5$.

13.     $H_2O$ has $2(1) + 6 = 8$ valence electrons.

$H_2O$ has a tetrahedral arrangement of the electron pairs about the O atom that requires $sp^3$ hybridization. Two of the four $sp^3$ hybrid orbitals are used to form bonds to the two hydrogen atoms, and the other two $sp^3$ hybrid orbitals hold the two lone pairs on oxygen. The two O–H bonds are formed from overlap of the $sp^3$ hybrid orbitals from oxygen with the 1s atomic orbitals from the hydrogen atoms. Each O–H covalent bond is called a sigma ($\sigma$) bond since the shared electron pair in each bond is centered in an area on a line running between the two atoms.

15.    Ethane, $C_2H_6$, has $2(4) + 6(1) = 14$ valence electrons. The Lewis structure is:

The carbon atoms are $sp^3$ hybridized. The six C–H sigma bonds are formed from overlap of the $sp^3$ hybrid orbitals on C with the 1s atomic orbitals from the hydrogen atoms. The carbon-carbon sigma bond is formed from overlap of an $sp^3$ hybrid orbital on each C atom.

Ethanol, $C_2H_6O$ has $2(4) + 6(1) + 6 = 20$ e$^-$

The two C atoms and the O atom are $sp^3$ hybridized. All bonds are formed from overlap with these $sp^3$ hybrid orbitals. The C–H and O–H sigma bonds are formed from overlap of $sp^3$ hybrid orbitals with hydrogen 1s atomic orbitals. The C–C and C–O sigma bonds are formed from overlap of the $sp^3$ hybrid orbitals on each atom.

17.    See Exercises 13.51, 13.52, and 13.54 for the Lewis structures. To predict the hybridization, first determine the arrangement of electron pairs about each central atom using the VSEPR model, then utilize the information in Figure 14.24 of the text to deduce the hybridization required for that arrangement of electron pairs.

13.51    a.    HCN;  C is sp hybridized.          b.    $PH_3$;  P is $sp^3$ hybridized.

c.    $CHCl_3$;  C is $sp^3$ hybridized.          d.    $NH_4^+$;  N is $sp^3$ hybridized.

e.    $H_2CO$;  C is $sp^2$ hybridized.          f.    $SeF_2$;  Se is $sp^3$ hybridized.

g.    $CO_2$;  C is sp hybridized.          h.    $O_2$;  Each O atom is $sp^2$ hybridized.

i.    HBr;  Br is $sp^3$ hybridized.

13.52   a.   All the central atoms are sp³ hybridized.

        b.   All the central atoms are sp³ hybridized.

        c.   All the central atoms are sp³ hybridized.

13.54   a.   In $NO_2^-$ and $NO_3^-$, N is sp² hybridized.  In $N_2O_4$, both central N atoms are also
             sp² hybridized.

        b.   In $OCN^-$ and $SCN^-$, the central carbon atoms in each ion are sp hybridized, and
             in $N_3^-$, the central N atom is also sp hybridized.

19.     a.

        tetrahedral          sp³
        109.5°               nonpolar

        b.

trigonal pyramid     sp³
< 109.5°             polar

The angles in $NF_3$ should be slightly less than 109.5° because the lone pair requires more
space than the bonding pairs.

        c.

V-shaped             sp³
< 109°.5             polar

        d.

trigonal planar      sp²
120°                 nonpolar

        e.

linear               sp
180°                 nonpolar

        f.

see-saw              dsp³
a)  ≈120°,  b) ≈90°  polar

g.

trigonal bipyramid        dsp³
a) 90°,    b) 120°        nonpolar

h.

linear        dsp³
180°          nonpolar

i.

square planar        d²sp³
90°                  nonpolar

j.

octahedral        d²sp³
90°               nonpolar

k.

square pyramid        d²sp³
≈90°                  polar

l.

T-shaped        dsp³
≈90°            polar

21.

For the p orbitals to properly line up to form the π bond, all six atoms are forced into the same plane.  If the atoms are not in the same plane, then the π bond could not form since the p orbitals would no longer be parallel to each other.

23.     To complete the Lewis structures, just add lone pairs of electrons to satisfy the octet rule for
        the atoms with fewer than eight electrons.

        Biacetyl ($C_4H_6O_2$) has $4(4) + 6(1) + 2(6) = 34$ valence electrons.

All CCO angles are 120°.  The six atoms are not forced to lie in
the same plane because of free rotation about the carbon-
carbon single (sigma) bonds.  There are 11 σ and 2 π bonds in
biacetyl.

        Acetoin ($C_4H_8O_2$) has $4(4) + 8(1) + 2(6) = 36$ valence electrons.

The carbon with the doubly bonded O is $sp^2$ hybridized.
The other 3 C atoms are $sp^3$ hybridized. Angle a = 120°
and angle b = 109.5°. There are 13 σ and 1 π bonds in
acetoin.

        *Note*: All single bonds are σ bonds, all double bonds are one σ and one π bond, and all triple
        bonds are one σ and two π bonds.

25.     To complete the Lewis structure, just add lone pairs of electrons to satisfy the octet rule for
        the atoms that have fewer than eight electrons.

a.  6               b.  4              c.  The center N in $-N=N=N$ group

d.  33 $\sigma$    e.  5 $\pi$        f.  180°

g.  ≈109.5°        h.  $sp^3$

27.     CO, $4 + 6 = 10 \; e^-$;    $CO_2$, $4 + 2(6) = 16 \; e^-$;    $C_3O_2$, $3(4) + 2(6) = 24 \; e^-$

:C≡O:                      Ö=C=Ö                      Ö=C=C=C=Ö

There is no molecular structure for the diatomic CO molecule. The carbon in CO is sp hybridized.  $CO_2$ is a linear molecule, and the central carbon atom is sp hybridized. $C_3O_2$ is a linear molecule with all the central carbon atoms exhibiting sp hybridization.

## The Molecular Orbital (MO) Model

29.     Bonding and antibonding molecular orbitals are both solutions to the quantum mechanical treatment of the molecule. Bonding orbitals form when in-phase orbitals combine to give constructive interference. This results in enhanced electron probability located between the two nuclei. The end result is that a bonding MO is lower in energy than the atomic orbitals from which it is composed. Antibonding orbitals form when out-of-phase orbitals combine. The mismatched phases produce destructive interference leading to a node of electron probability between the two nuclei. With electron distribution pushed to the outside, the energy of an antibonding orbital is higher than the energy of the atomic orbitals from which it is composed.

31.     a.   $H_2$ has two valence electrons to put in the MO diagram, whereas $He_2$ has four valence electrons.

$H_2$: $(\sigma_{1s})^2$            Bond order = B.O. = $(2-0)/2 = 1$
$He_2$: $(\sigma_{1s})^2(\sigma_{1s}*)^2$       B.O. = $(2-2)/2 = 0$

$H_2$ has a nonzero bond order, so MO theory predicts it will exist. The $H_2$ molecule is stable with respect to the two free H atoms. $He_2$ has a bond order of zero, so it should not form. The $He_2$ molecule is not more stable than the two free He atoms.

b.   See Figure 14.41 for the MO energy-level diagrams of $B_2$, $C_2$, $N_2$, $O_2$, and $F_2$. $B_2$ and $O_2$ have unpaired electrons in their electron configuration, so they are predicted to be paramagnetic. $C_2$, $N_2$, and $F_2$ have no unpaired electrons in the MO diagrams; they are all diamagnetic.

c.   From the MO energy diagram in Figure 14.41, $N_2$ maximizes the number of electrons in the lower-energy bonding orbitals and has no electrons in the antibonding 2p molecular orbitals. $N_2$ has the highest possible bond order of three, so it should be a very strong (stable) bond.

d.   $NO^+$ has $5 + 6 - 1 = 10$ valence electrons to place in the MO diagram, and $NO^-$ has $5 + 6 + 1 = 12$ valence electrons. The MO diagram for these two ions is assumed to be the same as that used for $N_2$.

$NO^+$: $(\sigma_{2s})^2(\sigma_{2s}{*})^2(\pi_{2p})^4(\sigma_{2p})^2$        B.O. $= (8-2)/2 = 3$

$NO^-$: $(\sigma_{2s})^2(\sigma_{2s}{*})^2(\pi_{2p})^4(\sigma_{2p})^2(\pi_{2p}{*})^2$      B.O. $= (8-4)/2 = 2$

$NO^+$ has a larger bond order than $NO^-$, so $NO^+$ should be more stable than $NO^-$.

33.

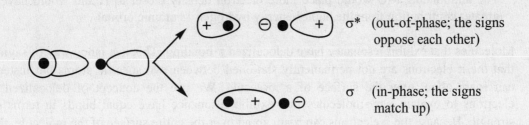

σ*   (out-of-phase; the signs oppose each other)

σ    (in-phase; the signs match up)

These molecular orbitals are sigma MOs because the electron density is cylindrically symmetric about the internuclear axis.

35.   $N_2$:   $(\sigma_{2s})^2(\sigma_{2s}{*})^2(\pi_{2p})^4(\sigma_{2p})^2$                  B.O. $= (8-2)/2 = 3$

We need to decrease the bond order from 3 to 2.5. There are two ways to do this. One is to add an electron to form $N_2^-$. This added electron goes into one of the $\pi_{2p}^*$ orbitals, giving a bond order of $(8-3)/2 = 2.5$. We could also remove a bonding electron to form $N_2^+$. The bond order for $N_2^+$ is also 2.5 [$= (7-2)/2$].

37.   CN:   $(\sigma_{2s})^2(\sigma_{2s}{*})^2(\pi_{2p})^4(\sigma_{2p})^1$

NO:   $(\sigma_{2s})^2(\sigma_{2s}{*})^2(\pi_{2p})^4(\sigma_{2p})^2(\pi_{2p}{*})^1$

$O_2^{2+}$:   $(\sigma_{2s})^2(\sigma_{2s}{*})^2(\sigma_{2p})^2(\pi_{2p})^4$

$N_2^{2+}$:   $(\sigma_{2s})^2(\sigma_{2s}{*})^2(\pi_{2p})^4$

If the added electron goes into a bonding orbital, the bond order would increase, making the species more stable and more likely to form. Between CN and NO, CN would most likely form $CN^-$ since the bond order increases (unlike $NO^-$, where the added electron goes into an antibonding orbital). Between $O_2^{2+}$ and $N_2^{2+}$, $N_2^+$ would most likely form since the bond order increases (unlike $O_2^{2+}$ going to $O_2^+$).

39.   The π bonds between S atoms and between C and S atoms are not as strong. The atomic orbitals do not overlap with each other as well as the smaller atomic orbitals of C and O overlap.

41.     Side-to-side overlap of these d orbitals would produce a π molecular orbital.  There would be no probability of finding an electron on the axis joining the two nuclei, which is characteristic of π MOs.

43.     a.   The electron density would be closer to F on average.  The F atom is more electronegative than the H atom, and the 2p orbital of F is lower in energy than the 1s orbital of H.

        b.   The bonding MO would have more fluorine 2p character since it is closer in energy to the fluorine 2p atomic orbital.

        c.   The antibonding MO would place more electron density closer to H and would have a greater contribution from the higher-energy hydrogen 1s atomic orbital.

45.     Molecules that exhibit resonance have delocalized π bonding. This is a fancy way of saying that the π electrons are not permanently stationed between two specific atoms but instead can roam about over the surface of a molecule. We use the concept of delocalized π electrons to explain why molecules that exhibit resonance have equal bonds in terms of strength. Because the π electrons can roam about over the entire surface of the molecule, the π electrons are shared by all the atoms in the molecule, giving rise to equal bond strengths.

        The classic example of delocalized π electrons is benzene ($C_6H_6$). Figure 14.50 shows the π molecular orbital system for benzene. Each carbon in benzene is $sp^2$ hybridized, leaving one unhybridized p atomic orbital. All six of the carbon atoms in benzene have an unhybridized p orbital pointing above and below the planar surface of the molecule. Instead of just two unhybridized p orbitals overlapping, we say all six of the unhybridized p orbitals overlap, resulting in delocalized π electrons roaming about above and below the entire surface of the benzene molecule.

        $SO_2$, $6 + 2(6) = 18$ $e^-$

        In $SO_2$ the central sulfur atom is $sp^2$ hybridized. The unhybridized p atomic orbital on the central sulfur atom will overlap with parallel p orbitals on each adjacent O atom. All three of these p orbitals overlap together, resulting in the π electrons moving about above and below the surface of the $SO_2$ molecule. With the delocalized π electrons, the S–O bond lengths in $SO_2$ are equal (and not different, as each individual Lewis structure indicates).

47.      The Lewis structures for $CO_3^{2-}$ are (24 e$^-$):

$$\left[ \underset{:\ddot{O}:\quad :\ddot{O}:}{\overset{:O:}{\underset{|}{C}}} \right]^{2-} \longleftrightarrow \left[ \underset{:\ddot{O}:\quad \ddot{O}:}{\overset{:\ddot{O}:}{\underset{||}{C}}} \right]^{2-} \longleftrightarrow \left[ \underset{:\ddot{O}:\quad \ddot{O}:}{\overset{:\ddot{O}:}{\underset{||}{C}}} \right]^{2-}$$

In the localized electron view, the central carbon atom is sp$^2$ hybridized; the sp$^2$ hybrid orbitals are used to form the three sigma bonds in $CO_3^{2-}$. The central C atom also has one unhybridized p atomic orbital that overlaps with another p atomic orbital from one of the oxygen atoms to form the π bond in each resonance structure. This localized π bond moves (resonates) from one position to another. In the molecular orbital model for $CO_3^{2-}$, all four atoms in $CO_3^{2-}$ have a p atomic orbital that is perpendicular to the plane of the ion. All four of these p orbitals overlap at the same time to form a delocalized π bonding system where the π electrons can roam above and below the entire surface of the ion. The π molecular orbital system for $CO_3^{2-}$ is analogous to that for $NO_3^-$ which is shown in Figure 14.51 of the text.

## Spectroscopy

49.      reduced mass $= \mu = \dfrac{m_1 m_2}{m_1 + m_2} = \dfrac{(1.0078)(78.918)}{1.0078 + 78.918}$ amu $\times \dfrac{1.66054 \times 10^{-27}\ kg}{amu}$

$$= 1.6524 \times 10^{-27}\ kg$$

$$\nu_o = \frac{1}{2\pi} \sqrt{\frac{k}{\mu}} = \frac{c}{\lambda} = 2.9979 \times 10^{10}\ cm\ s^{-1} \times 2650.\ cm^{-1} = 7.944 \times 10^{13}\ s^{-1}$$

$$7.944 \times 10^{13}\ s^{-1} = \frac{1}{2\pi} \sqrt{\frac{k}{\mu}} = \frac{1}{2\pi} \sqrt{\frac{k}{1.6524 \times 10^{-27}\ kg}}$$

Solving for k, the force constant:  k $= 411.7\ kg\ s^{-2} = 411.7\ N\ m^{-1}$

*Note*: one newton $= 1\ N = 1\ kg\ m\ s^{-2}$.

51.      a.    $\Delta E = 2hB(J_i + 1) = h\nu = \dfrac{hc}{\lambda}, \quad \dfrac{c}{\lambda} = 2B(J_i + 1)$

$$\frac{c}{\lambda} = 2B(0 + 1) = 2B = \frac{2.998 \times 10^8\ m\ s^{-1}}{2.60 \times 10^{-3}\ m} = 1.15 \times 10^{11}\ s^{-1}$$

$$B = \frac{1.15 \times 10^{-11}\ s^{-1}}{2} = 5.75 \times 10^{10}\ s^{-1}$$

$$I = \frac{h}{8\pi^2 B} = \frac{6.626 \times 10^{-34} \text{ J s}}{8\pi^2 (5.75 \times 10^{10} \text{ s}^{-1})} = 1.46 \times 10^{-46} \text{ kg m}^2$$

$$I = \mu R_e^2, \ \mu = \frac{m_1 m_2}{m_1 + m_2} = \frac{(12.000)(15.995)}{12.000 + 15.995} \text{ amu} \times \frac{1.66054 \times 10^{-27} \text{ kg}}{\text{amu}}$$

$$= 1.1385 \times 10^{-26} \text{ kg}$$

$$R_e^2 = \frac{I}{\mu} = \frac{1.46 \times 10^{-46} \text{ kg m}^2}{1.1385 \times 10^{-26} \text{ kg}} = 1.28 \times 10^{-20} \text{ m}^2, \ R_e = \text{bond length} = 1.13 \times 10^{-10} \text{ m}$$

$$= 113 \text{ pm}$$

b.   $v = \dfrac{\Delta E}{h} = 2B(J_i + 1) = 2B(2 + 1) = 6B$

From part a, $B = 5.75 \times 10^{10} \text{ s}^{-1}$, so: $v = 6(5.75 \times 10^{10} \text{ s}^{-1}) = 3.45 \times 10^{11} \text{ s}^{-1}$

53.   a.   The $-CH_2$ group neighbors a $-CH_3$ group, so a quartet of peaks should result (iv).

   b.   The $-CH_3$ H atoms are separated by more than three sigma bonds from other H atoms, so no spin-spin coupling should occur.  A singlet peak should result (i).

   c.   The $-CH_2$ H atoms neighbor two H atoms, so a triplet peak should result (iii).

   d.   The two H atoms in the $-CH_2F$ group neighbor one H atom, so a doublet peak should result (ii).

## Additional Exercises

55.   a.   $XeO_3$,  $8 + 3(6) = 26 \text{ e}^-$          b.   $XeO_4$,  $8 + 4(6) = 32 \text{ e}^-$

trigonal pyramid; $sp^3$                    tetrahedral; $sp^3$

   c.   $XeOF_4$,  $8 + 6 + 4(7) = 42 \text{ e}^-$          d.   $XeOF_2$,  $8 + 6 + 2(7) = 28 \text{ e}^-$

square pyramid; $d^2sp^3$                    T-shaped; $dsp^3$

e.   $XeO_3F_2$ has $8 + 3(6) + 2(7) = 40$ valence electrons.

trigonal bipyramid; $dsp^3$

57.    a.   No, some atoms are in different places.  Thus these are not resonance structures;  they are different compounds.

b.   For the first Lewis structure, all nitrogens are $sp^3$ hybridized and all carbons are $sp^2$ hybridized.  In the second Lewis structure, all nitrogens and carbons are $sp^2$ hybridized.

c.   For the reaction:

Bonds broken:

3 C=O (745 kJ/mol)
3 C–N (305 kJ/mol)
3 N–H (391 kJ/mol)

Bonds formed:

3 C=N (615 kJ/mol)
3 C–O (358 kJ/mol)
3 O–H (467 kJ/mol)

$\Delta H = 3(745) + 3(305) + 3(391) - [3(615) + 3(358) + 3(467)]$

$\Delta H = 4323$ kJ $- 4320$ kJ $= 3$ kJ

The bonds are slightly stronger in the first structure with the carbon-oxygen double bonds since $\Delta H$ for the reaction is positive.  However, the value of $\Delta H$ is so small that the best conclusion is that the bond strengths are comparable in the two structures.

59.

The two orbitals will overlap side-to-side, so when the orbitals are in-phase, a π bonding molecular orbital would form.

61.      Considering only the 12 valence electrons in $O_2$, the MO models would be:

O₂ ground state                    Arrangement of electrons consistent
                                   with the Lewis structure (double bond and
                                   no unpaired electrons).

It takes energy to pair electrons in the same orbital.  Thus the structure with no unpaired electrons is at a higher energy; it is an excited state.

63.      a.   $BH_3$ has $3 + 3(1) = 6$ valence electrons.

trigonal planar,  nonpolar,  120°,  $sp^2$

b.   $N_2F_2$ has $2(5) + 2(7) = 24$ valence electrons.

Can also be:

V-shaped about both N's;

$\approx 120°$ about both N's;

both N atoms:  $sp^2$

polar                                                                      nonpolar

These are distinctly different molecules.

c.   $C_4H_6$  has $4(4) + 6(1) = 22$ valence electrons.

All C atoms are trigonal planar with $120°$ bond angles and $sp^2$ hybridization. Because C and H have about equal electronegativities, the C–H bonds are essentially nonpolar, so the molecule is nonpolar. All neutral compounds composed of only C and H atoms are nonpolar.

d.   $ICl_3$ has $7 + 3(7) = 28$ valence electrons.

$\approx 90°$

$\approx 90°$

T-shaped, polar

$\approx 90°$,  $dsp^3$

65.    The Lewis structure for caffeine is:

The three C atoms each bonded to three H atoms are $sp^3$ hybridized (tetrahedral geometry); the other five C atoms with trigonal planar geometry are $sp^2$ hybridized. The one N atom with the double bond is $sp^2$ hybridized, and the other three N atoms are $sp^3$ hybridized. The answers to the questions are:

- 6 C and N atoms are $sp^2$ hybridized
- 6 C and N atoms are $sp^3$ hybridized
- 0 C and N atoms are sp hybridized (linear geometry)
- 25 σ bonds and 4 π bonds

67.     a.    The Lewis structures for NNO and NON are:

$$:N = N = \overset{\cdot\cdot}{\underset{\cdot\cdot}{O}}: \quad \longleftrightarrow \quad :N \equiv N - \overset{\cdot\cdot}{\underset{\cdot\cdot}{O}}: \quad \longleftrightarrow \quad :\overset{\cdot\cdot}{N} - N \equiv O:$$

$$:N = O = \overset{\cdot\cdot}{N}: \quad \longleftrightarrow \quad :N \equiv O - \overset{\cdot\cdot}{\underset{\cdot\cdot}{N}}: \quad \longleftrightarrow \quad :\overset{\cdot\cdot}{\underset{\cdot\cdot}{N}} - O \equiv N:$$

The NNO structure is correct.  From the Lewis structures we would predict both NNO and NON to be linear.  However, we would predict NNO to be polar and NON to be nonpolar.  Since experiments show $N_2O$ to be polar, then NNO is the correct structure.

b.    Formal charge = number of valence electrons of atoms - [(number of lone pair electrons) + 1/2 (number of shared electrons)].

$$:N = N = \overset{\cdot\cdot}{\underset{\cdot\cdot}{O}}: \quad \longleftrightarrow \quad :N \equiv N - \overset{\cdot\cdot}{\underset{\cdot\cdot}{O}}: \quad \longleftrightarrow \quad :\overset{\cdot\cdot}{\underset{\cdot\cdot}{N}} - N \equiv O:$$
$$\ \ -1 \quad +1 \quad\ 0 \qquad\qquad\quad 0 \quad +1 \quad -1 \qquad\qquad\quad -2 \quad +1 \quad +1$$

The formal charges for the atoms in the various resonance structures are below each atom.  The central N is sp hybridized in all the resonance structures.  We can probably ignore the third resonance structure on the basis of the relatively large formal charges as compared to the first two resonance structures.

c.    The sp hybrid orbitals from the center N overlap with atomic orbitals (or appropriate hybrid orbitals) from the other two atoms to form the two sigma bonds.  The remaining two unhybridized p orbitals from the center N overlap with two p orbitals from the peripheral N atom to form the two π bonds.

## Challenge Problems

69.   a.   $F_2^-(g) \rightarrow F(g) + F^-(g)$          $\Delta H = F_2^-$ bond energy

Using Hess's law:

$$F_2^-(g) \rightarrow F_2(g) + e^-  \quad \Delta H = \ \ 290. \text{ kJ (IE for } F_2^-)$$
$$F_2(g) \rightarrow 2\ F(g) \quad\quad\quad \Delta H = \ \ 154 \text{ kJ (BE for } F_2 \text{ from Table 13.6)}$$
$$F(g) + e^- \rightarrow F^-(g) \quad\quad \Delta H = -327.8 \text{ kJ (EA for F from Table 12.8)}$$

$$F_2^-(g) \rightarrow F(g) + F^-(g) \quad \Delta H = \ \ 116 \text{ kJ;  BE for } F_2^- = 116 \text{ kJ/mol}$$

Note that $F_2^-$ has a smaller bond energy than $F_2$.

b.   $F_2$:  $(\sigma_{2s})^2(\sigma_{2s}*)^2(\sigma_{2p})^2(\pi_{2p})^4(\pi_{2p}*)^4$          B.O. = (8 – 6)/2 = 1

$F_2^-$:  $(\sigma_{2s})^2(\sigma_{2s}*)^2(\sigma_{2p})^2(\pi_{2p})^4(\pi_{2p}*)^4(\sigma_{2p}*)^1$     B.O. = (8 – 7)/2 = 0.5

MO theory predicts that $F_2$ should have a stronger bond than $F_2^-$ because $F_2$ has the larger bond order. As determined in part a, $F_2$ indeed has a stronger bond because the $F_2$ bond energy (154 kJ/mol) is greater than the $F_2^-$ bond energy (116 kJ/mol).

71.   a.   $E = \dfrac{hc}{\lambda} = \dfrac{(6.626 \times 10^{-34} \text{ J s})(2.998 \times 10^8 \text{ m/s})}{25 \times 10^{-9} \text{ m}} = 7.9 \times 10^{-18} \text{ J}$

$7.9 \times 10^{-18} \text{ J} \times \dfrac{6.022 \times 10^{23}}{\text{mol}} \times \dfrac{1 \text{ kJ}}{1000 \text{ J}} = 4800 \text{ kJ/mol}$

Using $\Delta H$ values from the various reactions, 25-nm light has sufficient energy to ionize $N_2$ and N and to break the triple bond. Thus $N_2$, $N_2^+$, N, and $N^+$ will all be present, assuming excess $N_2$.

b.   To produce atomic nitrogen but no ions, the range of energies of the light must be from 941 kJ/mol to just below 1402 kJ/mol.

$\dfrac{941 \text{ kJ}}{\text{mol}} \times \dfrac{1 \text{ mol}}{6.022 \times 10^{23}} \times \dfrac{1000 \text{ J}}{1 \text{ kJ}} = 1.56 \times 10^{-18} \text{ J/photon}$

$\lambda = \dfrac{hc}{E} = \dfrac{(6.6261 \times 10^{-34} \text{ J s})(2.998 \times 10^8 \text{ m/s})}{1.56 \times 10^{-18} \text{ J}} = 1.27 \times 10^{-7} \text{ m} = 127 \text{ nm}$

$\dfrac{1402 \text{ kJ}}{\text{mol}} \times \dfrac{1 \text{ mol}}{6.0221 \times 10^{23}} \times \dfrac{1000 \text{ J}}{\text{kJ}} = 2.328 \times 10^{-18} \text{ J/photon}$

$\lambda = \dfrac{hc}{E} = \dfrac{(6.6261 \times 10^{-34} \text{ J s})(2.9979 \times 10^8 \text{ m/s})}{2.328 \times 10^{-18} \text{ J}} = 8.533 \times 10^{-8} \text{ m} = 85.33 \text{ nm}$

Light with wavelengths in the range of 85.33 nm $< \lambda \le$ 127 nm will produce N but no ions.

c.  $N_2$: $(\sigma_{2s})^2(\sigma_{2s}*)^2(\pi_{2p})^4(\sigma_{2p})^2$; the electron removed from $N_2$ is in the $\sigma_{2p}$ molecular orbital, which is lower in energy than the 2p atomic orbital from which the electron in atomic nitrogen is removed. Because the electron removed from $N_2$ is lower in energy than the electron removed from N, the ionization energy of $N_2$ is greater than that for N.

73.  The complete Lewis structure follows. All but two of the carbon atoms are sp$^3$ hybridized. The two carbon atoms that contain the double bond are sp$^2$ hybridized (see *).

No; most of the carbons are not in the same plane since a majority of carbon atoms exhibit a tetrahedral structure. *Note*: CH, CH$_2$, and CH$_3$ are shorthand for carbon atoms singly bonded to hydrogen atoms.

75.  a.  The CO bond is polar with the negative end around the more electronegative oxygen atom. We would expect metal cations to be attracted to and to bond to the oxygen end of CO on the basis of electronegativity.

b.

:C≡O:      FC (carbon) = 4 − 2 − 1/2(6) = −1

FC (oxygen) = 6 − 2 − 1/2(6) = +1

From formal charge, we would expect metal cations to bond to the carbon (with the negative formal charge).

c.  In molecular orbital theory, only orbitals with proper symmetry overlap to form bonding orbitals. The metals that form bonds to CO are usually transition metals, all of which have outer electrons in the d orbitals. The only molecular orbitals of CO that have proper symmetry to overlap with d orbitals are the $\pi_{2p}*$ orbitals, whose shape is similar to the d orbitals (see Figure 14.36). Since the antibonding molecular orbitals have more carbon character, one would expect the bond to form through carbon.

77.    One of the resonance structures for benzene is:

To break $C_6H_6(g)$ into $C(g)$ and $H(g)$ requires the breaking of 6 C–H bonds, 3 C=C bonds and 3 C–C bonds:

$$C_6H_6(g) \rightarrow 6\ C(g) + 6\ H(g) \quad \Delta H = 6D_{C-H} + 3D_{C=C} + 3D_{C-C}$$

$$\Delta H = 6(413\ kJ) + 3(614\ kJ) + 3(347\ kJ) = 5361\ kJ$$

The question asks for $\Delta H_f^{\circ}$ for $C_6H_6(g)$, which is $\Delta H$ for the reaction:

$$6\ C(s) + 3\ H_2(g) \rightarrow C_6H_6(g) \quad \Delta H = \Delta H_{f,\ C_6H_6(g)}^{\circ}$$

To calculate $\Delta H$ for this reaction, we will use Hess's law along with the value $\Delta H_f^{\circ}$ for $C(g)$ and the bond energy value for $H_2$ ($D_{H_2} = 432$ kJ/mol).

| | |
|---|---|
| $6\ C(g) + 6\ H(g) \rightarrow C_6H_6(g)$ | $\Delta H_1 = -5361\ kJ$ |
| $6\ C(s) \rightarrow 6\ C(g)$ | $\Delta H_2 = \ \ 6(717\ kJ)$ |
| $3\ H_2(g) \rightarrow 6\ H(g)$ | $\Delta H_3 = \ \ 3(432\ kJ)$ |

$6\ C(s) + 3\ H_2(g) \rightarrow C_6H_6(g) \qquad \Delta H = \Delta H_1 + \Delta H_2 + \Delta H_3 = 237\ kJ; \ \Delta H_{f,\ C_6H_6(g)}^{\circ} = 237\ kJ/mol$

The experimental $\Delta H_f^{\circ}$ for $C_6H_6(g)$ is more stable (lower in energy) by 154 kJ than the $\Delta H_f^{\circ}$ calculated from bond energies (83 – 237 = –154 kJ). This extra stability is related to benzene's ability to exhibit resonance. Two equivalent Lewis structures can be drawn for benzene. The π bonding system implied by each Lewis structure consists of three localized π bonds. This is not correct because all C–C bonds in benzene are equivalent. We say the π electrons in benzene are delocalized over the entire surface of $C_6H_6$ (see Section 14.5 of the text). The large discrepancy between $\Delta H_f^{\circ}$ values is due to the delocalized π electrons, whose effect was not accounted for in the calculated $\Delta H_f^{\circ}$ value. The extra stability associated with benzene can be called resonance stabilization. In general, molecules that exhibit resonance are usually more stable than predicted using bond energies.

79.    The two isomers having the formula $C_2H_6O$ are:

The first structure has three types of hydrogens ($a$, $b$, and $c$), so the signals should be seen in three different regions of the NMR spectra. The overall relative areas of the three signals should be in a 3:2:1 ratio due to the number of $a$, $b$, and $c$ hydrogens. The signal for the $a$ hydrogens will be split into a triplet signal due to the two neighboring $b$ hydrogens. The signal for the $b$ hydrogens should be split into a quintet signal due to the three neighboring $a$ hydrogens plus the neighboring $c$ hydrogen (four total protons). The $c$ hydrogen signal should be split into a triplet signal due to the two neighboring $b$ hydrogens.  In practice, however, the $c$ hydrogen bonded to the oxygen does not behave "normally".  This O–H hydrogen generally behaves as if it was more than three sigma bonds apart from the $b$ hydrogens.  Therefore, the spectrum will most likely have a triplet signal for the $a$ hydrogens, a quartet signal for the $b$ hydrogens, and a singlet signal for the $c$ hydrogen.

In the second structure, all six hydrogens are equivalent, so only one signal will appear in the NMR spectrum. This signal will be a singlet signal because the hydrogens in the two –CH$_3$ groups are more than three sigma bonds apart from each other (no splitting occurs).

# CHAPTER 15

# CHEMICAL KINETICS

## Reaction Rates

11.     0.0120/0.0080 = 1.5;  reactant B is used up 1.5 times faster than reactant A. This corresponds to a 3 to 2 mole ratio between B and A in the balanced equation. 0.0160/0.0080 = 2; product C is produced twice as fast as reactant A is used up. So the coefficient for C is twice the coefficient for A. A possible balanced equation is $2A + 3B \rightarrow 4C$.

13.     Using the coefficients in the balanced equation to relate the rates:

$$\frac{d[H_2]}{dt} = 3\frac{d[N_2]}{dt} \text{ and } \frac{d[NH_3]}{dt} = -2\frac{d[N_2]}{dt}$$

$$\text{So}: \frac{1}{3}\frac{d[H_2]}{dt} = -\frac{1}{2}\frac{d[NH_3]}{dt} \text{ or } \frac{d[NH_3]}{dt} = -\frac{2}{3}\frac{d[H_2]}{dt}$$

Ammonia is produced at a rate equal to 2/3 of the rate of consumption of hydrogen.

15.     Rate $= k[Cl]^{1/2}[CHCl_3]$, $\dfrac{mol}{L\,s} = k\left(\dfrac{mol}{L}\right)^{1/2}\left(\dfrac{mol}{L}\right)$;  k must have units of $L^{1/2}$ mol$^{-1/2}$ s$^{-1}$.

## Rate Laws from Experimental Data:  Initial Rates Method

17.     a.  In the first two experiments, [NO] is held constant and [Cl$_2$] is doubled.  The rate also doubled.  Thus the reaction is first order with respect to Cl$_2$.  Or mathematically:  Rate $= k[NO]^x[Cl_2]^y$

$$\frac{0.36}{0.18} = \frac{k(0.10)^x(0.20)^y}{k(0.10)^x(0.10)^y} = \frac{(0.20)^y}{(0.10)^y}, \ 2.0 = 2.0^y, \ y = 1$$

We can get the dependence on NO from the second and third experiments.  Here, as the NO concentration doubles (Cl$_2$ concentration is constant), the rate increases by a factor of four.  Thus, the reaction is second order with respect to NO.  Or mathematically:

285

$$\frac{1.45}{0.36} = \frac{k(0.20)^x(0.20)}{k(0.10)^x(0.20)} = \frac{(0.20)^x}{(0.10)^x}, \ 4.0 = 2.0^x, \ x = 2; \ \text{so}, \ \text{Rate} = k[NO]^2[Cl_2].$$

Try to examine experiments where only one concentration changes at a time.  The more variables that change, the harder it is to determine the orders.  Also, these types of problems can usually be solved by inspection.  In general, we will solve using a mathematical approach, but keep in mind that you probably can solve for the orders by simple inspection of the data.

b.   The rate constant k can be determined from the experiments.  From experiment 1:

$$\frac{0.18 \ \text{mol}}{\text{L min}} = k\left(\frac{0.10 \ \text{mol}}{\text{L}}\right)^2\left(\frac{0.10 \ \text{mol}}{\text{L}}\right), \ k = 180 \ \text{L}^2 \ \text{mol}^{-2} \ \text{min}^{-1}$$

From the other experiments:

$$k = 180 \ \text{L}^2 \ \text{mol}^{-2} \ \text{min}^{-1} \ (\text{second exp.}); \ k = 180 \ \text{L}^2 \ \text{mol}^{-2} \ \text{min}^{-1} \ (\text{third exp.})$$

The average rate constant is $k_{mean} = 1.8 \times 10^2 \ \text{L}^2 \ \text{mol}^{-2} \ \text{min}^{-1}$.

19.   a.   Rate = $k[NOCl]^n$;  using experiments two and three:

$$\frac{2.66 \times 10^4}{6.64 \times 10^3} = \frac{k(2.0 \times 10^{16})^n}{k(1.0 \times 10^{16})^n}, \ 4.01 = 2.0^n, \ n = 2; \ \text{Rate} = k[NOCl]^2$$

b.   $$\frac{5.98 \times 10^4 \ \text{molecules}}{\text{cm}^3 \ \text{s}} = k\left(\frac{3.0 \times 10^{16} \ \text{molecules}}{\text{cm}^3}\right)^2, \ k = 6.6 \times 10^{-29} \ \text{cm}^3 \ \text{molecules}^{-1} \ \text{s}^{-1}$$

The other three experiments give (6.7, 6.6, and 6.6) $\times 10^{-29} \ \text{cm}^3 \ \text{molecules}^{-1} \ \text{s}^{-1}$, respectively.  The mean value for k is $6.6 \times 10^{-29} \ \text{cm}^3 \ \text{molecules}^{-1} \ \text{s}^{-1}$.

c.   $$\frac{6.6 \times 10^{-29} \ \text{cm}^3}{\text{molecules s}} \times \frac{1 \ \text{L}}{1000 \ \text{cm}^3} \times \frac{6.022 \times 10^{23} \ \text{molecules}}{\text{mol}} = \frac{4.0 \times 10^{-8} \ \text{L}}{\text{mol s}}$$

21.   a.   Rate = $k[ClO_2]^x[OH^-]^y$;  From the first two experiments:

$$2.30 \times 10^{-1} = k(0.100)^x(0.100)^y \ \text{and} \ 5.75 \times 10^{-2} = k(0.0500)^x(0.100)^y$$

Dividing the two rate laws:  $4.00 = \dfrac{(0.100)^x}{(0.0500)^x} = 2.00^x, \ x = 2$

Comparing the second and third experiments:

$$2.30 \times 10^{-1} = k(0.100)(0.100)^y \ \text{and} \ 1.15 \times 10^{-1} = k(0.100)(0.0500)^y$$

Dividing: $2.00 = \dfrac{(0.100)^y}{(0.050)^y} = 2.0^y,\ y = 1$

The rate law is:  Rate $= k[ClO_2]^2[OH^-]$

$2.30 \times 10^{-1}$ mol $L^{-1}$ $s^{-1} = k(0.100$ mol/L$)^2(0.100$ mol/L$),\ k = 2.30 \times 10^2$ $L^2$ mol$^{-2}$ $s^{-1}$
$= k_{mean}$

b.  Rate $= \dfrac{2.30 \times 10^2\ L^2}{mol^2\ s} \times \left(\dfrac{0.175\ mol}{L}\right)^2 \times \dfrac{0.0844\ mol}{L} = 0.594$ mol $L^{-1}$ $s^{-1}$

23.  Rate $= k[N_2O_5]^x$;  the rate laws for the first two experiments are:

$2.26 \times 10^{-3} = k(0.190)^x$ and  $8.90 \times 10^{-4} = k(0.0750)^x$

Dividing: $2.54 = \dfrac{(0.190)^x}{(0.0750)^x} = (2.53)^x,\ x = 1$;  Rate $= k[N_2O_5]$

$k = \dfrac{Rate}{[N_2O_5]} = \dfrac{8.90 \times 10^{-4}\ mol\ L^{-1}\ s^{-1}}{0.0750\ mol/L} = 1.19 \times 10^{-2}\ s^{-1}$;  $k_{mean} = 1.19 \times 10^{-2}\ s^{-1}$

25.  Rate $= k[I^-]^x[OCl^-]^y[OH^-]^z$;  Comparing the first and second experiments:

$\dfrac{18.7 \times 10^{-3}}{9.4 \times 10^{-3}} = \dfrac{k(0.0026)^x(0.012)^y(0.10)^z}{k(0.0013)^x(0.012)^y(0.10)^z},\ 2.0 = 2.0^x,\ x = 1$

Comparing the first and third experiments:

$\dfrac{9.4 \times 10^{-3}}{4.7 \times 10^{-3}} = \dfrac{k(0.0013)(0.012)^y(0.10)^z}{k(0.0013)(0.0060)^y(0.10)^z},\ 2.0 = 2.0^y,\ y = 1$

Comparing the first and sixth experiments:

$\dfrac{4.8 \times 10^{-3}}{9.4 \times 10^{-3}} = \dfrac{k(0.0013)(0.012)(0.20)^z}{k(0.0013)(0.012)(0.10)^z},\ 1/2 = 2.0^z,\ z = -1$

Rate $= \dfrac{k[I^-][OCl^-]}{[OH^-]}$;  the presence of $OH^-$ decreases the rate of the reaction.

For the first experiment:

$\dfrac{9.4 \times 10^{-3}\ mol}{L\ s} = k\dfrac{(0.0013\ mol/L)(0.012\ mol/L)}{(0.10\ mol/L)},\ k = 60.3\ s^{-1} = 60.\ s^{-1}$

For all experiments, $k_{mean} = 60.\ s^{-1}$.

## Integrated Rate Laws

27.    Zero order: $t_{1/2} = \dfrac{[A]_0}{2k}$;    first order: $t_{1/2} = \dfrac{\ln 2}{k}$;   second order: $t_{1/2} = \dfrac{1}{k[A]_0}$

For a first-order reaction, if the first half-life equals 20. s, the second half-life will also be 20. s because the half-life for a first-order reaction is concentration-independent. The second half-life for a zero-order reaction will be 1/2(20.) = 10. s. This is because the half-life for a zero-order reaction has a direct relationship with concentration (as the concentration decreases by a factor of 2, the half-life decreases by a factor of 2). Because a second-order reaction has an inverse relationship between $t_{1/2}$ and $[A]_0$, the second half-life will be 40. s (twice the first half-life value).

29.    a.   Because the ln[A] versus time plot was linear, the reaction is first order in A.  The slope of the ln[A] versus time plot equals –k.  Therefore, the rate law, the integrated rate law, and the rate constant value are:

$$\text{Rate} = k[A]; \quad \ln[A] = -kt + \ln[A]_0; \quad k = 2.97 \times 10^{-2} \ \text{min}^{-1}$$

b.   The half-life expression for a first order rate law is:

$$t_{1/2} = \frac{\ln 2}{k} = \frac{0.6931}{k}, \quad t_{1/2} = \frac{0.6931}{2.97 \times 10^{-2} \ \text{min}^{-1}} = 23.3 \ \text{min}$$

c.   $2.50 \times 10^{-3}$ $M$ is 1/8 of the original amount of A present initially, so the reaction is 87.5% complete.  When a first-order reaction is 87.5% complete (or 12.5% remains), then the reaction has gone through 3 half-lives:

$$100\% \ \xrightarrow{\ t_{1/2}\ } \ 50.0\% \ \xrightarrow{\ t_{1/2}\ } \ 25.0\% \ \xrightarrow{\ t_{1/2}\ } \ 12.5\%; \quad t = 3 \times t_{1/2} = 3 \times 23.3 \ \text{min} = 69.9 \ \text{min}$$

Or we can use the integrated rate law:

$$\ln\!\left(\frac{[A]}{[A]_0}\right) = -kt, \quad \ln\!\left(\frac{2.50 \times 10^{-3} \ M}{2.00 \times 10^{-2} \ M}\right) = -(2.97 \times 10^{-2} \ \text{min}^{-1})t$$

$$t = \frac{\ln(0.125)}{-2.97 \times 10^{-2} \ \text{min}^{-1}} = 70.0 \ \text{min}$$

31.    The first assumption to make is that the reaction is first order.  For a first order reaction, a graph of $\ln[H_2O_2]$ versus time will yield a straight line.  If this plot is not linear, then the reaction is not first order, and we make another assumption.

| Time (s) | $[H_2O_2]$ (mol/L) | $\ln[H_2O_2]$ |
|---|---|---|
| 0 | 1.00 | 0.000 |
| 120. | 0.91 | −0.094 |
| 300. | 0.78 | −0.25 |
| 600. | 0.59 | −0.53 |
| 1200. | 0.37 | −0.99 |
| 1800. | 0.22 | −1.51 |
| 2400. | 0.13 | −2.04 |
| 3000. | 0.082 | −2.50 |
| 3600. | 0.050 | −3.00 |

*Note*:  We carried extra significant figures in some of the natural log values in order to reduce round-off error.  For the plots, we will do this most of the time when the natural log function is involved.

The plot of $\ln[H_2O_2]$ versus time is linear.  Thus the reaction is first order.  The differential rate law and integrated rate law are  $\text{Rate} = \dfrac{-d[H_2O_2]}{dt} = k[H_2O_2]$  and  $\ln[H_2O_2] = -kt + \ln[H_2O_2]_0$.

We determine the rate constant k by determining the slope of the $\ln[H_2O_2]$ versus time plot (slope = −k).  Using two points on the curve gives:

$$\text{slope} = -k = \frac{\Delta y}{\Delta x} = \frac{0 - (3.00)}{0 - 3600.} = -8.3 \times 10^{-4}\ s^{-1},\ \ k = 8.3 \times 10^{-4}\ s^{-1}$$

To determine $[H_2O_2]$ at 4000. s, use the integrated rate law, where $[H_2O_2]_0 = 1.00\ M$.

$$\ln[H_2O_2] = -kt + \ln[H_2O_2]_0\ \ \text{or}\ \ \ln\left(\frac{[H_2O_2]}{[H_2O_2]_0}\right) = -kt$$

$$\ln\left(\frac{[H_2O_2]}{1.00}\right) = -8.3 \times 10^{-4}\ s^{-1} \times 4000.\ s,\ \ \ln[H_2O_2] = -3.3,\ \ [H_2O_2] = e^{-3.3} = 0.037\ M$$

33.   Assume the reaction is first order and see if the plot of $\ln[NO_2]$ versus time is linear.  If this isn't linear, try the second order plot of $1/[NO_2]$ versus time.  The data and plots follow.

| Time (s) | $[NO_2]$ ($M$) | $\ln[NO_2]$ | $1/[NO_2]$ ($M^{-1}$) |
|---|---|---|---|
| 0 | 0.500 | −0.693 | 2.00 |
| $1.20 \times 10^3$ | 0.444 | −0.812 | 2.25 |
| $3.00 \times 10^3$ | 0.381 | −0.965 | 2.62 |
| $4.50 \times 10^3$ | 0.340 | −1.079 | 2.94 |
| $9.00 \times 10^3$ | 0.250 | −1.386 | 4.00 |
| $1.80 \times 10^4$ | 0.174 | −1.749 | 5.75 |

The plot of $1/[NO_2]$ versus time is linear. The reaction is second order in $NO_2$. The differential rate law and integrated rate law are $\text{Rate} = k[NO_2]^2$ and $\dfrac{1}{[NO_2]} = kt + \dfrac{1}{[NO_2]_0}$.

The slope of the plot $1/[NO_2]$ versus. t gives the value of k. Using a couple of points on the plot:

$$\text{slope} = k = \frac{\Delta y}{\Delta x} = \frac{(5.75 - 2.00)\ M^{-1}}{(1.80 \times 10^4 - 0)\ s} = 2.08 \times 10^{-4}\ \text{L mol}^{-1}\ \text{s}^{-1}$$

To determine $[NO_2]$ at $2.70 \times 10^4$ s, use the integrated rate law, where $1/[NO_2]_0 = 1/0.500\ M = 2.00\ M^{-1}$.

$$\frac{1}{[NO_2]} = kt + \frac{1}{[NO_2]_0},\ \frac{1}{[NO_2]} = \frac{2.08 \times 10^{-4}\ \text{L}}{\text{mol s}} \times 2.70 \times 10^4\ s + 2.00\ M^{-1}$$

$$\frac{1}{[NO_2]} = 7.62,\ [NO_2] = 0.131\ M$$

35.    From the data, the pressure of $C_2H_5OH$ decreases at a constant rate of 13 torr for every 100. s. Since the rate of disappearance of $C_2H_5OH$ is not dependent on concentration, the reaction is zero order in $C_2H_5OH$.

$$k = \frac{13\ \text{torr}}{100.\ s} \times \frac{1\ \text{atm}}{760\ \text{torr}} = 1.7 \times 10^{-4}\ \text{atm/s}$$

The rate law and integrated rate law are:

$$\text{Rate} = k = 1.7 \times 10^{-4} \text{ atm/s}; \quad P_{C_2H_5OH} = -kt + 250. \text{ torr}\left(\frac{1 \text{ atm}}{760 \text{ torr}}\right) = -kt + 0.329 \text{ atm}$$

At 900. s:

$$P_{C_2H_5OH} = -1.7 \times 10^{-4} \text{ atm/s} \times 900. \text{ s} + 0.329 \text{ atm} = 0.176 \text{ atm} = 0.18 \text{ atm} = 130 \text{ torr}$$

37.     a.   We check for first-order dependence by graphing ln[concentration] versus time for each set of data. The rate dependence on NO is determined from the first set of data since the ozone concentration is relatively large compared to the NO concentration, so it is effectively constant.

| Time (ms) | [NO] (molecules/cm$^3$) | ln[NO] |
|-----------|-------------------------|--------|
| 0         | $6.0 \times 10^8$       | 20.21  |
| 100.      | $5.0 \times 10^8$       | 20.03  |
| 500.      | $2.4 \times 10^8$       | 19.30  |
| 700.      | $1.7 \times 10^8$       | 18.95  |
| 1000.     | $9.9 \times 10^7$       | 18.41  |

Because ln[NO] versus t is linear, the reaction is first order with respect to NO.

We follow the same procedure for ozone using the second set of data. The data and plot are:

| Time (ms) | [O$_3$] (molecules/cm$^3$) | ln[O$_3$] |
|-----------|----------------------------|-----------|
| 0         | $1.0 \times 10^{10}$       | 23.03     |
| 50.       | $8.4 \times 10^9$          | 22.85     |
| 100.      | $7.0 \times 10^9$          | 22.67     |
| 200.      | $4.9 \times 10^9$          | 22.31     |
| 300.      | $3.4 \times 10^9$          | 21.95     |

The plot of $\ln[O_3]$ versus t is linear.  Hence the reaction is first order with respect to ozone.

b.   Rate $= k[NO][O_3]$ is the overall rate law.

c.   For NO experiment, Rate $= k'[NO]$ and $k' = -$(slope from graph of $\ln[NO]$ versus t).

$$k' = -\text{slope} = -\frac{18.41 - 20.21}{(1000. - 0) \times 10^{-3}\,\text{s}} = 1.8\,\text{s}^{-1}$$

For ozone experiment,  Rate $= k''[O_3]$ and $k'' = -$(slope from $\ln[O_3]$ versus t).

$$k'' = -\text{slope} = -\frac{(21.95 - 23.03)}{(300. - 0) \times 10^{-3}\,\text{s}} = 3.6\,\text{s}^{-1}$$

d.   From NO experiment, Rate $= k[NO][O_3] = k'[NO]$ where $k' = k[O_3]$.

$k' = 1.8\,\text{s}^{-1} = k(1.0 \times 10^{14}\ \text{molecules/cm}^3)$,  $k = 1.8 \times 10^{-14}\ \text{cm}^3\ \text{molecules}^{-1}\,\text{s}^{-1}$

We can check this from the ozone data.  Rate $= k''[O_3] = k[NO][O_3]$, where $k'' = k[NO]$.

$k'' = 3.6\,\text{s}^{-1} = k(2.0 \times 10^{14}\ \text{molecules/cm}^3)$,  $k = 1.8 \times 10^{-14}\ \text{cm}^3\ \text{molecules}^{-1}\,\text{s}^{-1}$

Both values of k agree.

39.    For a first-order reaction, the integrated rate law is  $\ln([A]/[A]_0) = -kt$.  Solving for k:

$$\ln\!\left(\frac{0.250\ \text{mol/L}}{1.00\ \text{mol/L}}\right) = -k \times 120.\ \text{s},\ \ k = 0.0116\,\text{s}^{-1}$$

$$\ln\!\left(\frac{0.350\ \text{mol/L}}{2.00\ \text{mol/L}}\right) = -0.0116\,\text{s}^{-1} \times t,\ \ t = 150.\ \text{s}$$

41.     Comparing experiments 1 and 2, as the concentration of AB is doubled, the initial rate increases by a factor of 4. The reaction is second order in AB.

Rate = $k[AB]^2$,  $3.20 \times 10^{-3}$ mol $L^{-1}s^{-1}$ = $k_1(0.200\ M)^2$

$k = 8.00 \times 10^{-2}$ L mol$^{-1}$s$^{-1}$ = $k_{mean}$

For a second-order reaction:

$$t_{1/2} = \frac{1}{k[AB]_0} = \frac{1}{8.00 \times 10^{-2}\ \text{L mol}^{-1}\text{s}^{-1} \times 1.00\ \text{mol}/\text{L}} = 12.5\ \text{s}$$

43.     a.   When a reaction is 75.0% complete (25.0% of reactant remains), this represents two half-lives (100% $\rightarrow$ 50% $\rightarrow$ 25%). The first-order half-life expression is $t_{1/2}$ = (ln 2)/k. Because there is no concentration dependence for a first-order half-life, 320. s = two half-lives, $t_{1/2}$ = 320./2 = 160. s. This is both the first half-life, the second half-life, etc.

        b.   $t_{1/2} = \dfrac{\ln 2}{k}$,  $k = \dfrac{\ln 2}{t_{1/2}} = \dfrac{\ln 2}{160.\ \text{s}} = 4.33 \times 10^{-3}\ \text{s}^{-1}$

             At 90.0% complete, 10.0% of the original amount of the reactant remains, so [A] = 0.100[A]$_0$.

             $\ln\left(\dfrac{[A]}{[A]_0}\right) = -kt$,  $\ln\dfrac{0.100[A]_0}{[A]_0} = -(4.33 \times 10^{-3}\ \text{s}^{-1})t$,  $t = \dfrac{\ln(0.100)}{-4.33 \times 10^{-3}\ \text{s}^{-1}} = 532\ \text{s}$

45.     a.   [A] = $-kt$ + [A]$_0$;  If k = $5.0 \times 10^{-2}$ mol L$^{-1}$ s$^{-1}$ and [A]$_0$ = $1.00 \times 10^{-3}\ M$, then:

             [A] = $-(5.0 \times 10^{-2}$ mol L$^{-1}$ s$^{-1})t$ + $1.00 \times 10^{-3}$ mol/L

        b.   $\dfrac{[A]_0}{2} = -(5.0 \times 10^{-2})t_{1/2}$ + [A]$_0$  because at $t = t_{1/2}$, [A] = [A]$_0$/2.

             $-0.50[A]_0 = -(5.0 \times 10^{-2})t_{1/2}$,  $t_{1/2} = \dfrac{0.50(1.00 \times 10^{-3})}{5.0 \times 10^{-2}} = 1.0 \times 10^{-2}\ \text{s}$

             *Note*:  We could have used the $t_{1/2}$ expression to solve ($t_{1/2} = \dfrac{[A]_0}{2k}$).

        c.   [A] = $-kt$ + [A]$_0$ = $-(5.0 \times 10^{-2}$ mol L$^{-1}$ s$^{-1})(5.0 \times 10^{-3}$ s) + $1.00 \times 10^{-3}$ mol/L

             [A] = $7.5 \times 10^{-4}$ mol/L

             [A]$_{reacted}$ = $1.00 \times 10^{-3}$ mol/L $- 7.5 \times 10^{-4}$ mol/L = $2.5 \times 10^{-4}$ mol/L

             [B]$_{produced}$ = [A]$_{reacted}$ = $2.5 \times 10^{-4}\ M$

47.    a.    Because $[A]_0 \ll [B]_0$ or $[C]_0$, the B and C concentrations remain constant at 1.00 $M$ for this experiment. Thus, rate = $k[A]^2[B][C] = k'[A]^2$ where $k' = k[B][C]$.

For this pseudo-second-order reaction:

$$\frac{1}{[A]} = k't + \frac{1}{[A]_0}, \quad \frac{1}{3.26 \times 10^{-5}\ M} = k'(3.00\ \text{min}) + \frac{1}{1.00 \times 10^{-4}\ M}$$

$$k' = 6890\ \text{L mol}^{-1}\ \text{min}^{-1} = 115\ \text{L mol}^{-1}\ \text{s}^{-1}$$

$$k' = k[B][C], \quad k = \frac{k'}{[B][C]}, \quad k = \frac{115\ \text{L mol}^{-1}\ \text{s}^{-1}}{(1.00\ M)(1.00\ M)} = 115\ \text{L}^3\ \text{mol}^{-3}\ \text{s}^{-1}$$

b.    For this pseudo-second-order reaction:

$$\text{Rate} = k'[A]^2, \quad t_{1/2} = \frac{1}{k'[A]_0} = \frac{1}{115\ \text{L mol}^{-1}\ \text{s}^{-1}(1.00 \times 10^{-4}\ \text{mol/L})} = 87.0\ \text{s}$$

c.    $\dfrac{1}{[A]} = k't + \dfrac{1}{[A]_0} = 115\ \text{L mol}^{-1}\ \text{s}^{-1} \times 600.\ \text{s} + \dfrac{1}{1.00 \times 10^{-4}\ \text{mol/L}} = 7.90 \times 10^4\ \text{L/mol}$

$[A] = 1/7.90 \times 10^4\ \text{L/mol} = 1.27 \times 10^{-5}\ \text{mol/L}$

From the stoichiometry in the balanced reaction, 1 mol of B reacts with every 3 mol of A.

Amount A reacted = $1.00 \times 10^{-4}\ M - 1.27 \times 10^{-5}\ M = 8.7 \times 10^{-5}\ M$

Amount B reacted = $8.7 \times 10^{-5}\ \text{mol/L} \times \dfrac{1\ \text{mol B}}{3\ \text{mol A}} = 2.9 \times 10^{-5}\ M$

$[B] = 1.00\ M - 2.9 \times 10^{-5}\ M = 1.00\ M$

As we mentioned in part a, the concentration of B (and C) remain constant because the A concentration is so small compared to the B (or C) concentration.

## Reaction Mechanisms

49.    In a unimolecular reaction, a single reactant molecule decomposes to products. In a bimolecular reaction, two molecules collide to give products. The probability of the simultaneous collision of three molecules with enough energy and the proper orientation is very small, making termolecular steps very unlikely.

51.    For elementary reactions, the rate law can be written using the coefficients in the balanced equation to determine the orders.

a.   Rate = k[CH₃NC]          b.   Rate = k[O₃][NO]

c.   Rate = k[O₃]             d.   Rate = k[O₃][O]

e.   Rate = $k\left[{}_{6}^{14}C\right]$ or Rate = kN, where N = the number of ${}_{6}^{14}C$ atoms (convention)

53.   A mechanism consists of a series of elementary reactions in which the rate law for each step can be determined using the coefficients in the balanced equations.  For a plausible mechanism, the rate law derived from a mechanism must agree with the rate law determined from experiment.  To derive the rate law from the mechanism, the rate of the reaction is assumed to equal the rate of the slowest step in the mechanism.

Because step 1 is the rate determining step, the rate law for this mechanism is Rate = k[C₄H₉Br].  To get the overall reaction, we sum all the individual steps of the mechanism.  Summing all steps gives:

$$C_4H_9Br \rightarrow C_4H_9^+ + Br^-$$
$$C_4H_9^+ + H_2O \rightarrow C_4H_9OH_2^+$$
$$C_4H_9OH_2^+ + H_2O \rightarrow C_4H_9OH + H_3O^+$$

$$\overline{C_4H_9Br + 2\ H_2O \rightarrow C_4H_9OH + Br^- + H_3O^+}$$

Intermediates in a mechanism are species that are neither reactants nor products but that are formed and consumed during the reaction sequence in the mechanism.  The intermediates for this mechanism are  C₄H₉⁺ and C₄H₉OH₂⁺.

55.   Let's determine the rate law for each mechanism.  If the rate law derived from the mechanism is the same as the experimental rate law, then the mechanism is possible (assuming the sum of all the steps in the mechanism gives the overall balanced equation).  When deriving rate laws from a mechanism, we must substitute for all intermediate concentrations.

a.   Rate = k₁[NO][O₂];   <u>not possible</u>

b.   Rate = k₂[NO₃][NO] and k₁[NO][O₂] = k₋₁[NO₃] or $[NO_3] = \dfrac{k_1}{k_{-1}}[NO][O_2]$

Rate = $\dfrac{k_2 k_1}{k_{-1}}[NO]^2[O_2]$;   <u>possible</u>

c.   Rate = k₁[NO]²;   <u>not possible</u>

d.   Rate = k₂[N₂O₂] and $[N_2O_2] = \dfrac{k_1}{k_{-1}}[NO]^2$

Rate = $\dfrac{k_2 k_1}{k_{-1}}[NO]^2$;   <u>not possible</u>

Only the mechanism in b is consistent, so only mechanism b is a possible mechanism for this reaction.

57.    Rate $= k_3[Br^-][H_2BrO_3^+]$; we must substitute for the intermediate concentration. Because steps 1 and 2 are fast-equilibrium steps, rate forward reaction = rate reverse reaction.

$$k_2[HBrO_3][H^+] = k_{-2}[H_2BrO_3^+]; \quad k_1[BrO_3^-][H^+] = k_{-1}[HBrO_3]$$

$$[HBrO_3] = \frac{k_1}{k_{-1}}[BrO_3^-][H^+]; \quad [H_2BrO_3^+] = \frac{k_2}{k_{-2}}[HBrO_3][H^+] = \frac{k_2 k_1}{k_{-2} k_{-1}}[BrO_3^-][H^+]^2$$

$$Rate = \frac{k_3 k_2 k_1}{k_{-2} k_{-1}}[Br^-][BrO_3^-][H^+]^2 = k[Br^-][BrO_3^-][H^+]^2$$

59.    Rate $= k_2[I^-][HOCl]$; from the fast-equilibrium first step:

$$k_1[OCl^-] = k_{-1}[HOCl][OH^-], \quad [HOCl] = \frac{k_1[OCl^-]}{k_{-1}[OH^-]}; \quad \text{substituting into the rate equation:}$$

$$Rate = \frac{k_2 k_1 [I^-][OCl^-]}{k_{-1}[OH^-]} = \frac{k[I^-][OCl^-]}{[OH^-]}$$

61.   a.   $MoCl_5^-$

      b.   Rate $= \dfrac{d[NO_2^-]}{dt} = k_2[NO_3^-][MoCl_5^-]$    (Only the last step contains $NO_2^-$.)

We use the steady-state assumption to substitute for the intermediate concentration in the rate law. The steady-state approximation assumes that the concentration of an intermediate remains constant; i.e., d[intermediate]/dt = 0. To apply the steady-state assumption, we write rate laws for all steps where the intermediate is produced and equate the sum of these rate laws to the sum of the rate laws where the intermediate is consumed. Applying the steady-state approximation to $MoCl_5^-$:

$$\frac{d[MoCl_5^-]}{dt} = 0, \text{ so } k_1[MoCl_6^{2-}] = k_{-1}[MoCl_5^-][Cl^-] + k_2[NO_3^-][MoCl_5^-]$$

$$[MoCl_5^-] = \frac{k_1[MoCl_6^{2-}]}{k_{-1}[Cl^-] + k_2[NO_3^-]}; \quad Rate = \frac{d[NO_2^-]}{dt} = \frac{k_1 k_2 [NO_3^-][MoCl_6^{2-}]}{k_{-1}[Cl^-] + k_2[NO_3^-]}$$

63.   a.   Rate $= \dfrac{d[E]}{dt} = k_2[B^*]$; assume $\dfrac{d[B^*]}{dt} = 0$, then $k_1[B]^2 = k_{-1}[B][B^*] + k_2[B^*]$.

$$[B^*] = \frac{k_1[B]^2}{k_{-1}[B] + k_2}; \text{ the rate law is: } Rate = \frac{d[E]}{dt} = \frac{k_1 k_2 [B]^2}{k_{-1}[B] + k_2}$$

b. When $k_2 \ll k_{-1}[B]$, then Rate $= \dfrac{d[E]}{dt} = \dfrac{k_1 k_2 [B]^2}{k_{-1}[B]} = \dfrac{k_1 k_2}{k_{-1}}[B]$.

The reaction is first order when the rate of the second step is very slow (when $k_2$ is very small).

c. Collisions between B molecules only transfer energy from one B to another. This occurs at a much faster rate than the decomposition of an energetic B molecule (B*).

## Temperature Dependence of Rate Constants and the Collision Model

65.    Two reasons are:

1) The collision must involve enough energy to produce the reaction; i.e., the collision energy must be equal to or exceed the activation energy.

2) The relative orientation of the reactants when they collide must allow formation of any new bonds necessary to produce products.

67.    $k = A \exp(-E_a/RT)$ or $\ln k = \dfrac{-E_a}{RT} + \ln A$ (the Arrhenius equation)

For two conditions:   $\ln\left(\dfrac{k_2}{k_1}\right) = \dfrac{E_a}{R}\left(\dfrac{1}{T_1} - \dfrac{1}{T_2}\right)$   (Assuming A is temperature independent.)

Let $k_1 = 3.52 \times 10^{-7}$ L mol$^{-1}$s$^{-1}$,  $T_1 = 555$ K;  $k_2 = ?$, $T_2 = 645$ K;  $E_a = 186 \times 10^3$ J/mol

$\ln\left(\dfrac{k_2}{3.52 \times 10^{-7}}\right) = \dfrac{1.86 \times 10^5 \text{ J/mol}}{8.3145 \text{ J K}^{-1}\text{ mol}^{-1}}\left(\dfrac{1}{555 \text{ K}} - \dfrac{1}{645 \text{ K}}\right) = 5.6$

$\dfrac{k_2}{3.52 \times 10^{-7}} = e^{5.6} = 270$,  $k_2 = 270(3.52 \times 10^{-7}) = 9.5 \times 10^{-5}$ L mol$^{-1}$ s$^{-1}$

69.    $\ln\left(\dfrac{k_2}{k_1}\right) = \dfrac{E_a}{R}\left(\dfrac{1}{T_1} - \dfrac{1}{T_2}\right)$;   $\dfrac{k_2}{k_1} = 7.00$, $T_1 = 295$ K, $E_a = 54.0 \times 10^3$ J/mol

$\ln(7.00) = \dfrac{54.0 \times 10^3 \text{ J/mol}}{8.3145 \text{ J K}^{-1}\text{ mol}^{-1}}\left(\dfrac{1}{295 \text{ K}} - \dfrac{1}{T_2}\right)$,   $\dfrac{1}{295 \text{ K}} - \dfrac{1}{T_2} = 3.00 \times 10^{-4}$

$\dfrac{1}{T_2} = 3.09 \times 10^{-3}$,  $T_2 = 324$ K $= 51°C$

71.    From the Arrhenius equation in logarithmic form ($\ln k = -E_a/RT + \ln A$), a graph of ln k versus. 1/T should yield a straight line with a slope equal to $-E_a/R$ and a $y$ intercept equal to ln A.

   a.    Slope = $-E_a/R$, $E_a = 1.10 \times 10^4 \text{ K} \times \dfrac{8.3145 \text{ J}}{\text{K mol}} = 9.15 \times 10^4 \text{ J/mol} = 91.5 \text{ kJ/mol}$

   b.    The units for A are the same as the units for k ($s^{-1}$).

      $y$ intercept = ln A, $A = e^{33.5} = 3.54 \times 10^{14} \text{ s}^{-1}$

   c.    $\ln k = -E_a/RT + \ln A$  or  $k = A \exp(-E_a/RT)$

      $k = 3.54 \times 10^{14} \text{ s}^{-1} \times \exp\left(\dfrac{-9.15 \times 10^{-4} \text{ J/mol}}{8.3145 \text{ J K}^{-1} \text{mol}^{-1} \times 298 \text{ K}}\right) = 3.24 \times 10^{-2} \text{ s}^{-1}$

73.    The Arrhenius equation is $k = A\exp(-E_a/RT)$ or, in logarithmic form, $\ln k = -E_a/RT + \ln A$. Hence a graph of ln k versus 1/T should yield a straight line with a slope equal to $-E_a/R$ since the logarithmic form of the Arrhenius equation is in the form of a straight-line equation, $y = mx + b$.

   *Note*:  We carried one extra significant figure in the following ln k values in order to reduce round-off error.

| T (K) | 1/T ($K^{-1}$) | k (L $mol^{-1}$ $s^{-1}$) | ln k |
|---|---|---|---|
| 195 | $5.13 \times 10^{-3}$ | $1.08 \times 10^9$ | 20.80 |
| 230. | $4.35 \times 10^{-3}$ | $2.95 \times 10^9$ | 21.81 |
| 260. | $3.85 \times 10^{-3}$ | $5.42 \times 10^9$ | 22.41 |
| 298 | $3.36 \times 10^{-3}$ | $12.0 \times 10^9$ | 23.21 |
| 369 | $2.71 \times 10^{-3}$ | $35.5 \times 10^9$ | 24.29 |

Using a couple of points on the plot:

$$\text{slope} = \frac{20.95 - 23.65}{5.00 \times 10^{-3} - 3.00 \times 10^{-3}} = \frac{-2.70}{2.00 \times 10^{-3}} = -1.35 \times 10^3 \text{ K} = \frac{-E_a}{R}$$

$E_a = 1.35 \times 10^3 \text{ K} \times 8.3145 \text{ J K}^{-1} \text{ mol}^{-1} = 1.12 \times 10^4 \text{ J/mol} = 11.2 \text{ kJ/mol}$

From the best straight line (by calculator):  slope $= -1.43 \times 10^3$ K and $E_a = 11.9$ kJ/mol

75.     In the following reaction profiles R = reactants, P = products, $E_a$ = activation energy, $\Delta E$ = overall energy change for the reaction, and RC = reaction coordinate, which is the same as reaction progress.

The second reaction profile represents a two-step reaction since an intermediate plateau appears between the reactants and the products.  This plateau (see I in plot) represents the energy of the intermediate.  The general reaction mechanism for this reaction is:

$$\begin{aligned} R &\rightarrow I \\ \underline{I \rightarrow P} \\ R &\rightarrow P \end{aligned}$$

In a mechanism, the rate of the slowest step determines the rate of the reaction.  The activation energy for the slowest step will be the largest energy barrier that the reaction must overcome.  Since the second hump in the diagram is at the highest energy,  the second step has the largest activation energy and will be the rate determining step (the slow step).

77.     When $\Delta E$ is positive (the products have higher energy than the reactants, as represented in the energy profile for Exercise 15.75), then $E_{a, \text{ forward}} > E_{a, \text{ reverse}}$.  Therefore, this reaction has a positive $\Delta E$ value.

## Catalysis

79.     A metal-catalyzed reaction can be dependent on the number of adsorption sites on the metal surface. Once the metal surface is saturated with reactant, the rate of reaction becomes independent of concentration.

81.    a.    W because it has a lower activation energy than the Os catalyst.

       b.    $k_w = A_w \exp[-E_a(W)/RT]$;   $k_{uncat} = A_{uncat} \exp[-E_a(uncat)/RT]$;   assume $A_w = A_{uncat}$.

$$\frac{k_w}{k_{uncat}} = \exp\left(\frac{-E_a(W)}{RT} + \frac{E_a(uncat)}{RT}\right)$$

$$\frac{k_w}{k_{uncat}} = \exp\left(\frac{-163,000\ J/mol + 335,000\ J/mol}{8.3145\ J\ K^{-1}\ mol^{-1} \times 298\ K}\right) = 1.41 \times 10^{30}$$

The W-catalyzed reaction is approximately $10^{30}$ times faster than the uncatalyzed reaction.

       c.    Because $[H_2]$ is in the denominator of the rate law, the presence of $H_2$ decreases the rate of the reaction.  For the decomposition to occur, $NH_3$ molecules must be adsorbed on the surface of the catalyst.  If $H_2$ is also adsorbed on the catalyst surface, then there are fewer sites for $NH_3$ molecules to be adsorbed, and the rate decreases.

83.    The mechanism for the chlorine catalyzed destruction of ozone is:

$$O_3 + Cl \rightarrow O_2 + ClO \quad \text{(slow)}$$
$$ClO + O \rightarrow O_2 + Cl \quad \text{(fast)}$$
$$\overline{\phantom{O_3 + O \rightarrow 2\ O_2}}$$
$$O_3 + O \rightarrow 2\ O_2$$

Because the chlorine atom-catalyzed reaction has a lower activation energy, the Cl-catalyzed rate is faster.  Hence Cl is a more effective catalyst.  Using the activation energy, we can estimate the efficiency that Cl atoms destroy ozone compared to NO molecules (see Exercise 15.82c).

At 25°C: $\dfrac{k_{Cl}}{k_{NO}} = \exp\left[\dfrac{-E_a(Cl)}{RT} + \dfrac{E_a(NO)}{RT}\right] = \exp\left[\dfrac{(-2100 + 11,900)\ J/mol}{(8.3145 \times 298)\ J/mol}\right] = e^{3.96} = 52$

At 25°C, the Cl catalyzed reaction is roughly 52 times faster than the NO-catalyzed reaction, assuming the frequency factor A is the same for each reaction.

85.    At high [S], the enzyme is completely saturated with substrate.  Once the enzyme is completely saturated, the rate of decomposition of ES can no longer increase, and the overall rate remains constant.

87.    $\text{Rate} = \dfrac{-d[A]}{dt} = k[A]^x$

Assuming the catalyzed and uncatalyzed reaction have the same form and orders and because concentrations are assumed equal, rate $\propto 1/\Delta t$, where $\Delta t = \Delta\text{time}$.

$$\frac{\text{Rate}_{cat}}{\text{Rate}_{un}} = \frac{\Delta t_{un}}{\Delta t_{cat}} = \frac{2400\ \text{yr}}{\Delta t_{cat}}\ \text{and}\ \frac{\text{rate}_{cat}}{\text{rate}_{un}} = \frac{k_{cat}}{k_{un}}$$

$$\frac{\text{Rate}_{cat}}{\text{Rate}_{un}} = \frac{k_{cat}}{k_{un}} = \frac{A\exp[-E_a(cat)/RT]}{A\exp[-E_a(un)/RT]} = \exp\left[\frac{-E_a(cat)+E_a(un)}{RT}\right]$$

$$\frac{k_{cat}}{k_{un}} = \exp\left(\frac{-5.90\times10^4\ \text{J/mol}+1.84\times10^5\ \text{J/mol}}{8.3145\ \text{J K}^{-1}\text{mol}^{-1}\times600.\ \text{K}}\right) = 7.62\times10^{10}$$

$$\frac{\Delta t_{un}}{\Delta t_{cat}} = \frac{\text{rate}_{cat}}{\text{rate}_{un}} = \frac{k_{cat}}{k_{un}},\quad \frac{2400\ \text{yr}}{\Delta t_{cat}} = 7.62\times10^{10},\quad \Delta t_{cat} = 3.15\times10^{-8}\ \text{yr} \approx 1\ \text{s}$$

## Additional Exercises

89.    The most common method to experimentally determine the differential rate law is the method of initial rates. Once the differential rate law is determined experimentally, the integrated rate law can be derived. However, sometimes it is more convenient and more accurate to collect concentration versus time data for a reactant. When this is the case, then we do "proof" plots to determine the integrated rate law. Once the integrated rate law is determined, the differential rate law can be determined. Either experimental procedure allows determination of both the integrated and the differential rate law; and which rate law is determined by experiment and which is derived is usually decided by which data are easiest and most accurately collected.

91.

| Heating Time | Untreated | | Deacidifying | | Antioxidant | |
|---|---|---|---|---|---|---|
| (days) | s | ln s | s | ln s | s | ln s |
| 0.00 | 100.0 | 4.605 | 100.1 | 4.606 | 114.6 | 4.741 |
| 1.00 | 67.9 | 4.218 | 60.8 | 4.108 | 65.2 | 4.177 |
| 2.00 | 38.9 | 3.661 | 26.8 | 3.288 | 28.1 | 3.336 |
| 3.00 | 16.1 | 2.779 | – | – | 11.3 | 2.425 |
| 6.00 | 6.8 | 1.92 | – | – | – | – |

a.   We used a calculator to fit the data by least squares.  The results follow.

Untreated:  ln s = –(0.465)t + 4.55,  k = 0.465 d⁻¹

Deacidifying agent:  ln s = –(0.659)t + 4.66,  k = 0.659 d⁻¹

Antioxidant:  ln s = –(0.779)t + 4.84,  k = 0.779 d⁻¹

b.   No, the silk degrades more rapidly with the additives since k increases.

c.   $t_{1/2}$ = (ln 2)/k;  untreated:  $t_{1/2}$ = 1.49 day;  deacidifying agent:  $t_{1/2}$ = 1.05 day;  antioxidant:  $t_{1/2}$ = 0.890 day.

93.   Carbon cannot form the fifth bond necessary for the transition state because of the small atomic size of carbon and because carbon doesn't have low-energy d orbitals available to expand the octet.

95.   a.

| t (s) | $[C_4H_6]$ (M) | $\ln[C_4H_6]$ | $1/[C_4H_6]$ ($M^{-1}$) |
|---|---|---|---|
| 0 | 0.01000 | –4.6052 | $1.000 \times 10^2$ |
| 1000. | 0.00629 | –5.069 | $1.59 \times 10^2$ |
| 2000. | 0.00459 | –5.384 | $2.18 \times 10^2$ |
| 3000. | 0.00361 | –5.624 | $2.77 \times 10^2$ |

The plot of $1/[C_4H_6]$ versus t is linear, thus the reaction is second order in butadiene.  From the plot (not included), the integrated rate law is:

$$\frac{1}{[C_4H_6]} = (5.90 \times 10^{-2} \text{ L mol}^{-1} \text{ s}^{-1})t + 100.0 \; M^{-1}$$

b.   When dimerization is 1.0% complete, 99.0% of $C_4H_6$ is left.

$$[C_4H_6] = 0.990(0.01000) = 0.00990 \ M; \quad \frac{1}{0.00990} = (5.90 \times 10^{-2})t + 100.0$$

$$t = 17.1 \ s \approx 20 \ s$$

c.   2.0% complete, $[C_4H_6] = 0.00980 \ M; \quad \dfrac{1}{0.00980} = (5.90 \times 10^{-2})t + 100.0,$

$$t = 34.6 \ s \approx 30 \ s$$

d.   $\dfrac{1}{[C_4H_6]} = kt + \dfrac{1}{[C_4H_6]_0}; \quad [C_4H_6]_0 = 0.0200 \ M; \ \text{at} \ t = t_{1/2}, \ [C_4H_6] = 0.0100 \ M.$

$$\frac{1}{0.0100} = (5.90 \times 10^{-2})t_{1/2} + \frac{1}{0.0200}, \quad t_{1/2} = 847 \ s = 850 \ s$$

Or:  $t_{1/2} = \dfrac{1}{k[A]_0} = \dfrac{1}{(5.90 \times 10^{-2} \ \text{L mol}^{-1} \ \text{s}^{-1})(2.00 \times 10^{-2} \ M)} = 847 \ s$

e.   From Exercise 15.32, $k = 1.4 \times 10^{-2} \ \text{L mol}^{-1} \ \text{s}^{-1}$ at 500. K.  From this problem, $k = 5.90 \times 10^{-2} \ \text{L mol}^{-1} \ \text{s}^{-1}$ at 620. K.

$$\ln\left(\frac{k_2}{k_1}\right) = \frac{E_a}{R}\left(\frac{1}{T_1} - \frac{1}{T_2}\right), \quad \ln\left(\frac{5.90 \times 10^{-2}}{1.4 \times 10^{-2}}\right) = \frac{E_a}{8.3145 \ \text{J K}^{-1} \ \text{mol}^{-1}}\left(\frac{1}{500. \ \text{K}} - \frac{1}{620. \ \text{K}}\right)$$

$$12 = E_a(3.9 \times 10^{-4}), \ E_a = 3.1 \times 10^4 \ \text{J/mol} = 31 \ \text{kJ/mol}$$

97.   $k = A \exp(-E_a/RT); \quad \dfrac{k_{cat}}{k_{uncat}} = \dfrac{A_{cat} \exp(-E_{a,\,cat}/RT)}{A_{uncat} \exp(-E_{a,\,uncat}/RT)} = \exp\left(\dfrac{-E_{a,\,cat} + E_{a,\,uncat}}{RT}\right)$

$$2.50 \times 10^3 = \frac{k_{cat}}{k_{uncat}} = \exp\left(\frac{-E_{a,\,cat} + 5.00 \times 10^4 \ \text{J/mol}}{8.3145 \ \text{J K}^{-1} \ \text{mol}^{-1} \times 310. \ \text{K}}\right)$$

$$\ln(2.50 \times 10^3) \ \times \ 2.58 \times 10^3 \ \text{J/mol} = -E_{a,\,cat} + 5.00 \times 10^4 \ \text{J/mol}$$

$$E_{a,\,cat} = 5.00 \times 10^4 \ \text{J/mol} - 2.02 \times 10^4 \ \text{J/mol} = 2.98 \times 10^4 \ \text{J/mol} = 29.8 \ \text{kJ/mol}$$

99.   a.

| T (K) | 1/T (K$^{-1}$) | k (min$^{-1}$) | ln k |
|-------|---------------|----------------|-------|
| 298.2 | $3.353 \times 10^{-3}$ | 178 | 5.182 |
| 293.5 | $3.407 \times 10^{-3}$ | 126 | 4.836 |
| 290.5 | $3.442 \times 10^{-3}$ | 100. | 4.605 |

A plot of ln k versus 1/T gives a straight line (plot not included).  The equation for the straight line is:

$$\ln k = -6.48 \times 10^3 (1/T) + 26.9$$

For the ln k versus 1/T plot, slope = $-E_a/R = -6.48 \times 10^3$ K.

$-6.48 \times 10^3$ K $= -E_a/8.3145$ J K$^{-1}$ mol$^{-1}$, $E_a = 5.39 \times 10^4$ J/mol $= 53.9$ kJ/mol

b.   ln k $= -6.48 \times 10^3 (1/288.2) + 26.9 = 4.42$, k $= e^{4.42} = 83$ min$^{-1}$

About 83 chirps per minute per insect.  Note:  We carried extra significant figures.

c.   k gives the number of chirps per minute.  The number or chirps in 15 s is k/4.

| T (°C) | T (°F) | k (min$^{-1}$) | 42 + 0.80(k/4) |
|--------|--------|----------------|----------------|
| 25.0 | 77.0 | 178 | 78° F |
| 20.3 | 68.5 | 126 | 67°F |
| 17.3 | 63.1 | 100. | 62°F |
| 15.0 | 59.0 | 83 | 59°F |

The rule of thumb appears to be fairly accurate, almost ±1°F.

101.   Rate $= \dfrac{d[Cl_2]}{dt} = k_2[NO_2Cl][Cl]$;  Assume $\dfrac{d[Cl]}{dt} = 0$, then:

$k_1[NO_2Cl] = k_{-1}[NO_2][Cl] + k_2[NO_2Cl][Cl]$,  $[Cl] = \dfrac{k_1[NO_2Cl]}{k_{-1}[NO_2] + k_2[NO_2Cl]}$

Rate $= \dfrac{d[Cl_2]}{dt} = \dfrac{k_1 k_2 [NO_2Cl]^2}{k_{-1}[NO_2] + k_2[NO_2Cl]}$

## Challenge Problems

103.   a.   Rate $= k[CH_3X]^x[Y]^y$;  for experiment 1, [Y] is in large excess, so its concentration will be constant.  Rate $= k'[CH_3X]^x$, where $k' = k(3.0 \, M)^y$.

A plot (not included) of ln[CH$_3$X] versus t is linear ($x = 1$).  The integrated rate law is:

ln[CH$_3$X] $= -(0.93)t - 3.99$;  $k' = 0.93$ h$^{-1}$

For experiment 2, [Y] is again constant, with Rate $= k''[CH_3X]^x$, where $k'' = k(4.5 \, M)^y$. The natural log plot is linear again with an integrated rate law:

ln[CH$_3$X] $= -(0.93)t - 5.40$;  $k'' = 0.93$ h$^{-1}$

Dividing the rate-constant values:  $\dfrac{k'}{k''} = \dfrac{0.93}{0.93} = \dfrac{k(3.0)^y}{k(4.5)^y}$,  $1.0 = (0.67)^y$, $y = 0$

Reaction is first order in $CH_3X$ and zero order in Y.  The overall rate law is:

Rate = $k[CH_3X]$, where $k = 0.93$ $h^{-1}$ at 25°C

b.   $t_{1/2} = (\ln 2)/k = 0.6931/(7.88 \times 10^8$ $h^{-1}) = 8.80 \times 10^{-10}$ hour

c.   $\ln\left(\dfrac{k_2}{k_1}\right) = \dfrac{E_a}{R}\left(\dfrac{1}{T_1} - \dfrac{1}{T_2}\right),$  $\ln\left(\dfrac{7.88 \times 10^8}{0.93}\right) = \dfrac{E_a}{8.3145\ J\ K^{-1}\ mol^{-1}}\left(\dfrac{1}{298\ K} - \dfrac{1}{358\ K}\right)$

$E_a = 3.0 \times 10^5$ J/mol = $3.0 \times 10^2$ kJ/mol

d.   From part a, the reaction is first order in $CH_3X$ and zero order in Y.  From part c, the activation energy is close to the C-X bond energy.  A plausible mechanism that explains the results in parts a and c is:

$CH_3X \rightarrow CH_3 + X$     (slow)

$CH_3 + Y \rightarrow CH_3Y$     (fast)

*Note*:  This is a possible mechanism because the derived rate law is the same as the experimental rate law (and the sum of the steps gives the overall balanced equation).

105.   $\dfrac{-d[A]}{dt} = k[A]^3,$  $\displaystyle\int_{[A]_0}^{[A]_t} \dfrac{d[A]}{[A]^3} = -\int_0^t k\ dt$

$\displaystyle\int x^n\ dx = \dfrac{x^{n+1}}{n+1};$  so:  $-\dfrac{1}{2[A]^2}\Bigg|_{[A]_0}^{[A]_t} = -kt,$  $-\dfrac{1}{2[A]_t^2} + \dfrac{1}{2[A]_0^2} = -kt$

For the half-life equation, $[A]_t = 1/2[A]_0$:

$-\dfrac{1}{2\left(\dfrac{1}{2}[A]_0\right)^2} + \dfrac{1}{2[A]_0^2} = -kt_{1/2},$  $-\dfrac{4}{2[A]_0^2} + \dfrac{1}{2[A]_0^2} = -kt_{1/2}$

$-\dfrac{3}{2[A]_0^2} = -kt_{1/2},$  $t_{1/2} = \dfrac{3}{2[A]_0^2\ k}$

The first half-life is $t_{1/2} = 40.$ s and corresponds to going from $[A]_0$ to $1/2[A]_0$.  The second half-life corresponds to going from $1/2\ [A]_0$ to $1/4\ [A]_0$.

First half-life = $\dfrac{3}{2[A]_0^2\ k};$  second half-life = $\dfrac{3}{2\left(\dfrac{1}{2}[A]_0\right)^2\ k} = \dfrac{6}{[A]_0^2\ k}$

$$\frac{\text{First half} - \text{life}}{\text{Second half} - \text{life}} = \frac{\dfrac{3}{2[A]_0^2\, k}}{\dfrac{6}{[A]_0^2\, k}} = 3/12 = 1/4$$

Because the first half-life is 40. s, the second half-life will be one-fourth of this, or 10. s.

107.    Rate = $k[A]^x[B]^y[C]^z$; during the course of experiment 1, [A] and [C] are essentially constant, and Rate = $k'[B]^y$, where $k' = k[A]_0^x[C]_0^z$.

| [B] ($M$) | time (s) | ln[B] | 1/[B] ($M^{-1}$) |
|---|---|---|---|
| $1.0 \times 10^{-3}$ | 0 | $-6.91$ | $1.0 \times 10^3$ |
| $2.7 \times 10^{-4}$ | $1.0 \times 10^5$ | $-8.22$ | $3.7 \times 10^3$ |
| $1.6 \times 10^{-4}$ | $2.0 \times 10^5$ | $-8.74$ | $6.3 \times 10^3$ |
| $1.1 \times 10^{-4}$ | $3.0 \times 10^5$ | $-9.12$ | $9.1 \times 10^3$ |
| $8.5 \times 10^{-5}$ | $4.0 \times 10^5$ | $-9.37$ | $12 \times 10^3$ |
| $6.9 \times 10^{-5}$ | $5.0 \times 10^5$ | $-9.58$ | $14 \times 10^3$ |
| $5.8 \times 10^{-5}$ | $6.0 \times 10^5$ | $-9.76$ | $17 \times 10^3$ |

A plot of 1/[B] versus t is linear (plot not included).  The reaction is second order in B, and the integrated rate equation is:

$$1/[B] = (2.7 \times 10^{-2}\ \text{L mol}^{-1}\ \text{s}^{-1})t + 1.0 \times 10^3\ M^{-1};\ k' = 2.7 \times 10^{-2}\ \text{L mol}^{-1}\ \text{s}^{-1}$$

For experiment 2, [B] and [C] are essentially constant, and Rate = $k''[A]^x$, where $k'' = k[B]_0^y[C]_0^z = k[B]_0^2[C]_0^z$.

| [A] ($M$) | time (s) | ln[A] | 1/[A] ($M^{-1}$) |
|---|---|---|---|
| $1.0 \times 10^{-2}$ | 0 | $-4.61$ | $1.0 \times 10^2$ |
| $8.9 \times 10^{-3}$ | 1.0 | $-4.72$ | 110 |
| $7.1 \times 10^{-3}$ | 3.0 | $-4.95$ | 140 |
| $5.5 \times 10^{-3}$ | 5.0 | $-5.20$ | 180 |
| $3.8 \times 10^{-3}$ | 8.0 | $-5.57$ | 260 |
| $2.9 \times 10^{-3}$ | 10.0 | $-5.84$ | 340 |
| $2.0 \times 10^{-3}$ | 13.0 | $-6.21$ | $5.0 \times 10^2$ |

A plot of ln[A] versus t is linear.  The reaction is first order in A, and the integrated rate law is:

$$\ln[A] = -(0.123\ \text{s}^{-1})t - 4.61;\ k'' = 0.123\ \text{s}^{-1}$$

*Note*: We will carry an extra significant figure in $k''$.

Experiment 3:   [A] and [B] are constant;  Rate = $k'''[C]^z$

The plot of [C] versus t is linear.  Thus $z = 0$.

The overall rate law is Rate = $k[A][B]^2$.

From experiment 1 (to determine k):

$$k' = 2.7 \times 10^{-2} \text{ L mol}^{-1} \text{ s}^{-1} = k[A]_0^x[C]_0^z = k[A]_o = k(2.0 \text{ } M), \ k = 1.4 \times 10^{-2} \text{ L}^2 \text{ mol}^{-2} \text{ s}^{-1}$$

From experiment 2:  $k'' = 0.123 \text{ s}^{-1} = k[B]_0^2, \quad k = \dfrac{0.123 \text{ s}^{-1}}{(3.0 \text{ } M)^2} = 1.4 \times 10^{-2} \text{ L}^2 \text{ mol}^{-2} \text{ s}^{-1}$

Thus Rate = $k[A][B]^2$ and $k = 1.4 \times 10^{-2} \text{ L}^2 \text{ mol}^{-2} \text{ s}^{-1}$.

109.    Rate = $\dfrac{-d[N_2O_5]}{dt} = k_1[M][N_2O_5] - k_{-1}[NO_3][NO_2][M]$

Assume $d[NO_3]/dt = 0$, so $k_1[N_2O_5][M] = k_{-1}[NO_3][NO_2][M] + k_2[NO_3][NO_2] +$
$$k_3[NO_3][NO].$$

$$[NO_3] = \frac{k_1[N_2O_5][M]}{k_{-1}[NO_2][M] + k_2[NO_2] + k_3[NO]}$$

Assume $\dfrac{d[NO]}{dt} = 0$, so $k_2[NO_3][NO_2] = k_3[NO_3][NO]$,  $[NO] = \dfrac{k_2}{k_3}[NO_2]$.

Substituting:  $[NO_3] = \dfrac{k_1[N_2O_5][M]}{k_{-1}[NO_2][M] + k_2[NO_2] + \dfrac{k_2 k_3}{k_3}[NO_2]} = \dfrac{k_1[N_2O_5][M]}{[NO_2](k_{-1}[M] + 2k_2)}$

Solving for the rate law:

$$\text{Rate} = \frac{-d[N_2O_5]}{dt} = k_1[N_2O_5][M] - \frac{k_{-1}k_1[NO_2][N_2O_5][M]^2}{[NO_2](k_{-1}[M] + 2k_2)} = k_1[N_2O_5][M]$$
$$- \frac{k_{-1}k_1[M]^2[N_2O_5]}{k_{-1}[M] + 2k_2}$$

$$\text{Rate} = \frac{-d[N_2O_5]}{dt} = \left(k_1 - \frac{k_{-1}k_1[M]}{k_{-1}[M] + 2k_2}\right)[N_2O_5][M]; \text{ simplifying:}$$

$$\text{Rate} = \frac{-d[N_2O_5]}{dt} = \frac{2k_1k_2[M][N_2O_5]}{k_{-1}[M] + 2k_2}$$

111.    a.    Rate $= (k_1 + k_2[H^+])[I^-]^m[H_2O_2]^n$

In all the experiments the concentration of $H_2O_2$ is small compared to the concentrations of $I^-$ and $H^+$. Therefore, the concentrations of $I^-$ and $H^+$ are effectively constant, and the rate law reduces to:

Rate $= k_{obs}[H_2O_2]^n$, where $k_{obs} = (k_1 + k_2[H^+])[I^-]^m$

Because all plots of $\ln[H_2O_2]$ versus time are linear, the reaction is first order with respect to $H_2O_2$ ($n = 1$). The slopes of the $\ln[H_2O_2]$ versus time plots equal $-k_{obs}$, which equals $-(k_1 + k_2[H^+])[I^-]^m$. To determine the order of $I^-$, compare the slopes of two experiments in which $I^-$ changes and $H^+$ is constant. Comparing the first two experiments:

$$\frac{\text{slope (exp. 2)}}{\text{slope (exp. 1)}} = \frac{-0.360}{-0.120} = \frac{-[k_1 + k_2(0.0400\ M)](0.3000\ M)^m}{-[k_1 + k_2(0.0400\ M)](0.1000\ M)^m}$$

$$3.00 = \left(\frac{0.3000\ M}{0.1000\ M}\right)^m = (3.000)^m, \quad m = 1$$

The reaction is also first order with respect to $I^-$.

b.    The slope equation has two unknowns, $k_1$ and $k_2$. To solve for $k_1$ and $k_2$, we must have two equations. We need to take one of the first set of three experiments and one of the second set of three experiments to generate the two equations in $k_1$ and $k_2$.

Experiment 1:  slope $= -(k_1 + k_2[H^+])[I^-]$

$-0.120\ \text{min}^{-1} = -[k_1 + k_2(0.0400\ M)](0.1000\ M)$ or $1.20 = k_1 + k_2(0.0400)$

Experiment 4:

$-0.0760\ \text{min}^{-1} = -[k_1 + k_2(0.0200\ M)](0.0750\ M)$ or $1.01 = k_1 + k_2(0.0200)$

Subtracting 4 from 1:

$$\begin{array}{r} 1.20 = k_1 + k_2(0.0400) \\ -1.01 = -k_1 - k_2(0.0200) \\ \hline 0.19 = k_2(0.0200), \quad k_2 = 9.5\ \text{L}^2\ \text{mol}^{-2}\ \text{min}^{-1} \end{array}$$

$1.20 = k_1 + 9.5(0.0400)$, $k_1 = 0.82\ \text{L mol}^{-1}\ \text{min}^{-1}$

c.    There are two pathways, one involving $H^+$ with rate $= k_2[H^+][I^-][H_2O_2]$ and another not involving $H^+$ with rate $= k_1[I^-][H_2O_2]$. The overall rate of reaction depends on which of these two pathways dominates, and this depends on the $H^+$ concentration.

# CHAPTER 16

# LIQUIDS AND SOLIDS

## Intermolecular Forces and Physical Properties

11.     Intermolecular forces are the relatively weak forces between molecules that hold the molecules together in the solid and liquid phases. Intramolecular forces are the forces within a molecule. These are the covalent bonds in a molecule. Intramolecular forces (covalent bonds) are much stronger than intermolecular forces.

Dipole forces are the forces that act between polar molecules. The electrostatic attraction between the partial positive end of one polar molecule and the partial negative end of another is the dipole force. Dipole forces are generally weaker than hydrogen bonding. Both of these forces are due to dipole moments in molecules. Hydrogen bonding is given a separate name from dipole forces because hydrogen bonding is a particularly strong dipole force. Any neutral molecule that has a hydrogen covalently bonded to N, O, or F exhibits the relatively strong hydrogen bonding intermolecular forces.

London dispersion forces are accidental-induced dipole forces. Like dipole forces, London dispersion forces are electrostatic in nature. Dipole forces are the electrostatic forces between molecules having a permanent dipole. London dispersion forces are the electrostatic forces between molecules having an accidental or induced dipole. All covalent molecules (polar and nonpolar) have London dispersion forces, but only polar molecules (those with permanent dipoles) exhibit dipole forces.

13.     Fusion refers to a solid converting to a liquid, and vaporization refers to a liquid converting to a gas. Only a fraction of the hydrogen bonds are broken in going from the solid phase to the liquid phase. Most of the hydrogen bonds are still present in the liquid phase and must be broken during the liquid to gas phase transition. Thus the enthalpy of vaporization is much larger than the enthalpy of fusion since more intermolecular forces are broken during the vaporization process.

15.     Ionic compounds have ionic forces. Covalent compounds all have London dispersion (LD) forces, whereas polar covalent compounds have dipole forces and/or hydrogen-bonding forces. For hydrogen bonding (H-bonding) forces, the covalent compound must have either a N–H, O–H, or F–H bond in the molecule.

a.  LD only           b.  dipole, LD           c.  H-bonding, LD

d.  ionic             e.  LD only ($CH_4$ in a nonpolar covalent compound.)

f.  dipole, LD        g.  ionic                h.  ionic

i.   LD mostly;  C–F bonds are polar, but polymers such as teflon are so large that LD forces are the predominant intermolecular forces.

j.   LD                    k.  dipole, LD                    l.   H-bonding, LD

m.  dipole, LD          n.   LD only

17.    a.   $H_2NCH_2CH_2NH_2$;  more extensive hydrogen bonding (H-bonding) is possible because two $NH_2$ groups are present.

b.   $H_2CO$; $H_2CO$ is polar, whereas $CH_3CH_3$ is nonpolar.  $H_2CO$ has dipole forces in addition to LD forces.  $CH_3CH_3$ only has LD forces.

c.   $CH_3OH$; $CH_3OH$ can form relatively strong H-bonding interactions, unlike $H_2CO$.

d.  HF; HF is capable of forming H-bonding interactions; HBr is not.

19.    a.   Neopentane is more compact than n-pentane.  There is less surface-area contact among neopentane molecules.  This leads to weaker LD forces and a lower boiling point.

b.  HF is capable of H-bonding; HCl is not.

c.   LiCl is ionic, and HCl is a molecular solid with only dipole forces and LD forces.  Ionic forces are much stronger than the forces for molecular solids.

d.  n-Hexane is a larger molecule, so it has stronger LD forces.

21.    Ar exists as individual atoms that are held together in the condensed phases by London dispersion forces. The molecule that will have a boiling point closest to Ar will be a nonpolar substance with about the same molar mass as Ar (39.95 g/mol); this same size nonpolar substance will have about equivalent strength of London dispersion forces. Of the choices, only $Cl_2$ (70.90 g/mol) and $F_2$ (38.00 g/mol) are nonpolar. Because $F_2$ has a molar mass closest to that of Ar, one would expect the boiling point of $F_2$ to be close to that of Ar.

23.    Ethanol, $C_2H_6O$, has $2(4) + 6(1) + 6 = 20$ valence electrons.

Exhibits H-bonding and London dispersion forces.

Dimethyl ether, $C_2H_6O$, also has 20 valence electrons. It has a Lewis structure of:

Exhibits dipole and London dispersion forces but no hydrogen bonding since it has no H covalently bonded to the O.

Propane, $C_3H_8$, has $3(4) + 8(1) = 20$ valence electrons.

Propane only has relatively nonpolar bonds, so it is nonpolar. Propane exhibits only London dispersion forces.

The three compounds have similar molar mass, so the strength of the London dispersion forces will be approximately equivalent. Because dimethyl ether has additional dipole forces, it will boil at a higher temperature than propane. The compound with the highest boiling point is ethanol since it exhibits relatively strong hydrogen-bonding forces. The correct matching of boiling points is:

ethanol, 78.5 °C;  dimethyl ether, −23 °C;  propane, −42.1 °C

25.    A single hydrogen bond in $H_2O$ has a strength of 21 kJ/mol.  Each $H_2O$ molecule forms two H-bonds.  Thus it should take 42 kJ/mol of energy to break all of the H-bonds in water. Consider the phase transitions:

$$\text{Solid} \xrightarrow{\text{6.0 kJ}} \text{liquid} \xrightarrow{\text{40.7 kJ}} \text{vapor} \qquad \Delta H_{sub} = \Delta H_{fus} + \Delta H_{vap}$$

It takes a total of 46.7 kJ/mol to convert solid $H_2O$ to vapor ($\Delta H_{sub}$).  This would be the amount of energy necessary to disrupt all of the intermolecular forces in ice.  Thus $(42 \div 46.7) \times 100 = 90.\%$ of the attraction in ice can be attributed to H-bonding.

## Properties of Liquids

27.    a.    Surface tension: the resistance of a liquid to an increase in its surface area.

b.    Viscosity: the resistance of a liquid to flow.

c.    Melting point:  the temperature (at constant pressure) where a solid converts entirely to a liquid as long as heat is applied. A more detailed definition is the temperature at which the solid and liquid states have the same vapor pressure under conditions where the total pressure is constant.

d.    Boiling point: the temperature (at constant pressure) where a liquid converts entirely to a gas as long as heat is applied. The detailed definition is the temperature at which the vapor pressure of the liquid is exactly equal to the external pressure.

e.    Vapor pressure: the pressure of the vapor over a liquid at equilibrium.

As the strengths of intermolecular forces increase, surface tension, viscosity, melting point, and boiling point increase, whereas vapor pressure decreases.

29.     Water is a polar substance, and wax is a nonpolar substance; they are not attracted to each other.  A molecule at the surface of a drop of water is subject to attractions only by water molecules below it and to each side.  The effect of this uneven pull on the surface water molecules tends to draw them into the body of the water and causes the droplet to assume the shape that has the minimum surface area, a sphere.

31.     The structure of $H_2O_2$ is $H - O - O - H$, which produces greater hydrogen bonding than in water.  Thus the intermolecular forces are stronger in $H_2O_2$ than in $H_2O$ resulting in a higher normal boiling point for $H_2O_2$ and a lower vapor pressure.

## Structures and Properties of Solids

33.     a.   Crystalline solid:              Regular, repeating structure

             Amorphous solid:               Irregular arrangement of atoms or molecules

        b.   Ionic solid:                   Made up of ions held together by ionic bonding.

             Molecular solid:               Made up of discrete covalently bonded molecules held together in the solid phase by weaker forces (LD, dipole or hydrogen bonds).

        c.   Molecular solid:               Discrete, individual molecules

             Covalent network solid:        No discrete molecules;  a covalent network solid is one large molecule; the interparticle forces are the covalent bonds between atoms.

        d.   Metallic solid:                Completely delocalized electrons, conductor of electricity (ions in a sea of electrons)

             Covalent network solid:        Localized electrons; insulator or semiconductor

35.     a.   Both forms of carbon are network solids. In diamond, each carbon atom is surrounded by a tetrahedral arrangement of other carbon atoms to form a huge molecule. Each carbon atom is covalently bonded to four other carbon atoms.

             The structure of graphite is based on layers of carbon atoms arranged in fused six-membered rings. Each carbon atom in a particular layer of graphite is surrounded by three other carbons in a trigonal planar arrangement. This requires $sp^2$ hybridization. Each carbon has an unhybridized p atomic orbital; all of these p orbitals in each six-membered ring overlap with each other to form a delocalized $\pi$ electron system.

        b.   Silica is a network solid having an empirical formula of $SiO_2$. The silicon atoms are singly bonded to four oxygens. Each silicon atom is at the center of a tetrahedral arrangement of oxygen atoms that are shared with other silicon atoms. The structure of silica is based on a network of $SiO_4$ tetrahedra with shared oxygen atoms rather than discrete $SiO_2$ molecules.

             Silicates closely resemble silica. The structure is based on interconnected $SiO_4$ tetrahedra. However, in contrast to silica, where the O/Si ratio is 2:1, silicates have O/Si ratios greater than 2:1 and contain silicon-oxygen anions. To form a neutral solid silicate, metal cations are needed to balance the charge. In other words, silicates are salts containing metal cations and polyatomic silicon-oxygen anions.

When silica is heated above its melting point and cooled rapidly, an amorphous (disordered) solid called glass results.  Glass more closely resembles a very viscous solution than it does a crystalline solid.  To affect the properties of glass, several different additives are thrown into the mixture.  Some of these additives are $Na_2CO_3$, $B_2O_3$, and $K_2O$, with each compound serving a specific purpose relating to the properties of glass.

37.   A crystalline solid will have the simpler diffraction pattern because a regular, repeating arrangement is necessary to produce planes of atoms that will diffract the X rays in regular patterns.  An amorphous solid does not have a regular repeating arrangement and will produce a complicated diffraction pattern.

39.   $\lambda = \dfrac{2d\sin\theta}{n} = \dfrac{2 \times 1.36 \times 10^{-10}\ m \times \sin 15.0°}{1} = 7.04 \times 10^{-11}\ m = 0.704\ \text{Å}$

41.   $n\lambda = 2d\sin\theta$,  $d = \dfrac{n\lambda}{2\sin\theta} = \dfrac{1 \times 1.54\ \text{Å}}{2 \times \sin 14.22°} = 3.13\ \text{Å} = 3.13 \times 10^{-10}\ m = 313\ pm$

43.   A cubic closest packed structure has a face-centered cubic unit cell.  In a face-centered cubic unit, there are:

$$8\ \text{corners} \times \frac{1/8\ \text{atom}}{\text{corner}} + 6\ \text{faces} \times \frac{1/2\ \text{atom}}{\text{face}} = 4\ \text{atoms}$$

The atoms in a face-centered cubic unit cell touch along the face diagonal of the cubic unit cell.  Using the Pythagorean formula, where $l$ = length of the face diagonal and r = radius of the atom:

$$l^2 + l^2 = (4r)^2$$

$$2l^2 = 16r^2$$

$$l = r\sqrt{8}$$

$l = r\sqrt{8} = 197 \times 10^{-12}\ m \times \sqrt{8} = 5.57 \times 10^{-10}\ m = 5.57 \times 10^{-8}\ cm$

Volume of a unit cell $= l^3 = (5.57 \times 10^{-8}\ cm)^3 = 1.73 \times 10^{-22}\ cm^3$

Mass of a unit cell $= 4\ \text{Ca atoms} \times \dfrac{1\ \text{mol Ca}}{6.022 \times 10^{23}\ \text{atoms}} \times \dfrac{40.08\ \text{g Ca}}{\text{mol Ca}} = 2.662 \times 10^{-22}\ \text{g Ca}$

Density $= \dfrac{\text{mass}}{\text{volume}} = \dfrac{2.662 \times 10^{-22}\ g}{1.73 \times 10^{-22}\ cm^3} = 1.54\ g/cm^3$

45.    There are four Ni atoms in each unit cell:  For a unit cell:

$$\text{density} = \frac{\text{mass}}{\text{volume}} = 6.84 \text{ g/cm}^3 = \frac{4 \text{ Ni atoms} \times \dfrac{1 \text{ mol Ni}}{6.022 \times 10^{23} \text{ atoms}} \times \dfrac{58.69 \text{ g Ni}}{\text{mol Ni}}}{l^3}$$

Solving:  $l = 3.85 \times 10^{-8}$ cm = cube edge length

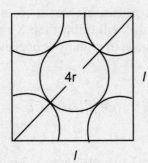

For a face centered cube:

$$(4r)^2 = l^2 + l^2 = 2l^2$$

$$r\sqrt{8} = l, \, r = l/\sqrt{8}$$

$$r = 3.85 \times 10^{-8} \text{ cm}/\sqrt{8}$$

$$r = 1.36 \times 10^{-8} \text{ cm} = 136 \text{ pm}$$

47.    For a body-centered unit cell:  8 corners $\times \dfrac{1/8 \text{ Ti}}{\text{corner}} + 1$ Ti at body center = 2 Ti atoms

All body-centered unit cells have two atoms per unit cell.  For a unit cell:

$$\text{density} = 4.50 \text{ g/cm}^3 = \frac{2 \text{ atoms Ti} \times \dfrac{1 \text{ mol Ti}}{6.022 \times 10^{23} \text{ atoms}} \times \dfrac{47.88 \text{ g Ti}}{\text{mol Ti}}}{l^3},$$

where $l$ = cube edge length

Solving:  $l$ = edge length of unit cell = $3.28 \times 10^{-8}$ cm = 328 pm

Assume Ti atoms just touch along the body diagonal of the cube, so body diagonal = 4 × radius of atoms = 4r.

The triangle we need to solve is:

$$(4r)^2 = (3.28 \times 10^{-8} \text{ cm})^2 + [(3.28 \times 10^{-8} \text{ cm})\sqrt{2}]^2, \quad r = 1.42 \times 10^{-8} \text{ cm} = 142 \text{ pm}$$

For a body-centered cubic unit cell, the radius of the atom is related to the cube edge length by $4r = l\sqrt{3}$ or $l = 4r/\sqrt{3}$.

49.    If gold has a face-centered cubic structure, then there are four atoms per unit cell, and from Exercise 16.43:

$$2l^2 = 16r^2$$

$$l = r\sqrt{8} = (144 \text{ pm})\sqrt{8} = 407 \text{ pm}$$

$$l = 407 \times 10^{-12} \text{ m} = 4.07 \times 10^{-8} \text{ cm}$$

$$\text{Density} = \frac{4 \text{ atoms Au} \times \dfrac{1 \text{ mol Au}}{6.022 \times 10^{23} \text{ atoms}} \times \dfrac{197.0 \text{ g Au}}{\text{mol Au}}}{(4.07 \times 10^{-8} \text{ cm})^3} = 19.4 \text{ g/cm}^3$$

If gold has a body-centered cubic structure, then there are two atoms per unit cell, and from Exercise 16.47:

$$16r^2 = l^2 + 2l^2$$

$$l = 4r/\sqrt{3} = 333 \text{ pm} = 333 \times 10^{-12} \text{ m}$$

$$l = 333 \times 10^{-10} \text{ cm} = 3.33 \times 10^{-8} \text{ cm}$$

$$\text{Density} = \frac{2 \text{ atoms Au} \times \dfrac{1 \text{ mol Au}}{6.022 \times 10^{23} \text{ atoms}} \times \dfrac{197.0 \text{ g Au}}{\text{mol Au}}}{(3.33 \times 10^{-8} \text{ cm})^3} = 17.7 \text{ g/cm}^3$$

The measured density of gold is consistent with a face-centered cubic unit cell.

51.    Conductor:    The energy difference between the filled and unfilled molecular orbitals is minimal. We call this energy difference the band gap. Because the band gap is minimal, electrons can easily move into the conduction bands (the unfilled molecular orbitals).

Insulator:    Large band gap; electrons do not move from the filled molecular orbitals to the conduction bands since the energy difference is large.

Semiconductor: Small band gap; the energy difference between the filled and unfilled molecular orbitals is smaller than in insulators, so some electrons can jump into the conduction bands. The band gap, however, is not as small as with conductors, so semiconductors have intermediate conductivity.

a. As the temperature is increased, more electrons in the filled molecular orbitals have sufficient kinetic energy to jump into the conduction bands (the unfilled molecular orbitals).

b. A photon of light is absorbed by an electron that then has sufficient energy to jump into the conduction bands.

c. An impurity either adds electrons at an energy near that of the conduction bands (n-type) or creates holes (unfilled energy levels) at energies in the previously filled molecular orbitals (p-type). Both n-type and p-type semiconductors increase conductivity by creating an easier path for electrons to jump from filled to unfilled energy levels.

In conductors, electrical conductivity is inversely proportional to temperature. Increases in temperature increase the motions of the atoms, which gives rise to increased resistance (decreased conductivity). In a semiconductor, electrical conductivity is directly proportional to temperature. An increase in temperature provides more electrons with enough kinetic energy to jump from the filled molecular orbitals to the conduction bands, increasing conductivity.

53. A rectifier is a device that produces a current that flows in one direction from an alternating current that flows in both directions. In a p-n junction, a p-type and an n-type semi-conductor are connected. The natural flow of electrons in a p-n junction is for the excess electrons in the n-type semiconductor to move to the empty energy levels (holes) of the p-type semiconductor. Only when an external electric potential is connected so that electrons flow in this natural direction will the current flow easily (forward bias). If the external electric potential is connected in reverse of the natural flow of electrons, no current flows through the system (reverse bias). A p-n junction only transmits a current under forward bias, thus converting the alternating current to direct current.

55. In has fewer valence electrons than Se. Thus Se doped with In would be a p-type semiconductor.

57. $E_{gap} = 2.5$ eV $\times 1.6 \times 10^{-19}$ J/eV $= 4.0 \times 10^{-19}$ J;  we want $E_{gap} = E_{light}$.

$$E_{light} = \frac{hc}{\lambda}, \ \lambda = \frac{hc}{E} = \frac{(6.63 \times 10^{-34} \text{ J s})(3.00 \times 10^8 \text{ m/s})}{4.0 \times 10^{-19} \text{ J}} = 5.0 \times 10^{-7} \text{ m} = 5.0 \times 10^2 \text{ nm}$$

59. The structures of most binary ionic solids can be explained by the closest packing of spheres. Typically, the larger ions, usually the anions, are packed in one of the closest packing arrangements, and the smaller cations fit into holes among the closest packed anions. There are different types of holes within the closest packed anions that are determined by the number of spheres that form them. Which of the three types of holes are filled usually depends on the relative size of the cation to the anion. Ionic solids will always try to maximize electrostatic attractions among oppositely charged ions and minimize the repulsions among ions with like charges.

The structure of sodium chloride can be described in terms of a cubic closest packed array of $Cl^-$ ions with $Na^+$ ions in all of the octahedral holes. An octahedral hole is formed between six $Cl^-$ anions. The number of octahedral holes is the same as the number of packed ions. So in the face-centered unit cell of sodium chloride, there are four net $Cl^-$ ions and four net octahedral holes. Because the stoichiometry dictates a 1:1 ratio between the number of $Cl^-$ anions and $Na^+$ cations, all of the octahedral holes must be filled with $Na^+$ ions.

In zinc sulfide, the sulfide anions also occupy the lattice points of a cubic closest packing arrangement. But instead of having the cations in octahedral holes, the $Zn^{2+}$ cations occupy tetrahedral holes. A tetrahedral hole is the empty space created when four spheres are packed together. There are twice as many tetrahedral holes as packed anions in the closest packed structure. Therefore, each face-centered unit cell of sulfide anions contains four net $S^{2-}$ ions and eight net tetrahedral holes. For the 1:1 stoichiometry to work out, only one-half of the tetrahedral holes are filled with $Zn^{2+}$ ions. This gives four $S^{2-}$ ions and four $Zn^{2+}$ ions per unit cell for an empirical formula of ZnS.

61.     There is one octahedral hole per closest packed anion in a closest packed structure.  If one-half of the octahedral holes are filled, then there is a 2:1 ratio of fluoride ions to cobalt ions in the crystal.  The formula is $CoF_2$.

63.     Mn ions at 8 corners:  8(1/8) = 1 Mn ion;  F ions at 12 edges:  12(1/4) = 3 F ions;  the formula is $MnF_3$.  Assuming fluoride is –1 charged, then the charge on Mn is +3.

65.     CsCl is a simple cubic array of $Cl^-$ ions with $Cs^+$ in the middle of each unit cell.  There is one $Cs^+$ and one $Cl^-$ ion in each unit cell.  $Cs^+$ and $Cl^-$ touch along the body diagonal.

Body diagonal $= 2r_{Cs^+} + 2r_{Cl^-} = \sqrt{3}\, l,\;\; l = $ length of cube edge

In each unit cell:

mass = 1 CsCl unit(1 mol/$6.022 \times 10^{23}$ units)(168.4 g/mol) = $2.796 \times 10^{-22}$ g

volume = $l^3$ = mass/density = $2.796 \times 10^{-22}$ g/3.97 g cm$^{-3}$ = $7.04 \times 10^{-23}$ cm$^3$

$l^3 = 7.04 \times 10^{-23}$ cm$^3$,  $l = 4.13 \times 10^{-8}$ cm = 413 pm = length of cube edge

$2r_{Cs^+} + 2r_{Cl^-} = \sqrt{3}\, l = \sqrt{3}\,(413 \text{ pm}) = 715$ pm

The distance between ion centers $= r_{Cs^+} + r_{Cl^-} = $ 715 pm/2 = 358 pm.

From ionic radius:  $r_{Cs^+} = 169$ pm and  $r_{Cl^-} = 181$ pm;  $r_{Cs^+} + r_{Cl^-} = 169 + 181 = 350.$ pm

The distance calculated from the density is 8 pm (2.3%) greater than that calculated from tables of ionic radii.

67.    For a cubic hole to be filled, the cation to anion radius ratio is between $0.732 < r_+/r_- < 1.00$.

CsBr:    $Cs^+$ radius = 169 pm, $Br^-$ radius = 195 pm; $r_+/r_- = 169/195 = 0.867$

From the radius ratio, $Cs^+$ should occupy cubic holes. The structure should be the CsCl structure. The actual structure is the CsCl structure.

KF:    $K^+$ radius = 133 pm, $F^-$ radius = 136 pm; $r_+/r_- = 133/136 = 0.978$

Again, we would predict a structure similar to CsCl, i.e., cations in the middle of a simple cubic array of anions. The actual structure is the NaCl structure.

The radius ratio rules fail for KF. Exceptions are common for crystal structures.

69.    a.    8 corners $\times \dfrac{1/8 \; Xe}{corner} + 1$ Xe inside cell = 2 Xe; 8 edges $\times \dfrac{1/4 \; F}{edge} + 2$ F inside cell = 4 F

Empirical formula is $XeF_2$. This is also the molecular formula.

b.    For a unit cell:

$$\text{mass} = 2 \; XeF_2 \text{ molecules} \times \frac{1 \; mol \; XeF_2}{6.022 \times 10^{23} \text{ molecules}} \times \frac{169.3 \; g \; XeF_2}{mol \; XeF_2} = 5.62 \times 10^{-22} \; g$$

$$\text{volume} = (7.02 \times 10^{-8} \; cm)(4.32 \times 10^{-8} \; cm)(4.32 \times 10^{-8} \; cm) = 1.31 \times 10^{-22} \; cm^3$$

$$\text{density} = d = \frac{\text{mass}}{\text{volume}} = \frac{5.62 \times 10^{-22} \; g}{1.31 \times 10^{-22} \; cm^3} = 4.29 \; g/cm^3$$

71.    With a cubic closest packed array of oxygen ions, we have 4 $O^{2-}$ ions per unit cell. We need to balance the total $-8$ charge of the anions with a $+8$ charge from the $Al^{3+}$ and $Mg^{2+}$ cations. The only combination of ions that gives a $+8$ charge is 2 $Al^{3+}$ ions and 1 $Mg^{2+}$ ion. The formula is $Al_2MgO_4$.

There are an equal number of octahedral holes as anions (4) in a cubic closest packed array and twice the number of tetrahedral holes as anions in a cubic closest packed array. For the stoichiometry to work out, we need 2 $Al^{3+}$ and 1 $Mg^{2+}$ per unit cell. Hence one-half the octahedral holes are filled with $Al^{3+}$ ions, and one-eighth the tetrahedral holes are filled with $Mg^{2+}$ ions.

73.    a.    Y: 1 Y in center; Ba: 2 Ba in center

Cu: 8 corners $\times \dfrac{1/8 \; Cu}{corner} = 1$ Cu, 8 edges $\times \dfrac{1/4 \; Cu}{edge} = 2$ Cu, total = 3 Cu atoms

O: 20 edges $\times \dfrac{1/4 \; O}{edge} = 5$ oxygen, 8 faces $\times \dfrac{1/2 \; O}{face} = 4$ oxygen, total = 9 O atoms

Formula: $YBa_2Cu_3O_9$

b.  The structure of this superconductor material follows the alternative perovskite structure described in Exercise 16.72b.  The $YBa_2Cu_3O_9$ structure is three of these cubic perovskite unit cells stacked on top of each other.  The oxygen atoms are in the same places, Cu takes the place of Ti, two of the calcium atoms are replaced by two barium atoms, and one Ca is replaced by Y.

c.  Y, Ba, and Cu are the same.  Some oxygen atoms are missing.

$$12 \text{ edges} \times \frac{1/4 \text{ O}}{\text{edge}} = 3 \text{ O}, \ \ 8 \text{ faces} \times \frac{1/2 \text{ O}}{\text{face}} = 4 \text{ O}, \ \text{ total} = 7 \text{ O atoms}$$

Superconductor formula is $YBa_2Cu_3O_7$.

## Phase Changes and Phase Diagrams

75.  A volatile liquid is one that evaporates relatively easily. Volatile liquids have large vapor pressures because the intermolecular forces that prevent evaporation are relatively weak.

77.  a.  As the intermolecular forces increase, the rate of evaporation decreases.

b.  As temperature increases, the rate of evaporation increases.

c.  As surface area increases, the rate of evaporation increases.

79.  $C_2H_5OH(l) \rightarrow C_2H_5OH(g)$ is an endothermic process.  Heat is absorbed when liquid ethanol vaporizes; the internal heat from the body provides this heat, which results in the cooling of the body.

81.  The mathematical equation that relates the vapor pressure of a substance to temperature is:

$$\ln P_{vap} = -\frac{\Delta H_{vap}}{R}\left(\frac{1}{T}\right) + C$$
$$\phantom{\ln P_{vap} =} y \phantom{xxxx} m \phantom{xxx} x \phantom{xx} + b$$

As shown above, this equation is in the form of the straight-line equation. If one plots $\ln P_{vap}$ versus $1/T$ with temperature in Kelvin, the slope of the straight line is $-\Delta H_{vap}/R$. Because $\Delta H_{vap}$ is always positive, the slope of the straight line will be negative.

83.  $\ln\left(\dfrac{P_1}{P_2}\right) = \dfrac{\Delta H_{vap}}{R}\left(\dfrac{1}{T_2} - \dfrac{1}{T_1}\right), \ \ln\left(\dfrac{836 \text{ torr}}{213 \text{ torr}}\right) = \dfrac{\Delta H_{vap}}{8.3145 \text{ J K}^{-1} \text{ mol}^{-1}}\left(\dfrac{1}{313 \text{ K}} - \dfrac{1}{353 \text{ K}}\right)$

Solving:  $\Delta H_{vap} = 3.1 \times 10^4$ J/mol;  for the normal boiling point, P = 1.00 atm = 760. torr.

$$\ln\left(\frac{760. \text{ torr}}{213 \text{ torr}}\right) = \frac{3.1 \times 10^4 \text{ J/mol}}{8.3145 \text{ J K}^{-1} \text{ mol}^{-1}}\left(\frac{1}{313 \text{ K}} - \frac{1}{T_1}\right), \ \frac{1}{313} - \frac{1}{T_1} = 3.4 \times 10^{-4}$$

$T_1 = 350.$ K = 77°C;  the normal boiling point of $CCl_4$ is 77°C.

85.     If we graph ln $P_{vap}$ versus $1/T$ with temperature in Kelvin, the slope of the resulting straight line will be $-\Delta H_{vap}/R$.

| $P_{vap}$ | ln $P_{vap}$ | T (Li) | $1/T$ | T (Mg) | $1/T$ |
|---|---|---|---|---|---|
| 1 torr | 0 | 1023 K | $9.775 \times 10^{-4}$ $K^{-1}$ | 893 K | $11.2 \times 10^{-4}$ $K^{-1}$ |
| 10. | 2.3 | 1163 | $8.598 \times 10^{-4}$ | 1013 | $9.872 \times 10^{-4}$ |
| 100. | 4.61 | 1353 | $7.391 \times 10^{-4}$ | 1173 | $8.525 \times 10^{-4}$ |
| 400. | 5.99 | 1513 | $6.609 \times 10^{-4}$ | 1313 | $7.616 \times 10^{-4}$ |
| 760. | 6.63 | 1583 | $6.317 \times 10^{-4}$ | 1383 | $7.231 \times 10^{-4}$ |

For Li:

We get the slope by taking two points $(x, y)$ that are on the line we draw. For a line, slope = $\Delta y/\Delta x$; or we can determine the straight-line equation using a calculator. The general straight-line equation is $y = mx + b$ where $m$ = slope and $b = y$ intercept.

The equation of the Li line is: ln $P_{vap} = -1.90 \times 10^4(1/T) + 18.6$, slope = $-1.90 \times 10^4$ K

Slope = $-\Delta H_{vap}/R$, $\Delta H_{vap} = -$slope $\times$ R = $1.90 \times 10^4$ K $\times$ 8.3145 J $K^{-1}$ $mol^{-1}$

$\Delta H_{vap} = 1.58 \times 10^5$ J/mol = 158 kJ/mol

For Mg:

The equation of the line is: ln $P_{vap} = -1.67 \times 10^4(1/T) + 18.7$, slope = $-1.67 \times 10^4$ K

$\Delta H_{vap} = -$slope $\times$ R = $1.67 \times 10^4$ K $\times$ 8.3145 J $K^{-1}$ $mol^{-1}$

$\Delta H_{vap} = 1.39 \times 10^5$ J/mol = 139 kJ/mol

The bonding is stronger in Li because $\Delta H_{vap}$ is larger for Li.

87.

Slope 5 > Slope 3 > Slope 1

Time 4 = 4 × Time 2

89.    To calculate $q_{total}$, break up the heating process into five steps.

$H_2O(s, -20.°C) \rightarrow H_2O(s, 0°C)$,  $\Delta T = 20.°C$

$$q_1 = s_{ice} \times m \times \Delta T = \frac{2.1\,J}{g\,°C} \times 5.00 \times 10^2\,g \times 20.°C = 2.1 \times 10^4\,J = 21\,kJ$$

$$H_2O(s, 0°C) \rightarrow H_2O(l, 0°C),\;\; q_2 = 5.00 \times 10^2\,g\,H_2O \times \frac{1\,mol}{18.02\,g} \times \frac{6.01\,kJ}{mol} = 167\,kJ$$

$$H_2O(l, 0°C) \rightarrow H_2O(l, 100.°C),\;\; q_3 = \frac{4.2\,J}{g\,°C} \times 5.00 \times 10^2\,g \times 100.°C = 2.1 \times 10^5\,J = 210\,kJ$$

$$H_2O(l, 100.°C) \rightarrow H_2O(g, 100.°C),\;\; q_4 = 5.00 \times 10^2\,g \times \frac{1\,mol}{18.02\,g} \times \frac{40.7\,kJ}{mol} = 1130\,kJ$$

$$H_2O(g, 100.°C) \rightarrow H_2O(g, 250.°C),\;\; q_5 = \frac{2.0\,J}{g\,°C} \times 5.00 \times 10^2\,g \times 150.°C = 1.5 \times 10^5\,J$$

$$= 150\,kJ$$

$$q_{total} = q_1 + q_2 + q_3 + q_4 + q_5 = 21 + 167 + 210 + 1130 + 150 = 1680\,kJ$$

91.    Heat released = $0.250\,g\,Na \times \dfrac{1\,mol}{22.99\,g} \times \dfrac{368\,kJ}{2\,mol} = 2.00\,kJ$

To melt 50.0 g of ice requires:  $50.0\,g\,ice \times \dfrac{1\,mol\,H_2O}{18.02\,g} \times \dfrac{6.01\,kJ}{mol} = 16.7\,kJ$

The reaction doesn't release enough heat to melt all of the ice.  The temperature will remain at 0°C.

93.     Total mass $H_2O$ = 18 cubes × $\dfrac{30.0\ g}{cube}$ = 540. g;  540. g $H_2O$ × $\dfrac{1\ mol\ H_2O}{18.02\ g}$ = 30.0 mol $H_2O$

Heat removed to produce ice at −5.0°C:

$$\left(\dfrac{4.18\ J}{g\ ^{\circ}C} \times 540.\ g \times 22.0\ ^{\circ}C\right) + \left(\dfrac{6.01 \times 10^3\ J}{mol} \times 30.0\ mol\right) + \left(\dfrac{2.08\ J}{g\ ^{\circ}C} \times 540.\ g \times 5.0\ ^{\circ}C\right)$$

$$= 4.97 \times 10^4\ J + 1.80 \times 10^5\ J + 5.6 \times 10^3\ J = 2.35 \times 10^5\ J$$

$$2.35 \times 10^5\ J \times \dfrac{1\ g\ CF_2Cl_2}{158\ J} = 1.49 \times 10^3\ g\ CF_2Cl_2\ \text{must be vaporized.}$$

95.     See Figures 16.55 and 16.58 for the phase diagrams of $H_2O$ and $CO_2$. Most substances exhibit only three different phases: solid, liquid, and gas. This is true for $H_2O$ and $CO_2$. Also typical of phase diagrams is the positive slopes for both the liquid-gas equilibrium line and the solid-gas equilibrium line. This is also true for both $H_2O$ and $CO_2$. The solid-liquid equilibrium line also generally has a positive slope. This is true for $CO_2$ but not for $H_2O$. In the $H_2O$ phase diagram, the slope of the solid-liquid line is negative. The determining factor for the slope of the solid-liquid line is the relative densities of the solid and liquid phases. The solid phase is denser than the liquid phase in most substances; for these substances, the slope of the solid-liquid equilibrium line is positive. For water, the liquid phase is denser than the solid phase, which corresponds to a negative-sloping solid-liquid equilibrium line. Another difference between $H_2O$ and $CO_2$ is the normal melting points and normal boiling points. The term normal just dictates a pressure of 1 atm. $H_2O$ has a normal melting point (0°C) and a normal boiling point (100°C), but $CO_2$ does not. At 1 atm pressure, $CO_2$ only sublimes (goes from the solid phase directly to the gas phase). There are no temperatures at 1 atm for $CO_2$ where the solid and liquid phases are in equilibrium or where the liquid and gas phases are in equilibrium. There are other differences, but those discussed above are the major ones.

The relationship between melting points and pressure is determined by the slope of the solid-liquid equilibrium line. For most substances ($CO_2$ included), the positive slope of the solid-liquid line shows a direct relationship between the melting point and pressure. As pressure increases, the melting point increases. Water is just the opposite since the slope of the solid-liquid line in water is negative. Here, the melting point of water is inversely related to the pressure.

For boiling points, the positive slope of the liquid-gas equilibrium line indicates a direct relationship between the boiling point and pressure. This direct relationship is true for all substances, including $H_2O$ and $CO_2$.

The critical temperature for a substance is defined as the temperature above which the vapor cannot be liquefied no matter what pressure is applied. The critical temperature, like the boiling-point temperature, is directly related to the strength of the intermolecular forces. Since $H_2O$ exhibits relatively strong hydrogen bonding interactions and $CO_2$ only exhibits London dispersion forces, one would expect a higher critical temperature for $H_2O$ than for $CO_2$.

97.    The critical temperature is the temperature above which the vapor cannot be liquefied no matter what pressure is applied. Since $N_2$ has a critical temperature below room temperature (~22°C), it cannot be liquefied at room temperature. $NH_3$, with a critical temperature above room temperature, can be liquefied at room temperature.

99.    A:  solid;              B:  liquid;                        C:  vapor

       D:  solid + vapor;       E:  solid + liquid + vapor

       F:  liquid + vapor;      G:  liquid + vapor;               H:  vapor

       triple point:  E;              critical point:  G

       Normal freezing point:  temperature at which solid-liquid line is at 1.0 atm (see following plot).

       Normal boiling point:  temperature at which liquid-vapor line is at 1.0 atm (see following plot).

       Because the solid-liquid equilibrium line has a positive slope, the solid phase is denser than the liquid phase.

101.   a.   two

       b.   Higher-pressure triple point:  graphite, diamond, and liquid;  lower-pressure triple point:  graphite, liquid, and vapor

       c.   It is converted to diamond (the more dense solid form).

       d.   Diamond is more dense, which is why graphite can be converted to diamond by applying pressure.

103.   Because the density of the liquid phase is greater than the density of the solid phase, the slope of the solid-liquid boundary line is negative (as in $H_2O$). With a negative slope, the melting points increase with a decrease in pressure, so the normal melting point of X should be greater than 225°C.

105.

As P is lowered, we go from a to b on the phase diagram. The water boils. The evaporation of the water is endothermic, and the water is cooled (b → c), forming some ice. If the pump is left on, the ice will sublime until none is left. This is the basis of freeze drying.

## Additional Exercises

107.    If $TiO_2$ conducts electricity as a liquid, then it is an ionic solid; if not, then $TiO_2$ is a network solid.

109.    $B_2H_6$: This compound contains only nonmetals, so it is probably a molecular solid with covalent bonding. The low boiling point confirms this.

   $SiO_2$: This is the empirical formula for quartz, which is a network solid.

   CsI: This is a metal bonded to a nonmetal, which generally form ionic solids. The electrical conductivity in aqueous solution confirms this.

   W:   Tungsten is a metallic solid as the conductivity data confirms.

111.

R = radius of sphere

r = radius of trigonal hole

For a right-angle triangle (opposite side not drawn in):

$$\cos 30° = \frac{adjacent}{hypotenuse} = \frac{R}{R + r}, \ 0.866 = \frac{R}{R + r}$$

$$(0.866)r = R - (0.866)R, \ r = \left(\frac{0.134}{0.866}\right) \times R = (0.155)R$$

The cation must have a radius that is 0.155 times the radius of the spheres to just fit into the trigonal hole.

113.   Ar is cubic closest packed.  There are four Ar atoms per unit cell and with a face-centered unit cell, the atoms touch along the face diagonal.  Let $l$ = length of cube edge.

Face diagonal = 4r = $l\sqrt{2}$ ,   $l$ = 4(190. pm)/$\sqrt{2}$ = 537 pm = $5.37 \times 10^{-8}$ cm

$$\text{Density} = \frac{\text{mass}}{\text{volume}} = \frac{4 \text{ atoms} \times \dfrac{1 \text{ mol}}{6.022 \times 10^{23} \text{ atoms}} \times \dfrac{39.95 \text{ g}}{\text{mol}}}{(5.37 \times 10^{-8} \text{ cm})^3} = 1.71 \text{ g/cm}^3$$

115.   Assuming 100.00 g: $28.31 \text{ g O} \times \dfrac{1 \text{ mol}}{15.999 \text{ g}} = 1.769 \text{ mol O}$;  $71.69 \text{ g Ti} \times \dfrac{1 \text{ mol}}{47.88 \text{ g}}$

$$= 1.497 \text{ mol Ti}$$

Dividing one mole quantity by the other, gives formulas of $TiO_{1.182}$ or $Ti_{0.8462}O$.

For $Ti_{0.8462}O$, let $x$ = $Ti^{2+}$ per mol $O^{2-}$  and  $y$ = $Ti^{3+}$ per mol $O^{2-}$.  Setting up two equations and solving:

$x + y = 0.8462$  and  $2x + 3y = 2$;  $2x + 3(0.8462 - x) = 2$

$x = 0.539$ mol $Ti^{2+}$/mol $O^{2-}$  and  $y = 0.307$ mol $Ti^{3+}$/mol $O^{2-}$

$\dfrac{0.539}{0.8462} \times 100 = 63.7\%$ of the titanium is $Ti^{2+}$ and 36.3% is $Ti^{3+}$.

117.   $24.7 \text{ g } C_6H_6 \times \dfrac{1 \text{ mol}}{78.11 \text{ g}} = 0.316 \text{ mol } C_6H_6$

$$P_{C_6H_6} = \frac{nRT}{V} = \frac{0.316 \text{ mol} \times \dfrac{0.08296 \text{ L atm}}{\text{K mol}} \times 293.2 \text{ K}}{100.0 \text{ L}} = 0.0760 \text{ atm, or } 57.8 \text{ torr}$$

119.   $w = -P\Delta V$;   assume constant P of 1.00 atm.

$$V_{373} = \frac{nRT}{P} = \frac{1.00 \, (0.08206)(373)}{1.00} = 30.6 \text{ L for 1 mol of water vapor}$$

Because the density of $H_2O(l)$ is 1.00 g/cm$^3$, 1.00 mol of $H_2O(l)$ occupies 18.0 cm$^3$ or 0.0180 L.

$w = -1.00$ atm (30.6 L − 0.0180 L) = −30.6 L atm

$w = -30.6$ L atm $\times$ 101.3 J L$^{-1}$ atm$^{-1}$ = $-3.10 \times 10^3$ J = −3.10 kJ

$\Delta E = q + w = 41.16$ kJ − 3.10 kJ = 38.06 kJ

$\dfrac{38.06}{41.16} \times 100 = 92.47\%$ of the energy goes to increase the internal energy of the water.

The remainder of the energy (7.53%) goes to do work against the atmosphere.

121.    The three sheets of B's form a cube. In this cubic unit cell there are B atoms at every face, B atoms at every edge, B atoms at every corner, and one B atom in the center. A representation of this cube is:

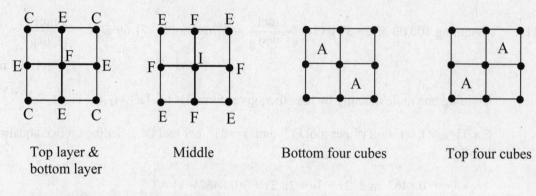

| Top layer &<br>bottom layer | Middle | Bottom four cubes | Top four cubes |

C = corner;    E = edge;    F = face;    I = inside

Each unit cell of B atoms contains four A atoms. The number of B atoms in each unit cell is:

(8 corners × 1/8) + (6 faces × 1/2) + (12 edges × 1/4) + (1 middle B) = 8 B atoms

The empirical formula is $AB_2$.

Each A atom is in a cubic hole of B atoms, so eight B atoms surround each A atom. This will also be true in the extended lattice. The structure of B atoms in the unit cell is a cubic arrangement with B atoms at every face, edge, corner, and center of the cube.

## Challenge Problems

123.    $E = \dfrac{h^2(n_x^2 + n_y^2 + n_z^2)}{8mL^2}$;  $E_{111} = \dfrac{3h^2}{8mL^2}$;  $E_{112} = \dfrac{6h^2}{8mL^2}$;  $\Delta E = E_{112} - E_{111} = \dfrac{3h^2}{8mL^2}$

$\Delta E = \dfrac{hc}{\lambda} = \dfrac{(6.626 \times 10^{-34}\,\text{J s})(2.998 \times 10^8\,\text{m/s})}{9.50 \times 10^{-9}\,\text{m}} = 2.09 \times 10^{-17}\,\text{J}$

$L^2 = \dfrac{3h^2}{8m\Delta E}$,  $L = \left(\dfrac{3h^2}{8m\Delta E}\right)^{1/2} = \left[\dfrac{3(6.626 \times 10^{-34}\,\text{J s})^2}{8(9.109 \times 10^{-31}\,\text{kg})(2.09 \times 10^{-17}\,\text{J})}\right]^{1/2}$

$L = 9.30 \times 10^{-11}\,\text{m} = 93.0\,\text{pm}$

The sphere that fits in this cube will touch the cube at the center of each face. The diameter of the sphere will equal the length of the cube.  So:

$$2r = L, \ r = 93.0/2 = 46.5 \text{ pm}$$

125.    Assuming 100.00 g of $MO_2$:

$$23.72 \text{ g O} \times \frac{1 \text{ mol O}}{16.00 \text{ g O}} = 1.483 \text{ mol O}$$

$$1.483 \text{ mol O} \times \frac{1 \text{ mol M}}{2 \text{ mol O}} = 0.7415 \text{ mol M}$$

$$100.00 \text{ g} - 23.72 \text{ g} = 76.28 \text{ g M}; \ \text{molar mass M} = \frac{76.28 \text{ g}}{0.7415 \text{ mol}} = 102.9 \text{ g/mol}$$

From the periodic table, element M is rhodium (Rh).

The unit cell for cubic closest packing is face-centered cubic (4 atoms/unit cell). The atoms for a face-centered cubic unit cell are assumed to touch along the face diagonal of the cube, so the face diagonal = 4r. The distance between the centers of touching Rh atoms will be the distance of 2r where r = radius of Rh atom.

Face diagonal = $\sqrt{2} \ l$, where $l$ = cube edge.

Face diagonal = 4r = 2 × 269.0 × $10^{-12}$ m = 5.380 × $10^{-10}$ m

$$\sqrt{2} \ l = 4r = 5.38 \times 10^{-10} \text{ m}, \ l = \frac{5.38 \times 10^{-10} \text{ m}}{\sqrt{2}} = 3.804 \times 10^{-10} \text{ m} = 3.804 \times 10^{-8} \text{ cm}$$

$$\text{Density} = \frac{4 \text{ atoms Rh} \times \dfrac{1 \text{ mol Rh}}{6.0221 \times 10^{23} \text{ atoms}} \times \dfrac{102.9 \text{ g Rh}}{\text{mol Rh}}}{(3.804 \times 10^{-8} \text{ cm})^3} = 12.42 \text{ g/cm}^3$$

127.    For water vapor at 30.0°C and 31.824 torr:

$$\text{density} = \frac{P(\text{molar mass})}{RT} = \frac{\left(\dfrac{31.824 \text{ atm}}{760}\right)\left(\dfrac{18.015 \text{ g}}{\text{mol}}\right)}{\dfrac{0.08206 \text{ L atm}}{\text{K mol}} \times 303.2 \text{ K}} = 0.03032 \text{ g/L}$$

The volume of one molecule is proportional to $d^3$, where d is the average distance between molecules.  For a large sample of molecules, the volume is still proportional to $d^3$.  So:

$$\frac{V_{gas}}{V_{liq}} = \frac{d_{gas}^3}{d_{liq}^3}$$

If we have 0.99567 g $H_2O$, then $V_{liq} = 1.0000$ cm$^3$ = $1.0000 \times 10^{-3}$ L.

$V_{gas} = 0.99567$ g $\times$ 1 L/0.03032 g = 32.84 L

$$\frac{d_{gas}^3}{d_{liq}^3} = \frac{32.84 \text{ L}}{1.0000 \times 10^{-3} \text{ L}} = 3.284 \times 10^4, \quad \frac{d_{gas}}{d_{liq}} = (3.284 \times 10^4)^{1/3} = 32.02, \quad \frac{d_{liq}}{d_{gas}} = 0.03123$$

129.    a.    Structure a:

Ba: 2 Ba inside unit cell;  Tl: 8 corners $\times \dfrac{1/8 \text{ Tl}}{\text{corner}} = 1$ Tl;  Cu: 4 edges $\times \dfrac{1/4 \text{ Cu}}{\text{edge}} = 1$ Cu

O: 6 faces $\times \dfrac{1/2 \text{ O}}{\text{face}} + 8$ edges $\times \dfrac{1/4 \text{ O}}{\text{edge}} = 5$ O

Formula = $TlBa_2CuO_5$.

Structure b:

Tl and Ba are the same as in structure a.

Ca: 1 Ca inside unit cell;  Cu: 8 edges $\times \dfrac{1/4 \text{ Cu}}{\text{edge}} = 2$ Cu

O: 10 faces $\times \dfrac{1/2 \text{ O}}{\text{edge}} + 8$ edges $\times \dfrac{1/4 \text{ O}}{\text{edge}} = 7$ O

Formula = $TlBa_2CaCu_2O_7$.

Structure c:

Tl and Ba are the same and two Ca atoms are located inside the unit cell.

Cu: 12 edges $\times \dfrac{1/4 \text{ Cu}}{\text{edge}} = 3$ Cu;  O: 14 faces $\times \dfrac{1/2 \text{ O}}{\text{face}} + 8$ edges $\times \dfrac{1/4 \text{ O}}{\text{edge}} = 9$ O

Formula = $TlBa_2Ca_2Cu_3O_9$.

Structure d:  Following similar calculations, formula = $TlBa_2Ca_3Cu_4O_{11}$.

b.    Structure a has one planar sheet of Cu and O atoms and the number increases by one for each of the remaining structures.  The order of superconductivity temperature from lowest to highest temperature is  a < b < c < d.

c.   $TlBa_2CuO_5$: $3 + 2(2) + x + 5(-2) = 0$,  $x = +3$
Only $Cu^{3+}$ is present in each formula unit.

$TlBa_2CaCu_2O_7$: $3 + 2(2) + 2 + 2(x) + 7(-2) = 0$,  $x = +5/2$
Each formula unit contains 1 $Cu^{2+}$ and 1 $Cu^{3+}$.

$TlBa_2Ca_2Cu_3O_9$: $3 + 2(2) + 2(2) + 3(x) + 9(-2) = 0$,  $x = +7/3$
Each formula unit contains 2 $Cu^{2+}$ and 1 $Cu^{3+}$.

$TlBa_2Ca_3Cu_4O_{11}$: $3 + 2(2) + 3(2) + 4(x) + 11(-2) = 0$,  $x = +9/4$
Each formula unit contains 3 $Cu^{2+}$ and 1 $Cu^{3+}$.

d.   This superconductor material achieves variable copper oxidation states by varying the numbers of Ca, Cu, and O in each unit cell.  The mixtures of copper oxidation states is discussed in part c.  The superconductor material in Exercise 16.73 achieves variable copper oxidation states by omitting oxygen at various sites in the lattice.

131.   For an octahedral hole, the geometry is:

$2R + 2r$     $45°$     $R = Cl^-$ radius

$2R$     $r = Li^+$ radius

From the diagram:  $\cos 45° = \dfrac{\text{adjacent}}{\text{hypotenuse}} = \dfrac{2R}{2R + 2r} = \dfrac{R}{R + r}$,  $0.707 = \dfrac{R}{R + r}$

$R = 0.707(R + r)$,  $r = (0.414)R$

LiCl unit cell:

length  of cube edge = $2R + 2r = 514$ pm

$2R + 2(0.414\,R) = 514$ pm,  $R = 182$ pm = $Cl^-$ radius

$2(182$ pm$) + 2r = 514$ pm,  $r = 75$ pm = $Li^+$ radius

From Figure 13.8, the $Li^+$ radius is 60 pm, and the $Cl^-$ radius is 181 pm.  The $Li^+$ ion is much smaller than calculated.  This probably means that the ions are not actually in contact with each other.  The octahedral holes are larger than the $Li^+$ ion.

# CHAPTER 17

# PROPERTIES OF SOLUTIONS

## Solution Composition

13.　　a.　$HNO_3(l) \rightarrow H^+(aq) + NO_3^-(aq)$　　　　b.　$Na_2SO_4(s) \rightarrow 2\ Na^+(aq) + SO_4^{2-}(aq)$

　　　　c.　$Al(NO_3)_3(s) \rightarrow Al^{3+}(aq) + 3\ NO_3^-(aq)$　　　d.　$SrBr_2(s) \rightarrow Sr^{2+}(aq) + 2\ Br^-(aq)$

　　　　e.　$KClO_4(s) \rightarrow K^+(aq) + ClO_4^-(aq)$　　　　f.　$NH_4Br(s) \rightarrow NH_4^+(aq) + Br^-(aq)$

　　　　g.　$NH_4NO_3(s) \rightarrow NH_4^+(aq) + NO_3^-(aq)$　　h.　$CuSO_4(s) \rightarrow Cu^{2+}(aq) + SO_4^{2-}(aq)$

　　　　i.　$NaOH(s) \rightarrow Na^+(aq) + OH^-(aq)$

15.　　$\text{Molality} = \dfrac{40.0\ \text{g EG}}{60.0\ \text{g H}_2\text{O}} \times \dfrac{1000\ \text{g}}{\text{kg}} \times \dfrac{1\ \text{mol EG}}{62.07\ \text{g}} = 10.7\ \text{mol/kg}$

$\text{Molarity} = \dfrac{40.0\ \text{g EG}}{100.0\ \text{g solution}} \times \dfrac{1.05\ \text{g}}{\text{cm}^3} \times \dfrac{1000\ \text{cm}^3}{\text{L}} \times \dfrac{1\ \text{mol}}{62.07\ \text{g}} = 6.77\ \text{mol/L}$

$40.0\ \text{g EG} \times \dfrac{1\ \text{mol}}{62.07\ \text{g}} = 0.644\ \text{mol EG};\ \ 60.0\ \text{g H}_2\text{O} \times \dfrac{1\ \text{mol}}{18.02\ \text{g}} = 3.33\ \text{mol H}_2\text{O}$

$\chi_{EG} = \dfrac{0.644}{3.33 + 0.644} = 0.162 = \text{mole fraction ethylene glycol}$

17.　　$25\ \text{mL C}_5\text{H}_{12} \times \dfrac{0.63\ \text{g}}{\text{mL}} = 16\ \text{g C}_5\text{H}_{12};\ \ 25\ \text{mL} \times \dfrac{0.63\ \text{g}}{\text{mL}} \times \dfrac{1\ \text{mol}}{72.15\ \text{g}} = 0.22\ \text{mol C}_5\text{H}_{12}$

$45\ \text{mL C}_6\text{H}_{14} \times \dfrac{0.66\ \text{g}}{\text{mL}} = 30.\ \text{g C}_6\text{H}_{14};\ \ 45\ \text{mL} \times \dfrac{0.66\ \text{g}}{\text{mL}} \times \dfrac{1\ \text{mol}}{86.17\ \text{g}} = 0.34\ \text{mol C}_6\text{H}_{14}$

$\text{Mass \% pentane} = \dfrac{\text{mass pentane}}{\text{total mass}} \times 100 = \dfrac{16\ \text{g}}{16\ \text{g} + 30.\ \text{g}} \times 100 = 35\%$

$$\chi_{pentane} = \frac{mol\ pentane}{total\ mol} = \frac{0.22\ mol}{0.22\ mol + 0.34\ mol} = 0.39$$

$$Molality = \frac{mol\ pentane}{kg\ hexane} = \frac{0.22\ mol}{0.030\ kg} = 7.3\ mol/kg$$

$$Molarity = \frac{mol\ pentane}{L\ solution} = \frac{0.22\ mol}{25\ mL + 45\ mL} \times \frac{1000\ mL}{1\ L} = 3.1\ mol/L$$

19.    If we have 1.00 L of solution:

$$1.37\ mol\ citric\ acid \times \frac{192.1\ g}{mol} = 263\ g\ citric\ acid$$

$$1.00 \times 10^3\ mL\ solution \times \frac{1.10\ g}{mL} = 1.10 \times 10^3\ g\ solution$$

$$Mass\ \%\ of\ citric\ acid = \frac{263\ g}{1.10 \times 10^3\ g} \times 100 = 23.9\%$$

In 1.00 L of solution, we have 263 g citric acid and $(1.10 \times 10^3 - 263) = 840$ g of $H_2O$.

$$Molality = \frac{1.37\ mol\ citric\ acid}{0.84\ kg\ H_2O} = 1.6\ mol/kg$$

$$840\ g\ H_2O \times \frac{1\ mol}{18.0\ g} = 47\ mol\ H_2O; \quad \chi_{citric\ acid} = \frac{1.37}{47 + 1.37} = 0.028$$

21.    Because the density of water is 1.00 g/mL, 100.0 mL of water has a mass of 100. g.

$$Density = \frac{mass}{volume} = \frac{10.0\ g\ H_3PO_4 + 100.\ g\ H_2O}{104\ mL} = 1.06\ g/mL = 1.06\ g/cm^3$$

$$Mol\ H_3PO_4 = 10.0\ g \times \frac{1\ mol}{97.99\ g} = 0.102\ mol\ H_3PO_4$$

$$Mol\ H_2O = 100.\ g \times \frac{1\ mol}{18.02\ g} = 5.55\ mol\ H_2O$$

$$Mole\ fraction\ of\ H_3PO_4 = \frac{0.102\ mol\ H_3PO_4}{(0.102 + 5.55)\ mol} = 0.0180$$

$$\chi_{H_2O} = 1.0000 - 0.0180 = 0.9820$$

$$\text{Molarity} = \frac{0.102 \text{ mol H}_3\text{PO}_4}{0.104 \text{ L soln}} = 0.981 \text{ mol/L}$$

$$\text{Molality} = \frac{0.102 \text{ mol H}_3\text{PO}_4}{0.100 \text{ kg solvent}} = 1.02 \text{ mol/kg}$$

## Thermodynamics of Solutions and Solubility

23. "Like dissolves like" refers to the nature of the intermolecular forces. Polar solutes and ionic solutes dissolve in polar solvents because the types of intermolecular forces present in solute and solvent are similar. When they dissolve, the strengths of the intermolecular forces in solution are about the same as in pure solute and pure solvent. The same is true for nonpolar solutes in nonpolar solvents. The strengths of the intermolecular forces (London dispersion forces) are about the same in solution as in pure solute and pure solvent. In all cases of like dissolves like, the magnitude of $\Delta H_{soln}$ is either a small positive number (endothermic) or a small negative number (exothermic), with a value close to zero. For polar solutes in nonpolar solvents and vice versa, $\Delta H_{soln}$ is a very large, unfavorable value (very endothermic). Because the energetics are so unfavorable, polar solutes do not dissolve in nonpolar solvents, and vice versa.

25. Using Hess's law:

$$\text{NaI(s)} \rightarrow \text{Na}^+(g) + \text{I}^-(g) \qquad\qquad \Delta H = -\Delta H_{LE} = -(-686 \text{ kJ/mol})$$
$$\text{Na}^+(g) + \text{I}^-(g) \rightarrow \text{Na}^+(aq) + \text{I}^-(aq) \qquad \Delta H = \Delta H_{hyd} = -694 \text{ kJ/mol}$$
_____
$$\text{NaI(s)} \rightarrow \text{Na}^+(aq) + \text{I}^-(aq) \qquad\qquad \Delta H_{soln} = -8 \text{ kJ/mol}$$

$\Delta H_{soln}$ refers to the heat released or gained when a solute dissolves in a solvent. Here, an ionic compound dissolves in water.

27. Both $Al(OH)_3$ and NaOH are ionic compounds. Because the lattice energy is proportional to the charge of the ions, the lattice energy of aluminum hydroxide is greater than that of sodium hydroxide. The attraction of water molecules for $Al^{3+}$ and $OH^-$ cannot overcome the larger lattice energy, and $Al(OH)_3$ is insoluble. For NaOH, the favorable hydration energy is large enough to overcome the smaller lattice energy, and NaOH is soluble.

29. Water is a polar molecule capable of hydrogen bonding. Polar molecules, especially molecules capable of hydrogen bonding, and ions can be hydrated. For covalent compounds, as polarity increases, the attraction to water (hydration) increases. For ionic compounds, as the charge of the ions increase and/or the size of the ions decrease, the attraction to water (hydration) increases.

    a.  $CH_3CH_2OH$; $CH_3CH_2OH$ is polar, whereas $CH_3CH_2CH_3$ is nonpolar.

    b.  $CHCl_3$; $CHCl_3$ is polar, whereas $CCl_4$ is nonpolar.

    c.  $CH_3CH_2OH$; $CH_3CH_2OH$ is much more polar than $CH_3(CH_2)_{14}CH_2OH$.

31.    As the length of the hydrocarbon chain increases, the solubility decreases.  The –OH end of the alcohols can hydrogen-bond with water.  The hydrocarbon chain, however, is basically nonpolar and interacts poorly with water.  As the hydrocarbon chain gets longer, a greater portion of the molecule cannot interact with the water molecules, and the solubility decreases; i.e., the effect of the –OH group decreases as the alcohols get larger.

33.    Structure effects refer to solute and solvent having similar polarities in order for solution formation to occur. Hydrophobic solutes are mostly nonpolar substances that are "water-fearing." Hydrophilic solutes are mostly polar or ionic substances that are "water-loving." Pressure has little effect on the solubilities of solids or liquids; it does significantly affect the solubility of a gas. Henry's law states that the amount of a gas dissolved in a solution is directly proportional to the pressure of the gas above the solution (C = kP). The equation for Henry's law works best for dilute solutions of gases that do not dissociate in or react with the solvent. HCl(g) does not follow Henry's law because it dissociates into $H^+$(aq) and $Cl^-$(aq) in solution (HCl is a strong acid). For $O_2$ and $N_2$, Henry's law works well since these gases do not react with the water solvent.

An increase in temperature can either increase or decrease the solubility of a solid solute in water. It is true that a solute dissolves more rapidly with an increase in temperature, but the amount of solid solute that dissolves to form a saturated solution can either decrease or increase with temperature. The temperature effect is difficult to predict for solid solutes. However, the temperature effect for gas solutes is easier to predict because the solubility of a gas typically decreases with increasing temperature.

35.    $P_{gas} = kC$,   $0.790 \text{ atm} = k \times \dfrac{8.21 \times 10^{-4} \text{ mol}}{L}$,   $k = 962$ L atm/mol

$P_{gas} = kC$,   $1.10 \text{ atm} = \dfrac{962 \text{ L atm}}{\text{mol}} \times C$,   $C = 1.14 \times 10^{-3}$ mol/L

37.    As the temperature increases the gas molecules will have a greater average kinetic energy.  A greater fraction of the gas molecules in solution will have a kinetic energy greater than the attractive forces between the gas molecules and the solvent molecules.  More gas molecules are able to escape to the vapor phase and the solubility of the gas decreases.

## Vapor Pressures of Solution

39.    $P_{soln} = \chi_{C_2H_5OH} \, P^o_{C_2H_5OH}$;   $\chi_{C_2H_5OH} = \dfrac{\text{mol } C_2H_5OH \text{ in solution}}{\text{total mol in solution}}$

$53.6 \text{ g } C_3H_8O_3 \times \dfrac{1 \text{ mol } C_3H_8O_3}{92.09 \text{ g}} = 0.582 \text{ mol } C_3H_8O_3$

$133.7 \text{ g } C_2H_5OH \times \dfrac{1 \text{ mol } C_2H_5OH}{46.07 \text{ g}} = 2.90 \text{ mol } C_2H_5OH$; total mol = 0.582 + 2.90 = 3.48 mol

$113 \text{ torr} = \dfrac{2.90 \text{ mol}}{3.48 \text{ mol}} \times P^o_{C_2H_5OH}$,   $P^o_{C_2H_5OH} = 136$ torr

41.     Solution d (methanol-water); methanol is more volatile than water, which will increase the
        total vapor pressure to a value greater than the vapor pressure of pure water at this
        temperature.

43.     $P_B = \chi_B P_B^o$, $\chi_B = P_B / P_B^o = 0.900 \text{ atm}/0.930 \text{ atm} = 0.968$

        $0.968 = \dfrac{\text{mol benzene}}{\text{total mol}}$; mol benzene $= 78.11 \text{ g } C_6H_6 \times \dfrac{1 \text{ mol}}{78.11 \text{ g}} = 1.000 \text{ mol}$

        Let $x$ = mol solute; then: $\chi_B = 0.968 = \dfrac{1.000 \text{ mol}}{1.000 + x}$, $\;0.968 + (0.968)x = 1.000$, $\;x = 0.033 \text{ mol}$

        Molar mass $= \dfrac{10.0 \text{ g}}{0.033 \text{ mol}} = 303 \text{ g/mol} \approx 3.0 \times 10^2 \text{ g/mol}$

45.     a.   $25 \text{ mL } C_5H_{12} \times \dfrac{0.63 \text{ g}}{\text{mL}} \times \dfrac{1 \text{ mol}}{72.15} = 0.22 \text{ mol } C_5H_{12}$

             $45 \text{ mL } C_6H_{14} \times \dfrac{0.66 \text{ g}}{\text{mL}} \times \dfrac{1 \text{ mol}}{86.17} = 0.34 \text{ mol } C_6H_{14}$; total mol $= 0.22 + 0.34 = 0.56 \text{ mol}$

             $\chi_{pen}^{L} = \dfrac{\text{mol pentane in solution}}{\text{total mol in solution}} = \dfrac{0.22 \text{ mol}}{0.56 \text{ mol}} = 0.39$, $\;\chi_{hex}^{L} = 1.00 - 0.39 = 0.61$

             $P_{pen} = \chi_{pen}^{L} P_{pen}^o = 0.39(511 \text{ torr}) = 2.0 \times 10^2 \text{ torr}$; $\;P_{hex} = 0.61(150. \text{ torr}) = 92 \text{ torr}$

             $P_{total} = P_{pen} + P_{hex} = 2.0 \times 10^2 + 92 = 292 \text{ torr} = 290 \text{ torr}$

        b.   From Chapter 5 on gases, the partial pressure of a gas is proportional to the number of
             moles of gas present. For the vapor phase:

             $\chi_{pen}^{V} = \dfrac{\text{mol pentane in vapor}}{\text{total mol vapor}} = \dfrac{P_{pen}}{P_{total}} = \dfrac{2.0 \times 10^2 \text{ torr}}{290 \text{ torr}} = 0.69$

        Note: In the Solutions Guide, we added V or L to the mole fraction symbol to emphasize
        which value we are solving. If the L or V is omitted, then the liquid phase is assumed.

47.     $P_{total} = P_{meth} + P_{prop}$, $\;174 \text{ torr} = \chi_{meth}^{L}(303 \text{ torr}) + \chi_{prop}^{L}(44.6 \text{ torr})$; $\;\chi_{prop}^{L} = 1.000 - \chi_{meth}^{L}$

        $174 = 303\chi_{meth}^{L} + (1.000 - \chi_{meth}^{L})44.6 \text{ torr}$, $\;\dfrac{129}{258} = \chi_{meth}^{L} = 0.500$

        $\chi_{prop}^{L} = 1.000 - 0.500 = 0.500$

49.    $50.0 \text{ g CH}_3\text{COCH}_3 \times \dfrac{1 \text{ mol}}{58.08 \text{ g}} = 0.861 \text{ mol acetone}$

$50.0 \text{ g CH}_3\text{OH} \times \dfrac{1 \text{ mol}}{32.04 \text{ g}} = 1.56 \text{ mol methanol}$

$\chi^L_{acetone} = \dfrac{0.861}{0.861 + 1.56} = 0.356; \quad \chi^L_{methanol} = 1.000 - \chi^L_{acetone} = 0.644$

$P_{total} = P_{methanol} + P_{acetone} = 0.644(143 \text{ torr}) + 0.356(271 \text{ torr}) = 92.1 \text{ torr} + 96.5 \text{ torr}$

$= 188.6 \text{ torr}$

Because partial pressures are proportional to the moles of gas present, in the vapor phase:

$\chi^V_{acetone} = \dfrac{P_{acetone}}{P_{total}} = \dfrac{96.5 \text{ torr}}{188.6 \text{ torr}} = 0.512; \quad \chi^V_{methanol} = 1.000 - 0.512 = 0.488$

The actual vapor pressure of the solution (161 torr) is less than the calculated pressure assuming ideal behavior (188.6 torr). Therefore, the solution exhibits negative deviations from Raoult's law. This occurs when the solute-solvent interactions are stronger than in pure solute and pure solvent.

51.    No, the solution is not ideal. For an ideal solution, the strengths of intermolecular forces in the solution are the same as in pure solute and pure solvent. This results in $\Delta H_{soln} = 0$ for an ideal solution. $\Delta H_{soln}$ for methanol-water is not zero. Because $\Delta H_{soln} < 0$, this solution exhibits negative deviation from Raoult's law.

53.    Solutions of A and B have vapor pressures less than ideal (see Figure 17.11 of the text), so this plot shows negative deviations from Rault's law. Negative deviations occur when the intermolecular forces are stronger in solution than in pure solvent and solute. This results in an exothermic enthalpy of solution. The only statement that is false is e. A substance boils when the vapor pressure equals the external pressure. Because $\chi_B = 0.6$ has a lower vapor pressure at the temperature of the plot than either pure A or pure B, one would expect this solution to require the highest temperature in order for the vapor pressure to reach the external pressure. Therefore, the solution with $\chi_B = 0.6$ will have a higher boiling point than either pure A or pure B. (Note that because $P^o_B > P^o_A$, B is more volatile than A.)

## Colligative Properties

55.    Osmotic pressure: the pressure that must be applied to a solution to stop osmosis; osmosis is the flow of solvent into the solution through a semipermeable membrane. The equation to calculate osmotic pressure $\pi$ is:

$\pi = MRT$

where $M$ is the molarity of the solution, R is the gas constant, and T is the Kelvin temperature. The molarity of a solution approximately equals the molality of the solution when 1 kg solvent $\approx$ 1 L solution. This occurs for dilute solutions of water because $d_{H_2O} = 1.00$ g/cm$^3$.

57.    Molality $= m = \dfrac{\text{mol solute}}{\text{kg solvent}} = \dfrac{27.0 \text{ g N}_2\text{H}_4\text{CO}}{150.0 \text{ g H}_2\text{O}} \times \dfrac{1000 \text{ g}}{\text{kg}} \times \dfrac{1 \text{ mol N}_2\text{H}_4\text{CO}}{60.06 \text{ g N}_2\text{H}_4\text{CO}} = 3.00$ molal

$\Delta T_b = K_b m = \dfrac{0.51 \,^\circ\text{C}}{\text{molal}} \times 3.00 \text{ molal} = 1.5^\circ\text{C}$

The boiling point is raised from 100.0°C to 101.5°C (assuming P = 1 atm).

59.    Molality $= m = \dfrac{50.0 \text{ g C}_2\text{H}_6\text{O}_2}{50.0 \text{ g H}_2\text{O}} \times \dfrac{1000 \text{ g}}{\text{kg}} \times \dfrac{1 \text{ mol}}{62.07 \text{ g}} = 16.1$ mol/kg

$\Delta T_f = K_f m = 1.86 \,^\circ\text{C/molal} \times 16.1 \text{ molal} = 29.9^\circ\text{C}; \quad T_f = 0.0^\circ\text{C} - 29.9^\circ\text{C} = -29.9^\circ\text{C}$

$\Delta T_b = K_b m = 0.51^\circ\text{C/molal} \times 16.1 \text{ molal} = 8.2^\circ\text{C}; \quad T_b = 100.0^\circ\text{C} + 8.2^\circ\text{C} = 108.2^\circ\text{C}$

61.    $m = \dfrac{24.0 \text{ g} \times \dfrac{1 \text{ mol}}{58.0 \text{ g}}}{0.600 \text{ kg}} = 0.690 \text{ mol/kg}; \quad \Delta T_b = K_b m = 0.51^\circ\text{C kg/mol} \times 0.690 \text{ mol/kg} = 0.35^\circ\text{C}$

$T_b = 99.725^\circ\text{C} + 0.35^\circ\text{C} = 100.08^\circ\text{C}$

63.    $\Delta T_b = 77.85^\circ\text{C} - 76.50^\circ\text{C} = 1.35^\circ\text{C}; \quad m = \dfrac{\Delta T_b}{K_b} = \dfrac{1.35 \,^\circ\text{C}}{5.03 \,^\circ\text{C kg/mol}} = 0.268$ mol/kg

Mol biomolecule $= 0.0150$ kg solvent $\times \dfrac{0.268 \text{ mol hydrocarbon}}{\text{kg solvent}} = 4.02 \times 10^{-3}$ mol

From the problem, 2.00 g biomolecule was used, which must contain $4.02 \times 10^{-3}$ mol biomolecule. The molar mass of the biomolecule is:

$\dfrac{2.00 \text{ g}}{4.02 \times 10^{-3} \text{ mol}} = 498$ g/mol

65.    $\Delta T_f = K_f m, \quad m = \dfrac{\Delta T_f}{K_f} = \dfrac{0.300^\circ\text{C}}{5.12 \,^\circ\text{C kg/mol}} = \dfrac{5.86 \times 10^{-2} \text{ mol thyroxine}}{\text{kg benzene}}$

The moles of thyroxine present are:

$0.0100$ kg benzene $\times \dfrac{5.86 \times 10^{-2} \text{ mol thyroxine}}{\text{kg benzene}} = 5.86 \times 10^{-4}$ mol thyroxine

From the problem, 0.455 g thyroxine was used; this must contain $5.86 \times 10^{-4}$ mol thyroxine. The molar mass of the thyroxine is:

$$\text{molar mass} = \frac{0.455 \text{ g}}{5.86 \times 10^{-4} \text{ mol}} = 776 \text{ g/mol}$$

67.    $M = \dfrac{1.0 \text{ g}}{\text{L}} \times \dfrac{1 \text{ mol}}{9.0 \times 10^4 \text{ g}} = 1.1 \times 10^{-5} \text{ mol/L}; \quad \pi = MRT$

At 298 K:  $\pi = \dfrac{1.1 \times 10^{-5} \text{ mol}}{\text{L}} \times \dfrac{0.08206 \text{ L atm}}{\text{K mol}} \times 298 \text{ K} \times \dfrac{760 \text{ torr}}{\text{atm}}, \quad \pi = 0.20 \text{ torr}$

Because d = 1.0 g/cm$^3$, 1.0 L of solution has a mass of 1.0 kg.  Because only 1.0 g of protein is present per liter solution, 1.0 kg of $H_2O$ is present, and molality equals molarity to the correct number of significant figures.

$\Delta T_f = K_f m = \dfrac{1.86 \text{ }^\circ\text{C}}{\text{molal}} \times 1.1 \times 10^{-5} \text{ molal} = 2.0 \times 10^{-5}\text{°C}$

69.    $\pi = MRT = \dfrac{0.1 \text{ mol}}{\text{L}} \times \dfrac{0.08206 \text{ L atm}}{\text{K mol}} \times 298 \text{ K} = 2.45 \text{ atm} \approx 2 \text{ atm}$

$\pi = 2 \text{ atm} \times \dfrac{760 \text{ mm Hg}}{\text{atm}} \approx 2000 \text{ mm} \approx 2 \text{ m}$

The osmotic pressure would support a mercury column of approximately 2 m.  The height of a fluid column in a tree will be higher because Hg is more dense than the fluid in a tree.  If we assume the fluid in a tree is mostly $H_2O$, then the fluid has a density of 1.0 g/cm$^3$.  The density of Hg is 13.6 g/cm$^3$.

Height of fluid ≈ 2 m × 13.6 ≈ 30 m

71.    With addition of salt or sugar, the osmotic pressure inside the fruit cells (and bacteria) is less than outside the cell.  Water will leave the cells, which will dehydrate any bacteria present, causing them to die.

## Properties of Electrolyte Solutions

73.    19.6 torr = $\chi_{H_2O}(23.8 \text{ torr})$,   $\chi_{H_2O} = 0.824$;   $\chi_{\text{solute}} = 1.000 - 0.824 = 0.176$

0.176 is the mole fraction of all the solute particles present.  Because NaCl dissolves to produce two ions in solution ($Na^+$ and $Cl^-$), 0.176 is the mole fraction of $Na^+$ and $Cl^-$ ions present.  The mole fraction of NaCl is 1/2 (0.176) = 0.0880 = $\chi_{\text{NaCl}}$.

At 45°C,  $P_{\text{soln}} = 0.824(71.9 \text{ torr}) = 59.2 \text{ torr}$.

75.    $Na_3PO_4(s) \rightarrow 3\ Na^+(aq) + PO_4^{3-}(aq)$,  i = 4.0;  $CaBr_2(s) \rightarrow Ca^{2+}(aq) + 2\ Br^-(aq)$,  i = 3.0

       $KCl(s) \rightarrow K^+(aq) + Cl^-(aq)$,   i = 2.0

       The effective particle concentrations of the solutions are (assuming complete dissociation):

> 4.0(0.010 molal) = 0.040 molal for the $Na_3PO_4$ solution;  3.0(0.020 molal) = 0.060 molal for the $CaBr_2$ solution;  2.0(0.020 molal) = 0.040 molal for the KCl solution;  slightly greater than 0.020 molal for the HF solution because HF only partially dissociates in water (it is a weak acid).

a.     The 0.010 $m$ $Na_3PO_4$ solution and the 0.020 $m$ KCl solution both have effective particle concentrations of 0.040 $m$ (assuming complete dissociation), so both of these solutions should have the same boiling point as the 0.040 $m$ $C_6H_{12}O_6$ solution (a nonelectrolyte).

b.     $P = \chi P°$;  as the solute concentration decreases, the solvent's vapor pressure increases because $\chi$ increases.  Therefore, the 0.020 $m$ HF solution will have the highest vapor pressure because it has the smallest effective particle concentration.

c.     $\Delta T = K_f m$; the 0.020 $m$ $CaBr_2$ solution has the largest effective particle concentration, so it will have the largest freezing point depression (largest $\Delta T$).

77.    The van't Hoff factor i is the number of moles of particles (ions) produced for every mole of solute dissolved. For NaCl, i = 2 since $Na^+$ and $Cl^-$ are produced in water; for $Al(NO_3)_3$, i = 4 since $Al^{3+}$ and 3 $NO_3^-$ ions are produced when $Al(NO_3)_3$ dissolves in water. In real life, the van't Hoff factor is rarely the value predicted by the number of ions a salt dissolves into; i is generally something less than the predicted number of ions. This is due to a phenomenon called ion pairing, where at any instant a small percentage of oppositely charged ions pair up and act like a single solute particle. Ion pairing occurs most when the concentration of ions is large. Therefore, dilute solutions behave most ideally; here, i is close to that determined by the number of ions in a salt.

79.    $NaCl(s) \rightarrow Na^+(aq) + Cl^-(aq)$,  i = 2.0

$$\pi = iMRT = 2.0 \times \frac{0.10\ mol}{L} \times \frac{0.08206\ L\ atm}{K\ mol} \times 293\ K = 4.8\ atm$$

       A pressure greater than 4.8 atm should be applied to ensure purification by reverse osmosis.

81.    a.    $MgCl_2$, i (observed) = 2.7

             $\Delta T_f = iK_f m = 2.7 \times 1.86\ °C/molal \times 0.050\ molal = 0.25°C$;  $T_f = -0.25°C$

             $\Delta T_b = iK_b m = 2.7 \times 0.51\ °C/molal \times 0.050\ molal = 0.069°C$;  $T_b = 100.069°C$

       b.    $FeCl_3$, i (observed) = 3.4

             $\Delta T_f = iK_f m = 3.4 \times 1.86\ °C/molal \times 0.050\ molal = 0.32°C$;  $T_f = -0.32°C$

             $\Delta T_b = iK_b m = 3.4 \times 0.51°C/molal \times 0.050\ molal = 0.087°C$;  $T_b = 100.087°C$

83.    $\pi = iMRT, \ M = \dfrac{\pi}{iRT} = \dfrac{2.50 \text{ atm}}{2.00 \times \dfrac{0.08206 \text{ L atm}}{\text{K mol}} \times 298 \text{ K}} = 5.11 \times 10^{-2} \text{ mol/L}$

Molar mass of compound $= \dfrac{0.500 \text{ g}}{0.1000 \text{ L} \times \dfrac{5.11 \times 10^{-2} \text{ mol}}{\text{L}}} = 97.8 \text{ g/mol}$

## Additional Exercises

85.    Both solutions and colloids have suspended particles in some medium.  The major difference between the two is the size of the particles.  A colloid is a suspension of relatively large particles compared to a solution.  Because of this, colloids will scatter light, whereas solutions will not.  The scattering of light by a colloidal suspension is called the Tyndall effect.

87.    Coagulation is the destruction of a colloid by the aggregation of many suspended particles to form a large particle that settles out of solution.

89.    A 92 proof ethanol solution is 46% $C_2H_5OH$ by volume.  Assuming 100.0 mL of solution:

mol ethanol $= 46 \text{ mL } C_2H_5OH \times \dfrac{0.79 \text{ g}}{\text{mL}} \times \dfrac{1 \text{ mol } C_2H_5OH}{46.07 \text{ g}} = 0.79 \text{ mol } C_2H_5OH$

molarity $= \dfrac{0.79 \text{ mol}}{0.1000 \text{ L}} = 7.9 \ M$ ethanol

91.    $\Delta T = K_f m, \ m = \dfrac{\Delta T}{K_f} = \dfrac{2.79°C}{1.86 \ °C/\text{molal}} = 1.50 \text{ molal}$

a.    $\Delta T = K_b m, \ \ \Delta T = (0.51°C/\text{molal})(1.50 \text{ molal}) = 0.77°C, \ \ T_b = 100.77°C$

b.    $P_{soln} = \chi_{water} P^o_{water}, \ \ \chi_{water} = \dfrac{\text{mol } H_2O}{\text{mol } H_2O + \text{mol solute}}$

Assuming 1.00 kg of water, we have 1.50 mol solute, and:

mol $H_2O = 1.00 \times 10^3 \text{ g } H_2O \times \dfrac{1 \text{ mol } H_2O}{18.02 \text{ g } H_2O} = 55.5 \text{ mol } H_2O$

$\chi_{water} = \dfrac{55.5 \text{ mol}}{1.50 + 55.5} = 0.974; \ \ P_{soln} = (0.974)(23.76 \text{ mm Hg}) = 23.1 \text{ mm Hg}$

c.    We assumed ideal behavior in solution formation, we assumed the solute was nonvolatile, and we assumed i = 1 (no ions formed).

93.

Benzoic acid is capable of hydrogen-bonding, but a significant part of benzoic acid is the nonpolar benzene ring.   In benzene, a hydrogen-bonded dimer forms.

The dimer is relatively nonpolar and thus more soluble in benzene than in water.  Because benzoic acid forms dimers in benzene, the effective solute particle concentration will be less than 1.0 molal. Therefore, the freezing-point depression would be less than 5.12°C   ($\Delta T_f = K_f m$).

95.     a.   $NH_4NO_3(s) \rightarrow NH_4^+(aq) + NO_3^-(aq)$  $\Delta H_{soln} = ?$

Heat gain by dissolution process = heat loss by solution;  We will keep all quantities positive in order to avoid sign errors.  Because the temperature of the water decreased, the dissolution of $NH_4NO_3$ is endothermic ($\Delta H$ is positive).  Mass of solution = 1.60 + 75.0 = 76.6 g

$$\text{Heat loss by solution} = \frac{4.18\,J}{°C\,g} \times 76.6\,g \times (25.00°C - 23.34°C) = 532\,J$$

$$\Delta H_{soln} = \frac{532\,J}{1.60\,g\,NH_4NO_3} \times \frac{80.05\,g\,NH_4NO_3}{mol\,NH_4NO_3} = 2.66 \times 10^4\,J/mol = 26.6\,kJ/mol$$

b.  We will use Hess's law to solve for the lattice energy.  The lattice energy equation is:

$NH_4^+(g) + NO_3^-(g) \rightarrow NH_4NO_3(s)$          $\Delta H$ = lattice energy

$NH_4^+(g) + NO_3^-(g) \rightarrow NH_4^+(aq) + NO_3^-(aq)$  $\Delta H = \Delta H_{hyd} = -630.\,kJ/mol$
$NH_4^+(aq) + NO_3^-(aq) \rightarrow NH_4NO_3(s)$   $\Delta H = -\Delta H_{soln} = -26.6\,kJ/mol$

───────────────────────────────────────────

$NH_4^+(g) + NO_3^-(g) \rightarrow NH_4NO_3(s)$   $\Delta H = \Delta H_{hyd} - \Delta H_{soln} = -657\,kJ/mol$

97.     $\chi_{pen}^V = 0.15 = \dfrac{P_{pen}}{P_{total}}$;   $P_{pen} = \chi_{pen}^L P_{pen}^o$;   $P_{total} = P_{pen} + P_{hex} = \chi_{pen}^L(511) + \chi_{hex}^L(150.)$

Because $\chi_{hex}^L = 1.000 - \chi_{pen}^L$ :   $P_{total} = \chi_{pen}^L(511) + (1.000 - \chi_{pen}^L)(150.) = 150. + 361\chi_{pen}^L$

$$\chi_{pen}^{V} = \frac{P_{pen}}{P_{total}}, \ 0.15 = \frac{\chi_{pen}^{L}(511)}{150. + 361\chi_{pen}^{L}}, \ 0.15(150. + 361\chi_{pen}^{L}) = 511\chi_{pen}^{L}$$

$$23 + 54\chi_{pen}^{L} = 511\chi_{pen}^{L}, \ \chi_{pen}^{L} = \frac{23}{457} = 0.050$$

99.    $14.2 \text{ mg CO}_2 \times \dfrac{12.01 \text{ mg C}}{44.01 \text{ mg CO}_2} = 3.88 \text{ mg C}; \ \% \text{ C} = \dfrac{3.88 \text{ mg}}{4.80 \text{ mg}} \times 100 = 80.8\% \text{ C}$

$1.65 \text{ mg H}_2\text{O} \times \dfrac{2.016 \text{ mg H}}{18.02 \text{ mg H}_2\text{O}} = 0.185 \text{ mg H}; \ \% \text{ H} = \dfrac{0.185 \text{ mg}}{4.80 \text{ mg}} \times 100 = 3.85\% \text{ H}$

Mass % O = 100.00 − (80.8 + 3.85) = 15.4% O

Out of 100.00 g:

$$80.8 \text{ g C} \times \frac{1 \text{ mol}}{12.01 \text{ g}} = 6.73 \text{ mol C}; \ \frac{6.73}{0.963} = 6.99 \approx 7$$

$$3.85 \text{ g H} \times \frac{1 \text{ mol}}{1.008 \text{ g}} = 3.82 \text{ mol H}; \ \frac{3.82}{0.963} = 3.97 \approx 4$$

$$15.4 \text{ g O} \times \frac{1 \text{ mol}}{16.00 \text{ g}} = 0.963 \text{ mol O}; \ \frac{0.963}{0.963} = 1.00$$

Therefore, the empirical formula is $C_7H_4O$.

$$\Delta T_f = K_f m, \ m = \frac{\Delta T_f}{K_f} = \frac{22.3 \ ^\circ\text{C}}{40. \ ^\circ\text{C} / \text{molal}} = 0.56 \text{ molal}$$

$$\text{Mol anthraquinone} = 0.0114 \text{ kg camphor} \times \frac{0.56 \text{ mol anthraquinone}}{\text{kg camphor}} = 6.4 \times 10^{-3} \text{ mol}$$

$$\text{Molar mass} = \frac{1.32 \text{ g}}{6.4 \times 10^{-3} \text{ mol}} = 210 \text{ g/mol}$$

The empirical mass of $C_7H_4O$ is $7(12) + 4(1) + 16 \approx 104$ g/mol.  Because the molar mass is twice the empirical mass, the molecular formula is $C_{14}H_8O_2$.

101.    a.   $m = \dfrac{\Delta T_f}{K_f} = \dfrac{1.32^\circ\text{C}}{5.12 \ ^\circ\text{C kg/mol}} = 0.258 \text{ mol/kg}$

$$\text{Mol unknown} = 0.01560 \text{ kg} \times \frac{0.258 \text{ mol unknown}}{\text{kg}} = 4.02 \times 10^{-3} \text{ mol}$$

$$\text{Molar mass of unknown} = \frac{1.22 \text{ g}}{4.02 \times 10^{-3} \text{ mol}} = 303 \text{ g/mol}$$

$$\text{Uncertainty in temperature} = \frac{0.04}{1.32} \times 100 = 3\%$$

A 3% uncertainty in 303 g/mol = 9 g/mol.

So molar mass = 303 ±9 g/mol.

b. No, codeine could not be eliminated since its molar mass is in the possible range including the uncertainty.

c. We would like the uncertainty to be ±1 g/mol. We need the freezing-point depression to be about 10 times what it was in this problem. Two possibilities are:

1. make the solution 10 times more concentrated (may be solubility problem)

2. use a solvent with a larger $K_f$ value, e.g., camphor

103.            $M_3X_2(s) \rightarrow 3 M^{2+}(aq) + 2 X^{3-}(aq)$        $K_{sp} = [M^{2+}]^3[X^{3-}]^2$

Initial    $s$ = solubility (mol/L)        0                0
Equil.                                    $3s$              $2s$

$K_{sp} = (3s)^3(2s)^2 = 108s^5$;  total ion concentration = $3s + 2s = 5s$.

$$\pi = iMRT, \quad iM = \text{total ion concentration} = \frac{\pi}{RT} = \frac{2.64 \times 10^{-2} \text{ atm}}{0.08206 \text{ L atm K}^{-1} \text{ mol}^{-1} \times 298 \text{ K}}$$

$$= 1.08 \times 10^{-3} \text{ mol/L}$$

$5s = 1.08 \times 10^{-3}$ mol/L,  $s = 2.16 \times 10^{-4}$ mol/L

$K_{sp} = 108s^5 = 108(2.16 \times 10^{-4})^5 = 5.08 \times 10^{-17}$

105.    $m = \dfrac{0.100 \text{ g} \times \dfrac{1 \text{ mol}}{100.0 \text{ g}}}{0.5000 \text{ kg}} = 2.00 \times 10^{-3}$ mol/kg $\approx 2.00 \times 10^{-3}$ mol/L    (dilute solution)

$\Delta T_f = iK_f m$,  $0.0056°C = i(1.86 °C/molal)(2.00 \times 10^{-3} \text{ molal})$,  i = 1.5

If i = 1.0, percent dissociation = 0%, and if i = 2.0, percent dissociation = 100%. Because i = 1.5, the weak acid is 50.% dissociated.

$$HA \rightleftharpoons H^+ + A^- \qquad K_a = \frac{[H^+][A^-]}{[HA]}$$

Because the weak acid is 50.% dissociated:

$$[H^+] = [A^-] = [HA]_0 \times 0.50 = 2.00 \times 10^{-3}\ M \times 0.50 = 1.0 \times 10^{-3}\ M$$

$$[HA] = [HA]_0 - \text{amount HA reacted} = 2.00 \times 10^{-3}\ M - 1.0 \times 10^{-3}\ M = 1.0 \times 10^{-3}\ M$$

$$K_a = \frac{[H^+][A^-]}{[HA]} = \frac{(1.0 \times 10^{-3})(1.0 \times 10^{-3})}{1.0 \times 10^{-3}} = 1.0 \times 10^{-3}$$

107.   $iM = \dfrac{\pi}{RT} = \dfrac{0.3950\ \text{atm}}{0.08206\ \text{L atm K}^{-1}\ \text{mol}^{-1}(298.2\ \text{K})} = 0.01614\ \text{mol/L} = \text{total ion concentration}$

$$0.01614\ \text{mol/L} = M_{Mg^{2+}} + M_{Na^+} + M_{Cl^-};\quad M_{Cl^-} = 2M_{Mg^{2+}} + M_{Na^+}\ \text{(charge balance)}$$

Combining:  $0.01614 = 3M_{Mg^{2+}} + 2M_{Na^+}$

Let $x$ = mass $MgCl_2$ and $y$ = mass NaCl;  then $x + y = 0.5000$ g.

$$M_{Mg^{2+}} = \frac{x}{95.218}\ \text{and}\ M_{Na^+} = \frac{y}{58.443}\ \ \text{(Because V = 1.000 L.)}$$

$$\text{Total ion concentration} = \frac{3x}{95.218} + \frac{2y}{58.443} = 0.01614\ \text{mol/L}$$

Rearranging:  $3x + (3.2585)y = 1.537$

Solving by simultaneous equations:

$$
\begin{aligned}
3x &+ (3.2585)y = 1.537 \\
-3(x &+ \quad\ y) = -3(0.5000) \\
\hline
(0.2585)y &= 0.037, \qquad y = 0.14\ \text{g NaCl}
\end{aligned}
$$

Mass $MgCl_2 = 0.5000$ g $- 0.14$ g $= 0.36$ g;  mass % $MgCl_2 = \dfrac{0.36\ \text{g}}{0.5000\ \text{g}} \times 100 = 72\%$

## Challenge Problems

109.   From the problem, $\chi_{C_6H_6}^L = \chi_{CCl_4}^L = 0.500$.  We need the pure vapor pressures $(P^\circ)$ in order to calculate the vapor pressure of the solution.

$$C_6H_6(l) \rightleftharpoons C_6H_6(g)\quad K = P_{C_6H_6} = P_{C_6H_6}^\circ\ \text{at } 25°C$$

$$\Delta G_{rxn}^\circ = \Delta G_{f, C_6H_6(g)}^\circ - \Delta G_{f, C_6H_6(l)}^\circ = 129.66\ \text{kJ/mol} - 124.50\ \text{kJ/mol} = 5.16\ \text{kJ/mol}$$

$$\Delta G^{\circ} = -RT \ln K, \ \ \ln K = \frac{-\Delta G^{\circ}}{RT} = \frac{-5.16 \times 10^3 \ \text{J/mol}}{(8.3145 \ \text{J K}^{-1} \ \text{mol}^{-1})(298 \ \text{K})} = -2.08$$

$$K = P^{\circ}_{C_6H_6} = e^{-2.08} = 0.125 \ \text{atm}$$

For CCl$_4$:  $\Delta G^{\circ}_{rxn} = \Delta G^{\circ}_{f, CCl_4(g)} - \Delta G^{\circ}_{f, CCl_4(l)} = -60.59 \ \text{kJ/mol} - (-65.21 \ \text{kJ/mol})$

$$= 4.62 \ \text{kJ/mol}$$

$$K = P^{\circ}_{CCl_4} = \exp\left(\frac{-\Delta G^{\circ}}{RT}\right) = \exp\left(\frac{-4620 \ \text{J/mol}}{8.3145 \ \text{J K}^{-1} \ \text{mol}^{-1} \times 298 \ \text{K}}\right) = 0.155 \ \text{atm}$$

$$P_{C_6H_6} = \chi^L_{C_6H_6} P^{\circ}_{C_6H_6} = 0.500(0.125 \ \text{atm}) = 0.0625 \ \text{atm}; \ \ P_{CCl_4} = 0.500(0.155 \ \text{atm})$$

$$= 0.0775 \ \text{atm}$$

$$\chi^V_{C_6H_6} = \frac{P_{C_6H_6}}{P_{total}} = \frac{0.0625 \ \text{atm}}{0.0625 \ \text{atm} + 0.0775 \ \text{atm}} = \frac{0.0625}{0.1400} = 0.446; \ \ \chi^V_{CCl_4} = 1.000 - 0.446$$

$$= 0.554$$

111.    For the second vapor collected, $\chi^V_{B,2} = 0.714$ and $\chi^V_{T,2} = 0.286$.  Let $\chi^L_{B,2}$ = mole fraction of benzene in the second solution and $\chi^L_{T,2}$ = mole fraction of toluene in the second solution.

$$\chi^L_{B,2} + \chi^L_{T,2} = 1.000$$

$$\chi^V_{B,2} = 0.714 = \frac{P_B}{P_{total}} = \frac{P_B}{P_B + P_T} = \frac{\chi^L_{B,2}(750.0 \ \text{torr})}{\chi^L_{B,2}(750.0 \ \text{torr}) + (1.000 - \chi^L_{B,2})(300.0 \ \text{torr})}$$

Solving:  $\chi^L_{B,2} = 0.500 = \chi^L_{T,2}$

This second solution came from the vapor collected from the first (initial) solution, so, $\chi^V_{B,1} = \chi^V_{T,1} = 0.500$.  Let $\chi^L_{B,1}$ = mole fraction benzene in the first solution and $\chi^L_{T,1}$ = mole fraction of toluene in first solution.  $\chi^L_{B,1} + \chi^L_{T,1} = 1.000$.

$$\chi^V_{B,1} = 0.500 = \frac{P_B}{P_{total}} = \frac{P_B}{P_B + P_T} = \frac{\chi^L_{B,1}(750.0 \ \text{torr})}{\chi^L_{B,1}(750.0 \ \text{torr}) + (1.000 - \chi^L_{B,1})(300.0 \ \text{torr})}$$

Solving:  $\chi^L_{B,1} = 0.286$

The original solution had $\chi_B = 0.286$ and $\chi_T = 0.714$.

113.   a.   Assuming $MgCO_3(s)$ does not dissociate, the solute concentration in water is:

$$\frac{560 \, \mu g \, MgCO_3(s)}{mL} = \frac{560 \, mg}{L} = \frac{560 \times 10^{-3} \, g}{L} \times \frac{1 \, mol \, MgCO_3}{84.32 \, g}$$

$$= 6.6 \times 10^{-3} \, mol \, MgCO_3/L$$

An applied pressure of 8.0 atm will purify water up to a solute concentration of:

$$M = \frac{\pi}{RT} = \frac{8.0 \, atm}{0.08206 \, L \, atm \, K^{-1} \, mol^{-1} \times 300. \, K} = \frac{0.32 \, mol}{L}$$

When the concentration of $MgCO_3(s)$ reaches 0.32 mol/L, the reverse osmosis unit can no longer purify the water.  Let V = volume (L) of water remaining after purifying 45 L of $H_2O$.  When V + 45 L of water has been processed, the moles of solute particles will equal:

$$6.6 \times 10^{-3} \, mol/L \times (45 \, L + V) = 0.32 \, mol/L \times V$$

Solving:  $0.30 = (0.32 - 0.0066) \times V$,  V = 0.96 L

The minimum total volume of water that must be processed is 45 L + 0.96 L = 46 L.

*Note:* If $MgCO_3$ does dissociate into $Mg^{2+}$ and $CO_3^{2-}$ ions, then the solute concentration increases to $1.3 \times 10^{-2} \, M$,  and at least 47 L of water must be processed.

   b.   No;  a reverse osmosis system that applies 8.0 atm can only purify water with a solute concentration of less than 0.32 mol/L.  Salt water has a solute concentration of 2(0.60 *M*) = 1.2 mol/L ions.  The solute concentration of salt water is much too high for this reverse osmosis unit to work.

115.   a.   Assuming no ion association between $SO_4^{2-}(aq)$ and $Fe^{3+}(aq)$, then i = 5 for $Fe_2(SO_4)_3$.

$\pi = iMRT = 5(0.0500 \, mol/L)(0.08206 \, L \, atm \, K^{-1} \, mol^{-1})(298 \, K) = 6.11 \, atm$

   b.   $Fe_2(SO_4)_3(aq) \rightarrow 2 \, Fe^{3+}(aq) + 3 \, SO_4^{2-}(aq)$

Under ideal circumstances, 2/5 of $\pi$ calculated above results from $Fe^{3+}$ and 3/5 results from $SO_4^{2-}$.  The contribution to $\pi$ from $SO_4^{2-}$ is $3/5 \times 6.11$ atm = 3.67 atm.  Because $SO_4^{2-}$ is assumed unchanged in solution, the $SO_4^{2-}$ contribution in the actual solution will also be 3.67 atm.  The contribution to the actual $\pi$ from the $Fe(H_2O)_6^{3+}$ dissociation reaction is $6.73 - 3.67 = 3.06$ atm.

The initial concentration of $Fe(H_2O)_6^{2+}$ is 2(0.0500) = 0.100 *M*.  The setup for the weak acid problem is:

$$Fe(H_2O)_6^{3+} \rightarrow H^+ + Fe(OH)(H_2O)_5^{2+} \qquad K_a = \frac{[H^+][Fe(OH)(H_2O)_5^{2+}]}{[Fe(H_2O)_6^{3+}]}$$

Initial   0.100 *M*          ~0                0
          $x$ mol/L of $Fe(H_2O)_6^{3+}$ reacts to reach equilibrium
Equil.   $0.100 - x$          $x$                $x$

$$\text{Total ion concentration} = iM = \frac{\pi}{RT} = \frac{3.06 \text{ atm}}{0.08206 \text{ L atm K}^{-1} \text{ mol}^{-1}(298 \text{ K})} = 0.125 \ M$$

$$0.125 \ M = 0.100 - x + x + x = 0.100 + x, \ x = 0.025 \ M$$

$$K_a = \frac{[H^+][Fe(OH)(H_2O)_5^{2+}]}{[Fe(H_2O)_6^{3+}]} = \frac{x^2}{0.100 - x} = \frac{(0.025)^2}{(0.100 - 0.025)} = \frac{(0.025)^2}{0.075}$$

$$K_a = 8.3 \times 10^{-3}$$

117.   a.   $\pi = iMRT, \ iM = \dfrac{\pi}{RT} = \dfrac{7.83 \text{ atm}}{0.08206 \text{ L atm K}^{-1} \text{ mol}^{-1} \times 298 \text{ K}} = 0.320 \text{ mol/L}$

Assuming 1.000 L of solution:

total mol solute particles = mol $Na^+$ + mol $Cl^-$ + mol NaCl = 0.320 mol

$$\text{mass solution} = 1000. \text{ mL} \times \frac{1.071 \text{ g}}{\text{mL}} = 1071 \text{ g solution}$$

mass NaCl in solution = 0.0100 × 1071 g = 10.7 g NaCl

$$\text{mol NaCl added to solution} = 10.7 \text{ g} \times \frac{1 \text{ mol}}{58.44 \text{ g}} = 0.183 \text{ mol NaCl}$$

Some of this NaCl dissociates into $Na^+$ and $Cl^-$ (two moles of ions per mole of NaCl), and some remains undissociated. Let $x$ = mol undissociated NaCl = mol ion pairs.

Mol solute particles = 0.320 mol = 2(0.183 − $x$) + $x$

$0.320 = 0.366 - x, \ x = 0.046$ mol ion pairs

$$\text{Fraction of ion pairs} = \frac{0.046}{0.183} = 0.25, \ \text{or } 25\%$$

b.   $\Delta T = K_f m$, where $K_f = 1.86$ °C kg/mol; from part a, 1.000 L of solution contains 0.320 mol of solute particles. To calculate the molality of the solution, we need the kilograms of solvent present in 1.000 L of solution.

Mass of 1.000 L solution = 1071 g; mass of NaCl = 10.7 g

Mass of solvent in 1.000 L solution = 1071 g − 10.7 g = 1060. g

$$\Delta T = 1.86 \text{ °C kg/mol} \times \frac{0.320 \text{ mol}}{1.060 \text{ kg}} = 0.562 \text{°C}$$

Assuming water freezes at 0.000°C, then $T_f = -0.562$°C.

# CHAPTER 18

# THE REPRESENTATIVE ELEMENTS

## Group 1A Elements

1. The gravity of the earth is not strong enough to keep the light $H_2$ molecules in the atmosphere.

3. a. $\Delta H° = 2(-46 \text{ kJ}) = -92 \text{ kJ}$; $\Delta S° = 2(193 \text{ J/K}) - [3(131 \text{ J/K}) + 192 \text{ J/K}] = -199 \text{ J/K}$

$\Delta G° = \Delta H° - T\Delta S° = -92 \text{ kJ} - 298 \text{ K}(-0.199 \text{ kJ/K}) = -33 \text{ kJ}$

b. Because $\Delta G°$ is negative, this reaction is spontaneous at standard conditions.

c. $\Delta G° = 0$ when $T = \dfrac{\Delta H°}{\Delta S°} = \dfrac{-92 \text{ kJ}}{-0.199 \text{ kJ/K}} = 460 \text{ K}$

At $T < 460$ K and standard pressures, the favorable $\Delta H°$ term dominates, and the reaction is spontaneous ($\Delta G° < 0$).

5. Ionic, covalent, and metallic (or interstitial); the ionic and covalent hydrides are true compounds obeying the law of definite proportions and differ from each other in the type of bonding. The interstitial hydrides are more like solid solutions of hydrogen with a transition metal and do not obey the law of definite proportions.

7. Alkali metals have a $ns^1$ valence shell electron configuration. Alkali metals lose this valence electron with relative ease to form $M^+$ cations when in ionic compounds. They all are easily oxidized. Therefore, in order to prepare the pure metals, alkali metals must be produced in the absence of materials ($H_2O$, $O_2$) that are capable of oxidizing them. The method of preparation is electrochemical processes, specifically, electrolysis of molten chloride salts and reduction of alkali salts with Mg and $H_2$. In all production methods, $H_2O$ and $O_2$ must be absent.

9. Hydrogen forms many compounds in which the oxidation state is +1, as do the Group 1A elements. For example, $H_2SO_4$ and HCl as compared to $Na_2SO_4$ and NaCl. On the other hand, hydrogen forms diatomic $H_2$ molecules and is a nonmetal, whereas the Group 1A elements are metals. Hydrogen also forms compounds with a -1 oxidation state, which is not characteristic of Group 1A metals, e.g., NaH.

11. $4 \text{ Li(s)} + O_2(g) \rightarrow 2 \text{ Li}_2O(s)$

$16 \text{ Li(s)} + S_8(s) \rightarrow 8 \text{ Li}_2S(s)$; $2 \text{ Li(s)} + Cl_2(g) \rightarrow 2 \text{ LiCl(s)}$

$12 \text{ Li(s)} + P_4(s) \rightarrow 4 \text{ Li}_3P(s); \quad 2 \text{ Li(s)} + H_2(g) \rightarrow 2 \text{ LiH(s)}$

$2 \text{ Li(s)} + 2 \text{ H}_2O(l) \rightarrow 2 \text{ LiOH(aq)} + H_2(g); \quad 2 \text{ Li(s)} + 2 \text{ HCl(aq)} \rightarrow 2 \text{ LiCl(aq)} + H_2(g)$

## Group 2A Elements

13.    Alkaline earth metals form +2 ions, so 2 moles of electrons are transferred to form the metal M.

$$\text{Mol M} = 748 \text{ s} \times \frac{5.00 \text{ C}}{\text{s}} \times \frac{1 \text{ mol e}^-}{96{,}485 \text{ C}} \times \frac{1 \text{ mol M}}{2 \text{ mol e}^-} = 1.94 \times 10^{-2} \text{ mol M}$$

$$\text{Molar mass of M} = \frac{0.471 \text{ g M}}{1.94 \times 10^{-2} \text{ mol M}} = 24.3 \text{ g/mol}; \quad MgCl_2 \text{ was electrolyzed.}$$

15.    The alkaline earth ions that give water the hard designation are $Ca^{2+}$ and $Mg^{2+}$. These ions interfere with the action of detergents and form unwanted precipitates with soaps. Large-scale water softeners remove $Ca^{2+}$ by precipitating out the calcium ions as $CaCO_3$. In homes, $Ca^{2+}$ and $Mg^{2+}$ (plus other cations) are removed by ion exchange. See Figure 18.6 for a schematic of a typical cation exchange resin.

17.    $CaCO_3(s) + H_2SO_4(aq) \rightarrow CaSO_4(aq) + H_2O(l) + CO_2(g)$

19.    $\dfrac{1 \text{ mg F}^-}{\text{L}} \times \dfrac{1 \text{ g}}{1000 \text{ mg}} \times \dfrac{1 \text{ mol F}^-}{19.0 \text{ g F}^-} = 5.3 \times 10^{-5} \, M \, \text{F}^- = 5 \times 10^{-5} \, M \, \text{F}^-$

$CaF_2(s) \rightleftharpoons Ca^{2+}(aq) + 2 \text{ F}^-(aq) \quad K_{sp} = [Ca^{2+}][F^-]^2 = 4.0 \times 10^{-11}$; precipitation will occur when $Q > K_{sp}$. Let's calculate $[Ca^{2+}]$ so that $Q = K_{sp}$.

$Q = 4.0 \times 10^{-11} = [Ca^{2+}]_0[F^-]_0^2 = [Ca^{2+}]_0(5 \times 10^{-5})^2, \quad [Ca^{2+}]_0 = 2 \times 10^{-2} \, M$

$CaF_2(s)$ will precipitate when $[Ca^{2+}]_0 > 2 \times 10^{-2} \, M$. Therefore, hard water should have a calcium ion concentration of less than $2 \times 10^{-2} \, M$ to avoid precipitate formation.

## Group 3A Elements

21.    The valence electron configuration of Group 3A elements is $ns^2np^1$. The lightest Group 3A element, boron, is a nonmetal because most of its compounds are covalent. Aluminum, although commonly thought of as a metal, does have some nonmetallic properties because its bonds to other nonmetals have significant covalent character. The other Group 3A elements have typical metal characteristics; their compounds formed with nonmetals are ionic. From this discussion, metallic character increases as the Group 3A elements get larger.

As discussed above, boron is a nonmetal in both properties and compounds formed. However aluminum has physical properties of metals, such as high thermal and electrical conductivities and a lustrous appearance. The compounds of aluminum with other nonmetals, however, do have some nonmetallic properties because the bonds have significant covalent character.

23.      $B_2H_6(g) + 3 O_2(g) \rightarrow 2 B(OH)_3(s)$

25.      $2 Ga(s) + 3 F_2(g) \rightarrow 2 GaF_3(s)$;  $4 Ga(s) + 3 O_2(g) \rightarrow 2 Ga_2O_3(s)$

         $16 Ga(s) + 3 S_8(s) \rightarrow 8 Ga_2S_3(s)$

         $2 Ga(s) + 6 HCl(aq) \rightarrow 2 GaCl_3(aq) + 3 H_2(g)$

## Group 4A Elements

27.      The valence electron configuration of Group 4A elements is $ns^2np^2$. The two most important elements on earth are both Group 4A elements. They are carbon, found in all biologically important molecules, and silicon, found in most of the compounds that make up the earth's crust. They are important because they are so prevalent in compounds necessary for life and the geologic world.

         As with Group 3A, Group 4A shows an increase in metallic character as the elements get heavier. Carbon is a typical nonmetal, silicon and germanium have properties of both metals and nonmetals so they are classified as semimetals, and tin and lead have typical metallic characteristics.

29.      White tin is stable at normal temperatures.  Gray tin is stable at temperatures below 13.2°C. Thus, for the phase change:  Sn(gray) → Sn(white),  ΔG is negative at T > 13.2°C and ΔG is positive at T < 13.2°C.  This is only possible if ΔH is positive and ΔS is positive.  Therefore, gray tin has the more ordered structure (has the smaller positional probability).

31.      $Sn(s) + 2F_2(g) \rightarrow SnF_4(s)$, tin(IV) fluoride;   $Sn(s) + F_2(g) \rightarrow SnF_2(s)$, tin(II) fluoride

33.      $Pb_3O_4$: we assign −2 for the oxidation state of O. The sum of the oxidation states of Pb must be +8.  We get this if two of the lead atoms are Pb(II) and one is Pb(IV).  Therefore, the mole ratio of lead(II) to lead(IV) is 2:1.

## Group 5A Elements

35.      $NO_4^{3-}$

Both $NO_4^{3-}$ and $PO_4^{3-}$ have 32 valence electrons so both have similar Lewis structures.  From the Lewis structure for $NO_4^{3-}$, the central N atom has a tetrahedral arrangement of electron pairs.  N is small.  There is probably not enough room for all four oxygen atoms around N.  P is larger; thus $PO_4^{3-}$ is stable.

$PO_3^-$

$PO_3^-$ and $NO_3^-$ both have 24 valence electrons, so both have similar Lewis structures. From the Lewis structure, $PO_3^-$ has a trigonal arrangement of electron pairs about the central P atom (two single bonds and one double bond). P=O bonds are not particularly stable, whereas N=O bonds are stable. Thus $NO_3^-$ is stable.

37.   N: $1s^2 2s^2 2p^3$; the extremes of the oxidation states for N can be rationalized by examining the electron configuration of N. Nitrogen is three electrons short of the stable Ne electron configuration of $1s^2 2s^2 2p^6$. Having an oxidation state of –3 makes sense. The +5 oxidation state corresponds to N "losing" its five valence electrons. In compounds with oxygen, the N–O bonds are polar covalent, with N having the partial positive end of the bond dipole. In the world of oxidation states, electrons in polar covalent bonds are assigned to the more electronegative atom; this is oxygen in N–O bonds. N can form enough bonds to oxygen to give it a +5 oxidation state. This loosely corresponds to losing all the valence electrons.

39.   This is due to nitrogen's ability to form strong π bonds, whereas heavier group 5A elements do not form strong π bonds. Therefore, $P_2$, $As_2$, and $Sb_2$ do not form since two π bonds are required to form these diatomic substances.

41.   a.   $NO_2$, $5 + 2(6) = 17$ $e^-$                       $N_2O_4$, $2(5) + 4(6) = 34$ $e^-$

plus other resonance structures          plus other resonance structures

b.   $BF_3$, $3 + 3(7) = 24$ $e^-$                         $NH_3$, $5 + 3(1) = 8$ $e^-$

$BF_3NH_3$, $24 + 8 = 32$ $e^-$

In reaction a, $NO_2$ has an odd number of electrons, so it is impossible to satisfy the octet rule. By dimerizing to form $N_2O_4$, the odd electron on two $NO_2$ molecules can pair up, giving a species whose Lewis structure can satisfy the octet rule.  In general, odd-electron species are very reactive.    In reaction b, $BF_3$ can be considered electron-deficient; boron has only six electrons around it.  By forming $BF_3NH_3$, the boron atom satisfies the octet rule by accepting a lone pair of electrons from $NH_3$ to form a fourth bond.

43.     As temperature increases, the value of K decreases. This is consistent with an exothermic reaction. In an exothermic reaction, heat is a product, and an increase in temperature shifts the equilibrium to the reactant side (as well as lowering the value of K).

45.

$$H_2N\text{—}NH_2 (l) + 2\ F\text{—}F\ (g) \longrightarrow 4\ H\text{—}F(g) + N\equiv N\ (g)$$

Bonds broken:                                                Bonds formed:

1 N–N (160. kJ/mol)                                4 H–F (565 kJ/mol)
4 N–H (391 kJ/mol)                                 1 N≡ N (941 kJ/mol)
2 F–F (154 kJ/mol)

$\Delta H$ = 160. + 4(391) + 2(154) – [4(565) + 941] = 2032 kJ – 3201 kJ = –1169 kJ

47.     Production of bismuth:

$$2\ Bi_2S_3(s) + 9\ O_2(g) \rightarrow 2\ Bi_2O_3(s) + 6\ SO_2(g);\ 2\ Bi_2O_3(s) + 3\ C(s) \rightarrow 4\ Bi(s) + 3\ CO_2(g)$$

Production of antimony:

$$2\ Sb_2S_3(s) + 9\ O_2(g) \rightarrow 2\ Sb_2O_3(s) + 6\ SO_2(g);\ 2\ Sb_2O_3(s) + 3\ C(s) \rightarrow 4\ Sb(s) + 3\ CO_2(g)$$

49.     TSP = $Na_3PO_4$; $PO_4^{3-}$ is the conjugate base of the weak acid $HPO_4^{2-}$ ($K_a = 4.8 \times 10^{-13}$).  All conjugate bases of weak acids are effective bases ($K_b = K_w/K_a = 1.0 \times 10^{-14}/4.8 \times 10^{-13} = 2.1 \times 10^{-2}$).  The weak base reaction of $PO_4^{3-}$ with $H_2O$ is $PO_4^{3-}(aq) + H_2O(l) \rightleftharpoons HPO_4^{2-}(aq) + OH^-(aq)$.

51.     $4\ As(s) + 3\ O_2(g) \rightarrow As_4O_6(s);\ 4\ As(s) + 5\ O_2(g) \rightarrow As_4O_{10}(s)$

$As_4O_6(s) + 6\ H_2O(l) \rightarrow 4\ H_3AsO_3(aq);\ As_4O_{10}(s) + 6\ H_2O(l) \rightarrow 4\ H_3AsO_4(aq)$

53.     $1/2\ N_2(g) + 1/2\ O_2(g) \rightarrow NO(g)\quad \Delta G° = \Delta G^{\circ}_{f,\,NO} = 87$ kJ/mol;  by definition, $\Delta G^{\circ}_f$ for a compound equals the free energy change that would accompany the formation of 1 mole of that compound from its elements in their standard states.  NO (and some other oxides of nitrogen) have weaker bonds than the triple bond of $N_2$ and the double bond of $O_2$.  Because of this, NO (and some other oxides of nitrogen) have higher (positive) standard free energies of formation than the relatively stable $N_2$ and $O_2$ molecules.

55.    The pollution provides nitrogen and phosphorus nutrients so that the algae can grow.  The algae consume oxygen, causing fish to die.

57.    $5 N_2O_4(l) + 4 N_2H_3CH_3(l) \rightarrow 12 H_2O(g) + 9 N_2(g) + 4 CO_2(g)$

$$\Delta H^\circ = \left[ 12 \text{ mol} \left( \frac{-242 \text{ kJ}}{\text{mol}} \right) + 4 \text{ mol} \left( \frac{-393.5 \text{ kJ}}{\text{mol}} \right) \right]$$

$$- \left[ 5 \text{ mol} \left( \frac{-20. \text{ kJ}}{\text{mol}} \right) + 4 \text{ mol} \left( \frac{54 \text{ kJ}}{\text{mol}} \right) \right] = -4594 \text{ kJ}$$

59.    Hydrazine also can hydrogen bond because it has covalent N–H bonds as well as having a lone pair of electrons on each N. The high boiling point for hydrazine's relatively small size supports this.

## Group 6A Elements

61.    The two allotropic forms of oxygen are $O_2$ and $O_3$.

$O_2$, 2(6) = 12 e⁻                    $O_3$, 3(6) = 18 e⁻

The MO electron configuration of $O_2$ has two unpaired electrons in the degenerate pi antibonding $(\pi^*_{2p})$ orbitals (see Figure 14.41). A substance with unpaired electrons is paramagnetic. Ozone has a V-shape molecular structure with a bond angle of 117°, slightly less than the predicted 120° trigonal planar bond angle.

63.    In the upper atmosphere, $O_3$ acts as a filter for ultraviolet (UV) radiation:

$O_3 \xrightarrow{h\nu} O_2 + O$   (See Exercise 18.62.)

$O_3$ is also a powerful oxidizing agent.  It irritates the lungs and eyes, and at high concentration, it is toxic.  The smell of a "fresh spring day" is $O_3$ formed during lightning discharges.  Toxic materials don't necessarily smell bad.  For example, HCN smells like almonds.

65.    +6 oxidation state:  $SO_4^{2-}$, $SO_3$, $SF_6$

+4 oxidation state:  $SO_3^{2-}$, $SO_2$, $SF_4$

+2 oxidation state:  $SCl_2$

 0 oxidation state:  $S_8$ and all other elemental forms of sulfur

−2 oxidation state:  $H_2S$, $Na_2S$

67.    $H_2SeO_4(aq) + 3\ SO_2(g) \rightarrow Se(s) + 3\ SO_3(g) + H_2O(l)$

## Group 7A Elements

69.    $O_2F_2$ has $2(6) + 2(7) = 26$ valence $e^-$.

$$:\ddot{F}—\ddot{O}—\ddot{O}—\ddot{F}:$$

Formal charge          0        0        0        0

Oxidation number      $-1$      $+1$      $+1$      $-1$

Oxidation numbers (states) are more useful.  We are forced to assign $+1$ as the oxidation number for oxygen.  Oxygen is very electronegative, and $+1$ is not a stable oxidation state for this element.

71.    Fluorine is the most reactive of the halogens because it is the most electronegative atom, and the bond in the $F_2$ molecule is very weak.

73.    $\begin{array}{ll} ClO^- + H_2O + 2\ e^- \rightarrow 2\ OH^- + Cl^- & E_c^\circ = 0.90\ V \\ 2\ NH_3 + 2\ OH^- \rightarrow N_2H_4 + 2\ H_2O + 2\ e^- & -E_a^\circ = 0.10\ V \end{array}$

$\overline{\phantom{ClO^-(aq) + 2\ NH_3(aq) \rightarrow Cl^-(aq) + N_2H_4(aq) + H_2O(l)}}$

$ClO^-(aq) + 2\ NH_3(aq) \rightarrow Cl^-(aq) + N_2H_4(aq) + H_2O(l) \qquad E_{cell}^\circ = 1.00\ V$

Because $E_{cell}^\circ$ is positive for this reaction, $ClO^-$, at standard conditions, can spontaneously oxidize $NH_3$ to the somewhat toxic $N_2H_4$.

75.    a.    $AgCl(s) \xrightarrow{h\nu} Ag(s) + Cl$; the reactive chlorine atom is trapped in the crystal.  When light is removed, $Cl$ reacts with silver atoms to re-form $AgCl$; that is, the reverse reaction occurs.  In pure $AgCl$, the $Cl$ atoms escape, making the reverse reaction impossible.

       b.    Over time chlorine is lost, and the presence of the dark silver metal is permanent.

## Group 8A Elements

77.    Helium is unreactive and doesn't combine with any other elements.  It is a very light gas and would easily escape the earth's gravitational pull as the planet was formed.

79.    $10.0\ m \times 10.0\ m \times 10.0\ m = 1.00 \times 10^3\ m^3$; from Table 18.22, volume % $Ar = 0.9\%$.

$$1.00 \times 10^3\ m^3 \times \left(\frac{10\ dm}{m}\right)^3 \times \frac{1\ L}{dm^3} \times \frac{0.9\ L\ Ar}{100\ L\ air} = 9 \times 10^3\ L\ of\ Ar\ in\ the\ room$$

$$PV = nRT, \quad n = \frac{PV}{RT} = \frac{(1.0 \text{ atm})(9 \times 10^3 \text{ L})}{(0.08206 \text{ L atm K}^{-1} \text{ mol}^{-1})(298 \text{ K})} = 4 \times 10^2 \text{ mol Ar}$$

$$4 \times 10^2 \text{ mol Ar} \times \frac{39.95 \text{ g}}{\text{mol}} = 2 \times 10^4 \text{ g Ar in the room}$$

$$4 \times 10^2 \text{ mol Ar} \times \frac{6.022 \times 10^{23} \text{ atoms}}{\text{mol}} = 2 \times 10^{26} \text{ atoms Ar in the room}$$

A 2-L breath contains: $2 \text{ L air} \times \dfrac{0.9 \text{ L Ar}}{100 \text{ L air}} = 2 \times 10^{-2} \text{ L Ar}$

$$n = \frac{PV}{RT} = \frac{(1.0 \text{ atm})(2 \times 10^{-2} \text{ L})}{(0.08206 \text{ L atm K}^{-1} \text{ mol}^{-1})(298 \text{ K})} = 8 \times 10^{-4} \text{ mol Ar}$$

$$8 \times 10^{-4} \text{ mol Ar} \times \frac{6.022 \times 10^{23} \text{ atoms}}{\text{mol}} = 5 \times 10^{20} \text{ atoms of Ar in a 2-L breath}$$

Because Ar and Rn are both noble gases, both species will be relatively unreactive. However, all nuclei of Rn are radioactive, unlike most nuclei of Ar. The radioactive decay products of Rn can cause biological damage when inhaled.

81.     One would expect $RnF_2$ and $RnF_4$ to form in fashion similar to $XeF_2$ and $XeF_4$. The chemistry of radon is difficult to study because radon isotopes are all radioactive. The hazards of dealing with radioactive materials are immense.

## Additional Exercises

83.     Solids have stronger intermolecular forces than liquids. In order to maximize the hydrogen bonding in the solid phase, ice is forced into an open structure. This open structure is why $H_2O(s)$ is less dense than $H_2O(l)$.

85.     Strontium and calcium are both alkaline earth metals, so both have similar chemical properties. Because milk is a good source of calcium, strontium could replace some calcium in milk without much difficulty.

87.
$$Tl^{3+} + 2 e^- \rightarrow Tl^+ \qquad\qquad E_c^\circ = 1.25 \text{ V}$$
$$3 \text{ I}^- \rightarrow I_3^- + 2 e^- \qquad\qquad -E_a^\circ = -0.55 \text{ V}$$

$$\overline{Tl^{3+} + 3 \text{ I}^- \rightarrow Tl^+ + I_3^- \qquad\qquad E_{cell}^\circ = 0.70 \text{ V}} \qquad (\text{Spontaneous because } E_{cell}^\circ > 0.)$$

In solution, $Tl^{3+}$ can oxidize $\text{I}^-$ to $I_3^-$. Thus we expect $TlI_3$ to be thallium(I) triiodide.

89.    The inert pair effect refers to the difficulty of removing the pair of s electrons from some of the elements in the fifth and sixth periods of the periodic table. As a result, multiple oxidation states are exhibited for the heavier elements of Groups 3A and 4A. $In^+$, $In^{3+}$, $Tl^+$, and $Tl^{3+}$ oxidation states are all important to the chemistry of In and Tl.

91.    $$15 \text{ kWh} = \frac{15000 \text{ J h}}{\text{s}} \times \frac{60 \text{ s}}{\text{min}} \times \frac{60 \text{ min}}{\text{h}} = 5.4 \times 10^7 \text{ J or } 5.4 \times 10^4 \text{ kJ} \quad \text{(Hall process)}$$

To melt 1.0 kg Al requires:  $1.0 \times 10^3 \text{ g Al} \times \dfrac{1 \text{ mol Al}}{26.98 \text{ g}} \times \dfrac{10.7 \text{ kJ}}{\text{mol Al}} = 4.0 \times 10^2 \text{ kJ}$

It is feasible to recycle Al by melting the metal because, in theory, it takes less than 1% of the energy required to produce the same amount of Al by the Hall process.

93.    Carbon cannot form the fifth bond necessary for the transition state because carbon doesn't have low-energy d orbitals available to expand its octet of valence electrons.

95.    a.    The Lewis structures for NNO and NON are (16 valence electrons each):

$$:N=N=O: \quad \longleftrightarrow \quad :N\equiv N-O: \quad \longleftrightarrow \quad :N-N\equiv O:$$

$$:N=O=N: \quad \longleftrightarrow \quad :N\equiv O-N: \quad \longleftrightarrow \quad :N-O\equiv N:$$

The NNO structure is correct. From the Lewis structures we would predict both NNO and NON to be linear. However, we would predict NNO to be polar and NON to be nonpolar. Becase experiments show $N_2O$ to be polar, then NNO is the correct structure.

b.    Formal charge = number of valence electrons of atoms - [(number of lone pair electrons) + 1/2 (number of shared electrons)].

$$:N=N=O: \quad \longleftrightarrow \quad :N\equiv N-O: \quad \longleftrightarrow \quad :N-N\equiv O:$$
$$\;\;{-1}\;\;{+1}\;\;\;\;{0} \qquad\qquad {0}\;\;\;{+1}\;\;\;{-1} \qquad\qquad {-2}\;\;\;{+1}\;\;\;{+1}$$

The formal charges for the atoms in the various resonance structures appear below each atom. The central N is sp hybridized in all the resonance structures. We can probably ignore the third resonance structure on the basis of the relatively large formal charges on the various atoms in $N_2O$ as compared with the first two resonance structures.

c.    The sp hybrid orbitals on the center N overlap with atomic orbitals (or hybrid orbitals) on the other two atoms to form the two sigma bonds. The remaining two unhybridized p orbitals on the center N overlap with two p orbitals on the peripheral N to form the two $\pi$ bonds.

97.    For a buffer solution: $pH = pK_a + \log\dfrac{[base]}{[acid]}$; $7.15 = -\log(6.2 \times 10^{-8}) + \log\dfrac{[HPO_4^{2-}]}{[H_2PO_4^-]}$

$7.15 = 7.21 + \log\dfrac{[HPO_4^{2-}]}{[H_2PO_4^-]}$, $\dfrac{[HPO_4^{2-}]}{[H_2PO_4^-]} = 10^{-0.06} = 0.9$, $\dfrac{[H_2PO_4^-]}{[HPO_4^{2-}]} = \dfrac{1}{0.9} = 1.1 \approx 1$

A best buffer has approximately equal concentrations of weak acid and conjugate base so that $pH \approx pK_a$ for a best buffer.   The $pK_a$ value for a $H_3PO_4/H_2PO_4^-$ buffer is $-\log (7.5 \times 10^{-3}) = 2.12$.  A pH of 7.1 is too high for a $H_3PO_4/H_2PO_4^-$ buffer to be effective. At this high a pH, there would be so little $H_3PO_4$ present that we could hardly consider it a buffer. This solution would not be effective in resisting pH changes, especially when a strong base is added.

## Challenge Problems

99.    $Mg^{2+} + P_3O_{10}^{5-} \rightleftharpoons MgP_3O_{10}^{3-}$   $pK = -8.60$; $[Mg^{2+}]_0 = \dfrac{50. \times 10^{-3}\ g}{L} \times \dfrac{1\ mol}{24.3\ g} = 2.1 \times 10^{-3}\ M$

$[P_3O_{10}^{5-}]_0 = \dfrac{40.\ g\ Na_5P_3O_{10}}{L} \times \dfrac{1\ mol}{367.9\ g} = 0.11\ M$

Assume the reaction goes to completion because K is large ($K = 10^{8.60} = 4.0 \times 10^8$).  Then solve the back-equilibrium problem to determine the small amount of $Mg^{2+}$ present.

|              | $Mg^{2+}$          | $+$ | $P_3O_{10}^{5-}$       | $\rightleftharpoons$ | $MgP_3O_{10}^{3-}$       |                    |
|--------------|--------------------|-----|------------------------|------|---------------------------|--------------------|
| Before       | $2.1 \times 10^{-3}\ M$ |     | $0.11\ M$              |      | $0$                       |                    |
| Change       | $-2.1 \times 10^{-3}$   |     | $-2.1 \times 10^{-3}$  | $\rightarrow$ | $+2.1 \times 10^{-3}$     | Reacts completely  |
| After        | $0$                |     | $0.11$                 |      | $2.1 \times 10^{-3}$      | New initial        |

$x$ mol/L $MgP_3O_{10}^{3-}$ dissociates to reach equilibrium

|              | $Mg^{2+}$ | $P_3O_{10}^{5-}$ |               | $MgP_3O_{10}^{3-}$     |
|--------------|-----------|------------------|---------------|------------------------|
| Change       | $+x$      | $+x$             | $\leftarrow$  | $-x$                   |
| Equil.       | $x$       | $0.11 + x$       |               | $2.1 \times 10^{-3} - x$ |

$K = 4.0 \times 10^8 = \dfrac{[MgP_3O_{10}^{3-}]}{[Mg^{2+}][P_3O_{10}^{5-}]} = \dfrac{2.1 \times 10^{-3} - x}{x(0.11 + x)}$   (assume $x \ll 2.1 \times 10^{-3}$)

$4.0 \times 10^8 \approx \dfrac{2.1 \times 10^{-3}}{x(0.11)}$, $x = [Mg^{2+}] = 4.8 \times 10^{-11}\ M$;  assumptions good.

101.    $3\ O_2(g) \rightleftharpoons 2\ O_3(g)$; $\Delta H° = 2(143) = 286\ kJ$; $\Delta G° = 2(163) = 326\ kJ$

$\ln K_p = \dfrac{-\Delta G°}{RT} = \dfrac{-326 \times 10^3\ J}{(8.3145\ J\ K^{-1}\ mol^{-1})(298\ K)} = -131.573$, $K_p = e^{-131.573} = 7.22 \times 10^{-58}$

*Note*: We carried extra significant figures for the $K_p$ calculation.

We need the value of K at 230. K. From Section 10.11 of the text: $\ln K = \dfrac{-\Delta H^\circ}{RT} + \dfrac{\Delta S^\circ}{R}$

For two sets of K and T:

$$\ln K_1 = \frac{-\Delta H^\circ}{R}\left(\frac{1}{T_1}\right) + \frac{\Delta S^\circ}{R}; \quad \ln K_2 = \frac{-\Delta H^\circ}{R}\left(\frac{1}{T_2}\right) + \frac{\Delta S^\circ}{R}$$

Subtracting the first expression from the second:

$$\ln K_2 - \ln K_1 = \frac{\Delta H^\circ}{R}\left(\frac{1}{T_1} - \frac{1}{T_2}\right) \quad \text{or} \quad \ln\frac{K_2}{K_1} = \frac{\Delta H^\circ}{R}\left(\frac{1}{T_1} - \frac{1}{T_2}\right)$$

Let $K_2 = 7.22 \times 10^{-58}$, $T_2 = 298$;  $K_1 = K_{230}$, $T_1 = 230.$ K;  $\Delta H^\circ = 286 \times 10^3$ J

$$\ln\frac{7.22 \times 10^{-58}}{K_{230}} = \frac{286 \times 10^3}{8.3145}\left(\frac{1}{230.} - \frac{1}{298}\right) = 34.13 \qquad \text{(Carrying extra sig. figs.)}$$

$$\frac{7.22 \times 10^{-58}}{K_{230}} = e^{34.13} = 6.6 \times 10^{14}, \ K_{230} = 1.1 \times 10^{-72}$$

$$K_{230} = 1.1 \times 10^{-72} = \frac{P_{O_3}^2}{P_{O_2}^3} = \frac{P_{O_3}^2}{(1.0 \times 10^{-3} \text{ atm})^3}, \ P_{O_3} = 3.3 \times 10^{-41} \text{ atm}$$

The volume occupied by one molecule of ozone is:

$$V = \frac{nRT}{P} = \frac{(1/6.022 \times 10^{23} \text{ mol})(0.08206 \text{ L atm K}^{-1}\text{ mol}^{-1})(230.\text{ K})}{(3.3 \times 10^{-41} \text{ atm})}, \ V = 9.5 \times 10^{17} \text{ L}$$

Equilibrium is probably not maintained under these conditions. When only two ozone molecules are in a volume of $9.5 \times 10^{17}$ L, the reaction is not at equilibrium. Under these conditions, Q > K, and the reaction shifts left. But with only two ozone molecules in this huge volume, it is extremely unlikely that they will collide with each other. At these conditions, the concentration of ozone is not large enough to maintain equilibrium.

103.   a.   NO is the catalyst. NO is present in the first step of the mechanism on the reactant side, but it is not a reactant since it is regenerated in the second step and does not appear in the overall balanced equation.

       b.   $NO_2$ is an intermediate. Intermediates also never appear in the overall balanced equation. In a mechanism, intermediates always appear first on the product side, whereas catalysts always appear first on the reactant side.

c.   $k = A \exp(-E_a/RT); \quad \dfrac{k_{cat}}{k_{un}} = \dfrac{A \exp[-E_a(cat)/RT]}{A \exp[-E_a(un)/RT]} = \exp\left[\dfrac{E_a(un) - E_a(cat)}{RT}\right]$

$\dfrac{k_{cat}}{k_{un}} = \exp\left(\dfrac{2100 \text{ J/mol}}{8.3145 \text{ J K}^{-1} \text{ mol}^{-1} \times 298 \text{ K}}\right) = e^{0.85} = 2.3$

The catalyzed reaction is approximately 2.3 times faster than the uncatalyzed reaction at 25°C.

d.   The mechanism for the chlorine catalyzed destruction of ozone is:

$\begin{array}{ll} O_3 + Cl \rightarrow O_2 + ClO & \text{slow} \\ ClO + O \rightarrow O_2 + Cl & \text{fast} \end{array}$

$\rule{6cm}{0.4pt}$

$O_3 + O \rightarrow 2\,O_2$

e.   Because the chlorine atom-catalyzed reaction has a lower activation energy, the Cl-catalyzed rate is faster.  Hence Cl is a more effective catalyst.  Using the activation energy, we can estimate the efficiency that Cl atoms destroy ozone compared to NO molecules.

At 25°C:  $\dfrac{k_{Cl}}{k_{NO}} = \exp\left[\dfrac{-E_a(Cl) + E_a(NO)}{RT}\right] = \exp\left[\dfrac{(-2100 + 11,900)\,\text{J/mol}}{(8.3145 \times 298)\,\text{J/mol}}\right]$

$= e^{3.96} = 52$

At 25°C, the Cl-catalyzed reaction is roughly 52 times faster (more efficient) than the NO-catalyzed reaction, assuming the frequency factor A is the same for each reaction and assuming similar rate laws.

105.                $Pb^{2+} \quad + \quad H_2EDTA^{2-} \quad \rightleftharpoons \quad PbEDTA^{2-} \quad + \quad 2\,H^+$

| | | | | | |
|---|---|---|---|---|---|
| Before | 0.0010 $M$ | 0.050 $M$ | 0 | $1.0 \times 10^{-6}$ $M$ | (Buffer, [H$^+$] constant) |
| Change | −0.0010 | −0.0010 | → +0.0010 | No change | Reacts completely |
| After | 0 | 0.049 | 0.0010 | $1.0 \times 10^{-6}$ | New initial conditions |

$x$ mol/L PbEDTA$^{2-}$ dissociates to reach equilibrium

| | | | | | |
|---|---|---|---|---|---|
| Change | +$x$ | +$x$ | ← −$x$ | | |
| Equil. | $x$ | 0.049 + $x$ | 0.0010 − $x$ | $1.0 \times 10^{-6}$ | (Buffer) |

$K = 1.0 \times 10^{23} = \dfrac{[PbEDTA^{2-}][H^+]^2}{[Pb^{2+}][H_2EDTA^{2-}]} = \dfrac{(0.0010 - x)(1.0 \times 10^{-6})^2}{(x)(0.049 + x)}$

$1.0 \times 10^{23} \approx \dfrac{(0.0010)(1.0 \times 10^{-12})}{(x)(0.049)}, \quad x = [Pb^{2+}] = 2.0 \times 10^{-37}\,M; \text{ assumptions good.}$

107.    a.    $K^+(blood) \rightleftharpoons K^+(muscle)$   $\Delta G° = 0$;   $\Delta G = RT \ln\left(\dfrac{[K^+]_m}{[K^+]_b}\right)$;   $\Delta G = w_{max}$

$\Delta G = \dfrac{8.3145\ J}{K\ mol}\,(310.\ K) \ln\left(\dfrac{0.15}{0.0050}\right)$,   $\Delta G = 8.8 \times 10^3\ J/mol = 8.8\ kJ/mol$

At least 8.8 kJ of work must be applied to transport 1 mol $K^+$.

b.    Other ions will have to be transported in order to maintain electroneutrality. Either anions must be transported into the cells, or cations ($Na^+$) in the cell must be transported to the blood. The latter is what happens:  $[Na^+]$ in blood is greater than $[Na^+]$ in cells as a result of this pumping.

c.    $\Delta G° = -RT \ln K = -(8.3145\ J\ K^{-1}\ mol^{-1})(310.\ K) \ln(1.7 \times 10^5) = -3.1 \times 10^4\ J/mol$
$$= -31\ kJ/mol$$

The hydrolysis of ATP (at standard conditions) provides 31 kJ/mol of energy to do work. We need 8.8 kJ of work to transport 1.0 mol of $K^+$.

$8.8\ kJ \times \dfrac{1\ mol\ ATP}{31\ kJ} = 0.28\ mol\ ATP$ must be hydrolyzed

# CHAPTER 19

# TRANSITION METALS AND COORDINATION CHEMISTRY

## Transition Metals

7. Cr and Cu are exceptions to the normal filling order of electrons.

   a. Cr: $[Ar]4s^1 3d^5$

      $Cr^{2+}$: $[Ar]3d^4$

      $Cr^{3+}$: $[Ar]3d^3$

   b. Cu: $[Ar]4s^1 3d^{10}$

      $Cu^+$: $[Ar]3d^{10}$

      $Cu^{2+}$: $[Ar]3d^9$

   c. V: $[Ar]4s^2 3d^3$

      $V^{2+}$: $[Ar]3d^3$

      $V^{3+}$: $[Ar]3d^2$

9. The lanthanide elements are located just before the 5d transition metals. The lanthanide contraction is the steady decrease in the atomic radii of the lanthanide elements when going from left to right across the periodic table. As a result of the lanthanide contraction, the sizes of the 4d and 5d elements are very similar (see the following exercise). This leads to a greater similarity in the chemistry of the 4d and 5d elements in a given vertical group.

11. a. molybdenum(IV) sulfide; molybdenum(VI) oxide

    b. $MoS_2$, +4; $MoO_3$, +6; $(NH_4)_2Mo_2O_7$, +6; $(NH_4)_6Mo_7O_{24} \cdot 4\ H_2O$, +6

13. $TiF_4$: ionic compound containing $Ti^{4+}$ ions and $F^-$ ions. $TiCl_4$, $TiBr_4$, and $TiI_4$: covalent compounds containing discrete, tetrahedral $TiX_4$ molecules. As these covalent molecules get larger, the boiling points and melting points increase because the London dispersion forces increase. $TiF_4$ has the highest boiling point because the interparticle forces are stronger in ionic compounds than in covalent compounds.

15. $H^+ + OH^- \rightarrow H_2O$; sodium hydroxide (NaOH) will react with the $H^+$ on the product side of the reaction. This effectively removes $H^+$ from the equilibrium, which will shift the reaction to the right to produce more $H^+$ and $CrO_4^{2-}$. As more $CrO_4^{2-}$ is produced, the solution turns yellow.

17. Because transition metals form bonds to species that donate lone pairs of electrons, transition metals are Lewis acids (electron pair acceptors). The Lewis bases in coordination compounds are the ligands, all of which have an unshared pair of electrons to donate. The coordinate covalent bond between the ligand and the transition metal just indicates that both electrons in the bond originally came from one of the atoms in the bond. Here, the electrons in the bond come from the ligand.

19. $Fe_2O_3(s) + 6\ H_2C_2O_4(aq) \rightarrow 2\ Fe(C_2O_4)_3^{3-}(aq) + 3\ H_2O(l) + 6\ H^+(aq)$; the oxalate anion forms a soluble complex ion with iron in rust ($Fe_2O_3$), which allows rust stains to be removed.

21.   See Section 19.3 for a good summary of the nomenclature rules.

a.   The correct name is tetraamminecopper(II) chloride. The complex ion is named incorrectly in several ways.

b.   The correct name is bis(ethylenediamine)nickel(II) sulfate. The ethylenediamine ligands are neutral, and sulfate has a 2– charge. Therefore, $Ni^{2+}$ is present, not $Ni^{4+}$.

c.   The correct name is potassium diaquatetrachlorochromate(III). Because the complex ion is an anion, the –ate suffix ending is added to the name of the metal. Also, the ligands were not in alphabetical order (*a* in aqua comes before *c* in chloro).

d.   The correct name is sodium tetracyanooxalatocobaltate(II). The only error is that tetra should be omitted in front of sodium. That four sodium ions are needed to balance charge is deduced from the name of the complex ion.

23.   To determine the oxidation state of the metal, you must know the charges of the various common ligands (see Tables 19.13 and 19.14 of the text).

a.   pentaamminechlororuthenium(III) ion

b.   hexacyanoferrate(II) ion

c.   tris(ethylenediamine)manganese(II) ion

d.   pentaamminenitrocobalt(III) ion

25.   a.   $K_2[CoCl_4]$

b.   $[Pt(H_2O)(CO)_3]Br_2$

c.   $Na_3[Fe(CN)_2(C_2O_4)_2]$

d.   $[Cr(NH_3)_3Cl(NH_2CH_2CH_2NH_2)]I_2$

27.   $BaCl_2$ gives no precipitate, so $SO_4^{2-}$ must be in the coordination sphere ($BaSO_4$ is insoluble). A precipitate with $AgNO_3$ means the $Cl^-$ is not in the coordination sphere, that is, $Cl^-$ is a counter ion. Because there are only four ammonia molecules in the coordination sphere, $SO_4^{2-}$ must be acting as a bidentate ligand. The structure is:

29.   Because each compound contains an octahedral complex ion, the formulas for the compounds are $[Co(NH_3)_6]I_3$, $[Pt(NH_3)_4I_2]I_2$, $Na_2[PtI_6]$, and $[Cr(NH_3)_4I_2]I$. Note that in some cases the $I^-$ ions are ligands bound to the transition metal ion as required for a coordination number of 6, whereas in other cases the $I^-$ ions are counter ions required to balance the charge of the complex ion. The $AgNO_3$ solution will only precipitate the $I^-$ counter ions and will not precipitate the $I^-$ ligands. Therefore, 3 mol AgI will precipitate per mole of $[Co(NH_3)_6]I_3$, 2 mol AgI will precipitate per mole of $[Pt(NH_3)_4I_2]I_2$, 0 mol AgI will precipitate per mole of $Na_2[PtI_6]$, and 1 mol AgI will precipitate per mole of $[Cr(NH_3)_4I_2]I$.

31.    a.    Isomers: species with the same formulas but different properties; they are different compounds. See the text for examples of the following types of isomers.

b.    Structural isomers: isomers that have one or more bonds that are different.

c.    Steroisomers: isomers that contain the same bonds but differ in how the atoms are arranged in space.

d.    Coordination isomers: structural isomers that differ in the atoms that make up the complex ion.

e.    Linkage isomers: structural isomers that differ in how one or more ligands are attached to the transition metal.

f.    Geometric isomers: (cis-trans isomerism); steroisomers that differ in the positions of atoms with respect to a rigid ring, bond, or each other.

g.    Optical isomers: steroisomers that are nonsuperimposable mirror images of each other; that is, they are different in the same way that our left and right hands are different.

33.    a.

cis                                        trans

*Note*:  $C_2O_4^{2-}$ is a bidentate ligand.  Bidentate ligands bond to the metal at two positions that are 90° apart from each other in octahedral complexes.  Bidentate ligands do not bond to the metal at positions 180° apart from each other.

b.

cis                                        trans

c.

cis                              trans

d.

*Note*:  N   N   is an abbreviation for the bidentate en ligand (ethylenediamine, $NH_2CH_2CH_2NH_2$).

35.

mirror

trans                                                          cis
(mirror image is
superimposable)                              The mirror image of the cis
                                             isomer is also superimposable.

No;  both the trans and the cis forms of $Co(NH_3)_4Cl_2^+$ have mirror images that are superimposable. For the cis form, the mirror image only needs a 90° rotation to produce the original structure. Hence neither the trans nor cis form is optically active.

37.

M = transition metal ion

and

39.    Linkage isomers differ in the way that the ligand bonds to the metal. $SCN^-$ can bond through the sulfur or through the nitrogen atom. $NO_2^-$ can bond through the nitrogen or through the oxygen atom. $OCN^-$ can bond through the oxygen or through the nitrogen atom. $N_3^-$, $NH_2CH_2CH_2NH_2$ and $I^-$ are not capable of linkage isomerism.

41.    Similar to the molecules discussed in Figures 19.16 and 19.17 of the text, $Cr(acac)_3$ and cis-$Cr(acac)_2(H_2O)_2$ are optically active. The mirror images of these two complexes are nonsuperimposable. There is a plane of symmetry in trans-$Cr(acac)_2(H_2O)_2$, so it is not optically active. A molecule with a plane of symmetry is never optically active because the mirror images are always superimposable. A plane of symmetry is a plane through a molecule where one side exactly reflects the other side of the molecule.

## Bonding, Color, and Magnetism in Coordination Compounds

43.    a.    Ligand that will give complex ions with the maximum number of unpaired electrons.

b.    Ligand that will give complex ions with the minimum number of unpaired electrons.

c.    Complex with a minimum number of unpaired electrons (low spin = strong field).

d.    Complex with a maximum number of unpaired electrons (high spin = weak field).

45.    $Sc^{3+}$ has no electrons in d orbitals. $Ti^{3+}$ and $V^{3+}$ have d electrons present. The color of transition metal complexes results from electron transfer between split d orbitals. If no d electrons are present, no electron transfer can occur, and the compounds are not colored.

47.    a.  $Fe^{2+}$: $[Ar]3d^6$

     ↑   ↑               —   —

↑↓  ↑  ↑           ↑↓  ↑↓  ↑↓

High spin, small Δ                Low spin, large Δ

b.  $Fe^{3+}$: $[Ar]3d^5$            c.  $Ni^{2+}$: $[Ar]3d^8$

    ↑  ↑                ↑   ↑

↑  ↑  ↑            ↑↓  ↑↓  ↑↓

High spin, small Δ

d.  $Zn^{2+}$: $[Ar]3d^{10}$

  ↑↓   ↑↓

↑↓  ↑↓  ↑↓

e.  $Co^{2+}$: $[Ar]3d^7$

   ↑   ↑               ↑  —

↑↓  ↑↓  ↑         ↑↓  ↑↓  ↑↓

High spin, small Δ           Low spin, large Δ

49.    To determine the crystal field diagrams, you need to determine the oxidation state of the transition metal, which can only be determined if you know the charges of the ligands (see Table 19.13).  The electron configurations and the crystal field diagrams follow.

   a.  $Ru^{2+}$: $[Kr]4d^6$, no unpaired $e^-$       b.  $Ni^{2+}$: $[Ar]3d^8$, 2 unpaired $e^-$

    —   —                   ↑   ↑

↑↓  ↑↓  ↑↓          ↑↓  ↑↓  ↑↓

Low spin, large Δ

c.  $V^{3+}$: $[Ar]3d^2$, 2 unpaired $e^-$

*Note*: $Ni^{2+}$ must have two unpaired electrons, whether high spin or low spin, and $V^{3+}$ must have two unpaired electrons, whether high spin or low spin.

51.    Replacement of water ligands by ammonia ligands resulted in shorter wavelengths of light being absorbed. Energy and wavelength are inversely related, so the presence of the $NH_3$ ligands resulted in a larger d-orbital splitting (larger $\Delta$). Therefore, $NH_3$ is a stronger-field ligand than $H_2O$.

53.    From Table 19.16 of the text, the violet complex ion absorbs yellow-green light ($\lambda \approx 570$ nm), the yellow complex ion absorbs blue light ($\lambda \approx 450$ nm), and the green complex ion absorbs red light ($\lambda \approx 650$ nm).  The spectrochemical series shows that $NH_3$ is a stronger-field ligand than $H_2O$, which is a stronger-field ligand than $Cl^-$.  Therefore, $Cr(NH_3)_6^{3+}$ will have the largest d-orbital splitting and will absorb the lowest wavelength electromagnetic radiation ($\lambda \approx 450$ nm) because energy and wavelength are inversely related ($\lambda = hc/E$).    Thus the yellow solution contains the $Cr(NH_3)_6^{3+}$ complex ion.  Similarly, we would expect the $Cr(H_2O)_4Cl_2^+$ complex ion to have the smallest d-orbital splitting because it contains the weakest-field ligands.  The green solution with the longest wavelength of absorbed light contains the $Cr(H_2O)_4Cl_2^+$ complex ion.  This leaves the violet solution, which contains the $Cr(H_2O)_6^{3+}$ complex ion.  This makes sense because we would expect $Cr(H_2O)_6^{3+}$ to absorb light of a wavelength between that of $Cr(NH_3)_6^{3+}$ and $Cr(H_2O)_4Cl_2^+$.

55.    a.  $Ru(phen)_3^{2+}$ exhibits optical isomerism [similar to $Co(en)_3^{3+}$ in Figure 19.16 of the text].

b.  $Ru^{2+}$: $[Kr]4d^6$;  because there are no unpaired electrons, $Ru^{2+}$ is a strong-field  (low-spin) case.

large $\Delta$

57.    The crystal field diagrams are different because the geometries of where the ligands point are different. The tetrahedrally oriented ligands point differently in relationship to the d orbitals than do the octahedrally oriented ligands. Also, we have more ligands in an octahedral complex.

See Figure 19.28 for the tetrahedral crystal field diagram. Notice that the orbitals are reverse of that in the octahedral crystal field diagram. The degenerate $d_{z^2}$ and $d_{x^2-y^2}$ are at a lower energy than the degenerate $d_{xy}$, $d_{xz}$, and $d_{yz}$ orbitals. Again, the reason for this is that tetrahedral ligands are oriented differently than octahedral field ligands, so the interactions with specifically oriented d orbitals are different. Also notice that the difference in magnitude of the d-orbital splitting for the two geometries. The d-orbital splitting in tetrahedral complexes is less than one-half the d-orbital splitting in octahedral complexes. There are no known ligands powerful enough to produce the strong-field case; hence all tetrahedral complexes are weak field or high spin.

59.    $CoBr_6^{4-}$ has an octahedral structure, and $CoBr_4^{2-}$ has a tetrahedral structure (as do most $Co^{2+}$ complexes with four ligands). Coordination complexes absorb electromagnetic radiation (EMR) of energy equal to the energy difference between the split d orbitals. Because the tetrahedral d-orbital splitting is less than one-half the octahedral d-orbital splitting, tetrahedral complexes will absorb lower-energy EMR, which corresponds to longer-wavelength EMR ($E = hc/\lambda$). Therefore, $CoBr_6^{4-}$ will absorb EMR having a wavelength shorter than $3.4 \times 10^{-6}$ m.

## Additional Exercises

61.    $Hg^{2+}(aq) + 2\ I^-(aq) \rightarrow HgI_2(s)$, orange precipitate

$HgI_2(s) + 2\ I^-(aq) \rightarrow HgI_4^{2-}(aq)$, soluble complex ion

$Hg^{2+}$ is a $d^{10}$ ion. Color is the result of electron transfer between split d orbitals. This cannot occur for the filled d orbitals in $Hg^{2+}$. Therefore, we would not expect $Hg^{2+}$ complex ions to form colored solutions.

63.    $Ni(CO)_4$ is composed of 4 CO molecules and Ni. Thus nickel has an oxidation state of zero.

65.    i.    $0.0203\ \text{g CrO}_3 \times \dfrac{52.00\ \text{g Cr}}{100.0\ \text{g CrO}_3} = 0.0106\ \text{g Cr};\quad \%\ \text{Cr} = \dfrac{0.0106\ \text{g}}{0.105\ \text{g}} \times 100 = 10.1\%\ \text{Cr}$

ii.    $32.93\ \text{mL HCl} \times \dfrac{0.100\ \text{mmol HCl}}{\text{mL}} \times \dfrac{1\ \text{mmol NH}_3}{\text{mmol HCl}} \times \dfrac{17.03\ \text{mg NH}_3}{\text{mmol}} = 56.1\ \text{mg NH}_3$

$\%\ \text{NH}_3 = \dfrac{56.1\ \text{mg}}{341\ \text{mg}} \times 100 = 16.5\%\ \text{NH}_3$

iii.   $73.53\% + 16.5\% + 10.1\% = 100.1\%$;  the compound must be composed of only Cr, $NH_3$, and I.

Out of 100.00 g of compound:

$$10.1 \text{ g Cr} \times \frac{1 \text{ mol}}{52.00 \text{ g}} = 0.194 \text{ mol}; \qquad \frac{0.194}{0.194} = 1.00$$

$$16.5 \text{ g NH}_3 \times \frac{1 \text{ mol}}{17.03 \text{ g}} = 0.969 \text{ mol}; \qquad \frac{0.969}{0.194} = 4.99$$

$$73.53 \text{ g I} \times \frac{1 \text{ mol}}{126.9 \text{ g}} = 0.5794 \text{ mol}; \qquad \frac{0.5794}{0.194} = 2.99$$

$Cr(NH_3)_5I_3$ is the empirical formula. $Cr^{3+}$ forms octahedral complexes. So compound A is made of the octahedral $[Cr(NH_3)_5I]^{2+}$ complex ion and two $I^-$ ions as counter ions; the formula is $[Cr(NH_3)_5I]I_2$. Lets check this proposed formula using the freezing-point data.

iv.  $\Delta T_f = iK_f m$;  for $[Cr(NH_3)_5I]I_2$,  i = 3.0 (assuming complete dissociation).

$$\text{Molality} = m = \frac{0.601 \text{ g complex}}{1.000 \times 10^{-2} \text{ kg H}_2\text{O}} \times \frac{1 \text{ mol complex}}{517.9 \text{ g complex}} = 0.116 \text{ mol/kg}$$

$\Delta T_f = 3.0 \times 1.86 \text{ °C kg/mol} \times 0.116 \text{ mol/kg} = 0.65\text{°C}$

Because $\Delta T_f$ is close to the measured value, this is consistent with the formula $[Cr(NH_3)_5I]I_2$.

67.    M = metal ion

69.    No; in all three cases, six bonds are formed between $Ni^{2+}$ and nitrogen, so $\Delta H$ values should be similar.  $\Delta S°$ for formation of the complex ion is most negative for 6 $NH_3$ molecules reacting with a metal ion (seven independent species become one).  For penten reacting with a metal ion, two independent species become one, so $\Delta S°$ is least negative for this reaction compared to the other reactions.  Thus the chelate effect occurs because the more bonds a chelating agent can form to the metal, the more favorable $\Delta S°$ is for the formation of the complex ion, and the larger is the formation constant.

71.     $(H_2O)_5Cr-Cl-Co(NH_3)_5 \rightarrow (H_2O)_5Cr-Cl-Co(NH_3)_5 \rightarrow Cr(H_2O)_5Cl^{2+} + Co(II)$ complex

Yes; this is consistent.  After the oxidation, the ligands on Cr(III) won't exchange. Because $Cl^-$ is in the coordination sphere, it must have formed a bond to Cr(II) before the electron transfer occurred (as proposed through the formation of the intermediate).

73.     a.                    $Fe(H_2O)_6^{3+} + H_2O \rightleftharpoons Fe(H_2O)_5(OH)^{2+} + H_3O^+$

| | | | |
|---|---|---|---|
| Initial | 0.10 $M$ | 0 | ~0 |
| Equil. | 0.10 $- x$ | $x$ | $x$ |

$$K_a = \frac{[Fe(H_2O)_5(OH)^{2+}][H_3O^+]}{[Fe(H_2O)_6^{3+}]}, \quad 6.0 \times 10^{-3} = \frac{x^2}{0.10 - x} \approx \frac{x^2}{0.10}$$

$x = 2.4 \times 10^{-2}$ ; assumption is poor ($x$ is 24% of 0.10).  Using successive approximations:

$$\frac{x^2}{0.10 - 0.024} = 6.0 \times 10^{-3}, \ x = 0.021$$

$$\frac{x^2}{0.10 - 0.021} = 6.0 \times 10^{-3}, \ x = 0.022; \quad \frac{x^2}{0.10 - 0.022} = 6.0 \times 10^{-3}, \ x = 0.022$$

$x = [H^+] = 0.022 \ M; \ pH = 1.66$

b.  Because of the lower charge, $Fe^{2+}(aq)$ will not be as strong an acid as $Fe^{3+}(aq)$. A solution of iron(II) nitrate will be less acidic (have a higher pH) than a solution with the same concentration of iron(III) nitrate.

75.     a.  In the lungs, there is a lot of $O_2$, and the equilibrium favors $Hb(O_2)_4$.  In the cells, there is a deficiency of $O_2$, and the equilibrium favors $HbH_4^{4+}$.

b.  $CO_2$ is a weak acid in water; $CO_2 + H_2O \rightleftharpoons HCO_3^- + H^+$.  Removing $CO_2$ essentially decreases $H^+$.   $Hb(O_2)_4$ is then favored, and $O_2$ is not released by hemoglobin in the cells.  Breathing into a paper bag increases $[CO_2]$ in the blood, thus increasing $[H^+]$, which shifts the reaction left.

c.  $CO_2$ builds up in the blood, and it becomes too acidic, driving the equilibrium to the left. Hemoglobin can't bind $O_2$ as strongly in the lungs.  Bicarbonate ion acts as a base in water and neutralizes the excess acidity.

## Challenge Problems

77.

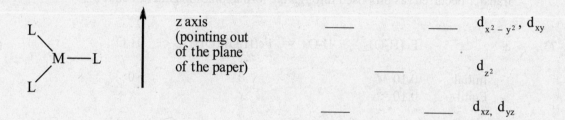

The $d_{x^2-y^2}$ and $d_{xy}$ orbitals are in the plane of the three ligands and should be destabilized the most. The amount of destabilization should be about equal when all the possible interactions are considered. The $d_{z^2}$ orbital has some electron density in the xy plane (the doughnut) and should be destabilized a lesser amount than the $d_{x^2-y^2}$ and $d_{xy}$ orbitals. The $d_{xz}$ and $d_{yz}$ orbitals have no electron density in the plane and should be lowest in energy.

79.

The $d_{z^2}$ orbital will be destabilized much more than in the trigonal planar case (see Exercise 19.77). The $d_{z^2}$ orbital has electron density on the z axis directed at the two axial ligands. The $d_{x^2-y^2}$ and $d_{xy}$ orbitals are in the plane of the three trigonal planar ligands and should be destabilized a lesser amount than the $d_{z^2}$ orbital; only a portion of the electron density in the $d_{x^2-y^2}$ and $d_{xy}$ orbitals is directed at the ligands. The $d_{xz}$ and $d_{yz}$ orbitals will be destabilized the least since the electron density is directed between the ligands.

81.    a.    Consider the following electrochemical cell:

$$Co^{3+} + e^- \rightarrow Co^{2+} \qquad\qquad E_c^\circ = 1.82 \text{ V}$$
$$Co(en)_3^{2+} \rightarrow Co(en)_3^{3+} + e^- \qquad\qquad -E_a^\circ = ?$$

$$\overline{Co^{3+} + Co(en)_3^{2+} \rightarrow Co^{2+} + Co(en)_3^{3+} \qquad\qquad E_{cell}^\circ = 1.82 - E_a^\circ}$$

The equilibrium constant for this overall reaction is:

$$Co^{3+} + 3\ en \rightarrow Co(en)_3^{3+} \qquad\qquad K_1 = 2.0 \times 10^{47}$$
$$Co(en)_3^{2+} \rightarrow Co^{2+} + 3\ en \qquad\qquad K_2 = 1/1.5 \times 10^{12}$$

$$Co^{3+} + Co(en)_3^{2+} \rightarrow Co(en)_3^{3+} + Co^{2+} \qquad K = K_1K_2 = \frac{2.0 \times 10^{47}}{1.5 \times 10^{12}} = 1.3 \times 10^{35}$$

From the Nernst equation for the overall reaction:

$$E_{cell}^{\circ} = \frac{0.0591}{n}\log K = \frac{0.0591}{1}\log(1.3 \times 10^{35}), \quad E_{cell}^{\circ} = 2.08\ V$$

$$E_{cell}^{\circ} = 1.82 - E_a^{\circ} = 2.08\ V,\ -E_a^{\circ} = 2.08\ V - 1.82\ V = 0.26\ V,\ so\ E_c^{\circ} = -0.26\ V$$

b.  The stronger oxidizing agent will be the more easily reduced species and will have the more positive standard reduction potential. From the reduction potentials, $Co^{3+}$ ($E^{\circ} = 1.82\ V$) is a much stronger oxidizing agent than $Co(en)_3^{3+}$ ($E^{\circ} = -0.26\ V$).

c.  In aqueous solution, $Co^{3+}$ forms the hydrated transition metal complex $Co(H_2O)_6^{3+}$. In both complexes, $Co(H_2O)_6^{3+}$ and $Co(en)_3^{3+}$, cobalt exists as $Co^{3+}$, which has six d electrons. Assuming a strong-field case for each complex ion, the d-orbital splitting diagram for each is:

When each complex gains an electron, the electron enters a higher-energy $e_g$ orbital. Because en is a stronger-field ligand than $H_2O$, the d-orbital splitting is larger for $Co(en)_3^{3+}$, and it takes more energy to add an electron to $Co(en)_3^{3+}$ than to $Co(H_2O)_6^{3+}$. Therefore, it is more favorable for $Co(H_2O)_6^{3+}$ to gain an electron than for $Co(en)_3^{3+}$ to gain an electron.

83.  a.          $AgBr(s)\ \rightleftharpoons\ Ag^+\ +\ Br^- \quad K_{sp} = [Ag^+][Br^-] = 5.0 \times 10^{-13}$

| | | | |
|---|---|---|---|
| Initial | $s$ = solubility (mol/L) | 0 | 0 |
| Equil. | | $s$ | $s$ |

$$K_{sp} = 5.0 \times 10^{-13} = s^2,\ s = 7.1 \times 10^{-7}\ mol/L$$

b.

$$AgBr(s) \rightleftharpoons Ag^+ + Br^- \qquad K_{sp} = 5.0 \times 10^{-13}$$
$$Ag^+ + 2\,NH_3 \rightleftharpoons Ag(NH_3)_2^+ \qquad K_f = 1.7 \times 10^7$$

---

$$AgBr(s) + 2\,NH_3(aq) \rightleftharpoons Ag(NH_3)_2^+(aq) + Br^-(aq) \quad K = K_{sp} \times K_f = 8.5 \times 10^{-6}$$

|  | AgBr(s) | + | 2 NH$_3$ | $\rightleftharpoons$ | Ag(NH$_3$)$_2^+$ | + | Br$^-$ |
|---|---|---|---|---|---|---|---|
| Initial |  |  | 3.0 $M$ |  | 0 |  | 0 |

$s$ mol/L of AgBr(s) dissolves to reach equilibrium = molar solubility

| Equil. |  |  | 3.0 − 2$s$ |  | $s$ |  | $s$ |

$$K = \frac{[Ag(NH_3)_2^+][Br^-]}{[NH_3]^2} = \frac{s^2}{(3.0 - 2s)^2}, \ 8.5 \times 10^{-6} \approx \frac{s^2}{(3.0)^2}, \ s = 8.7 \times 10^{-3} \ mol/L$$

Assumption good.

c.  The presence of NH$_3$ increases the solubility of AgBr. Added NH$_3$ removes Ag$^+$ from solution by forming the complex ion Ag(NH$_3$)$_2^+$. As Ag$^+$ is removed, more AgBr(s) will dissolve to replenish the Ag$^+$ concentration.

d.  Mass AgBr = $0.2500 \ L \times \dfrac{8.7 \times 10^{-3} \ mol \ AgBr}{L} \times \dfrac{187.8 \ g \ AgBr}{mol \ AgBr} = 0.41 \ g \ AgBr$

e.  Added HNO$_3$ will have no effect on the AgBr(s) solubility in pure water. Neither H$^+$ nor NO$_3^-$ reacts with Ag$^+$ or Br$^-$ ions. Br$^-$ is the conjugate base of the strong acid HBr, so it is a terrible base. Added H$^+$ will not react with Br$^-$ to any great extent. However, added HNO$_3$ will reduce the solubility of AgBr(s) in the ammonia solution. NH$_3$ is a weak base ($K_b = 1.8 \times 10^{-5}$). Added H$^+$ will react with NH$_3$ to form NH$_4^+$. As NH$_3$ is removed, a smaller amount of the Ag(NH$_3$)$_2^+$ complex ion will form, resulting in a smaller amount of AgBr(s) that will dissolve.

# CHAPTER 20

# THE NUCLEUS: A CHEMIST'S VIEW

## Radioactive Decay and Nuclear Transformations

1.  a.  Thermodynamic stability: the potential energy of a particular nucleus compared to the sum of the potential energies of its component protons and neutrons.

    b.  Kinetic stability: the probability that a nucleus will undergo decomposition to form a different nucleus.

    c.  Radioactive decay: a spontaneous decomposition of a nucleus to form a different nucleus.

    d.  Beta-particle production: a decay process for radioactive nuclides where an electron is produced; the mass number remains constant and the atomic number changes.

    e.  Alpha-particle production: a common mode of decay for heavy radioactive nuclides where a helium nucleus is produced, causing the atomic number and the mass number to change.

    f.  Positron production: a mode of nuclear decay in which a particle is formed having the same mass as an electron but opposite in charge.

    g.  Electron capture: a process in which one of the inner-orbital electrons in an atom is captured by the nucleus.

    h.  Gamma-ray emissions; the production of high-energy photons (gamma rays) that frequently accompany nuclear decays and particle reactions.

3.  All nuclear reactions must be charge-balanced and mass-balanced. To charge-balance, balance the sum of the atomic numbers on each side of the reaction, and to mass-balance, balance the sum of the mass numbers on each side of the reaction.

    a.  $^{51}_{24}Cr + ^{0}_{-1}e \rightarrow ^{51}_{23}V$　　　　b.  $^{131}_{53}I \rightarrow ^{0}_{-1}e + ^{131}_{54}Xe$

    c.  $^{32}_{15}P \rightarrow ^{0}_{-1}e + ^{32}_{16}S$　　　　d.  $^{235}_{92}U \rightarrow ^{4}_{2}He + ^{131}_{90}Th$

5.  a.  $^{68}_{31}Ga + ^{0}_{-1}e \rightarrow ^{68}_{30}Zn$　　　　b.  $^{62}_{29}Cu \rightarrow ^{0}_{+1}e + ^{62}_{28}Ni$

    c.  $^{212}_{87}Fr \rightarrow ^{4}_{2}He + ^{208}_{85}At$　　　　d.  $^{129}_{51}Sb \rightarrow ^{0}_{-1}e + ^{129}_{52}Te$

7. $^{247}_{97}Bk \rightarrow {}^{207}_{82}Pb + ? {}^{4}_{2}He + ? {}^{0}_{-1}e$; The change in mass number (247 - 207 = 40) is due exexclusively to the alpha particles. A change in mass number of 40 requires 10 ${}^{4}_{2}He$ particles to be produced. The atomic number only changes by 97 − 82 = 15. The 10 alpha particles change the atomic number by 20, so 5 ${}^{0}_{-1}e$ (five beta particles) are produced in the decay series of $^{247}Bk$ to $^{207}Pb$.

9. Reference Table 20.2 of the text for potential radioactive decay processes. $^{17}F$ and $^{18}F$ contain too many protons or too few neutrons. Electron capture and positron production are both possible decay mechanisms that increase the neutron-to-proton ratio. Alpha-particle production also increases the neutron-to-proton ratio, but it is not likely for these light nuclei. $^{21}F$ contains too many neutrons or too few protons. Beta-particle production lowers the neutron-to-proton ratio, so we expect $^{21}F$ to be a β-emitter.

11. a. $^{249}_{98}Cf + {}^{18}_{8}O \rightarrow {}^{263}_{106}Sg + 4{}^{1}_{0}n$      b. $^{259}_{104}Rf$; $^{263}_{106}Sg \rightarrow {}^{4}_{2}He + {}^{259}_{104}Rf$

## Kinetics of Radioactive Decay

13. $k = \dfrac{\ln 2}{t_{1/2}} = \dfrac{0.69315}{432.2 \text{ yr}} \times \dfrac{1 \text{ yr}}{365 \text{ d}} \times \dfrac{1 \text{ d}}{24 \text{ h}} \times \dfrac{1 \text{ h}}{3600 \text{ s}} = 5.086 \times 10^{-11} \text{ s}^{-1}$

Rate = kN = $5.086 \times 10^{-11} \text{ s}^{-1} \times 5.00 \text{ g} \times \dfrac{1 \text{ mol}}{241 \text{ g}} \times \dfrac{6.022 \times 10^{23} \text{ nuclei}}{\text{mol}}$

$$= 6.35 \times 10^{11} \text{ decays/s}$$

$6.35 \times 10^{11}$ alpha particles are emitted each second from a 5.00-g $^{241}Am$ sample.

15. a. $k = \dfrac{\ln 2}{t_{1/2}} = \dfrac{0.6931}{12.8 \text{ d}} \times \dfrac{1 \text{ d}}{24 \text{ h}} \times \dfrac{1 \text{ h}}{3600 \text{ s}} = 6.27 \times 10^{-7} \text{ s}^{-1}$

b. Rate = kN = $6.27 \times 10^{-7} \text{ s}^{-1} \left( 28.0 \times 10^{-3} \text{ g} \times \dfrac{1 \text{ mol}}{64.0 \text{ g}} \times \dfrac{6.022 \times 10^{23} \text{ nuclei}}{\text{mol}} \right)$

Rate = $1.65 \times 10^{14}$ decays/s

c. 25% of the $^{64}Cu$ will remain after 2 half-lives (100% decays to 50% after one half-life which decays to 25% after a second half-life). Hence 2(12.8 days) = 25.6 days is the time frame for the experiment.

17. $t = 63.0 \text{ yr}$; $k = \dfrac{\ln 2}{t_{1/2}}$; $\ln\left(\dfrac{N}{N_0}\right) = -kt = \dfrac{-0.6931 \times 63.0 \text{ yr}}{28.8 \text{ yr}} = -1.52$, $\left(\dfrac{N}{N_0}\right) = e^{-1.52}$

$$= 0.219$$

21.9% of the $^{90}Sr$ remains as of July 16, 2008.

19.    $175 \text{ mg Na}_3{}^{32}\text{PO}_4 \times \dfrac{32.0 \text{ mg }{}^{32}\text{P}}{165.0 \text{ mg Na}_3{}^{32}\text{PO}_4} = 33.9 \text{ mg }{}^{32}\text{P};\quad k = \dfrac{\ln 2}{t_{1/2}}$

$\ln\!\left(\dfrac{N}{N_0}\right) = -kt = \dfrac{-(0.6931)t}{t_{1/2}},\ \ln\!\left(\dfrac{m}{33.9 \text{ mg}}\right) = \dfrac{-0.6931(35.0 \text{ d})}{14.3 \text{ d}}$ ; carrying extra sig. figs.:

$\ln(m) = -1.696 + 3.523 = 1.827,\ m = e^{1.827} = 6.22 \text{ mg }{}^{32}\text{P}$ remains

21.    Plants take in $CO_2$ during the photosynthesis process, which incorporates carbon, including ${}^{14}\text{C}$, into its molecules. As long as the plant is alive, the ${}^{14}\text{C}/{}^{12}\text{C}$ ratio in the plant will equal the ratio in the atmosphere. When the plant dies, ${}^{14}\text{C}$ is not replenished because ${}^{14}\text{C}$ decays by beta-particle production. By measuring the ${}^{14}\text{C}$ activity today in the artifact and comparing this to the assumed ${}^{14}\text{C}$ activity when the plant died to make the artifact, an age can be determined for the artifact. The assumptions are that the ${}^{14}\text{C}$ level in the atmosphere is constant or that the ${}^{14}\text{C}$ level at the time the plant died can be calculated. A constant ${}^{14}\text{C}$ level is a pure assumption, and accounting for variation is complicated. Another problem is that some of the material must be destroyed to determine the ${}^{14}\text{C}$ level.

23.    $t_{1/2} = 5730 \text{ y};\ k = (\ln 2)/t_{1/2};\ \ln(N/N_0) = -kt;\ \ln\dfrac{15.1}{15.3} = \dfrac{-(\ln 2)t}{5730 \text{ yr}},\ t = 109$ years

No;  From ${}^{14}\text{C}$ dating, the painting was produced (at the earliest) during the late 1800s.

25.    $\ln\!\left(\dfrac{N}{N_0}\right) = -kt = \dfrac{-(\ln 2)t}{12.3 \text{ yr}},\ \ln\!\left(\dfrac{0.17 \times N_0}{N_0}\right) = -(5.64 \times 10^{-2})t,\ t = 31.4$ years

It takes 31.4 years for the tritium to decay to 17% of the original amount.  Hence the watch stopped fluorescing enough to be read in 1975 (1944 + 31.4).

27.    Assuming 1.000 g ${}^{238}\text{U}$ present in a sample, then 0.688 g ${}^{206}\text{Pb}$ is present.  Because 1 mol ${}^{206}\text{Pb}$ is produced per mol ${}^{238}\text{U}$ decayed:

$${}^{238}\text{U decayed} = 0.688 \text{ g Pb} \times \dfrac{1 \text{ mol Pb}}{206 \text{ g Pb}} \times \dfrac{1 \text{ mol U}}{\text{mol Pb}} \times \dfrac{238 \text{ g U}}{\text{mol U}} = 0.795 \text{ g }{}^{238}\text{U}$$

Original mass ${}^{238}\text{U}$ present $= 1.000 \text{ g} + 0.795 \text{ g} = 1.795 \text{ g }{}^{238}\text{U}$

$\ln\!\left(\dfrac{N}{N_0}\right) = -kt = \dfrac{-(\ln 2)t}{t_{1/2}},\ \ln\!\left(\dfrac{1.000 \text{ g}}{1.795 \text{ g}}\right) = \dfrac{-(0.693)t}{4.5 \times 10^9 \text{ yr}},\ t = 3.8 \times 10^9$ years

## Energy Changes in Nuclear Reactions

29.     $\Delta E = \Delta mc^2, \quad \Delta m = \dfrac{\Delta E}{c^2} = \dfrac{3.9 \times 10^{23} \text{ kg m}^2/\text{s}^2}{(3.00 \times 10^8 \text{ m/s})^2} = 4.3 \times 10^6 \text{ kg}$

The sun loses $4.3 \times 10^6$ kg of mass each second. *Note*: 1 J = 1 kg m$^2$/s$^2$.

31.     From the text, the mass of a proton = 1.00728 amu, the mass of a neutron = 1.00866 amu, and the mass of an electron = $5.486 \times 10^{-4}$ amu.

Mass of $^{56}_{26}$Fe nucleus = mass of atom − mass of electrons = 55.9349 − 26(0.0005486)

$= 55.9206$ amu

$26\,{}^1_1\text{H} + 30\,{}^1_0\text{n} \rightarrow {}^{56}_{26}\text{Fe}; \quad \Delta m = 55.9206 \text{ amu} - [26(1.00728) + 30(1.00866)] \text{ amu}$

$= -0.5285$ amu

$\Delta E = \Delta mc^2 = -0.5285 \text{ amu} \times \dfrac{1.6605 \times 10^{-27} \text{ kg}}{\text{amu}} \times (2.9979 \times 10^8 \text{ m/s})^2 = -7.887 \times 10^{-11} \text{ J}$

$\dfrac{\text{Binding energy}}{\text{Nucleon}} = \dfrac{7.887 \times 10^{-11} \text{ J}}{56 \text{ nucleons}} = 1.408 \times 10^{-12} \text{ J/nucleon}$

33.     Let $m_e$ = mass of electron;  for $^{12}$C (6e, 6p, 6n):  mass defect = $\Delta m$ = [mass of $^{12}$C nucleus] − [mass of 6 protons + mass of 6 neutrons]. *Note*: Atomic masses given include the mass of the electrons.

$\Delta m = 12.00000 \text{ amu} - 6m_e - [6(1.00782 - m_e) + 6(1.00866)]; \text{ mass of electrons cancel.}$

$\Delta m = 12.00000 - [6(1.00782) + 6(1.00866)] = -0.09888 \text{ amu}$

$\Delta E = \Delta mc^2 = -0.09888 \text{ amu} \times \dfrac{1.6605 \times 10^{-27} \text{ kg}}{\text{amu}} \times (2.9979 \times 10^8 \text{ m/s})^2$

$= -1.476 \times 10^{-11} \text{ J}$

$\dfrac{\text{Binding energy}}{\text{Nucleon}} = \dfrac{1.476 \times 10^{-11} \text{ J}}{12 \text{ nucleons}} = 1.230 \times 10^{-12} \text{ J/nucleon}$

For $^{235}$U (92e, 92p, 143n):

$\Delta m = 235.0439 - 92m_e - [92(1.00782 - m_e) + 143(1.00866)] = -1.9139 \text{ amu}$

$\Delta E = \Delta mc^2 = -1.9139 \times \dfrac{1.66054 \times 10^{-27} \text{ kg}}{\text{amu}} \times (2.99792 \times 10^8 \text{ m/s})^2 = -2.8563 \times 10^{-10} \text{ J}$

$$\frac{\text{Binding energy}}{\text{Nucleon}} = \frac{2.8563 \times 10^{-10} \text{ J}}{235 \text{ nucleons}} = 1.2154 \times 10^{-12} \text{ J/nucleon}$$

Because $^{56}$Fe is the most stable known nucleus, the binding energy per nucleon for $^{56}$Fe ($1.408 \times 10^{-12}$ J/nucleon) will be larger than that of $^{12}$C or $^{235}$U (see Figure 20.9 of the text).

35.    Binding energy $= \dfrac{1.326 \times 10^{-12} \text{ J}}{\text{nucleon}} \times 27 \text{ nucleons} = 3.580 \times 10^{-11}$ J for each $^{27}$Mg nucleus

$$\Delta E = \Delta mc^2, \quad \Delta m = \frac{\Delta E}{c^2} = \frac{-3.580 \times 10^{-11} \text{ J}}{(2.9979 \times 10^8 \text{ m/s})^2} = -3.983 \ 10^{-28} \text{ kg}$$

$$\Delta m = -3.983 \ 10^{-28} \text{ kg} \times \frac{1 \text{ amu}}{1.6605 \times 10^{-27} \text{ kg}} = -0.2399 \text{ amu} = \text{mass defect}$$

Let $m_{Mg}$ = mass of $^{27}$Mg nucleus;  an $^{27}$Mg nucleus has 12 p and 15 n.

$-0.2399$ amu $= m_{Mg} - (12m_p + 15m_n) = m_{Mg} - [12(1.00728 \text{ amu}) + 15(1.00866 \text{ amu})]$

$m_{Mg} = 26.9764$ amu

Mass of $^{27}$Mg atom = 26.9764 amu + 12$m_e$,  $26.9764 + 12(5.49 \times 10^{-4} \text{ amu}) = 26.9830$ amu (includes mass of 12 e$^-$)

37.    $_1^2\text{H} + _1^3\text{H} \rightarrow _2^4\text{He} + _0^1\text{n}$;  mass of electrons cancel when determining $\Delta m$ for this nuclear reaction.

$\Delta m = [4.00260 + 1.00866 - (2.01410 + 3.01605)]$ amu $= -1.889 \times 10^{-2}$ amu

For the production of 1 mol of $_2^4$He:   $\Delta m = -1.889 \times 10^{-2}$ g $= -1.889 \times 10^{-5}$ kg

$\Delta E = \Delta mc^2 = -1.889 \times 10^{-5}$ kg $\times (2.9979 \times 10^8 \text{ m/s})^2 = -1.698 \times 10^{12}$ J/mol

For one nucleus of $_2^4$He:

$$\frac{-1.698 \times 10^{12} \text{ J}}{\text{mol}} \times \frac{1 \text{ mol}}{6.0221 \times 10^{23} \text{ nuclei}} = -2.820 \times 10^{-12} \text{ J/nucleus}$$

## Detection, Uses, and Health Effects of Radiation

39.    The Geiger-Müller tube has a certain response time.  After the gas in the tube ionizes to produce a "count," some time must elapse for the gas to return to an electrically neutral state.  The response of the tube levels out because at high activities, radioactive particles are entering the tube faster than the tube can respond to them.

41.    Fission:    Splitting of a heavy nucleus into two (or more) lighter nuclei.

       Fusion:    Combining two light nuclei to form a heavier nucleus.

       The maximum binding energy per nucleon occurs at Fe.  Nuclei smaller than Fe become more stable by fusing to form heavier nuclei closer in mass to Fe.  Nuclei larger than Fe form more stable nuclei by splitting to form lighter nuclei closer in mass to Fe.

43.    Moderator:    Slows the neutrons to increase the efficiency of the fission reaction.

       Control rods:  Absorbs neutrons to slow or halt the fission reaction.

45.    In order to sustain a nuclear chain reaction, the neutrons produced by the fission process must be contained within the fissionable material so that they can go on to cause other fissions. The fissionable material must be closely packed together to ensure that neutrons are not lost to the outside.  The critical mass is the mass of material in which exactly one neutron from each fission event causes another fission event so that the process sustains itself.  A supercritical situation occurs when more than one neutron from each fission event causes another fission event.  In this case the process rapidly escalates, and the heat buildup causes a violent explosion.

47.    A nonradioactive substance can be put in equilibrium with a radioactive substance.  The two materials can then be checked to see whether all the radioactivity remains in the original material or if it has been scrambled by the equilibrium.

49.    All evolved oxygen in $O_2$ comes from water and not from carbon dioxide.

51.    Some factors for the biological effects of radiation exposure are:

       a.   The energy of the radiation. The higher the energy, the more damage it can cause.

       b.   The penetrating ability of radiation. The ability of specific radiation to penetrate human tissue where it can do damage must be considered.

       c.   The ionizing ability of the radiation. When biomolecules are ionized, their function is usually disturbed.

       d.   The chemical properties of the radiation source. Specifically, can the radioactive substance be readily incorporated into the body, or is the radiation source inert chemically so that it passes through the body relatively quickly.

       $^{90}$Sr will be incorporated into the body by replacing calcium in the bones. Once incorporated, $^{90}$Sr can cause leukemia and bone cancer. Krypton is chemically inert, so it will not be incorporated into the body.

53.    (i) and (ii) mean that Pu is not a significant threat outside the body.  Our skin is sufficient to keep out the α particles.  If Pu gets inside the body, it is easily oxidized to $Pu^{4+}$ (iv), which is chemically similar to $Fe^{3+}$ (iii).  Thus $Pu^{4+}$ will concentrate in tissues where $Fe^{3+}$ is found, including the bone marrow, where red blood cells are produced.  Once inside the body, α particles can cause considerable damage.

## Additional Exercises

55.     $20,000 \text{ ton TNT} \times \dfrac{4 \times 10^9 \text{ J}}{\text{ton TNT}} \times \dfrac{1 \text{ mol }^{235}\text{U}}{2 \times 10^{13} \text{ J}} \times \dfrac{235 \text{ g }^{235}\text{U}}{\text{mol }^{235}\text{U}} = 940 \text{ g }^{235}\text{U} \approx 900 \text{ g }^{235}\text{U}$

This assumes that all of the $^{235}\text{U}$ undergoes fission.

57.     The only product in the fast-equilibrium step is assumed to be $N^{16}O^{18}O_2$, where N is the central atom.  However, this is a reversible reaction where $N^{16}O^{18}O_2$ will decompose to NO and $O_2$.  Because any two oxygen atoms can leave $N^{16}O^{18}O_2$ to form $O_2$, we would expect (at equilibrium) one-third of the NO present in this fast equilibrium step to be $N^{16}O$ and two-thirds to be $N^{18}O$.  In the second step (the slow step), the intermediate $N^{16}O^{18}O_2$ reacts with the scrambled NO to form the $NO_2$ product, where N is the central atom in $NO_2$.  Any one of the three oxygen atoms can be transferred from $N^{16}O^{18}O_2$ to NO when the $NO_2$ product is formed.  The distribution of $^{18}O$ in the product can best be determined by forming a probability table.

|  | $N^{16}O$ (1/3) | $N^{18}O$ (2/3) |
|---|---|---|
| $^{16}O$ (1/3) from $N^{16}O^{18}O_2$ | $N^{16}O_2$ (1/9) | $N^{18}O^{16}O$ (2/9) |
| $^{18}O$ (2/3) from $N^{16}O^{18}O_2$ | $N^{16}O^{18}O$ (2/9) | $N^{18}O_2$ (4/9) |

From the probability table, 1/9 of the $NO_2$ is $N^{16}O_2$, 4/9 of the $NO_2$ is $N^{18}O_2$, and 4/9 of the $NO_2$ is $N^{16}O^{18}O$ (2/9 + 2/9 = 4/9).  *Note*:  $N^{16}O^{18}O$ is the same as $N^{18}O^{16}O$.  In addition, $N^{16}O^{18}O_2$ is not the only $NO_3$ intermediate formed; $N^{16}O_2{}^{18}O$ and $N^{18}O_3$ can also form in the fast-equilibrium first step.  However, the distribution of $^{18}O$ in the $NO_2$ product is the same as calculated above, even when these other $NO_3$ intermediates are considered.

59.     Assuming that the radionuclide is long lived enough that no significant decay occurs during the time of the experiment, the total counts of radioactivity injected are:

$$0.10 \text{ mL} \times \dfrac{5.0 \times 10^3 \text{ cpm}}{\text{mL}} = 5.0 \times 10^2 \text{ cpm}$$

Assuming that the total activity is uniformly distributed only in the rat's blood, the blood volume is:

$$V \times \dfrac{48 \text{ cpm}}{\text{mL}} = 5.0 \times 10^2 \text{ cpm}, \quad V = 10.4 \text{ mL} = 10. \text{ mL}$$

## Challenge Problems

61.     $_1^2H + _1^2H \rightarrow _2^4He$;  Q for $_1^2H = 1.6 \times 10^{-19}$ C;   mass of deuterium = 2 amu.

$$E = \frac{9.0 \times 10^9 \text{ J} \bullet \text{m}/C^2(Q_1Q_2)}{r} = \frac{9.0 \times 10^9 \text{ J} \bullet \text{m}/C^2(1.6 \times 10^{-19} \text{ C})^2}{2 \times 10^{-15} \text{ m}}$$

$$= 1 \times 10^{-13} \text{ J per alpha particle}$$

KE = 1/2 mv²;  $1 \times 10^{-13}$ J = 1/2 (2 amu × 1.66 × $10^{-27}$ kg/amu)v²,  v = 8 × $10^6$ m/s

From the kinetic molecular theory discussed in Chapter 5:

$$u_{rms} = \left(\frac{3RT}{M}\right)^{1/2} \text{ where M = molar mass in kilograms} = 2 \times 10^{-3} \text{ kg/mol for deuterium}$$

$$8 \times 10^6 \text{ m/s} = \left[\frac{3(8.3145 \text{ J K}^{-1} \text{ mol}^{-1})(T)}{2 \times 10^{-3} \text{ kg}}\right]^{1/2}, \quad T = 5 \times 10^9 \text{ K}$$

63.     $\text{mol I} = \frac{33 \text{ counts}}{\text{min}} \times \frac{1 \text{ mol I} \bullet \text{min}}{5.0 \times 10^{11} \text{ counts}} = 6.6 \times 10^{-11} \text{ mol I}$

$$[I^-] = \frac{6.6 \times 10^{-11} \text{ mol I}^-}{0.150 \text{ L}} = 4.4 \times 10^{-10} \text{ mol/L}$$

$$Hg_2I_2(s) \rightarrow Hg_2^{2+}(aq) \quad + \quad 2 I^-(aq) \qquad K_{sp} = [Hg_2^{2+}][I^-]^2$$

| | | | |
|---|---|---|---|
| Initial | s = solubility (mol/L) | 0 | 0 |
| Equil. | | s | 2s |

From the problem, $2s = 4.4 \times 10^{-10}$ mol/L,  $s = 2.2 \times 10^{-10}$ mol/L.

$K_{sp} = (s)(2s)^2 = (2.2 \times 10^{-10})(4.4 \times 10^{-10})^2 = 4.3 \times 10^{-29}$

65.     a.    For a gas, $u_{avg} = \sqrt{8RT/\pi M}$ where M is the molar mass in kg.  From the equation, the lighter the gas molecule, the faster is the average velocity.  Therefore, $^{235}UF_6$ will have the greater average velocity at a certain temperature because it is the lighter molecule.

        b.    From Graham's law (see Section 5.7 of the text):

$$\frac{\text{diffusion rate for } ^{235}UF_6}{\text{diffusion rate for } ^{238}UF_6} = \sqrt{\frac{M(^{238}UF_6)}{M(^{235}UF_6)}} = \sqrt{\frac{352.05 \text{ g/mol}}{349.03 \text{ g/mol}}} = 1.0043$$

Each diffusion step increases the $^{235}UF_6$ concentration by a factor of 1.0043.  To determine the number of steps to get to the desired 3.00% $^{235}U$, we use the following formula:

$$\frac{0.700 \ ^{235}UF_6}{99.3 \ ^{238}UF_6} \times (1.0043)^N = \frac{3.00 \ ^{235}UF_6}{97.0 \ ^{238}UF_6}$$

original ratio                                final ratio

where N represents the number of steps required.

Solving (and carrying extra sig. figs.):

$$(1.0043)^N = \frac{297.9}{67.9} = 4.387, \ \ N \log(1.0043) = \log(4.387)$$

$$N = \frac{0.6422}{1.863 \times 10^{-3}} = 345 \text{ steps}$$

Thus 345 steps are required to obtain the desired enrichment.

c.  $\dfrac{^{235}UF_6}{^{238}UF_6} \times (1.0043)^{100} = \dfrac{1526}{1.000 \times 10^5 - 1526}, \ \dfrac{^{235}UF_6}{^{238}UF_6} \times 1.5358 = \dfrac{1526}{98500}$

original ratio                                final ratio

$$\frac{^{235}UF_6}{^{238}UF_6} = 1.01 \times 10^{-2} = \text{initial } ^{235}U \text{ to } ^{238}U \text{ atom ratio}$$

# CHAPTER 21

# ORGANIC CHEMISTRY

## Hydrocarbons

1.  A hydrocarbon is a compound composed of only carbon and hydrogen. A saturated hydrocarbon has only carbon-carbon single bonds in the molecule. An unsaturated hydrocarbon has one or more carbon-carbon multiple bonds but may also contain carbon-carbon single bonds. A normal hydrocarbon has one chain of consecutively bonded carbon atoms. A branched hydrocarbon has at least one carbon atom not bonded to the end carbon of a chain of consecutively bonded carbon atoms. Instead, at least one carbon atom forms a bond to an inner carbon atom in the chain of consecutively bonded carbon atoms.

3.  In order to form, cyclopropane and cyclobutane are forced to form bond angles much smaller than the preferred 109.5° bond angles. Cyclopropane and cyclobutane easily react in order to obtain the preferred 109.5° bond angles.

5.  A difficult task in this problem is recognizing different compounds from compounds that differ by rotations about one or more C–C bonds (called conformations). The best way to distinguish different compounds from conformations is to name them. Different name = different compound; same name = same compound, so it is not an isomer but instead is a conformation.

    a.

    |  | | |
    | --- | --- | --- |
    | CH₃ | CH₃ | CH₃ |
    | CH₃CHCH₂CH₂CH₂CH₂CH₃ | CH₃CH₂CHCH₂CH₂CH₂CH₃ | CH₃CH₂CH₂CHCH₂CH₂CH₃ |
    | 2-methylheptane | 3-methylheptane | 4-methylheptane |

    b.

    |  | |
    | --- | --- |
    | CH₃ | CH₃ |
    | CH₃CCH₂CH₂CH₂CH₃ | CH₃CHCHCH₂CH₂CH₃ |
    | CH₃ | CH₃ |
    | 2,2-dimethylhexane | 2,3-dimethylhexane |

$$CH_3CHCH_2CHCH_2CH_3$$

2,4-dimethylhexane

$$CH_3CHCH_2CH_2CHCH_3$$

2,5-dimethylhexane

$$CH_3CH_2CCH_2CH_2CH_3$$

3,3-dimethylhexane

$$CH_3CH_2CHCHCH_2CH_3$$

3,4-dimethylhexane

$$CH_3CH_2CHCH_2CH_2CH_3$$

3-ethylhexane

c.

$$CH_3-C-CH-CH_2-CH_3$$

2,2,3-trimethylpentane

$$CH_3-C-CH_2-CH-CH_3$$

2,2,4-trimethylpentane

$$CH_3-CH-C-CH_2-CH_3$$

2,3,3-trimethylpentane

$$CH_3-CH-CH-CH-CH_3$$

2,3,4-trimethylpentane

$$CH_3-CH-CH-CH_2-CH_3$$

3-ethyl-2-methylpentane

$$CH_3-CH_2-C-CH_2-CH_3$$

3-ethyl-3-methylpentane

d.

$$CH_3-C-C-CH_3$$

2,2,3,3-tetramethylbutane

7.  London dispersion (LD) forces are the primary intermolecular forces exhibited by hydrocarbons. The strength of the LD forces depends on the surface-area contact among neighboring molecules. As branching increases, there is less surface-area contact among neighboring molecules, leading to weaker LD forces and lower boiling points.

9.  a.                                                      b.

c.                                                          d.  For 3-isobutylhexane, the longest chain is seven carbons long. The correct name is 4-ethyl-2-methyl-heptane. For 2-tert-butylpentane, the longest chain is six carbons long. The correct name is 2,2,3-trimethylhexane.

11. a.  2,2,4-trimethylhexane    b.  5-methylnonane    c.  2,2,4,4-tetramethylpentane

    d.  3-ethyl-3-methyloctane

    *Note*: For alkanes, always identify the longest carbon chain for the base name first, then number the carbons to give the lowest overall numbers for the substituent groups.

13. a.  1-butene                                b.  2-methyl-2-butene

    c.  2,5-dimethyl-3-heptene                   d.  2,3-dimethyl-1-pentene

    e.  1-ethyl-3-methylcyclopentene (double bond assumed between $C_1$ and $C_2$)

    f.  4-ethyl-3-methylcyclopentene             g.  4-methyl-2-pentyne

    Note: Multiple bonds are assigned the lowest number possible.

15. a.  1,3-dichlorobutane                      b.  1,1,1-trichlorobutane

    c.  2,3-dichloro-2,4-dimethylhexane         d.  1,2-difluoroethane

    e.  3-iodo-1-butene

    f.  2-bromotoluene (or o-bromotoluene or 1-bromo-2-methylbenzene)

    g.  1-bromo-2-methylcyclohexane

    h.  4-bromo-3-methylcyclohexene (double bond assumed between $C_1$ and $C_2$)

## Isomerism

19.  a.  1-sec-butylpropane

$$CH_2CH_2CH_3$$
$$|$$
$$CH_3CHCH_2CH_3$$

3-Methylhexane is correct.

b.  4-methylhexane

$$CH_3$$
$$|$$
$$CH_3CH_2CH_2CHCH_2CH_3$$

3-Methylhexane is correct.

c.  2-ethylpentane

$$CH_3CHCH_2CH_2CH_3$$
$$|$$
$$CH_2CH_3$$

3-Methylhexane is correct.

d.  1-ethyl-1-methylbutane

$$CH_2CH_3$$
$$|$$
$$CHCH_2CH_2CH_3$$
$$|$$
$$CH_3$$

3-Methylhexane is correct.

e.  3-methylhexane

$$CH_3CH_2CHCH_2CH_2CH_3$$
$$|$$
$$CH_3$$

This is a correct name.

f.  4-ethylpentane

$$CH_3CH_2CH_2CHCH_3$$
$$|$$
$$CH_2CH_3$$

3-Methylhexane is correct.

All six of these compounds are the same. They only differ from each other by rotations about one or more carbon-carbon single bonds. Only one isomer of $C_7H_{16}$ is present in all of these names, 3-methylhexane.

21.  To exhibit cis-trans isomerism, a compound must first have restricted rotation about a carbon-carbon bond. This occurs in compounds with double bonds and ring compounds. Second, the compound must have two carbons in the restricted rotation environment that each have two different groups bonded. For example, the compound in Exercise 21.13a has a double bond, but the first carbon in the double bond has two H atoms attached. This compound does not exhibit cis-trans isomerism. To see this, let's draw the potential cis-trans isomers:

These are the same compounds; they only differ by a simply rotation of the molecule. Therefore, they are not isomers of each other, but instead, they are the same compound. The only compounds that fulfill the restricted rotation requirement *and* have two different groups attached to carbons in the restricted rotation are compounds c and f. The cis-trans isomerism for these follows.

c.

f.

—— = out of plane of paper;   ----- = into plane of paper

23.   a.   All these structures have the formula $C_5H_8$. The compounds with the same physical properties will be the compounds that are identical to each other, i.e., compounds that only differ by rotations of C–C single bonds. To recognize identical compounds, name them. The names of the compounds are:

i.   trans-1,3-pentadiene          ii.   cis-1,3-pentadiene

iii.  cis-1,3-pentadiene           iv.   2-methyl-1,3-butadiene

Compounds ii and iii are identical compounds, so they would have the same physical properties.

b.   Compound i is a trans isomer because the bulkiest groups bonded to the carbon atoms in the $C_3=C_4$ double bond are as far apart as possible.

c.   Compound iv does not have carbon atoms in a double bond that each have two different groups attached. Compound iv does not exhibit cis-trans isomerism.

25.   The cis isomer has the $CH_3$ groups on the same side of the ring. The trans isomer has the $CH_3$ groups on opposite sides of the ring.

cis                              trans

27.    To help distinguish the different isomers, we will name them.

cis-1-chloro-1-propene          trans-1-chloro-1-propene

2-chloro-1-propene          3-chloro-1-propene          chlorocyclopropane

29.    $C_5H_{10}$ has the general formula for alkenes, $C_nH_{2n}$.  To distinguish the different isomers from each other, we will name them.  Each isomer must have a different name.

$CH_2$═$CHCH_2CH_2CH_3$              $CH_3CH$═$CHCH_2CH_3$

1-pentene                              2-pentene

$CH_2$═$CCH_2CH_3$                      $CH_3C$═$CHCH_3$
        |                                      |
       $CH_3$                                 $CH_3$

2-methyl-1-butene                    2-methyl-2-butene

$CH_3CHCH$═$CH_2$
        |
       $CH_3$

3-methyl-1-butene

Only 2-pentene exhibits cis-trans isomerism. The isomers are:

cis                                    trans

The other isomers of $C_5H_{10}$ do not contain carbons in the double bond to which two different groups are attached.

31.    a.   cis-1-bromo-1-propene      b.   cis-4-ethyl-3-methyl-3-heptene

       c.   trans-1,4-diiodo-2-propyl-1-pentene

*Note*: In general, cis-trans designations refer to the relative positions of the largest groups. In compound b, the largest group off the first carbon in the double bond is $CH_2CH_3$, and the largest group off the second carbon in the double bond is $CH_2CH_2CH_3$. Since their relative placement is on the same side of the double bond, this is the cis isomer.

33.    a.

ortho                         meta                         para

       b.   There are three trichlorobenzenes (1,2,3-trichlorobenzene, 1,2,4-trichlorobenzene, and 1,3,5-trichlorobenzene).

       c.   The meta isomer will be very difficult to synthesize.

       d.   1,3,5-Trichlorobenzene will be the most difficult to synthesize since all Cl groups are meta to each other in this compound.

## Functional Groups

35.

Carboxylic acid                                      Aldehyde

$$
R—\overset{\overset{\displaystyle :O:}{\|}}{C}—\overset{\cdot\cdot}{\underset{\cdot\cdot}{O}}—H
\qquad\qquad
R—\overset{\overset{\displaystyle :O:}{\|}}{C}—H
$$

RCOOH                                                   RCHO

The R designation refers to the rest of the organic molecule beyond the specific functional group indicated in the formula.  The R group may be a hydrogen but is usually a hydrocarbon fragment. The major point in the R group designation is that if the R group is an organic fragment, then the first atom in the R group is a carbon atom. What the R group has after the first carbon is not important to the functional group designation.

37.    Reference Table 21.4 for the common functional groups.

   a.  ketone          b.  aldehyde          c.  carboxylic acid          d.  amine

39.    a.

   b.  5 carbons in ring and the carbon in –$CO_2H$:  $sp^2$;  the other two carbons:  $sp^3$

   c.  24 sigma bonds; 4 pi bonds

41.    a.  3-chloro-1-butanol;  since the carbon containing the OH group is bonded to just one other carbon (one R group), this is a primary alcohol.

   b.  3-methyl-3-hexanol;  since the carbon containing the OH group is bonded to three other carbons (three R groups), this is a tertrary alcohol

   c.  2-methylcyclopentanol;  secondary alcohol (two R groups bonded to carbon containing the OH group). *Note*:  In ring compounds, the alcohol group is assumed to be bonded to $C_1$, so the number designation is commonly omitted for the alcohol group.

**43.**

OH
$CH_3CH_2CH_2CH_2CH_2$

1-pentanol

OH
$CH_3CH_2CH_2CHCH_3$

2-pentanol

OH
$CH_3CH_2CHCH_2CH_3$

3-pentanol

OH
$CH_3CH_2CHCH_2$
$CH_3$

2-methyl-1-butanol

OH
$CH_3CHCH_2CH_2$
$CH_3$

3-methyl-1-butanol

OH
$CH_3CH_2CCH_3$
$CH_3$

2-methyl-2-butanol

OH
$CH_3CHCHCH_3$
$CH_3$

3-methyl-2-butanol

$CH_3$  OH
$CH_3—C—CH_2$
$CH_3$

2,2-dimethyl-1-propanol

There are six isomeric ethers with formula $C_5H_{12}O$.  The structures follow.

$CH_3—O—CH_2CH_2CH_2CH_3$

$CH_3$
$CH_3—O—CHCH_2CH_3$

$CH_3$
$CH_3—O—CH_2CHCH_3$

$CH_3$
$CH_3—O—C—CH_3$
$CH_3$

$CH_3CH_2—O—CH_2CH_2CH_3$

$CH_3$
$CH_3CH_2—O—CH$
$CH_3$

**45.**    a.   4,5-dichloro-3-hexanone          b.   2,3-dimethylpentanal

c.   3-methylbenzaldehyde or m-methylbenzaldehyde

**47.**    a.                                        b.

O
$H—C—H$

O
$CH_3CH_2CH_2CCH_2CH_2CH_3$

c.                                        d.

O
$H—CCH_2CHCH_3$
$Cl$

O   $CH_3$
$CH_3$ $CCH_2CH_2CCH_3$
$CH_3$

49.

a.   trans-2-butene:

$$\begin{array}{c}CH_3\\ \phantom{xx}\end{array}\!\!C\!=\!C\!\!\begin{array}{c}H\\ \phantom{xx}\end{array}\text{,   formula} = C_4H_8$$

(structure shown with $CH_3$ and $H$ on left carbon, $H$ and $CH_3$ on right carbon)

(cyclobutane structure)   or   (methylcyclopropane structure)

b.   propanoic acid:   $CH_3CH_2\overset{\displaystyle O}{\overset{\|}{C}}\!-\!OH$,   formula = $C_3H_6O_2$

$CH_3\overset{\displaystyle O}{\overset{\|}{C}}\!-\!O\!-\!CH_3$   or   $H\overset{\displaystyle O}{\overset{\|}{C}}\!-\!O\!-\!CH_2CH_3$

c.   butanal:   $CH_3CH_2CH_2\overset{\displaystyle O}{\overset{\|}{C}}H$,   formula = $C_4H_8O$

$CH_3CH_2\overset{\displaystyle O}{\overset{\|}{C}}CH_3$

d.   butylamine:   $CH_3CH_2CH_2CH_2NH_2$,   formula = $C_4H_{11}N$:

A secondary amine has two R groups bonded to N.

$CH_3\!-\!\underset{\underset{\displaystyle CH_2CH_2CH_3}{|}}{N}\!-\!H$   or   $CH_3\!-\!\underset{\underset{\displaystyle CH_3CHCH_3}{|}}{N}\!-\!H$   or   $CH_3CH_2\!-\!\underset{\underset{\displaystyle CH_2CH_3}{|}}{N}\!-\!H$

e.   A tertiary amine has two R groups bonded to N.
     (See answer d for the structure of butylamine.)

$CH_3\!-\!\underset{\underset{\displaystyle CH_2CH_3}{|}}{N}\!-\!CH_3$

f.   2-methyl-2-propanol:   $CH_3\underset{\underset{\displaystyle OH}{|}}{\overset{\overset{\displaystyle CH_3}{|}}{C}}CH_3$ ,   formula = $C_4H_{10}O$

$CH_3\!-\!O\!-\!CH_2CH_2CH_3$   or   $CH_3\!-\!O\!-\!\underset{\underset{\displaystyle CH_3}{|}}{\overset{\overset{\displaystyle CH_3}{|}}{C}}H$   or   $CH_3CH_2\!-\!O\!-\!CH_2CH_3$

g.   A secondary alcohol has two R groups attached to the carbon bonded to the OH group.  (See answer f for the structure of 2-methyl-2-propanol.)

$$
\begin{array}{c}
\text{OH} \\
| \\
\text{CH}_3\text{CHCH}_2\text{CH}_3
\end{array}
$$

51.   a.   2-Chloro-2-butyne would have five bonds to the second carbon. Carbon never expands its octet.

$$
\begin{array}{c}
\text{Cl} \\
| \\
\text{CH}_3\text{---C}\equiv\text{CCH}_3
\end{array}
$$

b.   2-Methyl-2-propanone would have five bonds to the second carbon.

$$
\begin{array}{c}
\text{O} \\
\| \\
\text{CH}_3\text{---C---CH}_3 \\
| \\
\text{CH}_3
\end{array}
$$

c.   Carbon-1 in 1,1-dimethylbenzene would have five bonds.

d.   You cannot have an aldehyde functional group bonded to a middle carbon in a chain. Aldehyde groups, i.e.,

$$
\begin{array}{c}
\text{O} \\
\| \\
\text{---C---H}
\end{array}
$$

can only be at the beginning and/or the end of a chain of carbon atoms.

e.   You cannot have a carboxylic acid group bonded to a middle carbon in a chain. Carboxylic groups, i.e.,

$$
\begin{array}{c}
\text{O} \\
\| \\
\text{---C---OH}
\end{array}
$$

must be at the beginning and/or the end of a chain of carbon atoms.

f.  In cyclobutanol, the 1 and 5 positions refer to the same carbon atom. 5,5-Dibromo-1-cyclobutanol would have five bonds to carbon-1. This is impossible; carbon never expands its octet.

# Reactions of Organic Compounds

53.     Substitution:  An atom or group is replaced by another atom or group.

For example:  H in benzene is replaced by Cl.

$$C_6H_6 + Cl_2 \xrightarrow{\text{catalyst}} C_6H_5Cl + HCl$$

Addition:  Atoms or groups are added to a molecule.

For example:  $Cl_2$ adds to ethene.  $CH_2 = CH_2 + Cl_2 \rightarrow CH_2Cl–CH_2Cl$

55.   a.       $CH_2{=\!=}CH_2 + H_2O \xrightarrow{H^+}$

b.

c.

d.

57.

a.    CH$_3$CH—CHCH$_3$ (with H, H on the two central carbons)

b.    Cl Cl Cl Cl
      CH$_2$—CHCHCH—CH
              |        |
              CH$_3$    CH$_3$

c.    C$_6$H$_5$—Cl  +  HCl

d.    C$_4$H$_8$(g) + 6 O$_2$(g) → 4 CO$_2$(g) + 4 H$_2$O(g)


59.    When CH$_2$=CH$_2$ reacts with HCl, there is only one possible product, chloroethane. When Cl$_2$ is reacted with CH$_3$CH$_3$ (in the presence of light), there are six possible products because any number of the six hydrogens in ethane can be substituted for by Cl. The light-catalyzed substitution reaction is very difficult to control; hence it is not a very efficient method of producing monochlorinated alkanes.


61.    Primary alcohols (a, d, and f) are oxidized to aldehydes, which can be oxidized further to carboxylic acids.  Secondary alcohols (b, e, and f) are oxidized to ketones, and tertiary alcohols (c and f) do not undergo this type of oxidation reaction.  Note that compound f contains a primary, a secondary, and a tertiary alcohol.  For the primary alcohols (a, d, and f), we listed both the aldehyde and the carboxylic acid as possible products.


a.    H—C(=O)—CH$_2$CHCH$_3$   +   HO—C(=O)—CH$_2$CHCH$_3$
                    |                           |
                    CH$_3$                       CH$_3$

b.    CH$_3$—C(=O)—CHCH$_3$           c.    No reaction
                    |
                    CH$_3$

d.

e.

f.

63.    $KMnO_4$ will oxidize primary alcohols to aldehydes and then to carboxylic acids. Secondary alcohols are oxidized to ketones by $KMnO_4$. Tertiary alcohols and ethers are not oxidized by $KMnO_4$.

The three isomers and their reactions with $KMnO_4$ are:

The products of the reactions with excess $KMnO_4$ are 2-propanone and propanoic acid.

65.    a.    $CH_3CH = CH_2 + Br_2 \rightarrow CH_3CHBrCH_2Br$  (Addition reaction of $Br_2$ with propene)

b.

Oxidation of 2-propanol yields acetone (2-propanone).

c.

Addition of $H_2O$ to 2-methylpropene would yield tert-butyl alcohol (2-methyl-2-propanol) as the major product.

d.

Oxidation of 1-propanol would eventually yield propanoic acid. Propanal is produced first in this reaction and is then oxidized to propanoic acid.

67.

acetylsalicylic acid (aspirin)

methyl salicylate

## Polymers

69.    a.    Addition polymer: a polymer that forms by adding monomer units together (usually by reacting double bonds). Teflon, polyvinyl chloride, and polyethylene are examples of addition polymers.

       b.    Condensation polymer: a polymer that forms when two monomers combine by eliminating a small molecule (usually $H_2O$ or HCl). Nylon and Dacron are examples of condensation polymers.

       c.    Copolymer: a polymer formed from more than one type of monomer. Nylon and Dacron are copolymers.

       d.    Homopolymer: a polymer formed from the polymerization of only one type of monomer. Polyethylene, Teflon, and polystyrene are examples of homopolymers.

       e.    Polyester: a condensation polymer whose monomers link together by formation of the ester functional group.  Dacron is a polyester.

       f.    Polyamide: a condensation polymer whose monomers link together by formation of the amide functional group. Nylon is a polyamide as are proteins in the human body.

71.    a.    A polyester forms when an alcohol functional group reacts with a carboxylic acid functional group. The monomer for a homopolymer polyester must have an alcohol functional group and a carboxylic acid functional group present in the structure.

       b.    A polyamide forms when an amine functional group reacts with a carboxylic acid functional group. For a copolymer polyamide, one monomer would have at least two amine functional groups present, and the other monomer would have at least two carboxylic acid functional groups present. For polymerization to occur, each monomer must have two reactive functional groups present.

       c.    To form a typical addition polymer, a carbon-carbon double bond must be present. To form a polyester, the monomer would need the alcohol and carboxylic acid functional groups present. To form a polyamide, the monomer would need the amine and carboxylic acid functional groups present. The two possibilities are for the monomer to have a carbon-carbon double bond, an alcohol functional group, and a carboxylic acid functional group present or to have a carbon-carbon double bond, an amine functional group, and a carboxylic acid functional group present.

73.     The backbone of the polymer contains only carbon atoms, which indicates that Kel-F is an addition polymer.  The smallest repeating unit of the polymer and the monomer used to produce this polymer are:

$$
\begin{array}{cc}
\text{F F} & \text{F F}\\
\mid\ \mid & \mid\ \mid\\
-\!(\text{C}-\text{C})_n\!- & \text{C}=\text{C}\\
\mid\ \mid & \mid\ \mid\\
\text{Cl F} & \text{Cl F}
\end{array}
$$

75.    a.

$$H_2N-\!\!\!\bigcirc\!\!\!-NH_2 \quad \text{and} \quad HO_2C-\!\!\!\bigcirc\!\!\!-CO_2H$$

       b.   Repeating unit:

(structure of Nomex repeating unit)

       The two polymers differ in the substitution pattern on the benzene rings.  The Kevlar chain is straighter, and there is more efficient hydrogen bonding between Kevlar chains than between Nomex chains.

77.

$$
\begin{array}{c}
\text{CN} \quad\quad \text{CN}\\
\mid \quad\quad\quad \mid\\
-\!(\text{C}-\text{CH}_2-\text{C}-\text{CH}_2)_n\!-\\
\mid \quad\quad\quad \mid\\
\text{C}-\text{OCH}_3 \quad \text{C}-\text{OCH}_3\\
\parallel \quad\quad\quad \parallel\\
\text{O} \quad\quad\quad \text{O}
\end{array}
$$

       "Super glue" is an addition polymer formed by reaction of the C=C bond in methyl cyanoacrylate.

79.   Divinylbenzene is a crosslinking agent.  Divinylbenzene has two reactive double bonds that are both reacted when divinylbenzene inserts itself into two adjacent polymer chains during the polymerization process.  The chains cannot move past each other because the crosslinks bond adjacent polymer chains together, making the polymer more rigid.

81.   a.

      b.   Condensation;  HCl is eliminated when the polymer bonds form.

83.   Polyacrylonitrile:

      The CN triple bond is very strong and will not easily break in the combustion process.  A likely combustion product is the toxic gas hydrogen cyanide, HCN(g).

85.

      Two linkages are possible with glycerol.  A possible repeating unit with both types of linkages is shown above.  With either linkage, there are unreacted OH groups on the polymer chains.  These unreacted OH groups on adjacent polymer chains can react with the acid groups of phthalic acid to form crosslinks (bonds) between various polymer chains.

## Natural Polymers

87.

$$R - \underset{\underset{NH_2}{|}}{\overset{\overset{H}{|}}{C}} - \overset{\overset{O}{\|}}{C} - OH \quad = \text{general amino acid formula}$$

Hydrophilic ("water-loving") and hydrophobic ("water-fearing") refer to the polarity of the R group. When the R group consists of a polar group, then the amino acid is hydrophilic. When the R group consists of a nonpolar group, then the amino acid is hydrophobic.

89.    Denaturation changes the three-dimensional structure of a protein. Once the structure is affected, the function of the protein will also be affected.

91.    a.    Serine, tyrosine, and threonine contain the –OH functional group in the R group.

b.    Aspartic acid and glutamic acid contain the –COOH functional group in the R group.

c.    An amine group has a nitrogen bonded to other carbon and/or hydrogen atoms. Histidine, lysine, arginine, and tryptophan contain the amine functional group in the R group.

d.    The amide functional group is:

$$R - \overset{\overset{O}{\|}}{C} - \underset{}{N} - R''$$

with $R'$ on the nitrogen

This functional group is formed when individual amino acids bond together to form the peptide linkage. Glutamine and asparagine have the amide functional group in the R group.

93.    a.    Aspartic acid and phenylalanine make up aspartame.

amide bond forms here

b.    Aspartame contains the methyl ester of phenylalanine. This ester can hydrolyze to form methanol:

$$R\text{–}CO_2CH_3 + H_2O \rightleftharpoons RCO_2H + HOCH_3$$

95.

ser - ala                              ala - ser

97.    a.    Six tetrapeptides are possible.  From $NH_2$ to $CO_2H$ end:

phe-phe-gly-gly,  gly-gly-phe-phe,  gly-phe-phe-gly,

phe-gly-gly-phe,  phe-gly-phe-gly,  gly-phe-gly-phe

b.    Twelve tetrapeptides are possible.  From $NH_2$ to $CO_2H$ end:

phe-phe-gly-ala,  phe-phe-ala-gly,  phe-gly-phe-ala,

phe-gly-ala-phe,  phe-ala-phe-gly,  phe-ala-gly-phe,

gly-phe-phe-ala,  gly-phe-ala-phe,  gly-ala-phe-phe,

ala-phe-phe-gly,  ala-phe-gly-phe,  ala-gly-phe-phe

99.    a.    Covalent (forms a disulfide linkage)

b.    Hydrogen bonding (need N-H or O-H bond in side chain)

c.    Ionic (need $NH_2$ group on side chain of one amino acid with $CO_2H$ group on side chain of the other amino acid)

d.    London dispersion (need amino acids with nonpolar R groups)

101.    Glutamic acid: $R = -CH_2CH_2CO_2H$;  valine: $R = -CH(CH_3)_2$;  a polar side chain is replaced by a nonpolar side chain.  This could affect the tertiary structure of hemoglobin and the ability of hemoglobin to bind oxygen.

103.    Glutamic acid:                                      Monosodium glutamate:

One of the two acidic protons in the carboxylic acid groups is lost to form MSG. Which proton is lost is impossible for you to predict.

In MSG, the acidic proton from the carboxylic acid in the R group is lost, allowing formation of the ionic compound.

105.    See Figures 21.30 and 21.31 of the text for examples of the cyclization process.

D-Ribose

D-Mannose

107.    The aldohexoses contain six carbons and the aldehyde functional group.  Glucose, mannose, and galactose are aldohexoses.  Ribose and arabinose are aldopentoses since they contain five carbons with the aldehyde functional group.  The ketohexose (six carbons + ketone functional group) is fructose, and the ketopentose (five carbons + ketone functional group) is ribulose.

109.    A disaccharide is a carbohydrate formed by bonding two monosaccharides (simple sugars) together.  In sucrose, the simple sugars are glucose and fructose, and the bond formed between these two monosaccharides is called a glycoside linkage.

111.    The α and β forms of glucose differ in the orientation of a hydroxy group on one specific carbon in the cyclic forms (see Figure 21.31 of the text).  Starch is a polymer composed of only α-D-glucose, and cellulose is a polymer composed of only β-D-glucose.

113.    A chiral carbon has four different groups attached to it.  A compound with a chiral carbon is optically active.  Isoleucine and threonine contain more than the one chiral carbon atom (see asterisks).

isoleucine

threonine

115.    Only one of the isomers is optically active.  The chiral carbon in this optically active isomer is marked with an asterisk.

$$H - \overset{\displaystyle Cl}{\underset{\displaystyle CH = CH_2}{\overset{|}{\underset{|}{C^*}}}} - Br$$

117.    They all contain nitrogen atoms with lone pairs of electrons.

119.    The complementary base pairs in DNA are cytosine (C) and guanine (G), and thymine (T) and adenine (A).  The complementary sequence is C-C-A-G-A-T-A-T-G.

121.    Uracil will hydrogen bond to adenine.

123.    Base pair:

RNA        DNA

A . . . . . . . . T

G . . . . . . . . C

C . . . . . . . . G

U . . . . . . . . A

a.  Glu: CTT, CTC            Val:  CAA, CAG, CAT, CAC

Met: TAC                     Trp:  ACC

Phe: AAA, AAG                Asp:  CTA, CTG

b.  DNA sequence for trp-glu-phe-met:

ACC – CTT – AAA – TAC
or        or
CTC       AAG

c.  Due to glu and phe, there is a possibility of four different DNA sequences.  They are:

ACC–CTT–AAA–TAC  or  ACC–CTC–AAA–TAC or

ACC–CTT–AAG –TAC or ACC–CTC–AAG–TAC

d.

T—A—C—C—T—G—A—A—G

met        asp        phe

e.  TAC–CTA–AAG;  TAC–CTA–AAA;  TAC–CTG–AAA

125.   A deletion may change the entire code for a protein, thus giving an entirely different sequence of amino acids.  A substitution will change only one single amino acid in a protein.

## Additional Exercises

127.   $CH_3CH_2CH_2CH_2CH_2CH_2CH_2COOH + OH^- \rightarrow CH_3-(CH_2)_6-COO^- + H_2O$; octanoic acid is more soluble in 1 $M$ NaOH.  Added $OH^-$ will remove the acidic proton from octanoic acid, creating a charged species.  As is the case with any substance with an overall charge, solubility in water increases.  When morphine is reacted with $H^+$, the amine group is protonated, creating a positive charge on morphine ($R_3N + H^+ \rightarrow R_3NH^+$).  By treating morphine with HCl, an ionic compound results that is more soluble in water and in the bloodstream than is the neutral covalent form of morphine.

129.

To substitute for the benzene ring hydrogens, an iron(III) catalyst must be present. Without this special iron catalyst, the benzene ring hydrogens are unreactive. To substitute for an alkane hydrogen, specific wavelengths of light must be present. For toluene, the light-catalyzed reaction substitutes a chlorine for a hydrogen in the methyl group attached to the benzene ring.

131.    $85.63 \text{ g C} \times \dfrac{1 \text{ mol C}}{12.011 \text{ g C}} = 7.129 \text{ mol C}; \quad 14.37 \text{ g H} \times \dfrac{1 \text{ mol H}}{1.0079 \text{ g H}} = 14.26 \text{ mol H}$

Because the moles of H to moles of C ratio is 2:1 (14.26/7.129 = 2.000), the empirical formula is $CH_2$. The empirical formula mass $\approx 12 + 2(1) = 14$. Because $4 \times 14 = 56$ puts the molar mass between 50 and 60, the molecular formula is $C_4H_8$.

The isomers of $C_4H_8$ are:

CH$_2$═CHCH$_2$CH$_3$          CH$_3$CH═CHCH$_3$          CH═CHCH$_3$ (with CH$_3$ substituent)

1-butene                      2-butene                  2-methyl-1-propene

cyclobutane          methylcyclopropane

Only the alkenes will react with $H_2O$ to produce alcohols, and only 1-butene will produce a secondary alcohol for the major product and a primary alcohol for the minor product.

CH$_2$═CHCH$_2$CH$_3$ + H$_2$O ⟶ CH$_2$—CHCH$_2$CH$_3$ (H, OH)
2° alcohol, major product

CH$_2$═CHCH$_2$CH$_3$ + H$_2$O ⟶ CH$_2$—CHCH$_2$CH$_3$ (OH, H)
1° alcohol, minor product

2-Butene will produce only a secondary alcohol when reacted with $H_2O$, and 2-methyl-1-propene will produce a tertiary alcohol as the major product and a primary alcohol as the minor product.

133.    a.   The bond angles in the ring are about 60°. VSEPR predicts bond angles close to 109°. The bonding electrons are closer together than they prefer, resulting is strong electron-electron repulsions. Thus ethylene oxide is unstable (reactive).

        b.   The ring opens up during polymerization; the monomers link together through the formation of O—C bonds.

$$\left(-O-CH_2CH_2-O-CH_2CH_2-O-CH_2CH_2-\right)_n$$

**135.**

(Structure as drawn)

$$\left(-OCH_2CH_2OCN \cdots NCOCH_2CH_2OCN \cdots N-C-\right)_n$$

**137.**    The structures, the types of intermolecular forces exerted, and the boiling points for the compounds are:

CH₃CH₂CH₂COH

butanoic acid,  164°C
LD + dipole + H–bonding

CH₃CH₂CH₂CH₂CH₂OH

1-pentanol, 137°C
LD + H-bonding

CH₃CH₂CH₂CH₂CH

pentanal, 103°C
LD + dipole

CH₃CH₂CH₂CH₂CH₂CH₃

n-hexane, 69°C
LD only

All these compounds have about the same molar mass.  Therefore, the London dispersion (LD) forces in each are about the same.  The other types of forces determine the boiling-point order.  Because butanoic acid and 1-pentanol both exhibit hydrogen-bonding (H–bonding) interactions, these two compounds will have the two highest boiling points.  Butanoic acid has the highest boiling point since it exhibits H-bonding along with dipole-dipole forces due to the polar C=O bond.

**139.**    The isomers are:

HC—O—CH₃

CH₃—C—OH

HC=CH  (OH  OH)

C=CH₂  (HO   HO)

boils at lowest temperature
 (no H-bonding)

With the exception of the first isomer, the other isomers can form the relatively strong hydrogen-bonding interactions. The isomers that can hydrogen bond will boil at higher temperatures.

141.     a.

CH$_3$CHCH$_3$
|
CH$_2$CH$_3$

The longest chain is four carbons long. The correct name is 2-methylbutane.

b.

$$I \qquad CH_3$$
$$| \qquad |$$
CH$_3$CH$_2$CH$_2$CH$_2$C———CH$_2$
|
CH$_3$

The longest chain is seven carbons long and we would start the numbering system at the other end for lowest possible numbers.  The correct name is 3-iodo-3-methylheptane.

c.

CH$_3$
|
CH$_3$CH$_2$CH══C———CH$_3$

This compound cannot exhibit cis–trans isomerism since one of the double-bonded carbons has the same two groups (CH$_3$) attached. The numbering system should also start at the other end to give the double bond the lowest possible number. 2-Methyl-2-pentene is correct.

d.

Br OH
|   |
CH$_3$CHCHCH$_3$

The OH functional group gets the lowest number. 3-Bromo-2-butanol is correct.

143.     $\Delta G = \Delta H - T\Delta S$;  for the reaction, we break a P–O and an O–H bond and form a P–O and an O–H bond.   Thus $\Delta H \approx 0$.   $\Delta S < 0$ because two molecules are going to form one molecule (positional probability decreases).   Thus $\Delta G > 0$,  and the reaction is not spontaneous.

145.     a.   Even though this form of tartaric acid contains two chiral carbon atoms (see asterisks in the following structure), the mirror image of this form of tartaric acid is superimposable. Therefore, it is not optically active. One way to identify optical activity in molecules with two or more chiral carbon atoms is to look for a plane of symmetry in the molecule.  If a molecule has a plane of symmetry, then it is never optically active.  A plane of symmetry is a plane that bisects the molecule where one side exactly reflects the other side.

symmetry plane

b.  The optically active forms of tartaric acid have no plane of symmetry.  The structures of the optically active forms of tartaric acid are:

mirror

These two forms of tartaric acid are nonsuperimposable.

147.  For the reaction:

$$^+H_3NCH_2CO_2H \rightleftharpoons 2\,H^+ + H_2NCH_2CO_2^- \qquad K_{eq} = 7.3 \times 10^{-13} = K_a(-CO_2H) \times K_a(-NH_3^+)$$

$$7.3 \times 10^{-13} = \frac{[H^+]^2[H_2NCH_2CO_2^-]}{[^+H_3NCH_2CO_2H]} = [H^+]^2, \quad [H^+] = (7.3 \times 10^{-13})^{1/2}$$

$[H^+] = 8.5 \times 10^{-7}\,M; \quad pH = -\log[H^+] = 6.07 = $ isoelectric point

149.  $4.2 \times 10^{-3}\,g\,K_2CrO_7 \times \dfrac{1\,mol\,K_2Cr_2O_7}{294.20\,g} \times \dfrac{1\,mol\,Cr_2O_7^{2-}}{mol\,K_2Cr_2O_7} \times \dfrac{3\,mol\,C_2H_5OH}{2\,mol\,Cr_2O_7^{2-}}$

$$= 2.1 \times 10^{-5}\,mol\,C_2H_5OH$$

$$n_{\text{breath}} = \frac{PV}{RT} = \frac{\left(750. \text{ mm Hg} \times \dfrac{1 \text{ atm}}{760 \text{ mm Hg}}\right) \times 0.500 \text{ L}}{\dfrac{0.08206 \text{ L atm}}{\text{K mol}} \times 303 \text{ K}} = 0.0198 \text{ mol breath}$$

$$\text{Mol \% } C_2H_5OH = \frac{2.1 \times 10^{-5} \text{ mol } C_2H_5OH}{0.0198 \text{ mol total}} \times 100 = 0.11\% \text{ alcohol}$$

## Challenge Problems

151.    Assuming 1.000 L of the hydrocarbon ($C_xH_y$), then the volume of products will be 4.000 L and the mass of products ($H_2O + CO_2$) will be:

$$1.391 \text{ g/L} \times 4.000 \text{ L} = 5.564 \text{ g products}$$

$$\text{Mol } C_xH_y = n_{C_xH_y} = \frac{PV}{RT} = \frac{0.959 \text{ atm} \times 1.000 \text{ L}}{\dfrac{0.08206 \text{ L atm}}{\text{K mol}} \times 298 \text{ K}} = 0.0392 \text{ mol}$$

$$\text{Mol products} = n_p = \frac{PV}{RT} = \frac{1.51 \text{ atm} \times 4.000 \text{ L}}{\dfrac{0.08206 \text{ L atm}}{\text{K mol}} \times 375 \text{ K}} = 0.196 \text{ mol}$$

$C_xH_y + \text{oxygen} \rightarrow x \, CO_2 + y/2 \, H_2O$;    setting up two equations:

$(0.0392)x + 0.0392(y/2) = 0.196$   (moles of products)

$(0.0392)x(44.01 \text{ g/mol}) + 0.0392(y/2)(18.02 \text{ g/mol}) = 5.564 \text{ g}$   (mass of products)

Solving: $x = 2$ and $y = 6$, so the formula of the hydrocarbon is $C_2H_6$ which is ethane.

153.    a.    The three structural isomers of $C_5H_{12}$ are:

$$CH_3CH_2CH_2CH_2CH_3 \qquad\qquad CH_3CHCH_2CH_3 \qquad\qquad CH_3{-}\underset{\underset{CH_3}{|}}{\overset{\overset{CH_3}{|}}{C}}{-}CH_3$$
$$\underset{CH_3}{|}$$

$$\text{n-pentane} \qquad\qquad\qquad \text{2-methylbutane} \qquad\qquad \text{2,2-dimethylpropane}$$

n-Pentane will form three different monochlorination products:  1-chloropentane, 2-chloropentane, and 3-chloropentane (the other possible monochlorination products differ by a simple rotation of the molecule; they are not different products from the ones listed).

2,2-Dimethylpropane will only form one monochlorination product: 1-chloro-2,2-dimethylpropane. 2-Methylbutane is the isomer of $C_5H_{12}$ that forms four different monochlorination products: 1-chloro-2-methylbutane, 2-chloro-2-methylbutane, 3-chloro-2-methylbutane (or we could name this compound as 2-chloro-3-methylbutane), and 1-chloro-3-methylbutane.

b.   The isomers of $C_4H_8$ are:

The cyclic structures will not react with $H_2O$; only the alkenes will add $H_2O$ to the double bond. From Exercise 21.58, the major product of the reaction of 1-butene and $H_2O$ is 2-butanol (a 2° alcohol). 2-Butanol is also the major (and only) product when 2-butene and $H_2O$ react. 2-Methylpropene forms 2-methyl-2-propanol as the major product when reacted with $H_2O$; this product is a tertiary alcohol. Therefore, the $C_4H_8$ isomer is 2-methylpropene.

$$CH_2{=}\underset{\underset{CH_3}{|}}{C}{-}CH_3 + HOH \longrightarrow CH_3{-}\underset{\underset{OH}{|}}{\overset{\overset{CH_3}{|}}{C}}{-}CH_3$$

2-methyl-2-propanol (a 3° alcohol, 3 R groups)

c.   The structure of 1-chloro-1-methylcyclohexane is:

The addition reaction of HCl with an alkene is a likely choice for this reaction (see Exercise 21.58). The two isomers of $C_7H_{12}$ that produce 1-chloro-1-methylcyclohexane as the major product are:

d.  Working backwards, 2° alcohols produce ketones when they are oxidized (1° alcohols produce aldehydes, and then carboxylic acids).  The easiest way to produce the 2° alcohol from a hydrocarbon is to add $H_2O$ to an alkene.  The alkene reacted is 1-propene (or propene).

$$CH_2{=\!=}CHCH_3 \; + \; H_2O \; \longrightarrow \; CH_3\overset{\overset{\displaystyle OH}{|}}{C}CH_3 \; \xrightarrow{\text{oxidation}} \; CH_3\overset{\overset{\displaystyle O}{\|}}{C}CH_3$$

    propene                                                                    acetone

e.  The $C_5H_{12}O$ formula has too many hydrogens to be anything other than an alcohol (or an unreactive ether).  1° alcohols are first oxidized to aldehydes, and then to carboxylic acids.  Therefore, we want a 1° alcohol.  The 1° alcohols with formula $C_5H_{12}O$ are:

1-pentanol           2-methyl-1-butanol        3-methyl-1-butanol        2,2-dimethyl-1-propanol

    There are other alcohols with formula $C_5H_{12}O$, but they are all 2° or 3° alcohols, which do not produce carboxylic acids when oxidized.

155.   Treat this problem like a diprotic acid ($H_2A$) titration. The $K_{a_1}$ and $K_{a_2}$ reactions are:

$$^+NH_3CH_2COOH \rightleftharpoons \; ^+NH_3CH_2COO^- + H^+ \quad K_{a_1} = \frac{[^+NH_3CH_2COO^-][H^+]}{[^+NH_3CH_2COOH]} = 4.3 \times 10^{-3}$$

$$^+NH_3CH_2COO^- \rightleftharpoons NH_2CH_2COO^- + H^+ \quad K_{a_2} = \frac{[NH_2CH_2COO^-][H^+]}{[^+NH_3CH_2COO^-]}$$

$$K_{a_2} = K_w/K_b = 1.0 \times 10^{-14}/6.0 \times 10^{-5} = 1.7 \times 10^{-10}$$

As $OH^-$ is added, it reacts completely with the best acid present. From 0–50.0 mL of $OH^-$ added, the reaction is:

$$^+NH_3CH_2COOH + OH^- \rightarrow \; ^+NH_3CH_2COO^- + H_2O$$

At 50.0 mL $OH^-$ added (the first equivalence point), all of the $^+NH_3CH_2COOH$ has been converted into $^+NH_3CH_2COO^-$. This is an amphoteric species. To determine the pH when an amphoteric species is the major species present, we use the formula pH = $(pK_{a_1} + pK_{a_2})/2$. From 50.1–100.0 mL of $OH^-$ added, the reaction that occurs is:

$$^+NH_3CH_2COO^- + OH^- \rightarrow NH_2CH_2COO^- + H_2O$$

100.0 mL of $OH^-$ added represents the second equivalence point where $H_2NCH_2COO^-$ is the major amino acid species present.

a.   25.0 mL of $OH^-$ added represents the first halfway point to equivalence. Here, $[^+NH_3CH_2COOH] = [^+NH_3CH_2COO^-]$. This is a buffer solution where pH = $pK_{a_1}$.

At 25.0 mL $OH^-$ added:

$$pH = pK_{a_1} = -\log(4.3 \times 10^{-3}) = 2.37$$

50.0 mL of $OH^-$ added represents the first equivalence point. Here, $^+NH_3CH_2COO^-$ is the major amino acid species present. This is amphoteric species. At 50.0 mL $OH^-$ added:

$$pH = \frac{pK_{a_1} + pK_{a_2}}{2} = \frac{2.37 - \log(1.7 \times 10^{-10})}{2} = \frac{2.37 + 9.77}{2} = 6.07$$

75.0 mL of $OH^-$ added represents the second halfway point to equivalence. Here, $[^+NH_3CH_2COO^-] = [NH_2CH_2COO^-]$ and pH = $pK_{a_2}$. At 75.0 mL $OH^-$ added:

$$pH = pK_{a_2} = -\log(1.7 \times 10^{-10}) = 9.77$$

b.

Volume of OH⁻ added

The major amino acid species present are:

point A (0.0 mL OH⁻):        $^+NH_3CH_2COOH$

point B (25.0 mL OH⁻):        $^+NH_3CH_2COOH$ and $^+NH_3CH_2COO^-$

point C (50.0 mL OH⁻):        $^+NH_3CH_2COO^-$

point D (75.0 mL OH⁻):        $^+NH_3CH_2COO^-$ and $NH_2CH_2COO^-$

point E (100.0 mL OH⁻):      $NH_2CH_2COO^-$

c.   The various charged amino acid species are:

$^+NH_3CH_2COOH$              net charge = +1

$^+NH_3CH_2COO^-$              net charge = 0

$NH_2CH_2COO^-$              net charge = −1

The net charge is zero at the pH when the major amino acid species present is the $^+NH_3CH_2COO^-$ form;  this happens at the first equivalence point in our titration problem. From part a, this occurs at pH = 6.07 (the isoelectric point).

d.   The net charge is +1/2 when = [$^+NH_3CH_2COOH$] = [$^+NH_3CH_2COO^-$];  net charge = (+1 + 0)/2 = +1/2.  This occurs at the first halfway point to equivalence, where pH = $pK_{a_1}$ = 2.37.  The net charge is −1/2 when [$^+NH_3CH_2COO^-$] = [$NH_2CH_2COO^-$];  net charge = (0 − 1)/2 = −1/2.  This occurs at the second halfway point to equivalence, where pH = $pK_{a_2}$ = 9.77.